Controversies in Constitutional Law

Collections of Documents and Articles on Major Questions of American Law

General Editor

Paul Finkelman

University of Tulsa College of Law

A Garland Series

Immigration and the Constitution

Volume 1
The Origins of Constitutional Immigration Law

Edited with introductions by

Gabriel J. Chin
University of Cincinnati College of Law

Victor C. Romero
Pennsylvania State University, Dickinson School of Law

Michael A. Scaperlanda
University of Oklahoma College of Law

GARLAND PUBLISHING, INC.
A MEMBER OF THE TAYLOR & FRANCIS GROUP
New York & London
2000

Published in 2001 by
Garland Publishing, Inc.
29 West 35th Street
New York, NY 10001

Published in Great Britain by
Garland Publishing, Inc.
11 New Fetter Lane
London EC4 4EE

Garland is an imprint of the Taylor & Francis Group

10 9 8 7 6 5 4 3 2 1

The Library of Congress Cataloging-in-Publication Data

Immigration and the constitution / edited with introductions by
Gabriel J. Chin, Victor C. Romero, Michael A. Scaperlanda.
p. cm. — (Controversies in constitutional law)
Includes bibliographical references.
Contents: v. 1. The origins of constitutional immigration law — v. 2. Discrimination and equality in contemporary immigration law — v. 3. Shark infested waters : procedural due process in constitutional immigration law.
ISBN 0-8153-3346-3 (alk. paper)
1. Emigration and immigration law—United States. 2. Constitutional law—United States. 3. United States—Emigration and immigration—Government policy. I. Chin, Gabriel J. (Gabriel Jackson), 1964-. II. Romero, Victor C. III. Scaperlanda, Michael A. IV. Series
KF4819.A2 I4265 2000
342.73'082—dc21 00-064646
CIP

3 volume set ISBN:	**0-8153-3346-3**
Volume 1	0-8153-4059-1
Volume 2	0-8153-4060-5
Volume 3	0-8153-4061-3

Contents

General Introduction

This ongoing series, *Controversies in Constitutional Law*, provides teachers, scholars, and students convenient access to the debates and scholarly literature surrounding major questions of constitutional law. Each set of books—from two to four volumes—consists of four elements: an extended introduction to the issue by the editor; reprints of the significant cases and briefs on the subject; congressional testimony, and other primary documents on the issue; and a selection of the best scholarly articles on the subject. By conveniently gathering all this material in one place, each set of volumes allows users to become familiar quite quickly with the arguments and positions surrounding a particular constitutional controversy.

Scholars and students interested in constitutional law and public policy are often overwhelmed by the sheer amount of material published on controversial subjects. The topics in *Controversies in Constitutional Law* are timely, politically significant, and intellectually compelling. They are the issues that bring the legal academy, courts, politicians, and general public together—although not always in harmony.

Many scholars are barely able to keep up with the material in their own fields. Yet there is a need to learn about issues beyond our own specialties. A new course may require an introduction to a problem; our own research may lead us to a new issue. The series *Controversies in Constitutional Law* is designed to meet these needs, enabling scholars and teachers to come up to speed quickly on the topics in each volume.

This series also serves students at all levels of higher education. Students often encounter controversial topics without the background in how the problem developed. These volumes provide such a background. The important cases are reprinted in full; the introductions provide students with a map to the issues; the briefs, congressional testimony, and scholarly articles give convenient access to the arguments and debates.

The volumes bring together in one place a wide variety of sources and materials that are unlikely to be found in any one library. The primary documents include cases, briefs, testimony from legislative committee hearings, and, on occasion, executive branch publications. Few law libraries, university libraries, or public libraries will have all these materials. The secondary literature comes from the scholarly literature of law, history, political science, economics, criminology, and other relevant fields. Some of the articles also may be from serious nonacademic journals.

The related government publications, briefs, and congressional testimony are particularly important. Too often scholars overlook them. There is a tendency to focus on court opinions—what the Supreme Court said. Obviously this is important for understanding doctrine, but briefs, especially amicus briefs, are also an important source of information about the larger constitutional controversy the case represents. The politics of a case are often more apparent in these briefs than in either the oral agruments or the Court's opinion. Similarly, congressional hearings illustrate the social, philosophical, and political dimensions of these questions.

The goal of limiting each collection—two to four focused volumes—and the occasional inability to secure a reasonable copyright permission fee for some important articles often forces hard and judicious choices. Important articles that are not reprinted here are listed in the further reading sections at the end of the introductions.

The ongoing series *Controversies in Constitutional Law* provides in one single place comprehensive introductions to topics that will satisfy most lawyers, scholars, and students. By combining cases, briefs, legislative debates, and testimony from committee hearings with secondary articles, it is hoped to facilitate access to the issue presented in each set of volumes.

Series Introduction

It would be difficult to overstate the importance of immigration in American history and life. John F. Kennedy correctly called the United States "a nation of immigrants"; with the possible exception of Indians, America has always been composed of those who came from other lands, or descended of those who did. Immigration has made a monumental contribution to national wealth and culture; in almost every era, America has benefitted from the energy, industry and creativity of millions of newcomers. In almost every era, the country has also embraced, sometimes with difficulty, speakers of languages other than English, members of non-Caucasian races, and cultures.

As important as immigration has been to United States history, the ability to immigrate, may be more important to particular individuals. The course and quality of countless lives has been shaped by the opportunity to live in a free society, or not, a land with economic opportunities, or not, with one's parents, children and spouse, or not.

Immigration law and policy has been a forum for expression of the best and worst attitudes of politicians — and citizens. The country has been an open door for many, a refuge for millions. It has also passed laws which excluded aliens on grounds which reflect value judgments about the worth of various classes of people. The law of the land has excluded individuals on the basis of race, religion, sex, in-wedlock birth status, health, wealth and political beliefs. For reasons which have never been fully explained, the Supreme Court has held that Congress may do what it wishes with respect to immigration. Volume 1 of this set describes the history of United States immigration policy and the constitutional law of immigration as expressed by the Supreme Court and its critics.

While many of the most patently invidious forms of discrimination have been removed from immigration law, the law continues to make controversial class distinctions about who may enter and who cannot. These issues are discussed in Volume 2: *Discrimination and Equality in Contemporary Immigration Law*. The volume discusses various theories advanced to explain the nature and justification of authority to regulate immigration, and then explores the extent to which current law discriminates on the basis of race, gender and sexual orientation.

Substantive law — the formal rules for who may enter and who may not, who can stay and who must go — are critically important, but the procedures by which those rules are applied may be almost as significant. It makes a difference, for example, whether an individual has the right to judicial review of an administrative decision, or the right to hire counsel to assist with their case, and when. Volume 3: *Shark Infested Waters: Procedural Due Process* describes a number of procedural issues raised by current law, focusing on the constitutional rights, if any, of aliens seeking to enter the United States for the first time, and of resident aliens in the United States whom the government wishes to remove.

Volume Introduction

American immigration policy has been both remarkably magnanimous and shockingly cruel. The United States accepts more refugees and ordinary immigrants than any other country in the world; this tradition has created a "nation of immigrants." Yet, in some ways the law has been brutal. At various times, the law has excluded would-be immigrants on racial, religious and political grounds, and allowed for the deportation or exclusion of long-time resident aliens for relatively minor misconduct — even those who immigrated as infants or children.

The constitutional law of immigration fits firmly within the latter branch of this tradition. Most actions of the federal government are subject to judicial review, the supervision of the courts to ensure that official conduct conforms to the Constitution. As the first five entries in this volume demonstrate, things work differently in immigration. The Supreme Court has never declared unconstitutional an Act of Congress regulating immigration. The Court has frequently said that "over no conceivable subject is the legislative power of Congress more complete than it is over the admission of aliens. Our cases have long recognized the power to expel or exclude aliens as a fundamental sovereign attribute exercised by the Government's political departments largely immune from judicial control."[1] The special treatment of immigration statutes under the Constitution is called the "plenary power doctrine," the idea that federal power over immigration is simply beyond judicial review.[2]

The plenary power doctrine has proved quite durable. Over the years, Congress has discriminated in the immigration context on many grounds which are contrary to what are now widely accepted values. The first several entries in this volume describe the discriminatory background of American immigration law. Professor James F. Smith's article *A Nation That Welcomes Immigrants? An Historical Examination of United States Immigration Policy*,[3] offers a brief overview of almost two centuries of immigration policy. One aspect of that policy is a racial test on immigration; the Act of March 26, 1790 shows how deeply ingrained discrimination is in our immigration policy: The federal government believed that limiting citizenship by naturalization to free white persons was important enough that such a law was passed by the First Congress, and signed into law by President George Washington.

One important line of racially discriminatory statutes was in force in one form or another from 1882 to 1965. During this period, one of the primary goals of immigration policy was control of the "Yellow Peril." The Asian Exclusion Laws, beginning with the Chinese Exclusion Act in 1882 (later extended to members of all Asian races) carried out this purpose.

From 1921 to 1965, through the National Origins Quota System, immigration law discriminated against another group of racial undesirables, discouraging immigration of Africans, Southern and Eastern Europeans. It did this by allocating visas based on the national origin of people who were already in the United States in a way which disadvantaged "new immigrants," Catholics and Jews from southern and eastern Europe. Similarly, as Professor Bill Ong Hing explains in an article included in this volume, one of the ways Africans were discouraged from immigrating under the National Origins Quota System was that African Americans were not counted at all.[4] Latinos and Latinas were also not welcomed.[5]

Racial groups were not the only ones disfavored by the Immigration and Nationality Act. The law discriminated on the basis of religion, against Mormons as well as Catholics and Jews, for example. It discriminated on the basis of political affiliation, against Communists and Anarchists. It also discriminated on the basis of sexual orientation, sex, physical and mental disability, out-of-wedlock birth, and poverty.[6]

The racism, xenophobia and bigotry of these laws should not be entirely shocking, because many of them coincided with the eras of slavery and Jim Crow, when racial discrimination was carried out by state and federal governments. What is more remarkable is that congressional and executive decisions have gone almost entirely unchecked by the judicial branch, even in recent decades. The legal explanation for this oddity begins with two cases reprinted in this volume. Although more than a century old, the decisions of the Supreme Court in *Chae Chan Ping v. United States (The Chinese Exclusion Case)*[7] and *Fong Yue Ting v. United States*[8] remain the foundational cases of constitutional immigration law. In *Chae Chan Ping*, the Supreme Court upheld a portion of the Chinese Exclusion Act which held that Chinese could be excluded from the United States — even though they were U.S. residents who had left holding government-issued documents assuring them that they could return. *Fong Yue Ting* went one step further, holding that otherwise lawfully resident aliens of Chinese ancestry could be deported if they failed to register, a requirement imposed on no other racial group.

In both cases, the Court relied in part on racial reasoning. The Court in *Fong Yue Ting*, for example, explained that restrictions on Chinese immigration were developed because "the government of the United States was brought to the opinion that the presence within our territory of large numbers of Chinese laborers, of a distinct race and religion, remaining strangers in the land, residing apart by themselves, tenaciously adhering to the customs and usages of their own country, unfamiliar with our institutions, and apparently incapable of assimilating with our people, might endanger good order, and be injurious to the public interests . . . .[9]"

The Court also explained that judicial review was impossible because decisions about immigration had been structurally committed to branches of government other than the judiciary. Thus, said the Court in *Fong Yue Ting*:

> In exercising the great power which the people of the United States, by establishing a written constitution as the supreme and paramount law, have

> vested in this Court, of determining, whenever the question is properly brought before it, whether the acts of the legislative or of the executive are consistent with the constitution, it behooves the court to be careful that it does not undertake to pass upon political questions, the final decision of which has been committed by the constitution to the other departments of government.[10]

In *Nishimura Ekiu v. United States,*[11] a case from 1892, the Court seemed to say that not only was Congress free to legislate as it liked without substantive limitations, aliens had no right to raise constitutional due process objections to their treatment. The Supreme Court explained that as to unadmitted, non-resident aliens, "the decisions of executive or administrative officers, acting within powers conferred by Congress, are due process of law."[12]

A critical aspect of the Court's decisions is precisely where in the Constitution it found the authority for federal regulation of immigration. There were a number of plausible candidates, including the commerce clause,[13] the naturalization clause,[14] and the migration clause.[15] However, according to *Chae Chan Ping* and *Fong Yue Ting* federal authority over immigration did not flow from any particular language in the text of the Constitution. Instead it was inherent in sovereignty itself. This was an important conclusion, because enumerated powers were subject to the restrictions of the bill of rights, such as the requirement of due process of law. Inherent powers, such as the plenary power doctrine, by contrast, were understood to be unlimited.

The Court quickly recognized limited exceptions to the plenary power doctrine. In 1896, the Court held that it did not apply to criminal prosecutions, even ones which were immigration-related and in which aliens were defendants.[16] The Court also decided that aliens who had entered and developed attachments in the United States were entitled to some due process of law before they could be deported or excluded, explaining that "once an alien gains admission to our country and begins to develop the ties that go with permanent residence his constitutional status changes accordingly."[17]

Scholars and courts have struggled to explain precisely why immigration decisions should be beyond the scope of constitutional judicial review. One argument might be that the Constitution applies only domestically, and therefore aliens are not within its protection.[18] Even if true, that principle would not explain why resident aliens in the United States in deportation proceedings are subject to the plenary power doctrine. In any event, most aliens being excluded *are* physically in the United States, at an international airport or seaport, for example.

Another possibility might be that aliens who are not yet citizens are seeking a privilege, rather than exercising a right, when they apply to enter or seek to remain, in the United States.[19] At one time, the Supreme Court treated the due process clause as primarily applicable to "rights", not mere "privileges." This is at best a partial explanation, for the right-privilege distinction in constitutional law has lost much of its force in recent years, and in any event "even 'privileges' may not be denied for reasons which violate constitutional guarantees."[20]

Some courts have suggested that the plenary power doctrine is rooted in international law, and that international law authorizes nations to determine unilaterally who they will admit or exclude and on what terms.[21] Scholars have pointed out, though, that even if this was once true, as a result of changes in international law, nations are no longer free to act arbitrarily, and that if the plenary power doctrine evolved to reflect current international law principles, it would be much more protective of individual rights.[22]

The second group of materials deals with judicial and scholarly efforts to come to grips with classical immigration law. At one level, the criticisms of the plenary power doctrine and its internal inconsistencies have had little impact on it. As Professor T. Alexander Aleinikoff explains in his article *Federal Regulation of Aliens and the Constitution*, "[i]mmigration law has remained blissfully untouched by the virtual revolution in constitutional law since World War II, impervious to developments in due process, equal protection and criminal procedure."[23] How could the same justices who carried out the civil rights revolution domestically have determined that no change was needed in the immigration context? Professor Gabriel Chin's article *Segregation's Last Stronghold: Race Discrimination and the Constitutional Law of Immigration*,[24] underscores the potency of the plenary power doctrine. The article notes that even after the Supreme Court outlawed school segregation in *Brown v. Board of Education* in 1954, federal courts have continued to state that Congress remains free to discriminate on the basis of race in immigration. How can the weight of so many other changes in constitutional law, judicial review and respect for individual rights have failed to destroy this antique doctrine?

Ironically, one reason for the plenary power doctrine's survival is that it is becoming more flexible. Professor Peter Schuck's 1984 article *The Transformation of Immigration Law*,[25] argues that legal and social pressure was in the process of changing constitutional immigration law, and that cases were beginning to appear which would lead to reform of the classical approach. Immigrants often do win in the federal courts, not on constitutional claims, but on statutory grounds.

In addition, as Professor Hiroshi Motomura explains in his article *Immigration Law After a Century of Plenary Power: Phantom Constitutional Norms and Statutory Interpretation*,[26] the Supreme Court and lower federal courts effectively apply the Constitution to immigration in spite of the plenary power doctrine. Courts achieve this by assuming that statutes should be interpreted consistently with constitutional norms and values, even though a premise of the plenary power doctrine is that the Constitution is not directly applicable to immigration law. As a result, courts "produce results that are much more sympathetic to aliens than the results that would follow from the interpretation of statutes in light of the expressly applicable constitutional immigration law based on plenary power."[27]

Another reason for the plenary power doctrine's staying power may be its age. The 1954 decision in *Galvan v. Press*[28] illustrates the strength of the plenary power doctrine, and the Court's reluctance to break with stare decisis and repudiate it. *Galvan* involved a section of the INA which made deportable any alien who had been "a member of or affiliated with the Communist Party." Over the dissent of Justices Black and Douglas, Justice Frankfurter concluded that there was no constitutional infirmity in deporting a long-time resident alien on the basis of the statute, even though Communist Party membership was legal, even though the Act had not been in force during the individual's two year affiliation, and even though the alien was not given the opportunity to show that he was unaware of and did not share the illegal aims of the party.

The Court frankly acknowledged that deportation may "deprive a man of all that makes life worth living," and that "deportation is a drastic measure and at times the equivalent of banishment or exile." The Court admitted that the failure of the law to permit an alien to show that he or she acted with no bad purpose "strikes one with a sense of harsh incongruity." Yet, concluded the majority, respect for precedent required the result.

> In light of the expansion of the concept of substantive due process as a limitation upon all powers of Congress, even the war power, much could be said for the view, were we writing on a clean slate, that the Due Process Clause qualifies the scope of political discretion heretofore recognized as belonging to Congress in regulating the entry and deportation of aliens. And since the intrinsic consequences of deportation are so close to punishment for crime, it might fairly be said also that the ex post facto Clause, even though applicable only to punitive legislation, should be applied to deportation.
>
> But the slate is not clean. As to the extent of the power of Congress under review, there is not merely "a page of history" but a whole volume. Policies pertaining to the entry of aliens and their right to remain here are peculiarly concerned with the political conduct of government. In the enforcement of these policies, the Executive Branch of the Government must respect the procedural safeguards of due process. But that the formulation of these policies is entrusted exclusively to Congress has become about as firmly imbedded in the legislative and judicial tissues of our body politic as any aspect of our government. And whatever might have been said at an earlier date for applying the ex post facto Clause, it has been the unbroken rule of this Court that it has no application to deportation.
>
> We are not prepared to deem ourselves wiser or more sensitive to human rights than our predecessors, especially those who have been the most zealous in protecting civil rights under the Constitution and must therefore under our constitutional system recognize congressional power in dealing with aliens, on the basis of which we are unable to find [the statute] unconstitutional.[29]

The decision in *Galvan* is important because it correctly identifies stare decisis — obedience to precedent — as a major reason the Court has retained a constitutional doctrine which in other respects has little to recommend it.

Galvan is also an ironic decision, because although it gave great weight to stare decisis, the specific issue it addressed was handled so differently by the Court just a few years later that *Galvan* was effectively overruled. In *Rowoldt v. Perfetto*,[30] the Court read the provision making members of the Communist Party deportable to mean that aliens were not deportable simply because they filled out the application form, paid their dues and got Communist Party membership cards. The Court held that in spite of Rowoldt's 20-year party membership, which included a year as an employee of the Party, he did not have a "meaningful association;" the four dissenters were correct in arguing that that outcome "cannot be reconciled with *Galvan v. Press*."[31] So, in this and many other contexts, the Court has injected a sense of fairness into immigration law, not by saying the due process clause or equal protection principle rendered action unconstitutional, but by sometimes extravagantly interpreting statutes or facts to reach the conclusion that something immigration authorities wished to achieve had been prohibited by Congress.

Finally, there may be less than meets the eye to the plenary power doctrine. Although in a hundred years worth of cases, the Court has never struck down a Congressional statute as discriminatory or unfair, plenary power has always been controversial, even on the Court. In *Fong Yue Ting*, three justices dissented on constitutional grounds in favor of the aliens; many of the later cases also divided the justices. Even the majority decisions in plenary power cases are interesting. If immigration matters are committed to Congress, then decisions should be a sentence or two long: "Final decision with respect to immigration matters is confided to Congress. Therefore, whether we agree with it or not, we are not permitted to conclude that this statute is constitutionally invalid." However, the Court does not do that. They engage in some limited scrutiny of the reasonableness and rationality of Congressional action.

Even though, so far, the Court has invariably upheld congressional judgment, Congress knows that if they push too far, a statute might be held unconstitutional notwithstanding the rhetoric of the plenary power doctrine. Congress tends to change the law as domestic norms change, or even in advance of them. In 1952, for example, two years before *Brown v. Board of Education*, Congress repealed many of the racially discriminatory provisions of the Immigration and Nationality Act. In 1965, as it passed the Civil Rights Act and the Voting Rights Act, Congress got rid of the remaining discriminatory provisions, and established the current system which does not take into account race or religion at all.[32] So while in principle the plenary power doctrine says Congress can do anything, in recent years Congress has not really tested the extent of the Court's attachment to that idea.

The last article in this volume, Professor Stephen Legomsky's article *Ten More Years of Plenary Power: Immigration, Congress and the Courts*,[33] summarizes important features of the contemporary plenary power doctrine. One is that courts, even the Supreme Court, have suggested that the rational basis standard of judicial review, typically applied to economic regulation, applies to immigration classifications. These courts do not necessarily reject the plenary power doctrine, instead re-casting it as a limited form of judicial review. But if the courts are in fact engaging in judicial review, the premise of the plenary power doctrine — that for whatever reason the Constitution does not apply — seems undermined.

Professor Legomsky also notes that lower courts have used the recognized right to challenge deportation procedures as a way of expanding the ability of courts to review immigration decisions.

In sum, the constitutional law of immigration is in considerable disarray. The plenary power doctrine is unpopular with many scholars and judges, and there has been no convincing case made that it deserves to be part of our constitutional law today. Yet, in scores of cases, over twelve decades, it has invariably been sustained.

Further reading

Cases

Chae Chan Ping v. *United States* (The Chinese Exclusion Case), 130 U.S. 581, 9 S. Ct. 623, 32 L.Ed. 1068 (1889).
Nishimura Ekiu v. *United States*, 142 U.S. 651, 12 S. Ct. 336, 35 L.Ed. 1146 (1892).
Fong Yue Ting v. *United States*, 149 U.S. 698, 13 S. Ct. 1016, 37 L.Ed. 905 (1893).
Yamataya v. *Fisher* (The Japanese Immigrant Case), 189 U.S. 86, 23 S. Ct. 611, 47 L.Ed. 721 (1903).
Lem Moon Sing v. *United States*, 158 U.S. 538, 15 S. Ct. 967, 39 L.Ed. 1082 (1895).
United States v. *Ju Toy*, 198 U.S. 253, 25 S. Ct. 644 , 49 L.Ed. 1040 (1905).
United States EX REL. *Turner* v. *Williams*, 194 U.S. 279, 24 S. Ct. 719, 48 L.Ed 979 (1904) .
Oceanic Steam Navigation Co. v. *Stranahan*, 214 U.S. 320, 29 S. Ct. 671, 53 L.Ed. 1013 (1909).
Bugajewitz v. *Adams*, 228 U.S. 585, 33 S. Ct. 607, 57 L.Ed. 978 (1913).
Lapina v. *Williams*, 232 U.S. 78, 34 S. Ct. 196, 58 L.Ed. 515 (1914).
Mahler v. *Eby*, 264 U.S. 32, 44 S. Ct. 283, 68 L.Ed. 549 (1924).
United States EX REL. *Knauff* v. *Shaughnessy*, 338 U.S. 537, 70 S. Ct. 309, 94 L.Ed. 317 (1950).
(Discussed in Ellen Knauff, *The Ellen Knauff Story* (New York: Norton, 1952) (New York: DaCapo, reprint 1974)).
Shaughnessy v. *United States* EX REL. *Mezei*, 345 U.S. 206, 73 S. Ct. 625, 97 L.Ed. 956 (1953).
(Discussed in Charles Weisselberg, "The Exclusion and Detention of Aliens: Lessons from the lives of Ellen Knauff and Ignatz Mezei," *University of Pennsylvania Law Review* 143: 933 (1995)).
Harisiades v. *Shaughnessy*, 342 U.S. 580, 72 S. Ct. 512, 96 L.Ed. 586 (1952).
Galvan v. *Press*, 347 U.S. 522, 74 S. Ct. 737, 98 L.Ed. 911 (1954).
Boutilier v. *Immigration and Naturalization Service*, 387 U.S. 118, 87 S. Ct. 1563, 18 L.Ed.2d 661 (1967).
Fiallo v. *Bell*, 430 U.S. 787, 97 S. Ct.1473, 52 L.Ed.2d 50 (1977).
Kleindienst v. *Mandel*, 408 U.S. 753, 92 S. Ct. 2576, 33 L.Ed.2d 683 (1972).
(Discussed in John A. Scanlan, "Aliens in the Marketplace of Ideas: The Government, The Academy and the McCarran-Walter Act," *Texas Law Review* 66:1481 (1988). Reprinted in *Immigration and Nationality Law Review* 113 (1990)).
Landon v. *Plasencia*, 459 U.S. 21, 103 S. Ct. 321, 74 L.Ed.2d 21 (1982).
Immigration and Naturalization Service v. *Chadha*, 462 U.S. 919, 103 S. Ct. 2764, 77 L.Ed.2d 317 (1983).
(Discussed in Barbara Hinkson Craig, *Chadha: The Story of an Epic Constitutional Struggle*. (New York: Oxford, 1988)).
Arab American Anti-Discrimination Committee v. *Reno*, 525 U.S. 471, 119 S. Ct. 936, 142 L.Ed.2d 940 (1999).

General Materials

Aleinikoff, Thomas Alexander, David Martin, and Hiroshi Motomura. *Immigration and Citizenship: Process and Policy* (St. Paul: West Group, 4th ed., 1998).
Georgetown Immigration Law Journal (published by the Georgetown University Law Center).
Gordon, Charles, Stanley Mailman, and Stephen Yale-Loehr. *Immigration Law and Procedure* (multi-volume treatise).
Immigration and Nationality Law Review (Buffalo: William S. Hein and Co., 1976-Present) (annual anthology reprinting many leading law review articles).
Legomsky, Stephen H. *Immigration and Refugee Law and Policy* (Westbury: The Foundation Press, 2d ed., 1997).
Legomsky, Stephen H. *Immigration and the Judiciary: Law and Politics in Britain and America*. (Oxford: Clarendon Press, 1987).

Immigration History

Books

Daniels, Roger. *Coming to America: A History of Immigration and Ethnicity in American Life* (New York: HarperCollins, 1990).

Devine, Robert A. *American Immigration Policy, 1924–1952* (New Haven: Yale University Press, 1952).

Higham, John. *Strangers in the Land: Patterns of American Nativism, 1860–1925* (New Brunswick: Rutgers University Press, 1955).

Hutchinson, Edward P. *Legislative History of American Immigration Policy 1798–1965* (Philadelphia: University of Pennsyvlania Press, 1981).

Jones, Maldwyn Allen. *American Immigration* (Chicago and London: University of Chicago Press, 2d ed, 1992).

Salyer, Lucy E. *Laws Harsh as Tigers: Chinese Immigrants and the Shaping of Modern Immigration Law* (Chapel Hill and London: University of North Carolina Press, 1995).

Articles

Henkin, Louis. "The Constitution and United States Sovereignty: A Century of *Chinese Exclusion* and its Progeny," *Harvard Law Review* 100:853 (1987). Reprinted in *Immigration and Nationality Law Review* 115 (1988).

Journal of American Ethnic History (published by Transaction Periodicals Consortium, Rutgers University).

Nafziger, James A.R. "The General Admission of Aliens under International Law," *American Journal of International Law* 77: 804 (1983). Reprinted in *Immigration and Nationality Law Review* 275 (1986).

Neuman, Gerald. "The Lost Century of American Immigration Law (1776–1875)," *Columbia Law Review* 93:1833 (1993).

Legal, Economic and Policy Issues

Books

Borjas, George J. *Friends or Strangers: The Impact of Immigrants on the U.S. Economy* (New York: Basic Books, 1990).

Briggs, Vernon M., Jr. and Stephen Moore. *Still an Open Door? U.S. Immigration Policy and the American Economy* (Washington, D.C.: The American University Press, 1994).

Chin, Gabriel J. (ed.) *United States Commission on Immigration Reform: The Interim and Final Reports and Commentary* (Buffalo: William S. Hein & Co., 2000).

Konvitz, Milton R. *The Alien and the Asiatic in American Law* (Ithaca: Cornell University Press, 1946).

Neuman, Gerald L. *Strangers to the Constitution: Immigrants, Borders, and Fundamental Law* (Princeton: Princeton University Press, 1997).

Perea, Juan F. (ed.). *Immigrants Out! The New Nativism and the Anti-Immigrant Impulse in the United States* (New York and London: New York University Press, 1997).

Piatt, Bill. *Immigration Law: Cases and Materials* (Charlottesville: The Michie Co., 1994).

Pickus, Noah M. (ed.) *Immigration and Citizenship in the 21st Century* (Lanham: Rowman & Littlefield Publishers, 1998).

Reimers, David M. *Unwelcome Strangers: American Identity and the Turn Against Immigration* (New York: Columbia University Press, 1998).

Schrag, Philip G. *A Well-Founded Fear: The Congressional Battle to Save Political Asylum in America* (New York: Routledge, 2000).

Schuck, Peter. *Citizens, Strangers and In-Betweens: Essays on Immigration and Citizenship* (Boulder: Westview Press, 1998).

Smith, James P. and Barry Edmonston, (eds.) *The Immigration Debate: Studies on the Economic, Demographic and Fiscal Effects of Immigration* (Washington, D.C.: National Academy Press, 1998).

Weissbrodt, David. *Immigration Law and Procedure in a Nutshell* (St. Paul: West Publishing Co., Latest Edition).

Articles

Hahn, Richard F. Note, "Constitutional Limits on the Power to Exclude Aliens," *Columbia Law Review* 82: 957 (1982). Reprinted in *Immigration and Nationality Law Review* 7:207 (1983–84).

Legomsky, Stephen H. "Immigration Law and the Principle of Plenary Congressional Power," *Supreme Court Review* 1984: 255. Reprinted in *Immigration and Nationality Law Review* 81 (1986).

Motomura, Hiroshi. "The Curious Evolution of Immigration Law: Procedural Surrogates for Substantive Constitutional Rights," *Columbia Law Review* 92: 1625 (1992). Reprinted in *Immigration and Nationality Law Review* 14:3 (1991–92).

Note, "The Constitutional Requirement of Judicial Review for Administrative Deportation Decisions," *Harvard Law Review* 110:1850 (1997). Reprinted in *Immigration and Nationality Law Review* 18:342 (1997).

Wani, Ibrahim. Truth, "Strangers and Fiction: The Illegitimate Uses of Legal Fiction in Immigration Law," *Cardozo Law Review* 11: 51 (1989).

Notes

1 *Fiallo* v. *Bell*, 430 U.S. 787, 792, 97 S. Ct. 1473, 52 L.Ed.2d 50 (1977) (quoting *Oceanic Steam Navigation Co.* v. *Stranahan*, 214 U.S. 320, 339, 29 S. Ct. 671, 53 L.Ed. 1013 (1909); *Shaughnessy* v. *United States* EX REL. *Mezei*, 345 U.S. 206, 210, 73 S. Ct. 625, 97 L.Ed. 1956 (1953)).

2 *See* 5 Ronald D. Rotunda and John E. Nowak, *Treatise on Constitutional Law: Substance and Procedure*, § 22.2 at 191 (3d ed. 1999) ("In a long line of cases the Supreme Court has stated that the power of Congress over the admission of aliens to this country is absolute.").

3 *University of California, Davis Journal of International Law and Policy* 1: 227 (1995).

4 Immigration Policies: Messages of Exclusion to African Americans, *Howard Law Journal* 37:237 (1994), Reprinted in *Immigration and Nationality Law Review* 16:255 (1995–96).

5 *See, e.g.*, Gilbert Paul Carrasco, "Latinos in the United States." In *Immigrants Out! The New Nativism and the Anti-Immigrant Impulse in the United States* (Juan F. Perea ed., 1997).

6 *See generally* Committee on the Judiciary, House of Representatives, *Grounds for Exclusion of Aliens under the Immigration and Nationality Act: Historical Background and Analysis* (Washington, D.C.: U.S. Government Printing Office, 1988).

7 130 U.S. 581, 9 S. Ct. 623, 32 L.Ed. 1068 (1889).

8 149 U.S. 698, 13 S. Ct. 1016, 37 L.Ed. 905 (1893).

9 *Id.* at 717.

10 *Id.* at 712.

11 142 U.S. 651, 12 S. Ct. 336, 35 L.Ed. 1146 (1892).

12 *Id.* at 660, 12 S. Ct. 339.

13 Art. I, § 8, cl. 3 (Congress may "regulate Commerce with foreign nations."). Indeed, earlier cases had held that the commerce clause was the source of Congressional authority to regulate immigration. *See* Mary Sara Bilder, "The Struggle over Immigration: Indentured Servants, Slaves and Articles of Commerce," *Missouri Law Review* 61: 743 (1996). For a more detailed analysis of the constitutional basis of immigration law, see Stephen H. Legomsky, *Immigration and Refugee Law and Policy* (Foundation Press, 2d ed., 1997) 8–22.

14 Art. I, § 8, cl. 4 (Congress may "establish an uniform Rule of Naturalization.").

15 Art. 1, § 9, cl. 1 ("The Migration or Importation of such Persons as any of the States now existing shall think proper to admit, shall not be prohibited by the Congress prior to the year one thousand eight hundred and eight. . . ."). Of course, this was meant to refer to the slave trade.

16 *Wong Wing* v. *United States*, 163 U.S. 228, 16 S. Ct. 977, 41 L.Ed. 140 (1896).

17 *Landon* v. *Plasencia*, 459 U.S. 21, 32, 103 S. Ct. 321, 74 L.Ed.2d 21 (1982). *See also United States* EX REL. *Vajitaur* v. *Commissioner of Immigration*, 273 U.S. 103, 47 S. Ct. 302, 71 L.Ed. 560 (1927); *United States* EX REL. *Tisi* v. *Tod*, 264 U.S. 131, 44 S. Ct. 260, 68 L.Ed. 590 (1924); *Ng Fung Ho* v. *White*, 259 U.S. 276, 42 S. Ct. 492, 66 L.Ed. 938 (1922); *Yamataya* v. *Fisher* (The Japanese Immigrant Case), 189 U.S. 86, 23 S. Ct. 611, 47 L.Ed. 721 (1903).

18 *See United States* v. *Belmont*, 301 U.S. 324, 332, 57 S. Ct. 758, 81 L.Ed. 1134 (1937) ("our Constitution, laws and polices have no extraterritorial operation"); *Fong Yue Ting* v. *United States*, 149 U.S. 698, 738, 13 S. Ct. 1016, 37 L.Ed. 905 (1893) (Brewer, J., dissenting) (distinguishing deportation from interior of country from exclusion at the border because the "constitution has no extraterritorial effect").

[19] *See* 3 Ronald D. Rotunda and John E. Nowak, *Treatise on Constitutional Law: Substance and Procedure*, § 17.2, 4–7 (3d ed., 1999).

[20] *Id.* at 5.

[21] *See United States* v. *Curtiss-Wright Export Co.*, 299 U.S. 304, 318, 57 S. Ct. 216, 81 L.Ed. 255 (1936) (asserting that "the power to expel undesirable aliens . . . [is] found not in the provisions of the Constitution, but in the law of nations."); *Nishimura Ekiu* v. *United States*, 142 U.S. 651, 659, 12 S. Ct. 336, 35 L.Ed. 1146 (1892) (immigration power "is an accepted maxim of international law.").

[22] *See* Michael Scaperlanda, "Polishing the Tarnished Golden Door," *Wisconsin Law Review* 1993:965; Berta Esperanza Hernandez-Truyol, "Natives, Newcomers and Nativism: A Human Rights Model for the Twenty-first Century," *Fordham Urban Law Journal* 23: 1075 (1996)

[23] *American Journal of International Law* 83: 865 (1989).

[24] *UCLA Law Review* 46:1 (1998).

[25] *Columbia Law Review* 84:1 (1984).

[26] *Yale Law Journal* 100:545 (1990).

[27] *Id.* at 564-65.

[28] 347 U.S. 522, 74 S. Ct. 737, 98 L.Ed. 911 (1954).

[29] *Id.* at 530-32 (Citations omitted).

[30] 355 U.S. 115, 78 S. Ct. 180, 2 L.Ed.2d 140 (1957).

[31] 355 U.S. at 122, 78 S. Ct. at 184 (Harlan, J., dissenting).

[32] *See* Gabriel J. Chin, "The Civil Rights Revolution Comes to Immigration Law: A New Look at the Immigration and Nationality Act of 1965," *North Carolina Law Review* 75:263 (1996). Reprinted in *Immigration and Nationality Law Review* 17:87 (1995–96).

[33] *Hastings Constitutional Law Quarterly*, 22:925 (1995).

A NATION THAT WELCOMES IMMIGRANTS?
AN HISTORICAL EXAMINATION OF UNITED STATES IMMIGRATION POLICY

James F. Smith[*]

[*] Senior Lecturer, University of California, Davis School of Law. B.S. 1964, Arizona State University at Tempe; J.D. 1967, University of California at Berkeley.

INTRODUCTION

Give me your tired, your poor,
Your huddled masses, yearning to breathe free,
The wretched refuse of your teeming shore.
Send these, the homeless, tempest-tost to me,
I lift my lamp beside the golden door!

— Emma Lazarus, inscription on the Statue of Liberty

The United States has had one of the most generous and open immigration policies in the world. But since the ratification of the Constitution, United States immigration policy has been schizophrenic. Early Americans welcomed immigrants as a source of strength and vitality for a growing nation. However, from the 1830s on, when the Irish immigrants were castigated as unfit, some Americans began to view immigrants not as a resource, but as a cause of social and economic decline. These concerns have prompted a United States immigration policy that has at times been inconsistent, racist, and surprisingly un-American.

This Article examines the history of United States immigration policy. Part I discusses the history of United States immigration and naturalization legislation. Part II discusses United States immigration policy towards Mexico.

I. THE HISTORY OF UNITED STATES IMMIGRATION AND NATURALIZATION LEGISLATION

In the first century following ratification of the Constitution, Americans viewed immigration as a welcome source of national strength and wealth.[1] With few exceptions, immigration remained unregulated. The

---

[1] The Declaration of Independence itself cites restrictions on immigration as one of the causes for the American Revolution. *See also* EDWARD P. HUTCHINSON, LEGISLATIVE HISTORY OF AMERICAN IMMIGRATION POLICY 1798-1965, at 48 (1981) (quoting Abraham Lincoln, who encouraged immigration in a speech to Congress in 1863). Thomas Jefferson viewed the United States as an asylum for oppressed humanity. 11 ANNALS OF CONGRESS 16 (1801). President Tyler extended an invitation to the people of other countries to come and settle in the United States. CONG. GLOBE, 28th Cong., 1st Sess. 9 (1843).

nation's first immigration legislation, the Alien and Sedition Act of 1798, gave the president power to deport aliens he deemed "dangerous to the peace and safety of the United States." The Act also required registration of all aliens.[2] It expired in two years without having been used. The Alien Enemies and Alien Friends Act of 1798[3] permitted the deportation of all alien males fourteen and older who were natives, subjects, or citizens of any country at war with the United States.[4] President Thomas Jefferson criticized these measures for their failure to provide due process.

During the first three quarters of the 18th century, Congress resisted any further attempts to regulate immigration. In the 1830s and 1840s, a large influx of poor rural Catholics led to calls for "quality control" of immigrants by excluding criminals, paupers, and lunatics. This period witnessed the emergence of organized groups of "nativists" who favored restrictionist immigration policies. Such groups, which in the 1830s included the Secret Order of the Star Spangled Banner and the Know Nothing political party, have had a continuing influence in the United States.[5] But the dominant consensus favored an open immigration policy and Congress continued to defer to individual states to regulate immigration.[6]

A. *Selective Admissions*

United States possession of the former Mexican territories and discovery of gold in California lead to a large influx of Chinese and European immigrants. The rush to develop the West and build the transcontinental railroad demanded a quick supply of cheap labor. Both Chinese laborers and Mexican workers filled this need. Often, these laborers signed work contracts in return for transportation to the United States. Congress initially objected to such contracts trade as a form of slave trading.[7]

---

[2] 1 Stat. 570 (1798).

[3] 50 U.S.C. §§ 21-23 (1988).

[4] 1 Stat. 577 (1798); *see also* Lawrence H. Fuchs & Susan B. Forbes, *Select Commission on Immigration and Refugee Policy [SCIRP], U.S. Immigration Policy and the National Interest*, 92-93, 161-216, *reprinted in* THOMAS A. ALEINIKOFF & DAVID A. MARTIN, IMMIGRATION PROCESS AND POLICY 40-59 (2d ed. 1986).

[5] *See* Fuchs & Forbes, *supra* note 4, at 44 (discussing nativist groups).

[6] HUTCHINSON, *supra* note 1, at 46.

[7] *Id.* at 42; *see also* The Coolie Act of 1862, 12 Stat. 340 (1862).

The Immigration Act of 1875[8] prohibited immigrant labor contracts based on fraud and made contracting to supply Chinese laborers a felony. The measure also designated felony criminals and prostitutes as excludable classes of aliens. The exclusion of Chinese workers lead to a labor shortage in the West and to increased recruitment of Mexican workers.[9] In 1880, the United States negotiated a treaty with China permitting the former to regulate or temporarily suspend Chinese immigration.[10]

The Chinese Exclusion Act of 1882 was the United States' first comprehensive effort to restrict immigration. Congress responded to the economic depression of the time by scapegoating the Chinese as the source of unemployment and other economic dislocation. The hue and cry against the Chinese included the xenophobic complaints that they had failed to assimilate "our customs" and had "brought" with them prostitution. Some also feared that increased Chinese immigration could depress white settlement in the West.[11] The Act banned the immigration of Chinese laborers for ten years, provided for the deportation of illegal Chinese immigrants, and prohibited Chinese from becoming United States citizens.[12] These racist restrictions were later extended to all Asians, natives of India, and others.[13]

The Act instituted a head tax on all immigrants (except Canadians and Mexicans)[14] to defray the cost of regulating immigration and caring for immigrants. It provided for the exclusion of any convict, lunatic, idiot, or other person likely to become a public charge. Excluded aliens were to be returned to their country of origin. Immigrant convicts could be deported even after leaving the ship.[15]

This new policy of designating "excludable aliens" was later

---

[8] 18 Stat. 378 (1875).

[9] Gilberto Cardenas, *United States Immigration Policy Toward Mexico: An Historical Perspective*, 2 CHICANO L. REV. 66, 69 (1975).

[10] HUTCHINSON, *supra* note 1, at 76-77.

[11] *Id.* at 76.

[12] 22 Stat. 58 (1882).

[13] Immigration Act of 1917, 39 Stat. 874, 876 (1917).

[14] 23 Stat. 58 (1884); 34 Stat. 898 (1907).

[15] *Id.*

expanded to include aliens hired under labor contracts (1885),[16] paupers, polygamists, the insane and diseased (1891),[17] anarchists or subversives (1903),[18] the disabled (1907), stowaways, illiterates over age 16, alcoholics and psychotics (1917),[19] persons ineligible to become citizens (most Asians) and non-immigrants without proper documents (1924),[20] drug addicts, immoral sex-offenders, laborers not in demand in the United States, those previously deported for any reason and those aiding illegal aliens (1952).[21]

The 1917 enactment of the literacy test was an important symbolic victory for those nativist groups who favored a more restrictionist policy.[22] This controversial provision would have excluded a large portion of the nation's 19th century immigrants. In 1907, after the failure of the restrictionists to establish the test, Congress established the Dellingham Commission to study the impact of immigration on the United States. The Commission strongly recommended the literacy test to keep out inferior foreigners.[23] The measure became law in 1917 after four presidential vetoes.[24] However, it largely failed to control immigration in the manner that its proponents had intended because aspiring immigrants simply became literate.[25]

---

[16] Alien Contract Labor Act, ch. 164, 23 Stat. 332 (1885); Act of February 23, 1887, ch. 220, 24 Stat. 414 (1887).

[17] Act of March 3, 1891, ch. 551, § 1, 26 Stat. 1084 (1891).

[18] Act of March 3, 1903, ch. 1012, § 2, 32 Stat. 1214; Immigration Act of 1917, ch. 29, § 19, 39 Stat. 889; Anarchist Act of 1918, ch. 186, § 1, 40 Stat. 1012; Act of June 5, 1920, ch. 251, § 1, 41 Stat. 1009. Current versions are codified at 8 U.S.C. §§ 1182(a) (27)-(29), 1251(a) (15)-(16) (1988). *See generally* W. PRESTON JR., ALIENS AND DISSENTERS: FEDERAL SUPPRESSION OF RADICALS, 1903-1933, at 4 (1963).

[19] Act of February 5, 1917, ch. 29, § 3, 39 Stat. 875 (1917).

[20] Act of May 26, 1924, ch. 190, 43 Stat. 153 (1924).

[21] Immigration and Nationality Act, Pub. L. No. 414, § 212, 66 Stat. 163, 182-85 (1952).

[22] The 1917 Act codified the list of alien categories to be excluded and banned most immigration from Asia.

[23] Fuchs & Forbes, *supra* note 4, at 50.

[24] President Cleveland (1896), President Taft (1913), and President Wilson (1914 and 1917).

[25] Fuchs & Forbes, *supra* note 4, at 52.

B. National Origin Quotas (1921-1965)

Congressional concern about immigration was heightened immediately after World War I. Many Americans feared that millions of European refugees would immigrate to the United States and settle in urban areas that already faced unemployment and housing shortages.[26] As an alternative to complete prohibition of immigration,[27] Congress passed the first law establishing quotas for immigration.[28] The Act limited immigration to a three percent quota based on the number of aliens of that nationality living in the United States in 1910. Immigrants from the Western Hemisphere were exempted[29] despite concerns that "citizen" factories would be created.[30]

The quota used in the 1921 legislation disproportionately favored the most recent immigrants. The National Origins Act of 1924 ("Johnson-Reed") adopted a quota system that was designed to preserve the Northern European and British Isles composition of the population. Its racist formula was designed to maintain the national origin of the United States as it existed in 1890. The 1924 Act significantly decreased the quota allotment to Southern and Eastern Europeans (44% to 12%).[31] The total number of immigrants from the Eastern Hemisphere was limited to 150,000. The Western Hemisphere remained exempt. However, from 1928 on, bills were

---

[26] *See* The Immigration and Naturalization Systems of the United States, S. REP. NO. 1515, 91st Cong., 2d Sess. (1950). *See also* Esther Rosenfeld, *Fatal Lessons: United States Immigration Policy During the Holocaust,* 1 U.C. DAVIS J. INT'L L. & POL'Y 249 (1995) (detailing U.S. immigration policy during Holocaust).

[27] H. R. REP. NO. 14461, 66th Cong., 3d Sess. (1920).

[28] Act of May 19, 1921, ch. 8, § 2, 42 Stat. 5 (1921). Operation of the previous quota was extended by two years. Act of May 11, 1922, ch. 187, § 1, 42 Stat. 540 (1922).

[29] Congress's extreme concern with European refugee immigration best explains the Western Hemisphere exemption. World War I did not ravage the Western Hemisphere; therefore, Congress did not fear a huge influx of immigrants from the Western Hemisphere. Also, immigration from Europe constituted nearly 80% of total immigration between 1911 and 1920. Accordingly, Congress was less concerned with non-European immigration.

[30] 60 CONG. REC. 3443 (1921).

[31] 43 Stat. 153 (1924).

introduced in Congress to eliminate the exemption.[32] Abraham Lincoln's prophetic statement that should the nativists have their way, the Declaration of Independence would be rewritten to read: "All men are created equal, except Negroes, foreigners, and Catholics" seemed at hand.

C. *The Elimination of the Quota System and Illegalization of the Mexican Worker (1952, 1965-1976)*

During World War II, Congress repealed the exclusion legislation of 1882 because it felt that such legislation represented a continuing insult to its Chinese ally.[33] However, in 1952, the McCarran-Walter Act, which codified and revised United States immigration laws [hereafter the Immigration and Nationality Act or "INA"] and preserved the national origins quota system.[34] This provoked President Truman's veto, which Congress later overrode. The Act strengthened the exclusion provisions and made the naturalization process more rigorous.[35] It also eliminated all exclusions based on race and established a preference system favoring highly skilled workers.

In 1952, Congress also passed the so-called "Wetback Act," which was designed to discourage illegal immigration from Mexico.[36] The act permitted the Border Patrol to enter private lands within 25 miles of the border between the United States and Mexico.[37]

D. *The Immigration Act of 1965*

The United States civil rights movement of the 1960s greatly influenced the first major change in the Nation's immigration policy since

---

[32] H. R. REP. 10,995; H. R. REP. 6465; H. R. REP. 15,079; *see* HUTCHINSON, *supra* note 1, at 209 (noting that sugar beet growers blocked passage of bill in 1928).

[33] 60 Stat. 1353 (1943).

[34] 66 Stat. 163, 175-76 (1952).

[35] Lawrence A. Fuchs, *Immitration Policy and the Rule of Law*, 44 U. PITT. L. REV. 433, 434 (1983).

[36] Act of June 27, 1952, ch. 8, 66 Stat. 163, 228-29 (1952), *codified at* 8 U.S.C. § 1324(a) (1988).

[37] 60 Stat. 865 (1946); *see also infra*, notes 107-112 and accompanying text (discussing Wetback Act).

the National Origins Act of 1924. President John F. Kennedy had written a book, *A Nation of Immigrants,* which denounced the national origins quota system. Following Kennedy's assassination, President Johnson forcefully supported reform.[38] The Immigration Act of 1965 represented a step away from racially motivated quotas and toward a more humane system of immigration.[39] The 1965 Act abolished the national origin quotas system and also the Asian-Pacific triangle immigration restriction provisions of the 1952 Act. It replaced these restrictions with a per-country limit on every country outside of the Western Hemisphere. Eastern Hemisphere immigration was limited to 160,000 with a per-country limit of 20,000 immigrants per year.

The price for the abolition of the national origins system was, for the first time, to limit Western Hemisphere immigration (annual limit of 120,000).[40] The new system closed the last "Good Neighbor"[41] or open door policy of United States immigration. Although the Johnson Administration opposed the Western Hemisphere limitation, one researcher has written that:

> It can only be inferred that the . . . [limitations] . . . were to reassure those concerned at the possibility of large immigration from Latin America or fearful of ill effects on the labor market or to provide a *quid pro quo* to influential members of Congress or interest groups in return for their support of the bill.[42]

Future immigrants from the Western Hemisphere, without certain close relatives who were United States citizens or lawful permanent residents, would have to demonstrate their qualifications as skilled workers who were then in short supply.

---

[38] HUTCHINSON, *supra* note 1, at 435.

[39] 79 Stat. 911 (1965).

[40] Fuchs & Forbes, *supra* note 4, at 57.

[41] "Good Neighbor" is a phrase made popular by President Franklin Delano Roosevelt to describe friendly relations between the United States and Latin America.

[42] HUTCHINSON, *supra* note 1, at 378.

E. Immigration Legislation (1970-76)

By 1970, the end of the Bracero Program[43] and the imposition of the Western Hemisphere quotas and labor certification requirements had caused United States immigration policy toward Mexico to change from a generally open door policy to a more restrictive one.[44] Inevitably, "illegal immigration" from Mexico became a political issue. Congress had illegalized the Mexican worker and now had to deal with the "problem."[45] By 1971, it began to consider legislation to control the Mexican "illegal alien" problem. Representatives from the Mexican-American community argued that the United States should consider the special historic, geographic, social, and economic ties between the United States and Mexico. Certainly, the integrated economic and labor markets had existed before the creation of the international border and were continued for over a century under the open door policies.[46] Nonetheless, Congress has repeatedly rejected efforts to establish a preferential quota for Mexico, citing the distaste for any type of national origins quotas.[47]

---

[43] *See infra* notes 101-105 and accompanying text (discussing Bracero Program).

[44] In 1976, Western Hemisphere countries, including Mexico, were subjected to the 20,000 per country limit. Act of Oct. 20, 1976, Pub. L. 94-571, 90 Stat. 2703 (1976). While the legislation was neutral on its face, it had a uniquely prejudicial effect on Mexico, the only country to have exceeded that limit in the past several years. *The 'Rodino Bill' an Example of Prejudice Toward Mexican Immigration to the United States,* 2 CHICANO L. REV. 40, 46 (1975). Congress also noted that illegal immigration was so high that the proposed 35,000 limit for Mexico would still be insufficient. H. R. REP. NO. 94-1553, 94th Cong., 2d Sess. 9 (1975).

[45] The INS reported having deported 30,000 Mexicans in 1961, 108,000 in 1967 after the end of the Bracero program, and 450,000 in 1979. INS Annual Reports, 1961, 1967, and 1979. INS Commissioner Raymond F. Farrell directly attributed the rise in Mexican illegal immigration to the end of the Bracero program and the imposition of the Western Hemisphere quotas. *Hearings before House Subcomm. of the Judiciary,* 92nd Cong., 1st Sess. 19 (1971) (statement of Raymond F. Farrell, Commissioner, INS).

[46] H. R. REP. NO. 1, 92d Cong., 1st Sess., pt. 1, at 216, 269-70 1971.

[47] Both Presidents Gerald Ford and Jimmy Carter stated their support for a special quota for Mexico, but neither administrations took action to make this a reality. Fuchs & Forbes, *supra* note 4, at 58.

F. *The Refugee Act of 1980*

Throughout most of its history, the United States has treated refugees as simply another category of immigrants subject to the applicable admissions criteria and quotas. For example, in 1939 Congress defeated a bill to rescue 20,000 from Nazi Germany on the grounds that it would exceed the German quota. Following World War II, Congress was more generous and passed the Refugee Relief Act of 1948, under which 214,000 persons were admitted. However, the United States declined to join the 1951 United Nations Convention Relating to the Status of Refugees.[48]

In 1956 and 1957, Congress enacted measures to provide refugee status to those fleeing certain communist Eastern European countries and the Middle East. However, these provisions were also subject to the United States' refugee country quotas of the refugees. Throughout the 1960s and 1970s, the principal means for refugee admission was through the INS conditional entry or paroling power. The Refugee Act of 1980 was designed to bring the United States into conformity with the Refugee Protocol of 1968, which followed the 1951 Convention. It was also aimed to develop an ongoing mechanism for the admission of refugees abroad and the adjudication of political asylum claims in the United States.

G. *Immigration Reform and Control Act of 1986 ("Simpson-Rodino," hereafter "IRCA")*

The early 1980s saw a vigorous debate over United States immigration policies. Undocumented workers were seen as the foot soldiers of a "silent invasion." Many were concerned that the United States had lost control of its borders. Through employer sanctions, Senator Alan Simpson led the crusade to stop the "pull" factor of undocumented immigration. Liberals, however, were able to barter for a so-called amnesty or legalization law for certain aliens who were long-term residents despite their undocumented status.

Thus, the major reforms of the INA by the IRCA were: 1) to impose employer sanctions in the form of civil fines[49] and possible criminal

---

[48] 189 U.N.T.S. 137.

[49] Imposed fines will be not less than $250.00 nor more than $2000.00 for each unauthorized alien. 8 C.F.R. 274a.10 (1994). Employers may also be fined between $100.00 and $1,000.00 for each failure to complete I-9 forms (document verification). Immigration & Naturalization Act [INA], § 274A(e)(4), *codified at* 8 U.S.C. § 1324a(e)(4) (1988).

penalties[50] for the knowing hiring or continued employment of an "unauthorized alien";[51] 2) to create the so-called amnesty program under which aliens unlawfully in the country since before January 1, 1982 could apply for the newly created status of "temporary resident" (if successful, such aliens could later apply for permanent residence);[52] 3) to create a category of "Special Agricultural Workers" ("SAWs")[53] and "Replenishment Agricultural Workers" ("RAWs")[54] who could more easily qualify as temporary residents on the basis of having worked in "seasonal agricultural services,"[55] and 4) to create a new category of non-immigrant workers (H-2A).[56]

The legalization and SAW programs were short-term programs that ended in 1988.[57] However, the employer sanctions and agricultural worker

---

[50] INA § 274A(f), *codified at* 8 U.S.C. § 1324a(f) (1988). This code section provides for criminal penalties for a "pattern or practice" of unauthorized employment. *Id.*

[51] INA § 274A(a)(1), (h)(3), *codified at* 8 U.S.C. § 1324a(a)(1), (h)(3) (1988). Although the hiring of such aliens after November 6, 1986 is prohibited, employees hired before that date may be retained. IRCA § 101(a)(3); 8 C.F.R. § 274a (1994).

[52] INA § 245A, *codified at* 8 U.S.C. § 1255a (1988) (explaining that applications for permanent residence will be granted 18 months after temporary status is acquired if alien is otherwise eligible for immigrant visa and has "minimum understanding of ordinary English and knowledge or understanding of history and government of United States, or . . . is satisfactorily pursuing a course of study . . . to achieve . . . [these goals]." INA § 245A(b)(1)(D), *codified at* 8 U.S.C. § 1255a(b)(1)(D) (1988).

[53] INA § 210, *codified at* 8 U.S.C. § 1160 (1988). To be eligible for this program an alien must prove that he worked in a perishable crop for at least 90 days between May 1, 1985 and May 1, 1986.

[54] INA § 210A, *codified at* 8 U.S.C. § 1161 (1988).

[55] The legislation defines "seasonal agricultural service" as the performance of field work related to planting, cultural practices, cultivation, growing, and harvesting of fruits and vegetables or every kind and other perishable commodities, as defined in the regulations by the Secretary of Agriculture. INA § 210(h), *codified at* 8 U.S.C. § 1160(h) (1988).

[56] INA §§ 101(a)(15)(H)(ii), 216; *codified at* 8 U.S.C. §§ 1101(a)(15)(H)(ii), 1186 (1988).

[57] The period for legalization applications ended on May 4, 1988. INA § 245(a)(1)(A), *codified at* 8 U.S.C. § 1255(a)(1)(A) (1988). The SAW application period ended November 30, 1988. INA § 210(a)(1)(A), *codified at* 8 U.S.C.

provisions presented significant long-term changes in the immigration laws of the United States for all employers. Before the IRCA, the knowing employment of undocumented workers was not illegal. However, the act of transporting, concealing, harboring, or encouraging the entry of undocumented aliens has been punishable as a felony since 1952. The act of entering the United States without inspection has been a misdemeanor since the enactment of the INA in 1952. Aliens, however, were typically deported, not prosecuted.[58] Also, the smuggling provisions were not used against employers because of the famous "Texas Proviso" that exempted employers.[59]

The IRCA not only repealed the "Texas Proviso" but now requires all employers[60] to verify that each new employee possesses documents which demonstrate that she is authorized to work.[61] Both the employer and

---

§ 1160(a)(1)(A) (1988). Approximately 2 people million applied for legalization. About 10% of that number applied for the SAW program. Both programs have been the subject of considerable criticism and litigation. Critics charge that the INS administered the programs in such a rigorous fashion that thousands have been arbitrarily disqualified. For example, the INS regulations bar those who have left the United States for more than 45 days on any one occasion or for a total absence of more than 180 days. *See* 8 C.F.R. § 245a.2(h)(1) (1994). Also, only individuals were eligible to qualify, and not their spouses or children, which discouraged many from applying who feared that their family would be broken up. Although the legalization documents are confidential and may not be used to effect deportations, potential applicants were skeptical. SAW applicants did not receive the cooperation of the INS in obtaining the necessary proof of their employment from recalcitrant former employers.

[58] INA § 275, *codified at* 8 U.S.C. § 1325 (1988).

[59] The Texas Proviso stated: "Provided, however, that for the purpose of this section, employment (including the usual and normal practices incident to employment) shall not be deemed to constitute harboring." INS § 274, *codified at* 8 U.S.C. § 1255 (1988). *See also infra* notes 107-112 (discussing Wetback Act).

[60] The exemption covers "casual employment of an individual; who provided domestic service in a private home that is sporadic, irregular or intermittent." Also, independent contractors are not considered employees. 8 C.F.R. § 274a.1(f)-(h) (1994). The legislative history of the IRCA provides an exception to "casual hires (*i.e.* those that do not involve the existence of an employer/employee relationship)." H. R. REP. NO. 682, 99th Cong., 2d Sess., pt. 1, at 57 (1986).

[61] Employers must complete Form I-9, which states that they have inspected the required documents for all employees who were hired after November 6, 1986 and were still employed after May, 1987. INA § 272A(a)(1)(B), *codified at* 8 U.S.C.

the employee must sign a form under penalty of perjury.[62] The employer must swear that she has examined certain documents and that the employee is either a United States citizen, lawful permanent resident, or alien for three years or one year after the employment terminates, whichever occurs first.[63]

If the employer determines that "the document reasonably appears on its face to be genuine,"[64] such "good faith" examination establishes an "affirmative defense" to the sanctions.[65] These documents include those which establish citizenship,[66] lawful permanent residence,[67] a social security card plus proof of identity, or employment authorization by the INS.[68]

Because of the fears expressed that the IRCA would result in discrimination against "foreign appearing" workers, the IRCA prohibits discrimination on the basis of "national origin" or "citizenship status."[69] These provisions only protect citizens and permanent residents who demonstrate that they are "intending citizens" by formally declaring their intention to become naturalized in a timely fashion, once they are eligible to do so.[70] Also, an employer may hire a United States citizen in preference to an alien if the two are "equally qualified."[71] When President Reagan signed the IRCA, he signed a statement that these provisions only

---

§ 1324a(a)(1)(B) (1988); 8 C.F.R. § 274a.2 (1994).

[62] Form I-9 (Employer Eligibility Verification).

[63] INA § 274A(B), *codified at* 8 U.S.C. § 1324a(B) (1988); 8 C.F.R. § 274a.2(a) (1994).

[64] INA § 274A(b)(1)(A), *codified at* 8 U.S.C. § 1324a(b)(1)(A) (1988).

[65] INA § 274A(a)(3), *codified at* 8 U.S.C. § 1324a(a)(3) (1988); 8 C.F.R. § 274a.4 (1994).

[66] These documents have been held to include birth certificate, United States passport, certificate of citizenship or of naturalization.

[67] Form I-551 (green card).

[68] *See generally* INA § 274A(b)(1)(B)-(D), *codified at* 8 U.S.C. § 1324a(b)(1)(B)-(D) (1988).

[69] INA § 274(B)(a)(1), *codified at* 8 U.S.C. § 1324B(a)(1) (1988).

[70] INA § 274(B)(a)(3), *codified at* 8 U.S.C. § 1324B(a)(3) (1988).

[71] INA § 274(B)(a)(2), (4), *codified at* 8 U.S.C. § 1324B(a)(2), (4) (1988).

covered "discriminatory intent" and not actions having a "disparate impact."[72] If this interpretation is correct it would be difficult to prove such discrimination, as there is no requirement that records be kept of employment applicants who are not hired.[73]

There is little doubt that the passage of the IRCA has resulted in the proliferation of falsified documents.[74] In effect, the undocumented workers who were previously merely subject to deportation have now become the "falsely documented." In order to secure employment, these workers present false documents, which is a crime. If the primary goal of these provisions is to deter the worker's employment, then there is little indication that the employer's sanction law has been successful. Limited resources for enforcement and the employer's good faith defense for having inspected documents have reduced the program to little more than paperwork, except for the relatively few employers who have suffered civil penalties for violation.

The employer sanctions and legalization provisions appear relatively harsh when compared to the agricultural features of the IRCA. The necessary residence and labor requirements for the SAWs are more favorable because SAWs may effectively commute across the border.[75] SAWs may become permanent residents without meeting the "basic citizenship" requirements.[76] If the SAWs, who are not required to remain in agricultural work, and the non-immigrant agricultural workers in the new H-2A program do not meet the needs of agribusiness employers, the IRCA provides a safety valve. Here, the IRCA authorizes "replenishment agricultural workers" if the Secretaries of Agriculture and Labor determine that a shortage of available agricultural workers exist.[77] This generous provision for agricultural employers reflects the employer's political power.

---

[72] Statement by the President, reprinted in 63 Interp. Rel. 1036, 1037 (1986).

[73] 8 C.F.R. § 274a.2 (1994).

[74] Martha Brannigan, *Amnesty Program for Illegal Immigrants Spurs Increase in Fraudulent Documents*, WALL ST. J., Jan. 25, 1988, at 33.

[75] INA § 210(a)(4), (b)(3), (c) & (g), *codified at* 8 U.S.C. 1160(a)(4), (b)(3), (c) & (g) (1988).

[76] INA § 210(a)(2), *codified at* 8 U.S.C. § 1160(a)(2) (1988).

[77] INA § 210A, *codified at* 8 U.S.C. § 1161 (1988).

H. The Immigration Act of 1990

The Immigration Act of 1990[78] changed the landscape of the legal migration system.[79] The Act greatly increased employment-based immigration, allowed additional visas for some family-based categories, and created the new category of "diversity immigrants," that Congress designed to provide visas to aliens from "low admission" countries and regions.[80]

The conservative supporters of the Immigration Act of 1990 argued for fewer, or "better," immigrants.[81] The liberals, on the other hand, hoped that the Act would provide a more humane and generous immigration policy.[82] After years of trench warfare among the lobbyists, both sides seemed to have attained their objectives: Latinos and Asians received an increased number of visas for relatives, and businesses received easier access to imported skilled labor.[83] The changes aimed to restructure the ethnic mix of the immigrant pool by admitting more people who had no familial connections in the United States.[84] However, the practical effect of these changes was to give preferential treatment to the "better" immigrants such as millionaires and the highly-educated or highly-skilled.[85] Under the Act, wealthy foreigners can now buy their way into the United States as permanent residents. The price for the American Dream: $1 million. Congressman Bereuter observed that such a system of preferences is antithetical to the very words of the Statue of Liberty. Under the 1990 Immigration Act, the inscription on the Statue of Liberty should be changed from "Give me your tired, your poor" to "Give me a million bucks."[86] The poor and the less-educated immigrants are, simply stated, not as welcome as they have sometimes been.

---

[78] Immigration Act of 1990, Pub. L. No. 101-69, 104 Stat. 3537 (1990).

[79] ALEINIKOFF & MARTIN, *supra* note 4, at 61.

[80] *Id.*

[81] *See* Brae Canlen, *Who Gets In? A Breakdown of U.S. Immigration Numbers Reveals a Policy with No Rules and Many Exceptions,* CAL. LAW., Aug. 1994, at 2.

[82] *Id.*

[83] *Id.*

[84] *Id.*

[85] *Id.*

[86] 135 CONG. REC. H7911 (1990) (statement of Sen. Bereuter).

II. Policies Toward Mexico (1930-64): Repatriation, Braceros, "Operation Wetback," the "Wetback Act," "Operation Jobs" and IRCA

The Treaty of Guadalupe-Hidalgo (1848), which settled the Mexican-American War of 1847, sought to clarify the status of former Mexican citizens.[87] Mexicans remaining in the former Mexican territories were free to stay or to go to Mexico. Those who stayed could elect to be treated as either United States or Mexican citizens. If they did not elect within one year, they automatically became United States citizens. Most Mexicans remained rather than moving to Mexico.[88]

Little attention was paid to the newly created international border.[89] In the 1920s, the United States and Mexico entered into another treaty recognizing the citizenship rights of those traveling between the countries.[90] Most of the 20th century's Mexican immigrants have come to the United States in search of employment. This employment-based immigration has been continuous and significant but with varied legal and policy responses. One author has described the situation as follows:

> [I]llegal immigration from Mexico has been an integral part of a *de facto* U.S. policy with respect to the use of Mexican labor. Bluntly stated, the *de facto* policy has been — bring them in when they are needed, send them back when they aren't.[91]

It has been estimated that between 1900 and 1930 over 300,000 Mexican immigrants were legally admitted to the United States. Perhaps as many as one million undocumented immigrants settled in the same period. The Border Patrol was formed in 1924, but had little manpower on the southern border until much later.[92]

---

[87] 9 Stat. 922 (1848).

[88] *See* Gerald P. Lopez, *Undocumented Mexican Immigration: In Search of a Just Immigration Law and Policy*, 28 UCLA L. REV. 615, 642 (1981).

[89] *Id.* at 643. There were no records kept of immigration across the border until the early twentieth century. Fuchs & Forbes, *supra* note 4, at 53.

[90] 15 Stat. 687 (1868).

[91] Walter Fogel, *Illegal Alien Workers in the United States*, 16 INDUS. REL. 243, 246 (1977).

[92] *See generally* Fuchs & Forbes, *supra* note 4, at 53 (discussing Immigration Act of 1924).

The literacy and the contract labor exclusions of the 1917 Immigration Act contained a provision allowing the Commissioner of Immigration and the Secretary of Labor to waive immigration law provisions for temporary workers.[93] Under the Departmental Order of 1918, the Commissioner waived the head tax, contract labor laws and literacy requirements for Mexican laborers.[94] Attempts to amend those prior acts to exempt agricultural labor *expressly* or to prohibit aliens from commuting failed.[95] As a result, facing no serious restrictions, southwestern agricultural employers continued to rely on immigrant Mexican workers. These exceptions have been described as "the first Bracero program."[96]

A. Repatriation (1930)

During the depression years of 1921 and the 1930s, the Mexican government, like the United States government, directly participated in the repatriation of Mexican workers.[97] Mexicans were scapegoated as the cause of unemployment.[98] According to Mexican government statistics, 345,000 Mexicans — almost the number of immigrants counted in 1920 — returned to their homelands between 1929 and 1932, resulting in a net loss of Mexican immigration.[99] The State Department argued against placing a statutory ceiling on Mexican immigration because of potential for harm to the bilateral relation with Mexico. Congress deferred to administrative action which strictly enforced existing laws reducing Mexican immigration

---

[93] Act of Feb. 5, 1917, 39 Stat. 874, 878 (1917).

[94] Departmental Order No. 52641/202. *See* U.S. Imm. Serv. Bull. Vol. I, No. 3, at 1-4 (1918). *See also* Cardenas, *supra* note 9, at 68.

The admission of temporary contract laborers was authorized by section 3 of the 1917 Immigration Act; after 1951 it operated under Public Law 78, enacted that year.

[95] H. R. Rep. No. 7864 (1896).

[96] Cardenas, *supra* note 9, at 68.

[97] Manuel Garcia y Griego, *The Importation of Mexican Contract Laborers to the United States, 1942-64: Antecedents, Operation and Legacy*, WORKING PAPERS IN U.S.-MEXICAN STUDIES, U.C. SAN DIEGO 1, 6-11 (1981).

[98] Congressional Research Service, *Illegal Aliens: Analysis and Background, House Committee on the Judiciary*, 95th Cong., 1st Sess. 50-55, 48-49 (Comm. Print. 1977), *reprinted in* ALEINIKOFF & MARTIN, *supra* note 4, at 746-47.

[99] Garcia y Griego, *supra* note 97, at 11.

from 4000 to 250 per month.[100]

B. Braceros

The United States farm labor shortage caused by World War II led to a series of bilateral agreements with Mexico designed to alleviate the shortage.[101] The Bracero program, which was extended until 1964,[102] provided mechanisms for sending Mexican workers to United States agricultural areas. These understandings were a continuation of the pattern of recruitment and then repatriation of the Mexican workers. This pattern resulted from the active involvement of both governments. Nearly 500,000 Mexicans participated in the Bracero program at its peak in 1956.

After the United States and Mexico failed to renegotiate the Bracero Treaty in 1954, the Congress amended Public Law 78 to allow the United States to operate the program unilaterally.[103] As P.L. 78 illustrates, the two governments were not always in agreement. But by 1954, given the relative surplus of workers, the United States had the bargaining power to continue the program despite Mexican opposition. In two dramatic incidents in 1948 and 1954, the INS actively recruited workers in violation of existing agreements.[104]

With the rise of the civil rights movement in the United States in the early 1960s, Congress became increasingly concerned with the condition of both domestic and Mexican farm laborers. The 1956 publication of Ernesto Galarza's *Strangers in Our Fields,* documented the widespread abuses of the Bracero's rights under both international and domestic law and marked the beginning of increasing opposition (e.g., for labor unions and civil rights

---

[100] *See* Hutchinson, *supra* note 1, at 217. The INS administrators enforced the restrictive literacy, contract labor, and public charge exclusions rather than waiving them. Cardenas, *supra* note 9, at 68. The illegal denial of welfare benefits was another technique. *See* ALEINIKOFF & MARTIN, *supra* note 4, at 747.

[101] Act of April 29, 1943, Pub. L. No. 45, 57 Stat. 70 (1943); Act of Feb. 14, 1944, Pub. L. No. 229, 58 Stat. 11 (1944). The laws waived the head tax, contract labor provisions, and literacy requirements for the workers. The workers had to be photographed and fingerprinted in accordance with the recently passed Alien Registration Act. Act of June 29, 1940, 54 Stat. 670 (1940). Workers could be deported for failing to obey INS regulations.

[102] Act of Aug. 9, 1946, Pub. L. No. 707, 60 Stat. 969 (1946).

[103] 65 Stat. 119 (1951).

[104] Garcia y Griego, *supra* note 97, at 39.

groups) to P.L. 78.[105] Congress feared that use of Mexican workers depressed the wages and working conditions of United States workers and that farmers had little incentive to pay higher wages. P.L. 78 was allowed to expire in 1963.

While the Bracero program regulated legal migration, it also engendered illegal migration. Once workers became familiar with the United States and in many cases married and had children, they were not likely to simply return. In the meantime, their success in the United States became a beacon call for their former countrymen. Similarly, the SAW, RAW, and H-2A provisions of the IRCA, which encourage legal agricultural migration, will, undoubtedly, have the same effect.

California Governor Pete Wilson argued for these generous agricultural provisions then later complained about the use of social services by such workers. Although it is said that you cannot have it both ways, United States immigration policy had done just that. Indeed, the United States immigration policy toward Mexico may be fairly characterized as "bring in the workers when needed, whatever their status, and scapegoat them and drive them away when they are not." Such ebbs and flows have much to do with the economic cycle of the United States.[106] California's scapegoating of the undocumented worker in the Proposition 187 campaign may be seen in this context.

C. *Operation Wetback & Operation Jobs*

During the early 1950s, the INS reported that its increased apprehensions demonstrated "the greatest peacetime invasion complacently suffered by a country under open, flagrant, contemptuous violation of its laws."[107] This kind of rhetoric, along with an economic depression, resulted in the passage of the "Wetback Act," which provided criminal sanctions for the smuggling, harboring, and entry of aliens who had not

---

[105] *Id.* at 44-46.

[106] *See generally* Barbara Nesbet & Sherilyn K. Sellgren, *California's Proposition 187: A Painful History Repeats Itself*, 1 U.C. DAVIS J. INT'L L. & POL'Y 153 (1995) (discussing cyclical nature of U.S. economy and immigration policy); Minty Siu Chung, *Proposition 187: A Beginner's Tour Through a Recurring Nightmare*, 1 U.C. DAVIS J. INT'L L. & POL'Y 267 (1995) (comparing California's initiative to historical anti-Asian legislation).

[107] *Id.* at 23.

been legally admitted and inspected.[108] One consequence of the "Wetback Act" has been the practice of placing in custody those undocumented workers who could be "material witnesses" in the criminal prosecution of the smuggler.[109]

The "Wetback" provisions were not, however, vigorously enforced. This was due to political pressure, particularly from the Southwest, to cut back appropriations to the Border Patrol. Many overstay or otherwise deportable Braceros were legalized. Essentially, the employers and their political representatives decided that even though it increased apprehensions, the legalization of a steady supply of workers was the best solution. As legalization figures increased, apprehension rates went down.[110]

In 1954, the INS located and deported over 300,000 Mexicans under Operation Wetback. In a campaign reminiscent of the 1994 Proposition 187 campaign, there was widespread scapegoating of the Mexican immigrants for disease, loss of jobs, welfare, and border crime. The campaign was preceded by widespread scapegoating of the Mexican immigrants for disease, labor strikes, subversive and Communist infiltration, and border

---

[108] Act of March 20, 1952, Pub. L. No. 283, 66 Stat. 26 (1952); INA § 275, *codified at* 8 U.S.C. § 1325 (1988).

[109] 18 U.S.C. § 3144 (1988). These witnesses are often held in custody for weeks to months while a defendant charged with unlawful transportation of aliens is frequently released on bail and has fewer constitutional rights than accused smugglers. *See* United States v. Glasco, 488 F.2d 1068 (5th Cir. 1974); United States v. Anfield, 539 F.2d 674 (9th Cir. 1976).

In *In re* (Class Action Application for Habeas Corpus on Behalf of All Material Witnesses in the Western Dist. of Texas), 612 F. Supp 940, 944-945 (W.D.Tex. 1985), the court ruled that the statutory law of 8 U.S.C. §§ 3142 and 3144 and the Fifth Amendment required appointment of counsel to represent the alien material witnesses. The court later issued a standing order requiring that alien material witnesses be deposed and released within sixty days of the date of their detention. U.S. v. Guadian-Salazar, 824 F.2d 344 (5th Cir. 1987).

The INS regulations provide:

> No alien shall depart, or attempt to depart, from the United States if his departure would be prejudicial to the interests of the United States . . . as a witness in, or as a party to, any criminal case under investigation or pending in the United States: Provided, . . . [the] . . . alien . . . may be permitted to depart . . . with the consent of the appropriate prosecuting authority.

8 C.F.R. §§ 215.2(a), 215.3(g) (1994).

[110] *See* Garcia y Griego, *supra* note 97, at 24.

crimes.[111] In addition, in 1982 the INS undertook Operation Jobs, a well-publicized campaign to provide jobs for domestic workers. Under this campaign, the INS conducted raids on suspected employers of undocumented workers. Over 5,000 workers were arrested across the nation.[112]

CONCLUSION

The foregoing history is not a seamless web. An open immigration policy helped settle the West. Yet the Chinese and Mexicans who provided the cheap labor to do the job have on more than one occasion been blamed for most of the nation's ills. If the future is as the past, it seems likely that restrictionist immigration forces will be met by groups who take pride in our immigration history and continue to view immigrants as a resource, not a liability. The balance of this political jostling will depend on the state of the nation's economy and the generosity of its spirit. However, California's Proposition 187 campaign has a new and more sinister ring. Depriving children of their education or workers of their health care is not only mean-spirited, but terribly short sighted.

---

111 *Id.* at 28.

112 *See* ALEINIKOFF & MARTIN, *supra* note 4, at 40.

TITLE XXXI.

IMMIGRATION.

SECTION 1. No citizen of the United States, or foreigner coming into or residing within the same, shall, for himself or for any other person, either as master, factor, owner, or otherwise, build, equip, load, or otherwise prepare, any vessel, registered, enrolled, or licensed, in the United States, for the purpose of procuring from any port or place the subjects of China, Japan, or of any other oriental country, known as "coolies," to be transported to any foreign port, or place, to be disposed of, or sold, or transferred, for any time, as servants or apprentices, or to be held to service or labor.

Cooly-trade prohibited.

19 Feb., 1862, ch. 27, § 1, vol. 12, p. 340. 9 Feb., 1869, ch. 24, vol. 15, p. 269.

SEC. 2. If any vessel, belonging in whole or in part to a citizen of the United States, and registered, enrolled, or otherwise licensed therein, be employed in the "cooly-trade," so called, contrary to the provisions of the preceding section, such vessel, her tackle, apparel, furniture, and other appurtenances, shall be forfeited to the United States, and shall be liable to be seized, prosecuted, and condemned in any of the circuit courts or district courts of the United States for the district where the vessel may be found, seized, or carried.

Vessels employed in cooly-trade shall be forfeited.

Ibid.

SEC. 3. Every person who so builds, fits out, equips, loads, or otherwise prepares, or who sends to sea, or navigates, as owner, master, factor, agent, or otherwise, any vessel, belonging in whole or in part to a citizen of the United States, or registered, enrolled, or licensed within the same, knowing or intending that such vessel is to be or may be em-

Building vessels to engage in cooly-trade, how punished.

19 Feb., 1862, ch. 27, § 2, vol. 12, p. 340

ployed in that trade, contrary to the provisions of section one, shall be liable to a fine not exceeding two thousand dollars and be imprisoned not exceeding one year.

Punishment for violation of section one.

19 Feb., 1862, ch. 27, § 3, vol. 12, p. 340.

SEC. 4. Every citizen of the United States who, contrary to the provisions of section one, takes on board of any vessel, or receives or transports any such subjects as are described in that section, for the purpose of disposing of them in any way as therein prohibited, shall be liable to a fine not exceeding two thousand dollars and be imprisoned not exceeding one year.

This Title not to interfere with voluntary emigration.

19 Feb., 1862, ch. 27, § 4, vol. 12, p. 341.

SEC. 5. Nothing herein contained shall be deemed to apply to any voluntary emigration of the subjects specified in section one, or to any vessel carrying such person as passenger on board the same, but a certificate shall be prepared and signed by the consul or consular agent of the United States residing at the port from which such vessel may take her departure, containing the name of such person, and setting forth the fact of his voluntary emigration from such port, which certificate shall be given to the master of such vessel; and the same shall not be given until such consul or consular agent is first personally satisfied by evidence of the truth of the facts therein contained.

Examination of vessels.

19 Feb., 1862, ch. 27, § 6, vol. 12, p. 341.

SEC. 6. The President is empowered, in such way and at such time as he may judge proper, to direct the vessels of the United States, and the masters and commanders thereof, to examine all vessels navigated or owned in whole or in part by citizens of the United States, and registered, enrolled, or licensed under the laws thereof, whenever, in the judgment of such master or commanding officer, reasonable cause exists to believe that such vessel has on board any subjects of China, Japan, or other oriental country, known as "coolies;" and, upon sufficient proof that such vessel is employed in violation of the preceding provisions, to cause her to be carried, with her officers and crew, into any port or district within the United States, and delivered to the marshal of such district, to be held and disposed of according to law.

(130 U. S. 581)

CHAE CHAN PING v. UNITED STATES.[1]

(May 13, 1889.)

1. CHINESE—ACT OCT. 1, 1888—TREATY RIGHTS.
The fact that the Chinese exclusion act of Oct. 1, 1888, violates existing treaties with China, is no objection to its validity. Treaties are of no higher dignity than acts of congress, and may be modified or repealed by congress in like manner; and whether such modification or repeal is wise or just is not a judicial question.

2. SAME—CERTIFICATES OF IDENTITY.
The certificates of identity issued under the act of May 6, 1882, were mere licenses, revocable at the pleasure of congress.

3. ALIENS—POWER OF CONGRESS TO EXCLUDE.
Congress has power, even in times of peace, to exclude aliens from, or prevent their return to, the United States, for any reason it may deem sufficient.

Appeal from the Circuit Court of the United States for the Northern District of California.

Thos. D. Riordan, *Harvey S. Brown*, *George Hoadly*, and *Jas. C. Carter*, for appellant. *Sol. Gen. Jenks*, for the United States. *Geo. A. Johnson*, *John F. Swift*, and *Stephen M. White*, for the State of California.

FIELD, J. This case comes before us on appeal from an order of the circuit court of the United States for the Northern district of California, refusing to release the appellant, on a writ of *habeas corpus*, from his alleged unlawful detention by Capt. Walker, 582
*master of the steam-ship Belgic, lying within the harbor of San Francisco. The appellant is a subject of the emperor of China, and a laborer by occupation. He resided at San Francisco, Cal., following his occupation, from some time in 1875 until June 2, 1887, when he left for China on the steam-ship Gaelic, having in his possession a certificate in terms entitling him to return to the United States, bearing date on that day, duly issued to him by the collector of customs of the port of San Francisco, pursuant to the provisions of section 4 of the restriction act of May 6, 1882, as amended by the act of July 5, 1884, (22 St. p. 59, c. 126; 23 St. p. 115, c. 220.) On the 7th of September, 1888, the appellant, on his return to California, sailed from Hong Kong in the steam-ship Belgic, which arrived within the port of San Francisco on the 8th of October following. On his arrival he presented to the proper custom-house officers his certificate, and demanded permission to land. The collector of the port refused the permit, solely on the ground that under the act of congress approved October 1, 1888, supplementary to the restriction acts of 1882 and 1884, the certificate had been annulled, and his right to land abrogated, and he had been thereby forbidden again to enter the United States. 25 St. p. 504, c. 1064. The captain of the steam-ship, therefore, detained the appellant on board the steamer. Thereupon a petition on his behalf was presented to the circuit court of the United States for the Northern district of California, alleging that he was unlawfully restrained of his liberty, and praying that a writ of *habeas corpus* might be issued directed to the master of the steam-ship, commanding him to have the body of the appellant, with the cause of his detention, before the court at a time and place designated, to do and receive what might there be considered in the premises. A writ was accordingly issued, and in obedience to it the body of the appellant was produced before the court. Upon the hearing which followed, the court, after finding the facts substantially as stated, held as conclusions of law that the appellant was not entitled to enter the United States, and was not unlawfully restrained of his liberty, and ordered that he be remanded to the custody 583
of the master of the steam-ship from*which he had been taken under the writ. From this order an appeal was taken to this court. 584
*The appeal involves a consideration of the validity of the act of congress of October 1, 1888, prohibiting Chinese laborers from en-

[1] Affirming 36 Fed. Rep. 431.

tering the United States who had departed before its passage, having a certificate issued under the act of 1882 as amended by the act of 1884, granting them permission to return. The validity of the act is assailed as being in effect an expulsion from the country of Chinese laborers, in violation of existing treaties between the United States and the government of China, and of rights vested in them under the laws of congress.

It will serve to present with greater clearness the nature and force of the objections to the act if a brief statement be made of the general character of the treaties between the *590 two countries, and of the legislation of congress to carry them into execution. * The first treaty between the United States and the empire of China was concluded on the 3d of July, 1844, and ratified in December of the following year. 8 St. 592. Previous to that time there had been an extensive commerce between the two nations, that to China being confined to a single port. It was not, however, attended by any serious disturbances between our people there and the Chinese. In August, 1842, as the result of a war between England and China, a treaty was concluded stipulating for peace and friendship between them, and, among other things, that British subjects, with their families and establishments, should be allowed to reside for the purpose of carrying on mercantile pursuits at the five principal ports of the empire. 6 Hert. Treaties, 221. Actuated by a desire to establish by treaty friendly relations between the United States and the Chinese empire, and to secure to our people the same commercial privileges which had been thus conceded to British subjects, congress placed at the disposal of the president the means to enable him to establish future commercial relations between the two countries, "on terms of national equal reciprocity." Act March, 1843, c. 90, (5 St. 624.) A mission was accordingly sent by him to China, at the head of which was placed Mr. Caleb Cushing, a gentleman of large experience in public affairs. He found the Chinese government ready to concede by treaty to the United States all that had been reluctantly yielded to England through compulsion. As the result of his negotiations, the treaty of 1844 was concluded. It stipulated, among other things, that there should be a "perfect, permanent, and universal peace, and a sincere and cordial amity," between the two nations; that the five principal ports of the empire should be opened to the citizens of the United States, who should be permitted to reside with their families and trade there, and to proceed with their vessels and merchandise to and from any foreign port and either of said five ports; and while peaceably attending to their affairs should receive the protec- *591 tion of the Chinese authorities. Senate Document No. 138, 28th Cong. 2d Sess. *The treaty between England and China did not have the effect of securing permanent peace and friendship between those countries. British subjects in China were often subjected, not only to the violence of mobs, but to insults and outrages from local authorities of the country, which led to retaliatory measures for the punishment of the aggressors. To such an extent were these measures carried, and such resistance offered to them, that in 1856 the two countries were in open war. England then determined with the co-operation of France, between which countries there seemed to be perfect accord, to secure from the government of China, among other things, a recognition of the right of other powers to be represented there by accredited ministers, an extension of commercial intercourse with that country, and stipulations for religious freedom to all foreigners there, and for the suppression of piracy. England requested of the president the concurrence and active co-operation of the United States similar to that which France had accorded, and to authorize our naval and political authorities to act in concert with the allied forces. As this proposition involved a participation in existing hostilities, the request could not be acceded to, and the secretary of state in his communication to the English government explained that the war-making power of the United States was not vested in the president, but in congress, and that he had no authority, therefore, to order aggressive hostilities to be undertaken. But as the rights of citizens of the United States might be seriously affected by the results of existing hostilities, and commercial intercourse between the United States and China be disturbed, it was deemed advisable to send to China a minister plenipotentiary to represent our government, and watch our interests there. Accordingly, Mr. William B. Reed, of Philadelphia, was appointed such minister, and instructed, while abstaining from any direct interference, to aid by peaceful co-operation the objects the allied forces were seeking to accomplish. Senate Document No. 47, 35th Cong. 1st Sess. Through him a new treaty was negotiated with the Chinese government. It was concluded in June, 1858, and ratified in August of the following *592 year. *12 St. 1023. It reiterated the pledges of peace and friendship between the two nations, renewed the promise of protection to all citizens of the United States in China peaceably attending to their affairs, and stipulated for security to Christians in the profession of their religion.

Neither the treaty of 1844 nor that of 1858 touched upon the migration and emigration of the citizens and subjects of the two nations, respectively, from one country to the other. But in 1868 a great change in the relations of the two nations was made in that respect. In that year a mission from China, composed of distinguished functionaries of that empire, came to the United States with the professed object of establishing closer relations between the two countries and their peoples. At its head was placed Mr. Anson Burlingame, an eminent citizen of the United

States, who had at one time represented this country as commissioner to China. He resigned his office under our government to accept the position tendered to him by the Chinese government. The mission was hailed in the United States as the harbinger of a new era in the history of China,—as the opening up to free intercourse with other nations and peoples a country that for ages had been isolated and closed against foreigners, who were allowed to have intercourse and to trade with the Chinese only at a few designated places; and the belief was general, and confidently expressed, that great benefits would follow to the world generally, and especially to the United States. On its arrival in Washington, additional articles to the treaty of 1858 were agreed upon, which gave expression to the general desire that the two nations and their peoples should be drawn closer together. The new articles, eight in number, were agreed to on the 28th of July, 1868, and ratifications of them were exchanged at Pekin in November of the following year. 16 St. 739. Of these articles, the fifth, sixth, and seventh are as follows: "Art. 5. The United States of America and the emperor of China cordially recognize the inherent and inalienable right of man to change his home and allegiance, and also the mutual advantage of the free migration and emigration of their citizens and subjects respect- *593 ively from the one country*to the other for purposes of curiosity, of trade, or as permanent residents. The high contracting parties, therefore, join in reprobating any other than an entirely voluntary emigration for these purposes. They consequently agree to pass laws making it a penal offense for a citizen of the United States or Chinese subjects to take Chinese subjects either to the United States or to any other foreign country, or for a Chinese subject or citizen of the United States to take citizens of the United States to China or to any other foreign country, without their free and voluntary consent, respectively. Art. 6. Citizens of the United States visiting or residing in China shall enjoy the same privileges, immunities, or exemptions in respect to travel or residence as may there be enjoyed by the citizens or subjects of the most favored nation; and, reciprocally, Chinese subjects visiting or residing in the United shall enjoy the same privileges, immunities, and exemptions in respect to travel or residence as may there be enjoyed by the citizens or subjects of the most favored nation. But nothing herein contained shall be held to confer naturalization upon citizens of the United States in China, nor upon the subjects of China in the United States. Art. 7 Citizens of the United States shall enjoy all the privileges of the public educational institutions under the control of the government of China, and, reciprocally, Chinese subjects shall enjoy all the privileges of the public educational institutions under the control of the government of the United States, which are enjoyed in the respective countries by the citizens or subjects of the most favored nation. The citizens of the United States may freely establish and maintain schools within the empire of China at those places where foreigners are by treaty permitted to reside; and, reciprocally, Chinese subjects may enjoy the same privileges and immunities in the United States."

But notwithstanding these strong expressions of friendship and good will, and the desire they evince for free intercourse, events were transpiring on the Pacific coast which soon dissipated the anticipations indulged as to the benefits to follow the immigration of *594 Chinese to this country. The*previous treaties of 1844 and 1858 were confined principally to mutual declarations of peace and friendship, and to stipulations for commercial intercourse at certain ports in China, and for protection to our citizens while peaceably attending to their affairs. It was not until the additional articles of 1868 were adopted that any public declaration was made by the two nations that there were advantages in the free migration and emigration of their citizens and subjects, respectively, from one country to the other, and stipulations given that each should enjoy in the country of the other, with respect to travel or residence, the "privileges, immunities, and exemptions" enjoyed by citizens or subjects of the most favored nation. Whatever modifications have since been made to these general provisions have been caused by a well-founded apprehension—from the experience of years—that a limitation to the immigration of certain classes from China was essential to the peace of the community on the Pacific coast, and possibly to the preservation of our civilization there. A few words on this point may not be deemed inappropriate here, they being confined to matters of public notoriety, which have frequently been brought to the attention of congress. Report of Committee of H. R. No. 872, 46th Cong. 2d Sess.

The discovery of gold in California in 1848, as is well known, was followed by a large immigration thither from all parts of the world, attracted not only by the hope of gain from the mines, but from the great prices paid for all kinds of labor. The news of the discovery penetrated China, and laborers came from there in great numbers, a few with their own means, but by far the greater number under contract with employers, for whose benefit they worked. These laborers readily secured employment, and, as domestic servants, and in various kinds of outdoor work, proved to be exceedingly useful. For some years little opposition was made to them, except when they sought to work in the mines, but, as their numbers increased, they began to engage in various mechanical pursuits and trades, and thus came in competition with our artisans and mechanics, as well as our laborers in the field. The competition steadily increased as the laborers *595 came in*crowds on each steamer that arrived from China, or Hong Kong, an adjacent En-

v.9s.c.—40

glish port. They were generally industrious and frugal. Not being accompanied by families, except in rare instances, their expenses were small; and they were content with the simplest fare, such as would not suffice for our laborers and artisans. The competition between them and our people was for this reason altogether in their favor, and the consequent irritation, proportionately deep and bitter, was followed, in many cases, by open conflicts, to the great disturbance of the public peace. The differences of race added greatly to the difficulties of the situation. Notwithstanding the favorable provisions of the new articles of the treaty of 1868, by which all the privileges, immunities, and exemptions were extended to subjects of China in the United States which were accorded to citizens or subjects of the most favored nation, they remained strangers in the land, residing apart by themselves, and adhering to the customs and usages of their own country. It seemed impossible for them to assimilate with our people, or to make any change in their habits or modes of living. As they grew in numbers each year the people of the coast saw, or believed they saw, in the facility of immigration, and in the crowded millions of China, where population presses upon the means of subsistence, great danger that at no distant day that portion of our country would be overrun by them, unless prompt action was taken to restrict their immigration. The people there accordingly petitioned earnestly for protective legislation.

In December, 1878, the convention which framed the present constitution of California, being in session, took this subject up, and memorialized congress upon it, setting forth, in substance, that the presence of Chinese laborers had a baneful effect upon the material interests of the state, and upon public morals; that their immigration was in numbers approaching the character of an Oriental invasion, and was a menace to our civilization; that the discontent from this cause was not confined to any political party, or to any class or nationality, but was well nigh universal; that they retained the habits and customs of
*596 their own country, and in fact constituted a Chinese settlement within the state, without any interest in our country or its institutions; and praying congress to take measures to prevent their further immigration. This memorial was presented to congress in February, 1879. So urgent and constant were the prayers for relief against existing and anticipated evils, both from the public authorities of the Pacific coast and from private individuals, that congress was impelled to act on the subject. Many persons, however, both in and out of congress, were of opinion that, so long as the treaty remained unmodified, legislation restricting immigration would be a breach of faith with China. A statute was accordingly passed appropriating money to send commissioners to China to act with our minister there in negotiating and concluding by treaty a settlement of such matters of interest between the two governments as might be confided to them. 21 St. p. 133, c. 88. Such commissioners were appointed, and as the result of their negotiations the supplementary treaty of November 17, 1880, was concluded and ratified in May of the following year. 22 St. 826. It declares in its first article that "Whenever, in the opinion of the government of the United States, the coming of Chinese laborers to the United States, or their residence therein, affects or threatens to affect the interests of that country, or to endanger the good order of the said country or of any locality within the territory thereof, the government of China agrees that the government of the United States may regulate, limit, or suspend such coming or residence, but may not absolutely prohibit it. The limitation or suspension shall be reasonable, and shall apply only to Chinese who may go to the United States as laborers, other classes not being included in the limitations. Legislation taken in regard to Chinese laborers will be of such a character only as is necessary to enforce the regulation, limitation, or suspension of immigration, and immigrants shall not be subject to personal maltreatment or abuse." In its second article it declares that "Chinese subjects, whether proceeding to the United States as teachers, students, merchants, or from curiosity, together with their body and household servants, and Chinese laborers who are now in *597
the United States, shall *be allowed to go and come of their own free will and accord, and shall be accorded all the rights, privileges, immunities, and exemptions which are accorded to the citizens and subjects of the most favored nation."

The government of China thus agreed that, notwithstanding the stipulations of former treaties, the United States might regulate, limit, or suspend the coming of Chinese laborers, or their residence therein, without absolutely forbidding it, whenever in their opinion the interests of the country, or of any part of it, might require such action. Legislation for such regulation, limitation, or suspension was intrusted to the discretion of our government, with the condition that it should only be such as might be necessary for that purpose, and that the immigrants should not be maltreated or abused. On the 6th of May, 1882, an act of congress was approved, to carry this supplementary treaty into effect. 22 St. 58, c. 126. It is entitled "An act to execute certain treaty stipulations relating to Chinese." Its first section declares that after 90 days from the passage of the act, and for the period of 10 years from its date, the coming of Chinese laborers to the United States is suspended, and that it shall be unlawful for any such laborer to come, or, having come, to remain within the United States. The second makes it a misdemeanor, punishable by fine, to which imprisonment may be added, for the master of any vessel knowingly to bring within the United States from a foreign

country, and land, any such Chinese laborer. The third provides that those two sections shall not apply to Chinese laborers who were in the United States November 17, 1880, or who shall come within 90 days after the passage of the act. The fourth declares that, for the purpose of identifying the laborers who were here on the 17th of November, 1880, or who should come within the 90 days mentioned, and to furnish them with "the proper evidence" of their right to go from and come to the United States, the "collector of customs of the district from which any such Chinese laborer shall depart from the United States shall, in person or by deputy, go on board each vessel having on board any such Chinese laborer and cleared or about to sail
*598 from his district for a foreign port, and on such vessel make a list of all such Chinese laborers, which shall be entered in registry books to be kept for that purpose, in which shall be stated the name, age, occupation, last place of residence, physical marks or peculiarities, and all facts necessary for the identification of each of such Chinese laborers, which books shall be safely kept in the custom-house;" and each laborer thus departing shall be entitled to receive, from the collector or his deputy, a certificate containing such particulars, corresponding with the registry, as may serve to identify him. "The certificate herein provided for," says the section, "shall entitle the Chinese laborer to whom the same is issued to return to and re-enter the United States upon producing and delivering the same to the collector of customs of the district at which such Chinese laborer shall seek to re-enter."

The enforcement of this act with respect to laborers who were in the United States on November 17, 1880, was attended with great embarrassment, from the suspicious nature, in many instances, of the testimony offered to establish the residence of the parties, arising from the loose notions entertained by the witnesses of the obligation of an oath. This fact led to a desire for further legislation restricting the evidence receivable, and the amendatory act of July 5, 1884, was accordingly passed. 23 St. p. 115, c. 220. The committee of the house of representatives on foreign affairs, to whom the original bill was referred, in reporting it, recommending its passage, stated that there had been such manifold evasions, as well as attempted evasions, of the act of 1882, that it had failed to meet the demands which called it into existence. Report in H. R. No. 614, 48th Cong. 1st Sess. To obviate the difficulties attending its enforcement, the amendatory act of 1884 declared that the certificate which the laborer must obtain "shall be the only evidence permissible to establish his right of re-entry" into the United States. This act was held by this court not to require the certificate from laborers who were in the United States on the 17th of November, 1880, who had departed out of the country before May 6, 1882, and remained out until after July 5, 1884.
*Chew Heong v. U. S., 112 U. S. 536, 5 Sup. *599
Ct. Rep. 255. The same difficulties and embarrassments continued with respect to the proof of their former residence. Parties were able to pass successfully the required examination as to their residence before November 17, 1880, who, it was generally believed, had never visited our shores. To prevent the possibility of the policy of excluding Chinese laborers being evaded, the act of October 1, 1888, the validity of which is the subject of consideration in this case, was passed. It is entitled "An act a supplement to an act entitled 'An act to execute certain treaty stipulations relating to Chinese,' approved the 6th day of May, eighteen hundred and eigthy-two." 25 St. p. 504, c. 1064. It is as follows: "Be it enacted by the senate and house of representatives of the United States of America, in congress assembled, that from and after the passage of this act it shall be unlawful for any Chinese laborer who shall at any time heretofore have been, or who may now or hereafter be, a resident within the United States, and who shall have departed, or shall depart, therefrom, and shall not have returned before the passage of this act, to return to or remain in the United States. Sec. 2. That no certificates of identity provided for in the fourth and fifth sections of the act to which this is a supplement shall hereafter be issued; and every certificate heretofore issued in pursuance thereof is hereby declared void and of no effect, and the Chinese laborer claiming admission by virtue thereof shall not be permitted to enter the United States. Sec. 3. That all the duties prescribed, liabilities, penalties, and forfeitures imposed, and the powers conferred, by the second, tenth, eleventh, and twelfth sections of the act to which this is a supplement, are hereby extended, and made applicable to the provisions of this act. Sec. 4. That all such part or parts of the act to which this is a supplement as are inconsistent herewith are hereby repealed. Approved October 1, 1888."

The validity of this act, as already mentioned, is assailed, as being in effect an expulsion from the country of *Chinese laborers, *600
in violation of existing treaties between the United States and the government of China, and of rights vested in them under the laws of congress. The objection that the act is in conflict with the treaties was earnestly pressed in the court below, and the answer to it constitutes the principal part of its opinion. 36 Fed. Rep. 431. Here the objection made is that the act of 1888 impairs a right vested under the treaty of 1880, as a law of the United States, and the statutes of 1882 and of 1884 passed in execution of it. It must be conceded that the act of 1888 is in contravention of express stipulations of the treaty of 1868, and of the supplemental treaty of 1880, but it is not on that account invalid, or to be restricted in its enforcement. The treaties were of no greater legal obligation than the act of congress. By the constitution, laws made in pursuance there-

of, and treaties made under the authority of the United States, are both declared to be the supreme law of the land, and no paramount authority is given to one over the other. A treaty, it is true, is in its nature a contract between nations, and is often merely promissory in its character, requiring legislation to carry its stipulations into effect. Such legislation will be open to future repeal or amendment. If the treaty operates by its own force, and relates to a subject within the power of congress, it can be deemed in that particular only the equivalent of a legislative act, to be repealed or modified at the pleasure of congress. In either case the last expression of the sovereign will must control.

The effect of legislation upon conflicting treaty stipulations was elaborately considered in the Head-Money Cases, and it was there adjudged "that, so far as a treaty made by the United States with any foreign nation can become the subject of judicial cognizance in the courts of this country, it is subject to such acts as congress may pass for its enforcement, modification, or repeal." 112 U. S. 580, 599, 5 Sup. Ct. Rep. 247. This doctrine was affirmed and followed in Whitney v. Robertson, 124 U. S. 190, 195, 8 Sup. Ct. Rep. 456. It will not be presumed that the legislative department of the government will lightly pass laws which are in conflict with the treaties of the country; but
*601 that circumstances may arise which would not only justify the government in *disregarding their stipulations, but demand in the interests of the country that it should do so, there can be no question. Unexpected events may call for a change in the policy of the country. Neglect or violation of stipulations on the part of the other contracting party may require corresponding action on our part. When a reciprocal engagement is not carried out by one of the contracting parties, the other may also decline to keep the corresponding engagement. In 1798 the conduct towards this country of the government of France was of such a character that congress declared that the United States were freed and exonerated from the stipulations of previous treaties with that country. Its act on the subject was as follows: "An act to declare the treaties heretofore concluded with France no longer obligatory on the the United States. Whereas, the treaties concluded between the United States and France have been repeatedly violated on the part of the French government, and the just claims of the United States for reparation of the injuries so committed have been refused, and their attempts to negotiate an amicable adjustment of all complaints between the two nations have been repelled with indignity; and whereas, under authority of the French government, there is yet pursued against the United States a system of predatory violence, infracting the said treaties, and hostile to the rights of a free and independent nation: be it enacted by the senate and house of representatives of the United States of America, in congress assembled, that the United States are of right freed and exonerated from the stipulations of the treaties, and of the consular convention, heretofore concluded between the United States and France, and that the same shall not henceforth be regarded as legally obligatory on the government or citizens of the United States." 1 St. 578.

This act, as seen, applied in terms only to the future. Of course, whatever of a permanent character had been executed or vested under the treaties was not affected by it. In
that respect the abrogation of the obligations *602
of a treaty operates, *like the repeal of a law, only upon the future, leaving transactions executed under it to stand unaffected. The validity of this legislative release from the stipulations of the treaties was, of course, not a matter for judicial cognizance. The question whether our government is justified in disregarding its engagements with another nation is not one for the determination of the courts. This subject was fully considered by Mr. Justice Curtis, while sitting at the circuit, in Taylor v. Morton, 2 Curt. 454, 459, and he held that, while it would always be a matter of the utmost gravity and delicacy to refuse to execute a treaty, the power to do so was prerogative, of which no nation could be deprived without deeply affecting its independence; but whether a treaty with a foreign sovereign had been violated by him, whether the consideration of a particular stipulation of a treaty had been voluntarily withdrawn by one party so as to no longer be obligatory upon the other, and whether the views and acts of a foreign sovereign, manifested through his representative, had given just occasion to the political departments of our government to withhold the execution of a promise contained in a treaty or to act in direct contravention of such promise, were not judicial questions; that the power to determine them has not been confided to the judiciary, which has no suitable means to execute it, but to the executive and legislative departments of the government; and that it belongs to diplomacy and legislation, and not to the administration of existing laws. And the learned justice added, as a necessary consequence of these conclusions, that if congress has this power it is wholly immaterial to inquire whether it has, by the statute complained of, departed from the treaty or not; or, if it has, whether such departure was accidental or designed; and, if the latter, whether the reasons therefor were good or bad. These views were reasserted and fully adopted by this court in Whitney v. Robertson, 124 U. S. 190, 195, 8 Sup. Ct. Rep. 456. And we may add, to the concluding observation of the learned justice, that, if the power mentioned is vested in congress, any reflection upon its motives, or the motives of any of its members in exercising it, *603
would be entirely uncalled for. This court is not a censor of the morals *of other departments of the government; it is not invested

with any authority to pass judgment upon the motives of their conduct. When once it is established that congress possesses the power to pass an act, our province ends with its construction and its application to cases as they are presented for determination. Congress has the power under the constitution to declare war, and in two instances where the power has been exercised—in the war of 1812 against Great Britain, and in 1846 against Mexico—the propriety and wisdom and justice of its action were vehemently assailed by some of the ablest and best men in the country, but no one doubted the legality of the proceeding, and any imputation by this or any other court of the United States upon the motives of the members of congress who in either case voted for the declaration, would have been justly the cause of animadversion. We do not mean to intimate that the moral aspects of legislative acts may not be proper subjects of consideration. Undoubtedly they may be, at proper times and places, before the public, in the halls of congress, and in all the modes by which the public mind can be influenced. Public opinion thus enlightened, brought to bear upon legislation, will do more than all other causes to prevent abuses; but the province of the courts is to pass upon the validity of laws, not to make them, and, when their validity is established, to declare their meaning and apply their provisions. All else lies beyond their domain.

There being nothing in the treaties between China and the United States to impair the validity of the act of congress of October 1, 1888, was it on any other ground beyond the competency of congress to pass it? If so, it must be because it was not within the power of congress to prohibit Chinese laborers who had at the time departed from the United States, or should subsequently depart, from returning to the United States. Those laborers are not citizens of the United States; they are aliens. That the government of the United States, through the action of the legislative department, can exclude aliens from its territory is a proposition which we do not think open to controversy. Jurisdiction over its own territory to that extent is an incident of every independent nation. It is a part of
*604 its*independence. If it could not exclude aliens it would be to that extent subject to the control of another power. As said by this court in the case of The Exchange, 7 Cranch, 116, 136, speaking by Chief Justice MARSHALL: "The jurisdiction of the nation within its own territory is necessarily exclusive and absolute. It is susceptible of no limitation not imposed by itself. Any restriction upon it, deriving validity from an external source, would imply a diminution of its sovereignty to the extent of the restriction, and an investment of that sovereignty to the same extent in that power which could impose such restriction. All exceptions, therefore, to the full and complete power of a nation within its own territories, must be traced up to the consent of the nation itself. They can flow from no other legitimate source."

While under our constitution and form of government the great mass of local matters is controlled by local authorities, the United States, in their relation to foreign countries and their subjects or citizens, are one nation, invested with powers which belong to independent nations, the exercise of which can be invoked for the maintenance of its absolute independence and security throughout its entire territory. The powers to declare war, make treaties, suppress insurrection, repel invasion, regulate foreign commerce, secure republican governments to the states, and admit subjects of other nations to citizenship, are all sovereign powers, restricted in their exercise only by the constitution itself and considerations of public policy and justice which control, more or less, the conduct of all civilized nations. As said by this court in the case of Cohens v. Virginia, 6 Wheat. 264, 413, speaking by the same great chief justice: "That the United States form, for many, and for most important purposes, a single nation, has not yet been denied. In war, we are one people. In making peace, we are one people. In all commercial regulations, we are one and the same people. In many other respects, the American people are one; and the government which is alone capable of controlling and managing their interests in all these respects is the government of the Union. It is their government, and in that character they have no other. *605
America has chosen to*be in many respects, and to many purposes, a nation; and for all these purposes her government is complete; to all these objects, it is competent. The people have declared that in the exercise of all powers given for these objects it is supreme. It can, then, in effecting these objects, legitimately control all individuals or governments within the American territory. The constitution and laws of a state, so far as they are repugnant to the constitution and laws of the United States, are absolutely void. These states are constituent parts of the United States. They are members of one great empire,—for some purposes sovereign, for some purposes subordinate." The same view is expressed in a different form by Mr. Justice BRADLEY, in Knox v. Lee, 12 Wall. 457, 555, where he observes that "the United States is not only a government, but it is a national government, and the only government in this country that has the character of nationality. It is invested with power over all the foreign relations of the country, war, peace, and negotiations and intercourse with other nations; all of which are forbidden to the state governments. It has jurisdiction over all those general subjects of legislation and sovereignty which affect the interests of the whole people equally and alike, and which require uniformity of regulations and laws, such as the coinage, weights, and measures, bankruptcies, the postal system,

patent and copyright laws, the public lands, and interstate commerce; all which subjects are expressly or impliedly prohibited to the state governments. It has power to suppress insurrections, as well as to repel invasions, and to organize, arm, discipline, and call into service the militia of the whole country. The president is charged with the duty and invested with the power to take care that the laws be faithfully executed. The judiciary has jurisdiction to decide controversies between the states, and between their respective citizens, as well as questions of national concern; and the government is clothed with power to guaranty to every state a republican government, and to protect each of them against invasion and domestic violence."

The control of local matters being left to
*606 local authorities, and national matters being intrusted to the government of the *Union, the problem of free institutions existing over a widely extended country, having different climates and varied interests, has been happily solved. For local interests the several states of the Union exist, but for national purposes, embracing our relations with foreign nations, we are but one people, one nation, one power. To preserve its independence, and give security against foreign aggression and encroachment, is the highest duty of every nation, and to attain these ends nearly all other considerations are to be subordinated. It matters not in what form such aggression and encroachment come, whether from the foreign nation acting in its national character, or from vast hordes of its people crowding in upon us. The government, possessing the powers which are to be exercised for protection and security, is clothed with authority to determine the occasion on which the powers shall be called forth; and its determinations, so far as the subjects affected are concerned, are necessarily conclusive upon all its departments and officers. If, therefore, the government of the United States, through its legislative department, considers the presence of foreigners of a different race in this country, who will not assimilate with us, to be dangerous to its peace and security, their exclusion is not to be stayed because at the time there are no actual hostilities with the nation of which the foreigners are subjects. The existence of war would render the necessity of the proceeding only more obvious and pressing. The same necessity, in a less pressing degree, may arise when war does not exist, and the same authority which adjudges the necessity in one case must also determine it in the other. In both cases its determination is conclusive upon the judiciary. If the government of the country of which the foreigners excluded are subjects is dissatisfied with this action, it can make complaint to the executive head of our government, or resort to any other measure which, in its judgment, its interests or dignity may demand; and there lies its only remedy.

The power of the government to exclude foreigners from the country whenever, in its judgment, the public interests require such *607
exclusion, has been asserted in repeated *instances, and never denied by the executive or legislative departments. In a communication made in December, 1852, to Mr. A. Dudley Mann, at one time a special agent of the department of state in Europe, Mr. Everett, then secretary of state under President Fillmore, writes: "This government could never give up the right of excluding foreigners whose presence it might deem a source of danger to the United States." "Nor will this government consider such exclusion of American citizens from Russia necessarily a matter of diplomatic complaint to that country." In a dispatch to Mr. Fay, our minister to Switzerland, in March, 1856, Mr. Marcy, secretary of state under President Pierce, writes: "Every society possesses the undoubted right to determine who shall compose its members, and it is exercised by all nations, both in peace and war." "It may always be questionable whether a resort to this power is warranted by the circumstances, or what department of the government is empowered to exert it; but there can be no doubt that it is possessed by all nations, and that each may decide for itself when the occasion arises demanding its exercise." In a communication in September, 1869, to Mr. Washburne, our minister to France, Mr. Fish, secretary of state under President Grant, uses this language: "The control of the people within its limits, and the right to expel from its territory persons who are dangerous to the peace of the state, are too clearly within the essential attributes of sovereignty to be seriously contested. Strangers visiting or sojourning in a foreign country voluntarily submit themselves to its laws and customs, and the municipal laws of France, authorizing the expulsion of strangers, are not of such recent date, nor has the exercise of the power by the government of France been so infrequent, that sojourners within her territory can claim surprise when the power is put in force." In a communication to Mr. Foster, our minister to Mexico, in July, 1879, Mr. Evarts, secretary of state under President Hayes, referring to the power vested in the constitution of Mexico to expel objectionable foreigners, says: "The admission that, as that constitution now stands and is interpreted, foreigners who render themselves harmful or objectionable to the *608
general *government must expect to be liable to the exercise of the power adverted to, even in time of peace, remains, and no good reason is seen for departing from that conclusion now. But, while there may be no expedient basis on which to found objection, on principle and in advance of a special case thereunder, to the constitutional right thus asserted by Mexico, yet the manner of carrying out such asserted right may be highly objectionable. You would be fully justified in making earnest remonstrances should a cit-

izen of the United States be expelled from Mexican territory without just steps to assure the grounds of such expulsion, and in bringing the fact to the immediate knowledge of the department." In a communication to Mr. W. J. Stillman, under date of August 3, 1882, Mr. Frelinghuysen, secretary of state under President Arthur, writes: "This government cannot contest the right of foreign governments to exclude, on police or other grounds, American citizens from their shores." Whart. International Dig. § 206. The exclusion of paupers, criminals, and persons afflicted with incurable diseases, for which statutes have been passed, is only an application of the same power to particular classes of persons, whose presence is deemed injurious or a source of danger to the country. As applied to them, there has never been any question as to the power to exclude them. The power is constantly exercised; its existence is involved in the right of self-preservation. As to paupers, it makes no difference by whose aid they are brought to the country. As Mr. Fish, when secretary of state, wrote, in a communication under date of December 26, 1872, to Mr. James Moulding, of Liverpool, the government of the United States "is not willing and will not consent to receive the pauper class of any community who may be sent or may be assisted in their immigration at the expense of government or of municipal authorities." As to criminals, the power of exclusion has always been exercised, even in the absence of any statute on the subject. In a dispatch to Mr. Cramer, our minister to Switzerland, in December, 1881, Mr. Blaine, secretary
*609 of state under President Arthur, writes: "While, under the constitution and*the laws, this country is open to the honest and the industrious immigrant, it has no room outside of its prisons or almshouses for depraved and incorrigible criminals or hopelessly dependent paupers who may have become a pest or burden, or both, to their own country." Whart. International Dig., supra.

The power of exclusion of foreigners being an incident of sovereignty belonging to the government of the United States as a part of those sovereign powers delegated by the constitution, the right to its exercise at any time when, in the judgment of the government, the interests of the country require it, cannot be granted away or restrained on behalf of any one. The powers of government are delegated in trust to the United States, and are incapable of transfer to any other parties. They cannot be abandoned or surrendered. Nor can their exercise be hampered, when needed for the public good, by any considerations of private interest. The exercise of these public trusts is not the subject of barter or contract. Whatever license, therefore, Chinese laborers may have obtained, previous to the act of October 1, 1888, to return to the United States after their departure, is held at the will of the government, revocable at any time, at its pleasure. Whether a proper consideration by our government of its previous laws, or a proper respect for the nation whose subjects are affected by its action, ought to have qualified its inhibition, and made it applicable only to persons departing from the country after the passage of the act, are not questions for judicial determination. If there be any just ground of complaint on the part of China, it must be made to the political department of our government, which is alone competent to act upon the subject. The rights and interests created by a treaty, which have become so vested that its expiration or abrogation will not destroy or impair them, are such as are connected with and lie in property capable of sale and transfer, or other disposition, not such as are personal and untransferable in their character. Thus, in the Head-Money Cases, the court speaks of certain rights being in some instances conferred upon the citizens or subjects of one nation residing in the territorial limits of the other, which are "capable of enforcement *610
as*between private parties in the courts of the country." "An illustration of this character," it adds, "is found in treaties which regulate the mutual rights of citizens and subjects of the contracting nations in regard to rights of property by descent or inheritance, when the individuals concerned are aliens." 112 U. S. 580, 598, 5 Sup. Ct. Rep. 247. The passage cited by counsel from the language of Mr. Justice WASHINGTON in Society v. New Haven, 8 Wheat. 464, 493, also illustrates this doctrine. There the learned justice observes that, "if real estate be purchased or secured under a treaty, it would be most mischievous to admit that the extinguishment of the treaty extinguished the right to such estate. In truth, it no more affects such rights than the repeal of a municipal law affects rights acquired under it." Of this doctrine there can be no question in this court; but far different is this case, where a continued suspension of the exercise of a governmental power is insisted upon as a right, because, by the favor and consent of the government, it has not heretofore been exerted with respect to the appellant or to the class to which he belongs. Between property rights not affected by the termination or abrogation of a treaty, and expectations of benefits from the continuance of existing legislation, there is as wide a difference as between realization and hopes.

During the argument reference was made by counsel to the alien law of June 25, 1798, and to opinions expressed at the time by men of great ability and learning against its constitutionality. 1 St. 570, c. 58. We do not attach importance to those opinions in their bearing upon this case. The act vested in the president power to order all such aliens as he should judge dangerous to the peace and safety of the United States, or should have reasonable grounds to suspect were concerned in any treasonable or secret machination against the government, to depart out of the territory of the United States within

such time as should be expressed in his order. There were other provisions also distinguishing it from the act under consideration. The act was passed during a period of great
*717 political excitement, and it was attacked and defended with great zeal and ability. * It is enough, however, to say that it is entirely different from the act before us, and the validity of its provisions was never brought to the test of judicial decision in the courts of the United States. Order affirmed.

(149 U. S. 698)

FONG YUE TING v. UNITED STATES et al. WONG QUAN v. SAME. LEE JOE v. SAME.

(May 15, 1893.)

Nos. 1,345, 1,346, 1,347.

Chinese — Power of Exclusion and Expulsion — International and Constitutional Law — Treaties.

1. It is an accepted maxim of international law that every sovereign nation has the power, as inherent in sovereignty, and essential to self-preservation, to forbid the entrance of foreigners within its dominions, or to admit them only in such cases and upon such conditions as it may see fit to prescribe. Nishimura Ekiu v. U. S., 12 Sup. Ct. Rep. 336, 142 U. S. 651; Chae Chan Ping v. U. S., 9 Sup. Ct. Rep. 623, 130 U. S. 581; Knox v. Lee, 12 Wall. 457,—followed.

2. The right of a nation to expel or deport foreigners who have not been naturalized or taken any steps towards becoming citizens of the country rests upon the same grounds, and is as absolute and unqualified as the right to prohibit and prevent their entrance into the country.

3. The political department of the federal government, through the constitutional grant to it of control over international relations, has authority to expel aliens who have taken no steps to become citizens, even though they are subjects of a friendly power, and have acquired a domicile in this country. Mr. Chief Justice Fuller, Mr. Justice Field, and Mr. Justice Brewer, dissenting.

4. Chinese laborers who came to this country after the making of the Chinese treaties of July 28, 1868, and November 17, 1880, (16 Stat. 740; 22 Stat. 826,) acquired no right thereunder, or under the acts of congress in relation to the Chinese, as denizens or otherwise, to remain in this country, except by the license, permission, and sufferance of congress, to be withdrawn whenever, in its opinion, the public welfare might require it. Mr. Chief Justice Fuller, Mr. Justice Field, and Mr. Justice Brewer, dissenting. Chae Chan Ping v. U. S., 9 Sup. Ct. Rep. 623, 130 U. S. 581, followed.

5. Chinese laborers residing in the United States are entitled, like all other aliens, so long as they are permitted by the government to remain in the country, to all the safeguards of the constitution, and to the protection of the laws in regard to their rights of person and of property, and to their civil and criminal responsibility; but, as they have taken no steps to become citizens, and are incapable of becoming such under the naturalization laws, they remain subject to the power of congress to order their expulsion or deportation whenever, in its judgment, such a measure is necessary or expedient for the public interest. Mr. Chief Justice Fuller, Mr. Justice Field, and Mr. Justice Brewer, dissenting.

6. The act of May 5, 1892, requires in section 6 that all Chinese laborers entitled to remain in this country shall within one year from the date of the act obtain from the collector of internal revenue of the districts in which they reside, free of cost, a certificate of residence, which shall be in the nature of a passport enabling him to go into all parts of the United States, and which shall be recorded in the office of the collector; and provides that Chinese laborers who neglect to obtain such certificates, or are found in the United States without them, "shall be deemed and adjudged to be unlawfully within the United States, and may be arrested by any customs official, collector of internal revenue or his deputies, United States marshal or his deputies, and taken before a United States judge." The act makes it the duty of the judge to order that such laborer be deported from the United States, "unless he shall establish clearly, to the satisfaction of said judge, that by reason of accident, sickness, or other unavoidable cause he has been unable to procure his certificate, and to the satisfaction of the court, by at least one credible white witness, that he was a resident of the United States at the time of the passage of the act." *Held*, that the proceeding here provided for is in no proper sense a trial and sentence for crime, nor is the order of deportation a banishment in the technical sense, but the whole proceeding is merely a method of enforcing the return to his own country of an alien who fails to comply with the conditions prescribed for his continued residence here; and the provisions of the constitution requiring due process of law and trial by jury, and prohibiting unreasonable searches and seizures and cruel and unusual punishments, have no application. Mr. Chief Justice Fuller, Mr. Justice Field, and Mr. Justice Brewer, dissenting.

7. The provision which puts the burden of proof upon a Chinese laborer arrested for having no certificate, as well as the requirement of proof by one credible white witness that he was a resident of the United States at the time of the passage of the act, is within the acknowledged power of every legislature to prescribe the evidence which shall be received, and the effect of that evidence in the courts of its own government.

8. The provisions of an act of congress passed in the exercise of its constitutional authority must prevail even if they contravene the express stipulations of an earlier treaty.

Appeals from the circuit court of the United States in and for the southern district of New York. Affirmed. *699

*Statement by Mr. Justice GRAY:

These were three writs of habeas corpus, granted by the circuit court of the United States for the southern district of New York, upon petitions of Chinese laborers arrested and held by the marshal of the district for not having certificates of residence, under section 6 of the act of May 5, 1892, c. 60, which is copied in the margin.[1]

---

[1]An act to prohibit the coming of Chinese persons into the United States.

Be it enacted by the senate and house of representatives of the United States of America in congress assembled, that all laws now in force prohibiting and regulating the coming into this country of Chinese persons and persons of Chinese descent are hereby continued in force for a period of ten years from the passage of this act.

Sec. 2. That any Chinese person or person of Chinese descent, when convicted and adjudged under any of said laws to be not lawfully entitled to be or remain in the United States, shall be removed from the United States to China, unless he or they shall make it appear to the justice, judge, or commissioner before whom he or they are tried that he or they are subjects or citizens of some other country, in which case he or they shall be removed from the United States to such country: provided, that in any case where such other country, of which such Chinese person shall claim to be a citizen or subject, shall demand any tax as a condition of the removal of such person to that country, he or she shall be removed to China.

Sec. 3. That any Chinese person or person of Chinese descent arrested under the provisions of this act or the acts hereby extended shall be adjudged to be unlawfully within the United States, unless such person shall establish, by affirmative proof, to the satisfaction of such justice, judge, or commissioner, his lawful right to remain in the United States.

Sec. 4. That any such Chinese person or per-

*700 *701 * The rules and regulations made and promulgated by the secretary of the treasury under section 7 of that act prescribe*forms for applications for certificates of residence, for affidavits in support thereof, and for the certificates themselves; contain the provisions copied in the margin;[2] and also provide*for recording duplicates of the certificates in the office of the collector of internal revenue. *702

The first petition alleged that the petitioner was a person of the Chinese race, born in China, and not a naturalized citizen of the United States; that in or before 1879 he came to the United States, with the intention of remaining and taking up his residence therein, and with no definite intention of returning to China, and had ever since been a permanent resident of the United States, and for more than a year last past had resided in the city, county, and state of New York, and within the second district for the collection of internal revenue in that state; that he had not, since the passage of the act of 1892, applied to the collector of internal revenue of that district for a

son of Chinese descent convicted and adjudged to be not lawfully entitled to be or remain in the United States shall be imprisoned at hard labor for a period of not exceeding one year, and thereafter removed from the United States, as hereinbefore provided.

Sec. 5. That after the passage of this act, on an application to any judge or court of the United States in the first instance for a writ of habeas corpus, by a Chinese person seeking to land in the United States, to whom that privilege has been denied, no bail shall be allowed, and such application shall be heard and determined promptly, without unnecessary delay.

Sec. 6. And it shall be the duty of all Chinese laborers within the limits of the United States at the time of the passage of this act, and who are entitled to remain in the United States, to apply to the collector of internal revenue of their respective districts, within one year after the passage of this act, for a certificate of residence; and any Chinese laborer within the limits of the United States, who shall neglect, fail, or refuse to comply with the provisions of this act, or who, after one year from the passage hereof, shall be found within the jurisdiction of the United States without such certificate of residence, shall be deemed and adjudged to be unlawfully within the United States, and may be arrested by any United States customs official, collector of internal revenue or his deputies, United States marshal or his deputies, and taken before a United States judge, whose duty it shall be to order that he be deported from the United States, as hereinbefore provided, unless he shall establish clearly, to the satisfaction of said judge, that by reason of accident, sickness, or other unavoidable cause he has been unable to procure his certificate, and to the satisfaction of the court, and by at least one credible white witness, that he was a resident of the United States at the time of the passage of this act; and if upon the hearing it shall appear that he is so entitled to a certificate, it shall be granted, upon his paying the cost. Should it appear that said Chinaman had procured a certificate which has been lost or destroyed, he shall be detained, and judgment suspended a reasonable time to enable him to procure a duplicate from the officer granting it; and in such cases the cost of said arrest and trial shall be in the discretion of the court. And any Chinese person other than a Chinese laborer, having a right to be and remain in the United States, desiring such certificate as evidence of such right, may apply for and receive the same without charge.

Sec. 7. That immediately after the passage of this act the secretary of the treasury shall make such rules and regulations as may be necessary for the efficient execution of this act, and shall prescribe the necessary forms and furnish the necessary blanks to enable collectors of internal revenue to issue the certificates required hereby, and make such provisions that certificates may be procured in localities convenient to the applicants. Such certificates shall be issued without charge to the applicant, and shall contain the name, age, local residence, and occupation of the applicant, and such other description of the applicant as shall be prescribed by the secretary of the treasury; and a duplicate thereof shall be filed in the office of the collector of internal revenue for the district within which such Chinaman makes application.

Sec. 8. That any person who shall knowingly and falsely alter or substitute any name for the name written in such certificate, or forge such certificate, or knowingly utter any forged or fraudulent certificate, or falsely personate any person named in such certificate, shall be guilty of a misdemeanor, and upon conviction thereof shall be fined in a sum not exceeding one thousand dollars, or imprisoned in the penitentiary for a term of not more than five years.

Sec. 9. The secretary of the treasury may authorize the payment of such compensation in the nature of fees to the collectors of internal revenue, for services performed under the provisions of this act, in addition to salaries now allowed by law, as he shall deem necessary, not exceeding the sum of one dollar for each certificate issued.

[2] Collectors of internal revenue will receive applications on the following form, at their own offices, from such Chinese as are conveniently located thereto, and will cause their deputies to proceed to the towns or cities in their respective divisions where any considerable number of Chinese are residing, for the purpose of receiving applications. No application will be received later than May 5, 1893.

Collectors and deputies will give such notice, through leading Chinese, or by notices posted in the Chinese quarter of the various localities, as will be sufficient to apprise all Chinese residing in their districts of their readiness to receive applications, and the time and place where they may be made. All applications received by deputies must be forwarded to the collector's office, from whose office all certificates of residence will be issued, and sent to the deputy for delivery.

The affidavit of at least one credible witness of good character to the fact of residence and lawful status within the United States must be furnished with every application. If the applicant is unable to furnish such witness satisfactory to the collector or his deputy, his application will be rejected, unless he shall furnish other proof of his right to remain in the United States, in which case the application, with the proofs presented, shall be forwarded to the commissioner of internal revenue for his decision. The witness must appear before the collector or his deputy, and be fully questioned in regard to his testimony before being sworn.

In all cases of loss or destruction of original certificates of residence, where it can be established to the satisfaction of the collector of the district in which the certificate was issued that such loss or destruction was accidental, and without fault or negligence on the part of the applicant, a duplicate of the original may be issued under the same conditions that governed the original issue.

certificate of residence, as required by section 6, and was, and always had been, without such certificate of residence; and that he was arrested by the marshal, claiming authority to do so under that section, without any writ or warrant. The return of the marshal stated that the petitioner was found by him within the jurisdiction of the United States and in the southern district of New York, without the certificate of residence required by that section; that he had, therefore, arrested him, with the purpose and intention of taking him before a United States judge within that district; and that the petitioner admitted to the marshal, in reply to questions put through an interpreter, that he was a Chinese laborer, and was without the required certificate of residence.

The second petition contained similar allegations, and further alleged that the petitioner was taken by the marshal before the district judge for the southern district of
703 New York, and that "the said United States judge, without any hearing of any kind, thereupon ordered that your petitioner be*remanded to the custody of the marshal in and for the southern district of New York, and deported forthwith from the United States, as is provided in said act of May 5, 1892, all of which more fully appears by said order, a copy of which is hereto annexed and made a part hereof," and which is copied in the margin;[1] and that he was detained by virtue of the marshal's claim of authority and the judge's order. The marshal returned that he held the petitioner under that order.

In the third case the petition alleged, and the judge's order showed, the following state of facts: On April 11, 1893, the petitioner applied to the collector of internal revenue for a certificate of residence. The collector refused to give him a certificate, on the ground that the witnesses whom he produced to prove that he was entitled to the certificate were persons of the Chinese race, and not credible witnesses, and required of him to produce a witness other than a Chinaman to prove that he was entitled to the certificate, which he was unable to do, because 704
there was no person other than one of*the* Chinese race who knew and could truthfully swear that he was lawfully within the United States on May 5, 1892, and then entitled to remain therein; and because of such unavoidable cause he was unable to produce a certificate of residence, and was now without one. The petitioner was arrested by the marshal, and taken before the judge, and clearly established to the satisfaction of the judge that he was unable to procure a certificate of residence by reason of the unavoidable cause aforesaid; and also established to the judge's satisfaction, by the testimony of a Chinese resident of New York, that the petitioner was a resident of the United States at the time of the passage of the act; but, having failed to establish this fact clearly to the satisfaction of the court by at least one credible white witness, as required by the statute, the judge ordered the petitioner to be remanded to the custody of the marshal, and to be deported from the United States, as provided in the act.

Each petition alleged that the petitioner was arrested and detained without due process of law, and that section 6 of the act of May 5, 1892, was unconstitutional and void.

In each case the circuit court, after a hearing upon the writ of habeas corpus and the return of the marshal, dismissed the writ of habeas corpus, and allowed an appeal of the petitioner to this court, and admitted him to bail pending the appeal. All the proceedings, from the arrest to the appeal, took place on May 6th.

Jos. H. Choate, J. Hubley Ashton, and Maxwell Evarts, for appellants. Sol. Gen. Aldrich, for appellees.

Mr. Justice GRAY, after stating the facts, delivered the opinion of the court.

The general principles of public law which 705
lie at the foundation of these cases are clearly established by previous* judgments of this* court, and by the authorities therein referred to.

In the recent case of Nishimura Ekiu v. U. S., 142 U. S. 651, 659, 12 Sup. Ct. Rep. 336, the court, in sustaining the action of the executive department, putting in force an act of congress for the exclusion of aliens, said: "It is an accepted maxim of international law that every sovereign nation has the power, as inherent in sovereignty,

---

[1] In the matter of the arrest and deportation of Wong Quan, a Chinese laborer.

Wong Quan, a Chinese laborer, having been arrested in the city of New York on the 6th day of May, 1893, and brought before me, a United States judge, by John W. Jacobus, the marshal of the United States in and for the southern district of New York, as being a Chinese laborer found within the jurisdiction of the United States after the expiration of one year from the passage of the act of congress approved on the 5th day of May, 1892, and entitled "An act to prohibit the coming of Chinese persons into the United States," without having the certificate of residence required by said act; and the said Wong Quan having failed to clearly establish to my satisfaction that by reason of accident, sickness, or other unavoidable cause he had been unable to procure the said certificate, or that he had procured such certificate, and that the same had been lost or destroyed: Now, on motion of Edward Mitchell, the United States attorney in and for the southern district of New York, it is ordered that the said Wong Quan be, and he hereby is, remanded to the custody of the said John W. Jacobus, the United States marshal in and for the southern district of New York; and it is further ordered, that the said Wong Quan be deported from the United States of America in accordance with the provisions of said act of congress approved on the 5th day of May, 1892.

Dated New York, May 6, 1893.

Addison Brown,
United States District Judge for the Southern District of New York.

and essential to self-preservation, to forbid the entrance of foreigners within its dominions, or to admit them only in such cases and upon such conditions as it may see fit to prescribe. In the United States this power is vested in the national government, to which the constitution has committed the entire control of international relations, in peace as well as in war. It belongs to the political department of the government, and may be exercised either through treaties made by the president and senate or through statutes enacted by congress."

The same views were more fully expounded in the earlier case of Chae Chan Ping v. U. S., 130 U. S. 581, 9 Sup. Ct. Rep. 623, in which the validity of a former act of congress, excluding Chinese laborers from the United States, under the circumstances therein stated, was affirmed.

In the elaborate opinion delivered by Mr. Justice Field in behalf of the court it was said: "Those laborers are not citizens of the United States; they are aliens. That the government of the United States, through the action of the legislative department, can exclude aliens from its territory, is a proposition which we do not think open to controversy. Jurisdiction over its own territory to that extent is an incident of every independent nation. It is a part of its independence. If it could not exclude aliens, it would be to that extent subject to the control of another power." "The United States, in their relation to foreign countries and their subjects or citizens, are one nation, invested with powers which belong to independent nations, the exercise of which can be invoked for the maintenance of its absolute independence and security throughout its entire territory." 130 U. S. 603, 604, 9 Sup. Ct. Rep. 629.

*706 It was also said, repeating the language of Mr. Justice*Bradley in Knox v. Lee, 12 Wall. 457, 555: "The United States is not only a government, but it is a national government, and the only government in this country that has the character of nationality. It is invested with power over all the foreign relations of the country, war, peace, and negotiations and intercourse with other nations; all of which are forbidden to the state governments." 130 U. S. 605, 9 Sup. Ct. Rep. 629. And it was added: "For local interests, the several states of the Union exist; but for international purposes, embracing our relations with foreign nations, we are but one people, one nation, one power." 130 U. S. 606, 9 Sup. Ct. Rep. 630.

The court then went on to say: "To preserve its independence, and give security against foreign aggression and encroachment, is the highest duty of every nation; and to attain these ends nearly all other considerations are to be subordinated. It matters not in what form such aggression and encroachment come, whether from the foreign nation acting in its national character, or from vast hordes of its people crowding in upon us. The government, possessing the powers which are to be exercised for protection and security, is clothed with authority to determine the occasion on which the powers shall be called forth; and its determination, so far as the subjects affected are concerned, is necessarily conclusive upon all its departments and officers. If, therefore, the government of the United States, through its legislative department, considers the presence of foreigners of a different race in this country, who will not assimilate with us, to be dangerous to its peace and security, their exclusion is not to be stayed because at the time there are no actual hostilities with the nation of which the foreigners are subjects. The existence of war would render the necessity of the proceeding only more obvious and pressing. The same necessity, in a less pressing degree, may arise when war does not exist, and the same authority which adjudges the necessity in one case must also determine it in the other. In both cases its determination is conclusive upon the judiciary. If the government of the country of which the foreigners excluded are subjects is dissatisfied with this action, it can make *707 complaint to the*executive head of our government, or resort to any other measure which, in its judgment, its interests or dignity may demand; and there lies its only remedy. The power of the government to exclude foreigners from the country, whenever, in its judgment, the public interests require such exclusion, has been asserted in repeated instances, and never denied by the executive or legislative departments." 130 U. S. 606, 607, 9 Sup. Ct. Rep. 631. This statement was supported by many citations from the diplomatic correspondence of successive secretaries of state, collected in Whart. Int. Law Dig. § 206.

The right of a nation to expel or deport foreigners who have not been naturalized, or taken any steps towards becoming citizens of the country, rests upon the same grounds, and is as absolute and unqualified, as the right to prohibit and prevent their entrance into the country.

This is clearly affirmed in dispatches referred to by the court in Chae Chan Ping's Case. In 1856, Mr. Marcy wrote: "Every society possesses the undoubted right to determine who shall compose its members, and it is exercised by all nations, both in peace and war. A memorable example of the exercise of this power in time of peace was the passage of the alien law of the United States in the year 1798." In 1869, Mr. Fish wrote: "The control of the people within its limits, and the right to expel from its territory persons who are dangerous to the peace of the state, are too clearly within the essential attributes of sovereignty to be seriously contested." Whart. Int. Law Dig. § 206; 130 U. S. 607, 9 Sup. Ct. Rep. 630.

The statements of leading commentators on the law of nations are to the same effect.

Vattel says: "Every nation has the right to refuse to admit a foreigner into the country, when he cannot enter without putting the nation in evident danger, or doing it a manifest injury. What it owes to itself, the care of its own safety, gives it this right; and, in virtue of its natural liberty, it belongs to the nation to judge whether its circumstances will or will not justify the admission of the foreigner." "Thus, also, it
*708 has a right to send them elsewhere, if it has just cause to*fear that they will corrupt the manners of the citizens; that they will create religious disturbances, or occasion any other disorder, contrary to the public safety. In a word, it has a right, and is even obliged, in this respect, to follow the rules which prudence dictates." Vatt. Law Nat. lib. 1, c. 19, §§ 230, 231.

Ortolan says: "The government of each state has always the right to compel foreigners who are found within its territory to go away, by having them taken to the frontier. This right is based on the fact that, the foreigner not making part of the nation, his individual reception into the territory is matter of pure permission, of simple tolerance, and creates no obligation. The exercise of this right may be subjected, doubtless, to certain forms by the domestic laws of each country; but the right exists none the less, universally recognized and put in force. In France no special form is now prescribed in this matter; the exercise of this right of expulsion is wholly left to the executive power." Ortolan, Diplomatie de la Mer, (4th Ed.) lib. 2, c. 14, p. 297.

Phillimore says: "It is a received maxim of international law that the government of a state may prohibit the entrance of strangers into the country, and may, therefore, regulate the conditions under which they shall be allowed to remain in it, or may require and compel their departure from it." 1 Phillim. Int. Law, (3d Ed.) c. 10, § 220.

Bar says: "Banishment and extradition must not be confounded. The former is simply a question of expediency and humanity, since no state is bound to receive all foreigners, although, perhaps, to exclude all would be to say good-bye to the international union of all civilized states; and although in some states, such as England, strangers can only be expelled by means of special acts of the legislative power, no state has renounced its right to expel them, as is shown by the alien bills which the government of England has at times used to invest itself with the right of expulsion." "Banishment is regulated by rules of expediency and humanity, and is a matter for the police of the state. No doubt the police can appre-
*709 hend any foreigner who refuses to quit the country in*spite of authoritative orders to do so, and convey him to the frontier." Bar, Int. Law, (Gillespie's Ed. 1883,) 708, note, 711.

In the passages just quoted from Gillespie's translation of Bar, "banishment" is evidently used in the sense of expulsion or deportation by the political authority on the ground of expediency, and not in the sense of transportation or exile by way of punishment for crime. Strictly speaking, "transportation," "extradition," and "deportation," although each has the effect of removing a person from the country, are different things, and have different purposes. "Transportation" is by way of punishment of one convicted of an offense against the laws of the country. "Extradition" is the surrender to another country of one accused of an offense against its laws, there to be tried, and, if found guilty, punished. "Deportation" is the removal of an alien out of the country simply because his presence is deemed inconsistent with the public welfare, and without any punishment being imposed or contemplated, either under the laws of the country out of which he is sent or under those of the country to which he is taken.

In England, the only question that has ever been made in regard to the power to expel aliens has been whether it could be exercised by the king without the consent of parliament. It was formerly exercised by the king, but in later times by parliament, which passed several acts on the subject between 1793 and 1848. 2 Inst. 57; 1 Chalm. Op. 26; 1 Bl. Comm. 260; Chit. Prerog. 49; 1 Phillim. Int. Law, c. 10, § 220, and note; 30 Parl. Hist. 157, 167, 188, 217, 229; 34 Hans. Deb. (1st Series) 441, 445, 471, 1065–1071; 6 Law Rev. Quar. 27.

Eminent English judges, sitting in the judicial committee of the privy council, have gone very far in supporting the exclusion or expulsion, by the executive authority of a colony, of aliens having no absolute right to enter its territory or to remain therein.

In 1837, in a case arising in the island of Mauritius, which had been conquered by Great Britain from France in 1810, and in
which the law of France continued in force, *710
Lord*Lyndhurst, Lord Brougham, and Justices Bosanquet and Erskine, although considering it a case of great hardship, sustained the validity of an order of the English governor, deporting a friendly alien, who had long resided and carried on business in the island, and had enjoyed the privileges and exercised the rights of a person duly domiciled, but who had not, as required by the French law, obtained from the colonial government formal and express authority to establish a domicile there. In re Adam, 1 Moore, P. C. (N. S.) 460.

In a recent appeal from a judgment of the supreme court of the colony of Victoria, a collector of customs, sued by a Chinese immigrant for preventing him from landing in the colony, had pleaded a justification under the order of a colonial minister claiming to exercise an alleged prerogative of the crown to exclude alien friends, and denied the right of a court of law to examine his action, on the ground that what he had done was an act of state; and the plaintiff had demurred to the plea. Lord Chancellor Halsbury, speaking for himself, for Lord Herschell, (now lord chancellor,) and for other lords, after deciding against the plaintiff on a question of statutory construction, took occasion to observe: "The facts appearing on the record raise, quite apart from the statutes referred to, a grave question as to the plaintiff's right to maintain the action. He can only do so if he can establish that an alien has a legal right, enforceable by action, to enter British territory. No authority exists for the proposition that an alien has any such right. Circumstances may occur in which the refusal to permit an alien to land might be such an interference with international comity as would properly give rise to diplomatic remonstrance from the country of which he was a native; but it is quite another thing to assert that an alien, excluded from any part of her majesty's dominions by the executive government there, can maintain an action in a British court, and raise such questions as were argued before their lordships on the present appeal,—whether the proper officer for giving or refusing access to the country has been duly authorized by his own colonial government, whether the colonial government has received sufficient delegated author- *711 ity*from the crown to exercise the authority which the crown had a right to exercise through the colonial government if properly communicated to it, and whether the crown has the right, without parliamentary authority, to exclude an alien. Their lordships cannot assent to the proposition that an alien refused permission to enter British territory can, in an action in a British court, compel the decision of such matters as these, involving delicate and difficult constitutional questions affecting the respective rights of the crown and parliament, and the relations of this country to her self-governing colonies. When once it is admitted that there is no absolute and unqualified right of action on behalf of an alien refused admission to British territory, their lordships are of opinion that it would be impossible, upon the facts which the demurrer admits, for an alien to maintain an action." Musgrove v. Chun Teeong Toy, [1891] App. Cas. 272, 282, 283.

The right to exclude or to expel all aliens, or any class of aliens, absolutely or upon certain conditions, in war or in peace, being an inherent and inalienable right of every sovereign and independent nation, essential to its safety, its independence, and its welfare, the question now before the court is whether the manner in which congress has exercised this right in sections 6 and 7 of the act of 1892 is consistent with the constitution.

The United States are a sovereign and independent nation, and are vested by the constitution with the entire control of international relations, and with all the powers of government necessary to maintain that control, and to make it effective. The only government of this country which other nations recognize or treat with is the government of the Union, and the only American flag known throughout the world is the flag of the United States.

The constitution of the United States speaks with no uncertain sound upon this subject. That instrument, established by the people of the United States as the fundamental law of the land, has conferred upon the president the executive power; has made him the commander in chief of the army and navy; has authorized him, by and *712 with the consent of the *senate, to make treaties, and to appoint ambassadors, public ministers, and consuls; and has made it his duty to take care that the laws be faithfully executed. The constitution has granted to congress the power to regulate commerce with foreign nations, including the entrance of ships, the importation of goods, and the bringing of persons into the ports of the United States; to establish a uniform rule of naturalization; to define and punish piracies and felonies committed on the high seas, and offenses against the law of nations; to declare war, grant letters of marque and reprisal, and make rules concerning captures on land and water; to raise and support armies, to provide and maintain a navy, and to make rules for the government and regulation of the land and naval forces; and to make all laws necessary and proper for carrying into execution these powers, and all other powers vested by the constitution in the government of the United States, or in any department or officer thereof. And the several states are expressly forbidden to enter into any treaty, alliance, or confederation; to grant letters of marque and reprisal; to enter into any agreement or compact with another state, or with a foreign power; or to engage in war, unless actually invaded, or in such imminent danger as will not admit of delay.

In exercising the great power which the people of the United States, by establishing a written constitution as the supreme and paramount law, have vested in this court, of determining, whenever the question is properly brought before it, whether the acts of the legislature or of the executive are consistent with the constitution, it behooves the court to be careful that it does not undertake to pass upon political questions, the final decision of which has been committed by the constitution to the other departments of the government.

As long ago said by Chief Justice Marshall, and since constantly maintained by this court: "The sound construction of the constitution must allow to the national legislature that discretion, with respect to the means by which the powers it confers are to be carried into execution, which will enable that body to perform the high duties *713 assigned to it, in the* manner most beneficial to the people. Let the end be legitimate, let it be within the scope of the constitution; and all means which are appropriate, which are plainly adapted to that end, which are not prohibited, but consistent with the letter and spirit of the constitution, are constitutional." "Where the law is not prohibited, and is really calculated to effect any of the objects intrusted to the government, to undertake here to inquire into the degree of its necessity would be to pass the line which circumscribes the judicial department, and to tread on legislative ground. This court disclaims all pretensions to such a power." McCulloch v. Maryland, 4 Wheat. 316, 421, 423; Juilliard v. Greenman, 110 U. S. 421, 440, 450, 4 Sup. Ct. Rep. 122; Ex parte Yarbrough, 110 U. S. 651, 658, 4 Sup. Ct. Rep. 152; In re Rapier, 143 U. S. 110, 134, 12 Sup. Ct. Rep. 374; Logan v. U. S., 144 U. S. 263, 283, 12 Sup. Ct. Rep. 617.

The power to exclude or to expel aliens, being a power affecting international relations, is vested in the political departments of the government, and is to be regulated by treaty or by act of congress, and to be executed by the executive authority according to the regulations so established, except so far as the judicial department has been authorized by treaty or by statute, or is required by the paramount law of the constitution, to intervene.

In Nishimura Ekiu's Case, it was adjudged that, although congress might, if it saw fit, authorize the courts to investigate and ascertain the facts upon which the alien's right to land was made by the statutes to depend, yet congress might intrust the final determination of those facts to an executive officer; and that, if it did so, his order was due process of law, and no other tribunal, unless expressly authorized by law to do so, was at liberty to re-examine the evidence on which he acted, or to controvert its sufficiency. 142 U. S. 660, 12 Sup. Ct. Rep. 336.

The power to exclude aliens, and the power to expel them, rest upon one foundation, are derived from one source, are supported by the same reasons, and are in truth but parts of one and the same power.

The power of congress, therefore, to expel, like the power to exclude, aliens, or *714 any specified class of aliens, from the* country, may be exercised entirely through executive officers; or congress may call in the aid of the judiciary to ascertain any contested facts on which an alien's right to be in the country has been made by congress to depend.

Congress, having the right, as it may see fit, to expel aliens of a particular class, or to permit them to remain, has undoubtedly the right to provide a system of registration and identification of the members of that class within the country, and to take all proper means to carry out the system which it provides.

It is no new thing for the lawmaking power, acting either through treaties made by the president and senate, or by the more common method of acts of congress, to submit the decision of questions, not necessarily of judicial cognizance, either to the final determination of executive officers, or to the decision of such officers in the first instance, with such opportunity for judicial review of their action as congress may see fit to authorize or permit.

For instance, the surrender, pursuant to treaty stipulations, of persons residing or found in this country, and charged with crime in another, may be made by the executive authority of the president alone, when no provision has been made by treaty or by statute for an examination of the case by a judge or magistrate. Such was the case of Jonathan Robbins, under article 27 of the treaty with Great Britain of 1794, in which the president's power in this regard was demonstrated in the masterly and conclusive argument of John Marshall in the house of representatives. 8 Stat. 129; Whart. State Tr. 392; U. S. v. Nash, Bee, 286, 5 Wheat. append. 3. But provision may be made, as it has been by later acts of congress, for a preliminary examination before a judge or commissioner; and in such case the sufficiency of the evidence on which he acts cannot be reviewed by any other tribunal, except as permitted by statute. Act Aug. 12, 1848, c. 167, (9 Stat. 302;) Rev. St. §§ 5270–5274; Ex parte Metzger, 5 How. 176; Benson v. McMahon, 127 U. S. 457, 8 Sup. Ct. Rep. 1240; In re Luis Oteiza y Cortes, 136 U. S. 330, 10 Sup. Ct. Rep. 1031.

So claims to recover back duties illegally exacted on imports may, if congress so provides, be finally determined by the *715 secretary* of the treasury. Cary v. Curtis, 3 How. 236; Curtis v. Fiedler, 2 Black, 461, 478, 479; Arnson v. Murphy, 109 U. S. 238, 240, 3 Sup. Ct. Rep. 184. But congress may, as it did for long periods, permit them to be tried by suit against the collector of customs; or it may, as by the existing statutes, provide for their determination by a board of general appraisers, and allow the decisions of that board to be reviewed by the courts in such particulars only as may be prescribed by law. Act June 10, 1890, c. 407, §§ 14, 15, 25, (26 Stat. 137, 138, 141;) In re Fassett, 142 U. S. 479, 486, 487, 12 Sup. Ct. Rep. 295; Passavant v. U. S., 148 U. S. 214, 13 Sup. Ct. Rep. 572.

To repeat the careful and weighty words uttered by Mr. Justice Curtis in delivering a unanimous judgment of this court upon the question what is due process of law:

"To avoid misconstruction upon so grave a subject, we think it proper to state that we do not consider congress can either withdraw from judicial cognizance any matter which, from its nature, is the subject of a suit at the common law or in equity or admiralty, nor, on the other hand, can it bring under the judicial power a matter which, from its nature, is not a subject for judicial determination. At the same time there are matters involving public rights, which may be presented in such form that the judicial power is capable of acting on them, and which are susceptible of judicial determination, but which congress may or may not bring within the cognizance of the courts of the United States, as it may deem proper." Murray v. Hoboken, etc., Co., 18 How. 272, 284.

Before examining in detail the provisions of the act of 1892, now in question, it will be convenient to refer to the previous statutes, treaties, and decisions upon the subject.

The act of congress of July 27, 1868, c. 249, (re-enacted in sections 1999-2001, Rev. St.,) began with these recitals: "Whereas, the right of expatriation is a natural and inherent right of all people, indispensable to the enjoyment of the rights of life, liberty, and the pursuit of happiness; and whereas, in the recognition of this principle this government has freely received emigrants from all nations, and invested them
*716 with the rights of citizenship." It then declared that *any order or decision of any officer of the United States to the contrary was inconsistent with the fundamental principles of this government; enacted that "all naturalized citizens of the United States, while in foreign states, shall be entitled to and shall receive from this government the same protection of persons and property that is accorded to native-born citizens in like situations and circumstances;" and made it the duty of the president to take measures to protect the rights in that respect of "any citizen of the United States." 15 Stat. 223, 224.

That act, like any other, is subject to alteration by congress whenever the public welfare requires it. The right of protection which it confers is limited to citizens of the United States. Chinese persons, not born in this country, have never been recognized as citizens of the United States, nor authorized to become such under the naturalization laws. Rev. St. (2d Ed.) §§ 2165, 2169; Acts April 14, 1802, c. 28, (2 Stat. 153;) May 26, 1824, c. 186, (4 Stat. 69;) July 14, 1870, c. 254, § 7, (16 Stat. 256;) Feb. 18, 1875, c. 80, (18 Stat. 318;) In re Ah Yup, 5 Sawy. 155; Act of May 6, 1882, c. 126, § 14, (22 Stat. 61.)

The treaty made between the United States and China on July 28, 1868, contained the following stipulations:

"Art. 5. The United States of America and the emperor of China cordially recognize the inherent and inalienable right of man to change his home and allegiance, and also the mutual advantage of the free migration and emigration of their citizens and subjects, respectively, from one country to the other, for purposes of curiosity, of trade, or as permanent residents.

"Art. 6. Citizens of the United States visiting or residing in China, * * * and, reciprocally, Chinese subjects visiting or residing in the United States, shall enjoy the same privileges, immunities, and exemptions, in respect to travel or residence, as may there be enjoyed by the citizens or subjects of the most favored nation. But nothing herein contained shall be held to confer naturalization upon citizens of the United States in China, nor upon the subjects of China in the United States." 16 Stat. 740. *717

*After some years' experience under that treaty, the government of the United States was brought to the opinion that the presence within our territory of large numbers of Chinese laborers, of a distinct race and religion, remaining strangers in the land, residing apart by themselves, tenaciously adhering to the customs and usages of their own country, unfamiliar with our institutions, and apparently incapable of assimilating with our people, might endanger good order, and be injurious to the public interests, and therefore requested and obtained from China a modification of the treaty. Chew Heong v. U. S., 112 U. S. 536, 542, 543, 5 Sup. Ct. Rep. 255; Chae Chan Ping v. U. S., 130 U. S. 581, 595, 596, 9 Sup. Ct. Rep. 623.

On November 17, 1880, a supplemental treaty was accordingly concluded between the two countries, which contained the following preamble and stipulations:

"Whereas, the government of the United States, because of the constantly increasing immigration of Chinese laborers to the territory of the United States, and the embarrassments consequent upon such immigration, now desires to negotiate a modification of the existing treaties which shall not be in direct contravention of their spirit:

"Article 1. Whenever, in the opinion of the government of the United States, the coming of Chinese laborers to the United States, or their residence therein, affects or threatens to affect the interests of that country, or to endanger the good order of the said country, or of any locality within the territory thereof, the government of China agrees that the government of the United States may regulate, limit, or suspend such coming or residence, but may not absolutely prohibit it. The limitation or suspension shall be reasonable, and shall apply only to Chinese who may go to the United States as laborers, other classes not being included in the limitations. Legislation taken in regard to Chinese laborers will be of such a character only as is necessary to enforce the regulation, limitation, or suspension of immigration, and immigrants shall not be subject to personal maltreatment or abuse.

"Art. 2. Chinese subjects, whether proceeding to the *718 United States as teachers, students, merchants, or from curiosity, together with their body and household servants, and Chinese laborers who are now in the United States, shall be allowed to go and come of their own free will and accord, and shall be accorded all the rights, privileges, immunities, and exemptions which are accorded to the citizens and subjects of the most favored nation.

"Art. 3. If Chinese laborers, or Chinese of any other class, now either permanently or temporarily residing in the territory of the United States, meet with ill treatment at the hands of any other persons, the government of the United States will exert all its power to devise measures for their protection, and to secure to them the same rights, privileges, immunities, and exemptions as may be enjoyed by the citizens or subjects of the most favored nation, and to which they are entitled by treaty." 22 Stat. 826, 827.

The act of May 6, 1882, c. 126, entitled "An act to execute certain treaty stipulations relating to Chinese," and amended by the act of July 5, 1884, c. 220, began with the recital that, "in the opinion of the government of the United States, the coming of Chinese laborers to this country endangers the good order of certain localities within the territories thereof;" and, in section 1, suspended their coming for 10 years, and enacted that it should "not be lawful for any Chinese laborer to come from any foreign port or place, or, having so come, to remain within the United States;" in section 3, that this provision should not apply to Chinese laborers who were in the United States on November 17, 1880, or who came here within 90 days after the passage of the act of 1882, and who should produce evidence of that fact, as afterwards required by the act, to the master of the vessel and to the collector of the port; and, in section 4, that "for the purpose of properly identifying Chinese laborers who were in the United States" at such time, "and in order to furnish them with the proper evidence of their right to go from and come to the United States," as provided by that act and by the treaty of November 17, 1880, the collector of customs of the district from which any Chinese laborers should depart from *719 the United States by sea should go on board the vessel, and make and register a list of them, with all facts necessary for their identification, and should give to each a corresponding certificate, which should entitle him "to return to and re-enter the United States, upon producing and delivering the same to the collector of customs" to be canceled. The form of certificate prescribed by the act of 1884 differed in some particulars from that prescribed by the act of 1882, and the act of 1884 added that "said certificate shall be the only evidence to establish his right of re-entry." Each act further enacted, in section 5, that any such Chinese laborer, being in the United States, and desiring to depart by land, should be entitled to a like certificate of identity; and, in section 12, that no Chinese person should be permitted to enter the United States by land without producing such a certificate, and that "any Chinese person found unlawfully within the United States shall be caused to be removed therefrom to the country from whence he came, and at the cost of the United States, after being brought before some justice, judge, or commissioner of a court of the United States, and found to be one not lawfully entitled to be or remain in the United States." The act of 1884 further enacted, in section 16, that a violation of any of the provisions of the act, the punishment of which was not therein otherwise provided for, should be deemed a misdemeanor, and be punishable by fine not exceeding $1,000, or by imprisonment for not more than one year, or by both such fine and imprisonment. 22 Stat. 58–60; 23 Stat. 115–118.

Under those acts this court held, in Chew Heong v. U. S., 112 U. S. 536, 5 Sup. Ct. Rep. 255, that the clause of section 4 of the act of 1884, making the certificate of identity the only evidence to establish a right to re-enter the United States, was not applicable to a Chinese laborer who resided in the United States at the date of the treaty of 1880, departed by sea before the passage of the act of 1882, remained out of the United States until after the passage of the act of 1884, and then returned by sea; and in U. S. v. Jung Ah Lung, 124 U. S. 621, 8 Sup. Ct. Rep. 663, that a Chinese laborer, who resided in the United *720 States at the date of the treaty of 1880, and until 1883, when he left San Francisco for China, taking with him a certificate of identity from the collector of the port in the form provided by the act of 1882, which was stolen from him in China, was entitled to land again in the United States in 1885, on proving by other evidence these facts, and his identity with the person described in the register kept by the collector of customs as the one to whom that certificate was issued.

Both those decisions proceeded upon a consideration of the various provisions of the acts of 1882 and 1884, giving weight to the presumption that they should not, unless unavoidably, be construed as operating retrospectively, or as contravening the stipulations of the treaty. In the first of those cases Justices Field and Bradley, and in the second case Justices Field, Harlan, and Lamar, dissented from the judgment, being of opinion that the necessary construction of those acts was against the Chinese laborer; and in none of the opinions in either case was it suggested that the acts in question, if construed as contended by the United States, and so as to contravene the treaty, would be unconstitutional or inoperative.

In our jurisprudence it is well settled that the provisions of an act of congress, passed in the exercise of its constitutional authority, on this, as on any other, subject, if clear and explicit, must be upheld by the courts, even in contravention of express stipulations in an earlier treaty. As was said by this court in Chae Chan Ping's Case, following previous decisions: "The treaties were of no greater legal obligation than the act of congress. By the constitution, laws made in pursuance thereof, and treaties made under authority of the United States, are both declared to be the supreme law of the land, and no paramount authority is given to one over the other. A treaty, it is true, is in its nature a contract between nations, and is often merely promissory in its character, requiring legislation to carry its stipulations into effect. Such legislation will be open to future repeal or amendment. If the treaty operates by its own force, and relates to a subject within the power of congress, it can be deemed in that particular
*721 only the equivalent of a legislative act, to be repealed or*modified at the pleasure of congress. In either case the last expression of the sovereign will must control." "So far as a treaty made by the United States with any foreign nation can become the subject of judicial cognizance in the courts of this country it is subject to such acts as congress may pass for its enforcement, modification, or repeal." 130 U. S. 600, 9 Sup. Ct. Rep. 623. See, also, Foster v. Neilson, 2 Pet. 253, 314; Edye v. Robertson, 112 U. S. 580, 597–599, 5 Sup. Ct. Rep. 247; Whitney v. Robertson, 124 U. S. 190, 8 Sup. Ct. Rep. 456.

By the supplementary act of October 1, 1888, c. 1064, it was enacted, in section 1, that "from and after the passage of this act it shall be unlawful for any Chinese laborer, who shall at any time heretofore have been, or who may now or hereafter be, a resident within the United States, and who shall have departed or shall depart therefrom, and shall not have returned before the passage of this act, to return to, or remain in, the United States;" and, in section 2, that "no certificates of identity, provided for in the fourth and fifth sections of the act to which this is a supplement, shall hereafter be issued; and every certificate heretofore issued in pursuance thereof is hereby declared void and of no effect, and the Chinese laborer claiming admission by virtue thereof shall not be permitted to enter the United States." 25 Stat. 504.

In the case of Chae Chan Ping, already often referred to, a Chinese laborer, who had resided in San Francisco from 1875 until June 2, 1887, when he left that port for China, having in his possession a certificate issued to him on that day by the collector of customs, according to the act of 1884, and in terms entitling him to return to the United States, returned to the same port on October 8, 1888, and was refused by the collector permission to land, because of the provisions of the act of October 1, 1888, above cited. It was strongly contended in his behalf that by his residence in the United States for 12 years preceding June 2, 1887, in accordance with the fifth article of the treaty of 1868, he had now a lawful right to be in the United States, and had a vested right to return to the United States, which could
not be taken from him by any exercise of *722
mere legislative power by*congress; that he had acquired such a right by contract between him and the United States, by virtue of his acceptance of the offer contained in the acts of 1882 and 1884, to every Chinese person then here, if he should leave the country, complying with specified conditions, to permit him to return; that, as applied to him, the act of 1888 was unconstitutional, as being a bill of attainder and an ex post facto law; and that the depriving him of his right to return was punishment, which could not be inflicted except by judicial sentence. The contention was thus summed up at the beginning of the opinion: "The validity of the act is assailed as being in effect an expulsion from the country of Chinese laborers, in violation of existing treaties between the United States and the government of China, and of rights vested in them under the laws of congress." 130 U. S. 584–589, 9 Sup. Ct. Rep. 624.

Yet the court unanimously held that the statute of 1888 was constitutional, and that the action of the collector in refusing him permission to land was lawful; and, after the passages already quoted, said: "The power of exclusion of foreigners being an incident of sovereignty belonging to the government of the United States, as a part of those sovereign powers delegated by the constitution, the right to its exercise at any time when, in the judgment of the government, the interests of the country require it, cannot be granted away or restrained on behalf of any one. The powers of government are delegated in trust to the United States, and are incapable of transfer to any other parties. They cannot be abandoned or surrendered. Nor can their exercise be hampered, when needed for the public good, by any considerations of private interest. The exercise of these public trusts is not the subject of barter or contract. Whatever license, therefore, Chinese laborers may have obtained, previous to the act of October 1, 1888, to return to the United States after their departure, is held at the will of the government, revocable at any time, at its pleasure." "The rights and interests created by a treaty, which have become so vested that its expiration or abrogation will not destroy or impair them, are such as are connected with and lie in property, ca-

723 pable of sale and transfer or other disposition; not such as are personal and untransferable in their character." "But far different is this case, where a continued suspension of the exercise of a governmental power is insisted upon as a right, because, by the favor and consent of the government, it has not heretofore been exerted with respect to the appellant, or to the class to which he belongs. Between property rights not affected by the termination or abrogation of a treaty, and expectations of benefits from the continuance of existing legislation, there is as wide a difference as between realization and hopes." 130 U. S. 609, 610, 9 Sup. Ct. Rep. 631.

It thus appears that in that case it was directly adjudged, upon full argument and consideration, that a Chinese laborer, who had been admitted into the United States while the treaty of 1868 was in force, by which the United States and China "cordially recognize the inherent and inalienable right of man to change his home and allegiance, and also the mutual advantage of the free migration and emigration of their citizens and subjects, respectively, from one country to the other," not only for the purpose of curiosity or of trade, but "as permanent residents," and who had continued to reside here for 12 years, and who had then gone back to China, after receiving a certificate, in the form provided by act of congress, entitling him to return to the United States, might be refused readmission into the United States, without judicial trial or hearing, and simply by reason of another act of congress, passed during his absence, and declaring all such certificates to be void, and prohibiting all Chinese laborers who had at any time been residents in the United States, and had departed therefrom and not returned before the passage of this act, from coming into the United States.

In view of that decision, which, as before observed, was a unanimous judgment of the court, and which had the concurrence of all the justices who had delivered opinions in the cases arising under the acts of 1882 and 1884, it appears to be impossible to hold that a Chinese laborer acquired, under any of the treaties or acts of congress, any right, as a denizen, or otherwise, to be and remain in
724 this country, except by the license, permission, and sufferance of congress, to be withdrawn, whenever, in its opinion, the public welfare might require it.

By the law of nations, doubtless, aliens residing in a country, with the intention of making it a permanent place of abode, acquire, in one sense, a domicile there; and, while they are permitted by the nation to retain such a residence and domicile, are subject to its laws, and may invoke its protection against other nations. This is recognized by those publicists who, as has been seen, maintain in the strongest terms the right of the nation to expel any or all aliens at its pleasure. Vatt. Law Nat. lib. 1, c. 19, § 213; 1 Phillim. Int. Law, c. 18, § 321; Mr. Marcy, in Koszta's Case, 2 Whart. Int. Law Dig. § 198. See, also, Lau Ow Bew v. U. S., 144 U. S. 47, 62, 12 Sup. Ct. Rep. 517; Merl. Repert. "Domicile," § 13, quoted in the case above cited, of In re Adam, 1 Moore, P. C. (N. S.) 460, 472, 473.

Chinese laborers, therefore, like all other aliens residing in the United States for a shorter or longer time, are entitled, so long as they are permitted by the government of the United States to remain in the country, to the safeguards of the constitution, and to the protection of the laws, in regard to their rights of person and of property, and to their civil and criminal responsibility. But they continue to be aliens, having taken no steps towards becoming citizens, and incapable of becoming such under the naturalization laws; and therefore remain subject to the power of congress to expel them, or to order them to be removed and deported from the country, whenever, in its judgment, their removal is necessary or expedient for the public interest.

Nothing inconsistent with these views was decided or suggested by the court in Chy Lung v. Freeman, 92 U. S. 275, or in Yick Wo v. Hopkins, 118 U. S. 356, 6 Sup. Ct. Rep. 1064, cited for the appellants.

In Chy Lung v. Freeman, a statute of the state of California, restricting the immigration of Chinese persons, was held to be unconstitutional and void, because it contravened the grant in the constitutional congress of the power to regulate commerce with foreign nations. *725

* In Yick Wo v. Hopkins the point decided* was that the fourteenth amendment of the constitution of the United States, forbidding any state to deprive any person of life, liberty, or property without due process of law, or to deny to any person within its jurisdiction the equal protection of the laws, was violated by a municipal ordinance of San Francisco, which conferred upon the board of supervisors arbitrary power, without regard to competency of persons or to fitness of places, to grant or refuse licenses to carry on public laundries, and which was executed by the supervisors by refusing licenses to all Chinese residents, and granting them to other persons under like circumstances. The question there was of the power of a state over aliens continuing to reside within its jurisdiction, not of the power of the United States to put an end to their residence in the country.

The act of May 5, 1892, c. 60, is entitled "An act to prohibit the coming of Chinese persons into the United States;" and provides, in section 1, that "all laws now in force, prohibiting and regulating the coming into this country of Chinese persons and persons of Chinese descent, are hereby continued in force for a period of ten years from the passage of this act."

The rest of the act (laying aside, as immaterial, section 5, relating to an application for a writ of habeas corpus "by a Chinese person seeking to land in the United States, to whom that privilege has been denied") deals with two classes of Chinese persons: First, those "not entitled to be or remain in the United States;" and, second, those "entitled to remain in the United States." These words of description neither confer nor take away any right, but simply designate the Chinese persons who were not, or who were, authorized or permitted to remain in the United States under the laws and treaties existing at the time of the passage of this act, but subject, nevertheless, to the power of the United States, absolutely or conditionally, to withdraw the permission, and to terminate the authority to remain.

Sections 2-4 concern Chinese "not lawfully *726 entitled to be or remain in the United States," and provide that, after trial*before a justice, judge, or commissioner, a "Chinese person, or person of Chinese descent, convicted and adjudged to be not lawfully entitled to be or remain in the United States," shall be imprisoned at hard labor for not more than a year, and be afterwards removed to China, or other country of which he appears to be a citizen or subject.

The subsequent sections relate to Chinese laborers "entitled to remain in the United States" under previous laws. Sections 6 and 7 are the only sections which have any bearing on the cases before us, and the only ones, therefore, the construction or effect of which need now be considered.

The manifest objects of these sections are to provide a system of registration and identification of such Chinese laborers, to require them to obtain certificates of residence, and, if they do not do so within a year, to have them deported from the United States.

Section 6, in the first place, provides that "it shall be the duty of all Chinese laborers, within the limits of the United States at the time of the passage of this act, and who are entitled to remain in the United States, to apply to the collector of internal revenue of their respective districts, within one year after the passage of this act, for a certificate of residence." This provision, by making it the duty of the Chinese laborer to apply to the collector of internal revenue of the district for a certificate, necessarily implies a correlative duty of the collector to grant him a certificate, upon due proof of the requisite facts. What this proof shall be is not defined in the statute, but is committed to the supervision of the secretary of the treasury by section 7, which directs him to make such rules and regulations as may be necessary for the efficient execution of the act, to prescribe the necessary forms, and to make such provisions that certificates may be procured in localities convenient to the applicants, and without charge to them; and the secretary of the treasury has, by such rules and regulations, provided that the fact of residence shall be proved by "at least one credible witness of good character," or, in case of necessity, by other proof. The statute and the regulations, in order to make sure that every such Chinese*laborer may have a certificate, *727 in the nature of a passport, with which he may go into any part of the United States, and that the United States may preserve a record of all such certificates issued, direct that a duplicate of each certificate shall be recorded in the office of the collector who granted it, and may be issued to the laborer upon proof of loss or destruction of his original certificate. There can be no doubt of the validity of these provisions and regulations, unless they are invalidated by the other provisions of section 6.

This section proceeds to enact that any Chinese laborer within the limits of the United States, who shall neglect, fail, or refuse to apply for a certificate of residence within the year, or who shall afterwards be found within the jurisdiction of the United States without such a certificate, "shall be deemed and adjudged to be unlawfully within the United States." The meaning of this clause, as shown by those which follow, is not that this fact shall thereupon be held to be conclusively established against him, but only that the want of a certificate shall be prima facie evidence that he is not entitled to remain in the United States; for the section goes on to direct that he "may be arrested by any customs official, collector of internal revenue or his deputies, United States marshal or his deputies, and taken before a United States judge;" and that it shall thereupon be the duty of the judge to order that the laborer "be deported from the United States" to China, (or to any other country which he is a citizen or subject of, and which does not demand any tax as a condition of his removal to it,) "unless he shall establish clearly, to the satisfaction of said judge, that by reason of accident, sickness, or other unavoidable cause he has been unable to procure his certificate, and to the satisfaction of the court, and by at least one credible white witness, that he was a resident of the United States at the time of the passage of this act; and if, upon the hearing, it shall appear that he is so entitled to a certificate, it shall be granted upon his paying the cost. Should it appear that said Chinaman had procured a certificate which has been lost or destroyed, he shall be detained, and judgment suspended a reasonable time, to enable him to procure a duplicate from the*officer granting it; and in such *728 cases the cost of said arrest and trial shall be in the discretion of the court."

For the reasons stated in the earlier part of this opinion, congress, under the power to exclude or expel aliens, might have directed any Chinese laborer found in the United States without a certificate of residence to be removed out of the country by executive officers, without judicial trial or examination, just as it might have authorized such officers absolutely to prevent his entrance into the country. But congress has not undertaken to do this.

The effect of the provisions of section 6 of the act of 1892 is that, if a Chinese laborer, after the opportunity afforded him to obtain a certificate of residence within a year, at a convenient place, and without cost, is found without such a certificate, he shall be so far presumed to be not entitled to remain within the United States that an officer of the customs, or a collector of internal revenue, or a marshal, or a deputy of either, may arrest him, not with a view to imprisonment or punishment, or to his immediate deportation without further inquiry, but in order to take him before a judge, for the purpose of a judicial hearing and determination of the only facts which, under the act of congress, can have a material bearing upon the question whether he shall be sent out of the country, or be permitted to remain.

The powers and duties of the executive officers named being ordinarily limited to their own districts, the reasonable inference is that they must take him before a judge within the same judicial district; and such was the course pursued in the cases before us.

The designation of the judge, in general terms, as "a United States judge," is an apt and sufficient description of a judge of a court of the United States, and is equivalent to or synonymous with the designation, in other statutes, of the judges authorized to issue writs of habeas corpus, or warrants to arrest persons accused of crime. Rev. St. §§ 752, 1014.

When, in the form prescribed by law, the executive officer, acting in behalf of the United States, brings the Chinese laborer
*729 before the judge, in order that he may be heard, and *the facts upon which depends his right to remain in the country be decided, a case is duly submitted to the judicial power; for here are all the elements of a civil case,—a complainant, a defendant, and a judge,—actor, reus, et judex. 3 Bl. Comm. 25; Osborn v. Bank, 9 Wheat. 738, 819. No formal complaint or pleadings are required, and the want of them does not affect the authority of the judge or the validity of the statute.

If no evidence is offered by the Chinaman, the judge makes the order of deportation as upon a default. If he produces competent evidence to explain the fact of his not having a certificate, it must be considered by the judge; and if he thereupon appears to be entitled to a certificate, it is to be granted to him. If he proves that the collector of internal revenue has unlawfully refused to give him a certificate, he proves an "unavoidable cause," within the meaning of the act, for not procuring one. If he proves that he had procured a certificate, which has been lost or destroyed, he is to be allowed a reasonable time to procure a duplicate thereof.

The provision which puts the burden of proof upon him of rebutting the presumption arising from his having no certificate, as well as the requirement of proof "by at least one credible white witness that he was a resident of the United States at the time of the passage of this act," is within the acknowledged power of every legislature to prescribe the evidence which shall be received, and the effect of that evidence, in the courts of its own government. Ogden v. Saunders, 12 Wheat. 213, 262, 349; Pillow v. Roberts, 13 How. 472, 476; Cliquot's Champagne, 3 Wall. 114, 143; Ex parte Fisk, 113 U. S. 713, 721, 5 Sup. Ct. Rep. 724; Holmes v. Hunt, 122 Mass. 505, 516–519. The competency of all witnesses, without regard to their color, to testify in the courts of the United States, rests on acts of congress, which congress may, at its discretion, modify or repeal. Rev. St. §§ 858, 1977. The reason for requiring a Chinese alien, claiming the privilege of remaining in the United States, to prove the fact of his residence here at the time of the passage of the act "by at least one credible white witness,"
may have been the experience of congress, as *730
*mentioned by Mr. Justice Field in Chae Chan Ping's Case, that the enforcement of former acts, under which the testimony of Chinese persons was admitted to prove similar facts, "was attended with great embarrassment, from the suspicious nature, in many instances, of the testimony offered to establish the residence of the parties, arising from the loose notions entertained by the witnesses of the obligation of an oath." 130 U. S. 598, 9 Sup. Ct. Rep. 627. And this requirement, not allowing such a fact to be proved solely by the testimony of aliens in a like situation, or of the same race, is quite analogous to the provision, which has existed for 77 years in the naturalization laws, by which aliens applying for naturalization must prove their residence within the limits and under the jurisdiction of the United States, for five years next preceding, "by the oath or affirmation of citizens of the United States." Acts March 22, 1816, c. 32, § 2, (3 Stat. 259;) May 24, 1828, c. 116, § 2, (4 Stat. 311;) Rev. St. § 2165, cl. 6; 2 Kent, Comm. 65.

The proceeding before a United States judge, as provided for in section 6 of the act of 1892, is in no proper sense a trial and sentence for a crime or offense. It is simply the ascertainment, by appropriate and lawful means, of the fact whether the conditions exist upon which congress has enacted that an alien of this class may remain within the country. The order of deportation is not a punishment for crime. It is not a banishment, in the sense in which that word is often applied to the expulsion of a citizen from his country by way of punishment. It is but a method of enforcing the return to his own country of an alien who has not complied with the conditions upon the performance of which the government of the nation, acting within its constitutional authority, and through the proper departments, has determined that his continuing to reside here shall depend. He has not, therefore, been deprived of life, liberty, or property without due process of law; and the provisions of the constitution, securing the right of trial by jury, and prohibiting unreasonable searches and seizures and cruel and unusual punishments, have no application.

*731 *The question whether, and upon what conditions, these aliens shall be permitted to remain within the United States being one to be determined by the political departments of the government, the judicial department cannot properly express an opinion upon the wisdom, the policy, or the justice of the measures enacted by congress in the exercise of the powers confided to it by the constitution over this subject.

The three cases now before us do not differ from one another in any material particular.

In the first case the petitioner had wholly neglected, failed, and refused to apply to the collector of internal revenue for a certificate of residence, and, being found without such a certificate after a year from the passage of the act of 1892, was arrested by the United States marshal, with the purpose, as the return states, of taking him before a United States judge within the district; and thereupon, before any further proceeding, sued out a writ of habeas corpus.

In the second case the petitioner had likewise neglected, failed, and refused to apply to the collector of internal revenue for a certificate of residence, and, being found without one, was arrested by the marshal, and taken before the district judge of the United States, who ordered him to be remanded to the custody of the marshal, and to be deported from the United States, in accordance with the provisions of the act. The allegation in the petition that the judge's order was made "without any hearing of any kind" is shown to be untrue by the recital in the order itself (a copy of which is annexed to and made part of the petition) that he had failed to clearly establish to the judge's satisfaction that by reason of accident, sickness, or other unavoidable cause he had been unable to procure a certificate, or that he had procured one, and it had been lost or destroyed.

In the third case the petitioner had, within the year, applied to a collector of internal revenue for a certificate of residence, and had been refused it, because he produced, and could produce, none but Chinese witnesses, to prove the residence necessary to entitle him to a certificate. Being found without a certificate of residence, he was arrested by the *marshal, and taken before the *732 United States district judge, and established to the satisfaction of the judge that, because of the collector's refusal to give him a certificate of residence, he was without one by an unavoidable cause; and also proved, by a Chinese witness only, that he was a resident of the United States at the time of the passage of the act of 1892. Thereupon the judge ordered him to be remanded to the custody of the marshal, and to be deported from the United States, as provided in that act.

It would seem that the collector of internal revenue, when applied to for a certificate, might properly decline to find the requisite fact of residence upon testimony which, by an express provision of the act, would be insufficient to prove that fact at a hearing before the judge. But if the collector might have received and acted upon such testimony, and did, upon any ground, unjustifiably refuse a certificate of residence, the only remedy of the applicant was to prove by competent and sufficient evidence at the hearing before the judge the facts requisite to entitle him to a certificate. To one of those facts—that of residence—the statute, which, for the reasons already stated, appears to us to be within the constitutional authority of congress to enact, peremptorily requires at that hearing the testimony of a credible white witness; and it was because no such testimony was produced that the order of deportation was made.

Upon careful consideration of the subject, the only conclusion which appears to us to be consistent with the principles of international law, with the constitution and laws of the United States, and with the previous decisions of this court, is that in each of these cases the judgment of the circuit court dismissing the writ of habeas corpus is right, and must be affirmed.

Mr. Justice BREWER, dissenting.

I dissent from the opinion and judgment of the court in these cases, and, the questions being of importance, I deem it not improper to briefly state my reasons therefor. *733

*I rest my dissent on three propositions: First, that the persons against whom the penalties of section 6 of the act of 1892 are directed are persons lawfully residing within the United States; secondly, that as such they are within the protection of the constitution, and secured by its guaranties against oppression and wrong; and, third, that section 6 deprives them of liberty, and imposes punishment, without due process of law, and in disregard of constitutional guaranties, especially those found in the 4th, 5th, 6th, and 8th articles of the amendments.

And, first, these persons are lawfully residing within the limits of the United States. By the treaty of July 28, 1868, (16 Stat. 740,) commonly known as the "Burlingame Treaty," it was provided, (article 5:) "The United States of America and the emperor of China cordially recognize the inherent and inalienable right of man to change his home and allegiance, and also the mutual advantage of the free migration and emigration of their citizens and subjects, respectively, from the one country to the other, for purposes of curiosity, of trade, or as permanent residents." And, (article 6:) "Citizens of the United States visiting or residing in China shall enjoy the same privileges, immunities, or exemptions in respect to travel or residence as may there be enjoyed by the citizens or subjects of the most favored nation; and, reciprocally, Chinese subjects visiting or residing in the United States shall enjoy the same privileges, immunities, and exemptions in respect to travel or residence as may there be enjoyed by the citizens or subjects of the most favored nation." At that time we sought Chinese emigration. The subsequent treaty of November 17, 1880, (22 Stat. 826,) which looked to a restriction of Chinese emigration, nevertheless contained, in article 2, this provision:

"Art. 2. Chinese subjects, whether proceeding to the United States as teachers, students, merchants, or from curiosity, together with their body and household servants, and Chinese laborers who are now in the United States, shall be allowed to go and come of their own free will and accord, and shall
*734 be accorded all the rights, privileges, immunities,*and exemptions which are accorded to the citizens and subjects of the most favored nation."

While, subsequently to this treaty, congress passed several acts—May 6, 1882, (22 Stat. 58;) July 5, 1884, (23 Stat. 115;) October 1, 1888, (25 Stat. 504)—to restrict the entrance into this country of Chinese laborers, and while the validity of this restriction was sustained in the Chinese Exclusion Case, 130 U. S. 581, 9 Sup. Ct. Rep. 623, yet no act has been passed, denying the right of those laborers who had once lawfully entered the country to remain, and they are here not as travelers, or only temporarily. We must take judicial notice of that which is disclosed by the census, and which is also a matter of common knowledge. There are 100,000 and more of these persons living in this country, making their homes here, and striving by their labor to earn a livelihood. They are not travelers, but resident aliens.

But, further, this section 6 recognizes the fact of a lawful residence, and only applies to those who have such; for the parties named in the section, and to be reached by its provisions, are "Chinese laborers within the limits of the United States at the time of the passage of this act, and who are entitled to remain in the United States." These appellants, therefore, are lawfully within the United States, and are here as residents, and not as travelers. They have lived in this country, respectively, since 1879, 1877, and 1874,—almost as long a time as some of those who were members of the congress that passed this act of punishment and expulsion.

That those who have become domiciled in a country are entitled to a more distinct and larger measure of protection than those who are simply passing through, or temporarily in, it, has long been recognized by the law of nations. It was said by this court in the case of The Venus, 8 Cranch, 253, 278: "The writers upon the law of nations distinguish between a temporary residence in a foreign country, for a special purpose, and a residence accompanied with an intention to make it a permanent place of abode. The latter is styled by Vattel 'domicile,' which he defines to be 'a habitation fixed in any place, with an intention of always staying there.' Such*a person, says *735
this author, becomes a member of the new society, at least as a permanent inhabitant, and is a kind of citizen of an inferior order from the native citizens; but is, nevertheless, united and subject to the society, without participating in all its advantages. This right of domicile, he continues, is not established unless the person makes sufficiently known his intention of fixing there, either tacitly or by an express declaration. Vatt. Law Nat. pp. 92, 93. Grotius nowhere uses the word 'domicile,' but he also distinguishes between those who stay in a foreign country by the necessity of their affairs, or from any other temporary cause, and those who reside there from a permanent cause. The former he denominates 'strangers,' and the latter, 'subjects.'" The rule is thus laid down by Sir Robert Phillimore: "There is a class of persons which cannot be, strictly speaking, included in either of these denominations of naturalized or native citizens, namely, the class of those who have ceased to reside in their native country, and have taken up a permanent abode in another. These are domiciled inhabitants. They have not put on a new citizenship through some formal mode enjoined by the law of the new country. They are de facto, though not de jure, citizens of the country of their domicile." 1 Phillim. Int. Law, c. 18, p. 347.

In the Koszta Case it was said by Secretary Marcy: "This right to protect persons having a domicile, though not native-born or naturalized citizens, rests on the firm foundation of justice, and the claim to be protected is earned by considerations which the protecting power is not at liberty to disregard. Such domiciled citizen pays the same price for his protection as native-born or naturalized citizens pay for theirs. He is under the bonds of allegiance to the country of his residence, and, if he breaks them, incurs the same penalties. He owes the same obedience to the civil laws. His property is in the same way, and to the same extent, as theirs, liable to contribute to the support of the government. In nearly all respects his and their condition as to the duties and burdens of government are undistinguishable." 2 Whart. Int. Law Dig. § 198.

*736 *And in Lau Ow Bew v. U. S., 144 U. S. 47, 61, 12 Sup. Ct. Rep. 521, this court declared that, "by general international law, foreigners who have become domiciled in a country other than their own acquire rights, and must discharge duties, in many respects the same as possessed by and imposed upon the citizens of the country, and no restriction on the footing upon which such persons stand, by reason of their domicile, is to be presumed."

Indeed, there is force in the contention of counsel for appellants that these persons are "denizens," within the true meaning and spirit of that word as used in the common law. The old definition was this:

"A denizen of England by letters patent for life, entayl or in fee, whereby he becomes a subject in regard of his person." Craw v. Ramsey, Vaughan, 278.

And again:

"A denizen is an alien born, but who has obtained ex donatione regis letters patent to make him an English subject. * * * A denizen is in a kind of middle state between an alien and a natural-born subject, and partakes of both of them." 1 Bl. Comm. 374.

In respect to this, after quoting from some of the early constitutions of the states, in which the word "denizen" is found, counsel say: "It is claimed that the appellants in this case come completely within the definition quoted above. They are alien born, but they have obtained the same thing as letters patent from this country. They occupy a middle state between an alien and a native. They partake of both of them. They cannot vote, or, as it is stated in Bacon's Abridgment, they have no 'power of making laws,' as a native-born subject has, nor are they here as ordinary aliens. An ordinary alien within this country has come here under no prohibition and no invitation, but the appellants have come under the direct request and invitation, and under the 'patent,' of the federal government. They have been guarantied 'the same privileges, immunities, and exemptions in respect to * * * residence' (Burlingame Treaty, concluded July 28, 1868) as that enjoyed in
*737 the United States by the citizens and *subjects of the most favored nation. They have been told that if they would come here they would be treated just the same as we treat an Englishman, an Irishman, or a Frenchman. They have been invited here, and their position is much stronger than that of an alien, in regard to whom there is no guaranty from the government, and who has come not in response to any invitation, but has simply drifted here because there is no prohibition to keep him out. They certainly come within the meaning of 'denizen,' as used in the constitutions of the states."

But, whatever rights a resident alien might have in any other nation, here he is within the express protection of the constitution, especially in respect to those guaranties which are declared in the original amendments. It has been repeated so often as to become axiomatic that this government is one of enumerated and delegated powers; and, as declared in article 10 of the amendments, "the powers not delegated to the United States by the constitution, nor prohibited by it to the states, are reserved to the states, respectively, or to the people."

It is said that the power here asserted is inherent in sovereignty. This doctrine of powers inherent in sovereignty is one both indefinite and dangerous. Where are the limits to such powers to be found, and by whom are they to be pronounced? Is it within legislative capacity to declare the limits? If so, then the mere assertion of an inherent power creates it, and despotism exists. May the courts establish the boundaries? Whence do they obtain the authority for this? Shall they look to the practices of other nations to ascertain the limits? The governments of other nations have elastic powers. Ours are fixed and bounded by a written constitution. The expulsion of a race may be within the inherent powers of a despotism. History, before the adoption of this constitution, was not destitute of examples of the exercise of such a power; and its framers were familiar with history, and wisely, as it seems to me, they gave to this government no general power to banish. Banishment may be resorted to as *738
punishment for crime; but *among the powers reserved to the people, and not delegated to the government, is that of determining whether whole classes in our midst shall, for no crime but that of their race and birthplace, be driven from our territory.

Whatever may be true as to exclusion,—and as to that see Chinese Exclusion Case, 130 U. S. 581, 9 Sup. Ct. Rep. 623, and Nishimura Ekiu v. United States, 142 U. S 651, 12 Sup. Ct. Rep. 336,—I deny that there is any arbitrary and unrestrained power to banish residents, even resident aliens. What, it may be asked, is the reason for any difference? The answer is obvious. The constitution has no extraterritorial effect, and those who have not come lawfully within our territory cannot claim any protection from its provisions; and it may be that the national government, having full control of all matters relating to other nations, has the power to build, as it were, a Chinese wall around our borders, and absolutely forbid aliens to enter. But the constitution has potency everywhere within the limits of our territory, and the powers which the national government may exercise within such limits are those, and only those, given to it by that instrument. Now, the power to remove resident aliens is, confessedly, not expressed. Even if it be among the powers implied, yet still it can be exercised only in subordination to the limitations and restrictions imposed by the constitution. In the case of Monongahela Navigation Company v. United States, 148 U. S. 312, 336, 13 Sup. Ct. Rep. 630, it was said: "But, like the other powers granted to congress by the constitution, the power to regulate commerce is subject to all the limitations imposed by such instrument, and among them is that of the fifth amendment we have heretofore quoted. Congress has supreme control over the regulation of commerce; but if, in exercising that supreme control, it deems it necessary to take private property, then it must proceed subject to the limitations imposed by this fifth amendment, and can take only on payment of just compensation." And, if that be true of the powers expressly granted, it must as certainly be true of those that are only granted by implication.

When the first 10 amendments were presented for adoption, *739 they were preceded by a preamble stating that the conventions of many states had at the time of their adopting the constitution expressed a desire, "in order to prevent misconception or abuse of its powers, that further declaratory and restrictive clauses should be added." It is worthy of notice that in them the word "citizen" is not found. In some of them the descriptive word is "people," but in the fifth it is broader, and the word is "person," and in the sixth it is the "accused," while in the third, seventh, and eighth there is no limitation as to the beneficiaries suggested by any descriptive word.

In the case of Yick Wo v. Hopkins, 118 U. S. 356, 369, 6 Sup. Ct. Rep. 1070, it was said: "The fourteenth amendment of the constitution is not confined to the protection of citizens. It says: 'Nor shall any state deprive any person of life, liberty, or property without due process of law, nor deny to any person within its jurisdiction the equal protection of the laws.' These provisions are universal in their application to all persons within the territorial jurisdiction, without regard to any differences of race, of color, or of nationality; and the equal protection of the laws is a pledge of the protection of equal laws." The matter considered in that case was of a local nature, a municipal ordinance for regulating the carrying on of public laundries, something fairly within the police power of a state; and yet, because its provisions conflicted with the guaranties of the fourteenth amendment, the ordinance was declared void.

If the use of the word "person" in the fourteenth amendment protects all individuals lawfully within the state, the use of the same word, "person," in the fifth must be equally comprehensive, and secures to all persons lawfully within the territory of the United States the protection named therein; and a like conclusion must follow as to the sixth.

I pass, therefore, to the consideration of my third proposition: Section 6 deprives of "life, liberty, and property without due process of law." It imposes punishment without a trial, and punishment cruel and severe. It places the liberty of one individual subject to the unrestrained control of *740 another. Notice its provisions: It first commands all to register. He who does not register violates that law, and may be punished; and so the section goes on to say that one who has not complied with its requirements, and has no certificate of residence, "shall be deemed and adjudged to be unlawfully within the United States," and then it imposes as a penalty his deportation from the country. Deportation is punishment. It involves—First, an arrest, a deprival of liberty; and, second, a removal from home, from family, from business, from property. In 1 Rap. & L. Law Dict. p. 109, "banishment" is thus defined: "A punishment by forced exile, either for years or for life, inflicted principally upon political offenders; 'transportation' being the word used to express a similar punishment of ordinary criminals." In 4 Bl. Comm. 377, it is said: "Some punishments consist in exile or banishment, by abjuration of the realm, or transportation." In Vattel we find that "banishment is only applied to condemnation in due course of law." Note to section 228 in 1 Vattel.

But it needs no citation of authorities to support the proposition that deportation is punishment. Every one knows that to be forcibly taken away from home and family and friends and business and property, and sent across the ocean to a distant land, is punishment, and that oftentimes most severe and cruel. Apt and just are the words of one of the framers of this constitution,—President Madison,—when he says, (4 Elliot Deb. 555:) "If the banishment of an alien from a country into which he has been invited as the asylum most auspicious to his happiness,—a country where he may have formed the most tender connections; where he may have invested his entire property, and acquired property of the real and permanent, as well as the movable and temporary, kind; where he enjoys, under the laws, a greater share of the blessings of personal security and personal liberty than he can elsewhere hope for; * * * if, moreover, in the execution of the sentence against him, he is to be exposed, not only to the ordinary dangers of the sea, but to the peculiar casualties incident to a crisis of war and of unusual licentiousness on *741 that element, and possibly to vindictive purposes, which his immigration itself may have provoked,—if a banishment of this sort be not a punishment, and among the severest of punishments, it will be difficult to imagine a doom to which the name can be applied."

But punishment implies a trial: "No person shall be deprived of life, liberty, or property without due process of law." Due process requires that a man be heard before he is condemned, and both heard and condemned in the due and orderly procedure of a trial, as recognized by the common law from time immemorial. It was said by this court in Hagar v. Reclamation Dist. No. 108, 111 U. S. 701, 708, 4 Sup. Ct. Rep. 667: "Undoubtedly, where life and liberty are involved, due process requires that there be a regular course of judicial proceedings, which imply that the party to be affected shall have notice, and an opportunity to be heard." And by Mr. Justice Bradley, in defining "due process of law" in Davidson v. New Orleans, 96 U. S. 97, 107: "If found to be suitable or admissible in the special case, it will be adjudged to be 'due process of law;' but if found to be arbitrary, oppressive, and unjust, it may be declared to be not 'due process of law.'" And no person who has once come within the protection of the constitution can be punished without a trial. It may be summary, as for petty offenses and in cases of contempt, but still a trial, as known to the common law. It is said that a person may be extradited without a previous trial, but extradition is simply one step in the process of arresting and securing for trial. He may be removed by extradition from California to New York, or from this country to another, but such proceeding is not oppressive or unjust, but suitable and necessary, and therefore due process of law. But here the Chinese are not arrested and extradited for trial, but arrested, and, without a trial, punished by banishment.

Again, it is absolutely within the discretion of the collector to give or refuse a certificate to one who applies therefor. Nowhere is it provided what evidence shall be furnished to the collector, and nowhere is it made mandatory upon him to grant a certificate on the production of such evidence. It *742 cannot be due process of law to impose punishment on any person for failing to have that in his possession, the possession of which he can obtain only at the arbitrary and unregulated discretion of any official. It will not do to say that the presumption is that the official will act reasonably, and not arbitrarily. When the right to liberty and residence is involved, some other protection than the mere discretion of any official is required. Well was it said by Mr. Justice Matthews in the case of Yick Wo v. Hopkins, 118 U. S., on page 369, 6 Sup. Ct. Rep., on page 1071: "When we consider the nature and the theory of our institutions of government, the principles upon which they are supposed to rest, and review the history of their development, we are constrained to conclude that they do not mean to leave room for the play and action of purely personal and arbitrary power."

Again, a person found without such certificate may be taken before a United States judge. What judge? A judge in the district in which the party resides or is found? There is no limitation in this respect. A Chinese laborer in San Francisco may be arrested by a deputy United States marshal, and taken before a judge in Oregon; and, when so taken before that judge, it is made his duty to deport such laborer, unless he proves his innocence of any violation of the law, and that, too, by at least one credible white witness. And how shall he obtain that witness? No provision is made in the statute therefor. Will it be said that article 6 of the amendments gives to the accused a right to have a compulsory process for obtaining witnesses in his favor? The reply is that if he is entitled to one part of that article he is entitled to all, and among them is the right to a speedy and public trial by an impartial jury of the state and district. The only theory upon which this proceeding can be sustained is that he has no right to any benefits of this article 6; and if he has no right thereto, and the statute has made no provision for securing his witnesses, or limiting the proceeding to a judge of the district where he resides, the results follow inevitably, as stated, that he may be arrested by any one of the numerous officials named in the statute, and carried before any judge in *743 the United States that such official may select, and then, unless he proves that which he is given no means of proving, be punished by removal from home, friends, family, property, business, to another country.

It is said that these Chinese are entitled while they remain to the safeguards of the constitution, and to the protection of the laws in regard to their rights of person and of property, but that they continue to be aliens, subject to the absolute power of congress to forcibly remove them. In other words, the guaranties of "life, liberty, and property," named in the constitution, are theirs by sufferance, and not of right. Of what avail are such guaranties?

Once more: Supposing a Chinaman from San Francisco, having obtained a certificate, should go to New York or other place in pursuit of work, and on the way his certificate be lost or destroyed. He is subject to arrest and detention, the cost of which is in the discretion of the court, and judgment of deportation will be suspended a reasonable time to enable him to obtain a duplicate from the officer granting it. In other words, he cannot move about in safety without carrying with him this certificate. The situation was well described by Senator Sherman in the debate in the senate: "They are here ticket-of-leave men. Precisely as, under the Australian law, a convict is allowed to go at large, upon a ticket-of-leave, these people are to be allowed to go at large, and earn their livelihood, but they must have their tickets-of-leave in their possession." And he added: "This inaugurates in our system of government a new departure; one, I believe, never before practiced, although it was suggested in conference that some such rules had been adopted in slavery times to secure the peace of society."

It is true this statute is directed only against the obnoxious Chinese, but, if the power exists, who shall say it will not be exercised to-morrow against other classes and other people? If the guaranties of these amendments can be thus ignored in order to get rid of this distasteful class, what security have others that a like disregard of its provisions may not be resorted to? Profound and wise were the *744 observations of Mr. Justice Bradley, speaking for the court in Boyd v. U. S., 116 U. S. 616, 635, 6 Sup. Ct. Rep. 535: "Illegitimate and unconstitutional practices get their first footing in that way, namely, by silent approaches, and slight deviations from legal modes of procedure. This can only be obviated by adhering to the rule that constitutional provisions for the security of person and property should be liberally construed. A close and literal construction deprives them of half their efficacy, and leads to gradual depreciation of the right, as if it consisted more in sound than in substance. It is the duty of the courts to be watchful for the constitutional rights of the citizen, and against any stealthy encroachments thereon. Their motto should be, 'obsta principiis.' "

In the Yick Wo Case, in which was presented a municipal ordinance fair on its face, but contrived to work oppression to a few engaged in a single occupation, this court saw no difficulty in finding a constitutional barrier to such injustice. But this greater wrong, by which a hundred thousand people are subject to arrest and forcible deportation from the country, is beyond the reach of the protecting power of the constitution. Its grievous wrong suggests this declaration of wisdom coming from the dawn of English history: "Verily, he who dooms a worse doom to the friendless and the comer from afar than to his fellow, injures himself." The Laws of King Cnut, 1 Thorp, Anc. Laws Eng. p. 397.

In view of this enactment of the highest legislative body of the foremost Christian nation, may not the thoughtful Chinese disciple of Confucius fairly ask, "Why do they send missionaries here?"

Mr. Justice FIELD, dissenting.

I also wish to say a few words upon these cases, and upon the extraordinary doctrines announced in support of the orders of the court below.

*745 With the treaties between the United States and China, and the subsequent legislation adopted by congress to prevent the immigration of Chinese laborers into this country, resulting in the exclusion act of October 1, 1888, the court is familiar. They have often been before us, and have been considered in almost every phase. The act of 1888 declared that after its passage it should be unlawful for any Chinese laborer—who might then or thereafter be a resident of the United States, who should depart therefrom, and not return before the passage of the act—to return or remain in the United States. The validity of this act was sustained by this court. Chinese Exclusion Case, 130 U. S. 581, 9 Sup. Ct. Rep. 623. In the opinion announcing the decision we considered the treaties with China, and also the legislation of congress, and the causes which led to its enactment. The court cited numerous instances in which statesmen and jurists of eminence had held that it was the undoubted right of every independent nation to exclude foreigners from its limits whenever, in its judgment, the public interests demanded such exclusion.

"The power of exclusion of foreigners," said the court, "being an incident of sovereignty belonging to the government of the United States as a part of those sovereign powers delegated by the constitution, the right to its exercise at any time when, in the judgment of the government, the interests of the country require it, cannot be granted away or restrained on behalf of any one. The powers of government are delegated in trust to the United States, and are incapable of transfer to any other parties. They cannot be abandoned or surrendered. Nor can their exercise be hampered, when needed for the public good, by any considerations of private interest. The exercise of these public trusts is not the subject of barter or contract. Whatever license, therefore, Chinese laborers may have obtained previous to the act of October 1, 1888, to return to the United States after their departure, is held at the will of the government, revocable at any time at its pleasure. Whether a proper consideration by our government of its previous laws, or a proper respect for the nation whose subjects are affected by its action, ought to have qualified its inhibition, and made it applicable only to *746 persons departing from the country after the passage of the act, are not questions for judicial determination. If there be any just ground of complaint on the part of China, it must be made to the political department of our government, which is alone competent to act upon the subject."

I had the honor to be the organ of the court in announcing this opinion and judgment. I still adhere to the views there expressed, in all particulars; but between legislation for the exclusion of Chinese persons,—that is, to prevent them from entering the country,—and legislation for the deportation of those who have acquired a residence in the country under a treaty with China, there is a wide and essential difference. The power of the government to exclude foreigners from this country,—that is, to prevent them from entering it,—whenever the public interests, in its judgment, require such exclusion, has been repeatedly asserted by the legislative and executive departments of our government, and never denied; but its power to deport from the country persons lawfully domiciled therein by its consent, and engaged in the ordinary pursuits of life, has never been asserted by the legislative or executive departments, except for crime, or as an act of war, in view of existing or anticipated hostilities, unless the alien act of 1798 can be considered as recognizing that doctrine. 1 Stat. p. 570, c. 58. That act vested in the president power to order all such aliens as he should adjudge dangerous to the peace and safety of the United States, or should have reasonable grounds to suspect were concerned in any treasonable or secret machinations against the government, to depart out of the territory of the United States within such time as should be expressed in his order; and in case any alien when thus ordered to depart should be found at large within the United States after the term limited in the order, not having obtained a license from the president to reside therein, or, having obtained such license, should not have conformed thereto, he should, on conviction thereof, be imprisoned for a term not exceeding three years, and should never afterwards be admitted to become a citizen of the United States, with a proviso that if
*747 the alien thus ordered to depart should prove to the *satisfaction of the president, by evidence to be taken before such person or persons as he should direct, that no injury or danger to the United States would arise from suffering him to reside therein, the president might grant a license to him to remain within the United States for such time as he should judge proper, and at such place as he should designate. The act also provided that the president might require such alien to enter into a bond to the United States, in such penal sum as he might direct, with one or more sureties, to the satisfaction of the person authorized by the president to take the same, conditioned for his good behavior during his residence in the United States, and not to violate his license, which the president might revoke whenever he should think proper. The act also provided that it should be lawful for the president, whenever he deemed it necessary for the public safety, to order to be removed out of the territory of the United States any alien in prison in pursuance of the act, and to cause to be arrested, and sent out of the United States, such aliens as may have been ordered to depart, and had not obtained a license, in all cases where, in the opinion of the president, the public safety required a speedy removal, and that if any alien thus removed or sent out of the United States should voluntarily return, unless by permission of the president, such alien, being convicted thereof, should be imprisoned so long as, in the opinion of the president, the public safety might require.

The passage of this act produced great excitement throughout the country, and was severely denounced by many of its ablest statesmen and jurists as unconstitutional and barbarous, and among them may be mentioned the great names of Jefferson and Madison, who are throughout our country honored and revered for their lifelong devotion to principles of constitutional liberty. It was defended by its advocates as a war measure. John Adams, the president of the United States at the time, who approved the bill, and against whom the responsibility for its passage was charged, states in his correspondence that the bill was intended as a measure of that character. Volume 9 of his works, p. 291. The state of Virginia denounced it in severe terms. Its general as- *748
sembly*passed resolutions upon the act, and another act of the same session of congress, known as the "Sedition Act." Upon the first —the alien act—one of the resolutions declared that it exercised a power nowhere delegated to the federal government, and which, by uniting legislative and judicial powers to those of executive, subverted the general principles of free government, as well as the particular organization and positive provisions of the federal constitution. 4 Elliot, Deb. 528. The resolutions upon both acts were transmitted to the legislatures of different states, and their communications in answer to them were referred to a committee of the general assembly of Virginia, of which Mr. Madison was a member, and upon them his celebrated report was made. With reference to the alien act, after observing that it was incumbent in this, as in every other exercise of power by the federal government, to prove from the constitution that it granted the particular power exercised, and also that much confusion and fallacy had been thrown into the question to be considered by blending the two cases of aliens, members of a hostile nation, and aliens, members of friendly nations, he said: "With respect to alien enemies, no doubt has been intimated as to the federal authority over them; the constitution having expressly delegated to congress

the power to declare war against any nation, and, of course, to treat it and all its members as enemies. With respect to aliens who are not enemies, but members of nations in peace and amity with the United States, the power assumed by the act of congress is denied to be constitutional, and it is accordingly against this act that the protest of the general assembly is expressly and exclusively directed." Id. 554.

"Were it admitted, as is contended, that the 'act concerning aliens' has for its object, not a penal, but a preventive, justice, it would still remain to be proved that it comes within the constitutional power of the federal legislature, and, if within its power, that the legislature has exercised it in a constitutional manner. * * * But it can never be admitted that the removal of aliens, authorized by the act, is to be considered, not as punishment for an offense, but as a measure of *749 precaution and prevention. If the banishment of an alien from a country into which he has been invited as the asylum most auspicious to his happiness,—a country where he may have formed the most tender connections; where he may have invested his entire property, and acquired property of the real and permanent as well as the movable and temporary kind; where he enjoys, under the laws, a greater share of the blessings of personal security and personal liberty than he can elsewhere hope for; * * * if a banishment of this sort be not a punishment, and among the severest of punishments, it would be difficult to imagine a doom to which the name can be applied. And, if it be a punishment, it will remain to be inquired whether it can be constitutionally inflicted, on mere suspicion, by the single will of the executive magistrate, on persons convicted of no personal offense against the laws of the land, nor involved in any offense against the law of nations, charged on the foreign state of which they are members." 4 Elliot, Deb. 555. "It does not follow because aliens are not parties to the constitution, as citizens are parties to it, that, whilst they actually conform to it, they have no right to its protection. Aliens are not more parties to the laws than they are parties to the constitution, yet it will not be disputed that, as they owe, on one hand, a temporary obedience, they are entitled, in return, to their protection and advantage. If aliens had no rights under the constitution, they might not only be banished, but even capitally punished, without a jury, or the other incidents to a fair trial. But, so far has a contrary principle been carried, in every part of the United States, that, except on charges of treason, an alien has, besides all the common privileges, the special one of being tried by a jury, of which one-half may be also aliens.

"It is said, further, that, by the law and practice of nations, aliens may be removed, at discretion, for offenses against the law of nations; that congress are authorized to define and punish such offenses; and that to be dangerous to the peace of society is, in aliens, one of those offenses.

"The distinction between alien enemies and alien friends is *750 a clear and conclusive answer to this argument. Alien enemies are under the law of nations, and liable to be punished for offenses against it. Alien friends, except in the single case of public ministers, are under the municipal law, and must be tried and punished according to that law only." 4 Elliot, Deb. 556.

Massachusetts, evidently considering the alien act as a war measure, adopted in anticipation of probable hostilities, said, in answer to the resolutions of Virginia, among other things, that "the removal of aliens is the usual preliminary of hostility, and is justified by the invariable usages of nations. Actual hostility had, unhappily, been long experienced, and a formal declaration of it the government had reason daily to expect." Id. 535.

The duration of the act was limited to two years, and it has ever since been the subject of universal condemnation. In no other instance, until the law before us was passed, has any public man had the boldness to advocate the deportation of friendly aliens in time of peace. I repeat the statement that in no other instance has the deportation of friendly aliens been advocated as a lawful measure by any department of our government. And it will surprise most people to learn that any such dangerous and despotic power lies in our government,—a power which will authorize it to expel at pleasure, in time of peace, the whole body of friendly foreigners of any country domiciled herein by its permission; a power which can be brought into exercise whenever it may suit the pleasure of congress, and be enforced without regard to the guaranties of the constitution intended for the protection of the rights of all persons in their liberty and property. Is it possible that congress can, at its pleasure, in disregard of the guaranties of the constitution, expel at any time the Irish, German, French, and English who may have taken up their residence here on the invitation of the government, while we are at peace with the countries from which they came, simply on the ground that they have not been naturalized?

Notwithstanding the activity of the public authorities in enforcing the exclusion act of 1888, it was constantly evaded. *751 Chinese laborers came into the country by water and by land; they came through the open ports, and by rivers reaching the seas, and they came by the way of the Canadas and Mexico. New means of ingress were discovered, and, in spite of the vigilance of the police and customs officers, great numbers clandestinely found their way into the country. Their resemblance to each other rendered it difficult, and often impossible, to prevent this evasion of the laws. It was under these circumstances that the act of May 5, 1892, was passed. It had two objects in view. There were two classes of Chinese persons in the country,—those who had evaded the laws excluding them and entered clandestinely, and those who had entered lawfully, and resided therein under the treaty with China.

The act of 1892 extended, for the period of 10 years from its passage, all laws then in force, prohibiting and regulating the coming into the country of Chinese persons, or persons of Chinese descent; and it provided that any person, when convicted or adjudged under any of those laws of not legally being or remaining in the United States, should be removed therefrom to China, or to such other country as it might appear he was a subject of, unless such other country should demand a tax as a condition of his removal thereto, in which case he should be removed to China. The act also provided that a Chinese person arrested under its provisions, or the provisions of the acts extended, should be adjudged to be unlawfully within the United States, unless he should establish by affirmative proof his lawful right to remain within the United States, and that any Chinese person, or persons of Chinese descent, "convicted and adjudged not lawfully entitled to be or remain in the United States, should be imprisoned at hard labor for a period not exceeding one year, and thereafter removed from the United States." With this class of Chinese, and with the provisions of law applicable to them, we have no concern in the present case. We have only to consider the provisions of the act applicable to the second class of Chinese persons,—those who had a lawful right to remain in the United *752 States. By the additional articles to the treaty of 1858, adopted in 1868, generally called the "Burlingame Treaty," the governments of the two countries recognized "the inherent and inalienable right of man to change his home and allegiance, and also the mutual advantage of free migration and emigration of their citizens and subjects, respectively, from the one country to the other, for purposes of curiosity, of trade, or as permanent residents;" and accordingly the treaty, in the additional articles, provided that citizens of the United States visiting or residing in China, and Chinese subjects visiting or residing in the United States, should reciprocally enjoy the same privileges, immunities, and exemptions in respect to travel or residence as should be enjoyed by citizens or subjects of the most favored nation, in the country in which they should, respectively, be visiting or residing. 16 Stat. 739, 740. The supplemental treaty of November 17, 1880, providing for the limitation or suspension of the emigration of Chinese laborers, declared that "the limitation or suspension shall be reasonable, and apply only to Chinese who may go to the United States as laborers,—other classes not being included in the limitation,"—and that "Chinese subjects, whether residing in the United States as teachers, students, merchants, or from curiosity, together with their body and household servants, and Chinese laborers who were then in the United States, shall be allowed to go and come of their own free will and accord, and shall be accorded all rights, privileges, immunities, and exemptions which are accorded to the citizens and subjects of the most favored nation."

There are many thousands of Chinese laborers who came to the country, and resided in it, under the additional articles of the treaty adopted in 1868, and were in the country at the time of the adoption of the supplemental treaty of November, 1880. To these laborers, thus lawfully within the limits of the United States, section 6 of the act of May 5, 1892, relates. That section, so far as applicable to the present cases, is as follows:

"Sec. 6. And it shall be the duty of all Chinese laborers within the limits of the United States at the time of the passage of this act, and who are entitled to remain in the *753 United States, to apply to the collector of internal revenue of their respective districts, within one year after the passage of this act, for a certificate of residence, and any Chinese laborer within the United States, who shall neglect, fail, or refuse to comply with the provisions of this act, or who, after one year from the passage hereof, shall be found within the jurisdiction of the United States without such certificate of residence, shall be deemed and adjudged to be unlawfully within the United States, and may be arrested by any United States customs official, collector of internal revenue or his deputies, United States marshal or his deputies, and taken before a United States judge, whose duty it shall be to order that he be deported from the United States, as hereinbefore provided, unless he shall establish clearly, to the satisfaction of the said judge, that by reason of accident, sickness, or other unavoidable cause, he has been unable to procure his certificate, and to the satisfaction of the court, and by at least one credible white witness, that he was a resident of the United States at the time of the passage of this act; and if, upon the hearing, it shall appear that he is so entitled to a certificate, it shall be granted, upon his paying the cost. Should it appear that said Chinaman had procured a certificate which had been lost or destroyed, he shall be detained, and judgment suspended, a reasonable time, to enable him to procure a duplicate from the officer granting it, and in such cases the cost of said arrest and trial shall be in the discretion of the court."

The purpose of this section was to secure

the means of readily identifying the Chinese laborers present in the country, and entitled to remain, from those who may have clandestinely entered the country in violation of its laws. Those entitled to remain, by having a certificate of their identification, would enable the officers of the government to readily discover, and bring to punishment, those not entitled to enter, but who are excluded. To procure such a certificate was not a hardship to the laborers, but a means to secure full protection to them, and at the same time prevent an evasion of the law.

This object being constitutional, the only *754 question for our *consideration is the lawfulness of the procedure provided for its accomplishment; and this must be tested by the provisions of the constitution and laws intended for the protection of all persons against encroachment upon their rights. Aliens from countries at peace with us, domiciled within our country by its consent, are entitled to all the guaranties for the protection of their persons and property which are secured to native-born citizens. The moment any human being from a country at peace with us comes within the jurisdiction of the United States, with their consent,—and such consent will always be implied when not expressly withheld, and, in the case of the Chinese laborers before us, was, in terms, given by the treaty referred to,—he becomes subject to all their laws, is amenable to their punishment, and entitled to their protection. Arbitrary and despotic power can no more be exercised over them, with reference to their persons and property, than over the persons and property of native-born citizens. They differ only from citizens in that they cannot vote, or hold any public office. As men having our common humanity, they are protected by all the guaranties of the constitution. To hold that they are subject to any different law, or are less protected in any particular, than other persons, is, in my judgment, to ignore the teachings of our history, the practice of our government, and the language of our constitution. Let us test this doctrine by an illustration: If a foreigner who resides in the country by its consent commits a public offense, is he subject to be cut down, maltreated, imprisoned, or put to death by violence, without accusation made, trial had, and judgment of an established tribunal, following the regular forms of judicial procedure? If any rule in the administration of justice is to be omitted or discarded in his case, what rule is it to be? If one rule may lawfully be laid aside in his case, another rule may also be laid aside, and all rules may be discarded. In such instances a rule of evidence may be set aside in one case, a rule of pleading in another; the testimony of eye-witnesses may be rejected, and hearsay adopted; or no evidence *755 at all may be received, but simply an inspection of the accused, as is often*the case in tribunals of Asiatic countries, where personal caprice and not settled rules prevail. That would be to establish a pure, simple, undisguised despotism and tyranny, with respect to foreigners resident in the country by its consent, and such an exercise of power is not permissible, under our constitution. Arbitrary and tyrannical power has no place in our system. As said by this court, speaking by Mr. Justice Matthews, in Yick Wo v. Hopkins, 118 U. S. 366, 369, 6 Sup. Ct. Rep. 1064: "When we consider the nature and the theory of our institutions of government, the principles upon which they are supposed to rest, and review the history of their development, we are constrained to conclude they do not mean to leave room for the play and action of purely personal and arbitrary power. * * * The fundamental rights to life, liberty, and the pursuit of happiness, considered as individual possessions, are secured by those maxims of constitutional law which are the monuments showing the victorious progress of the race in securing to man the blessings of civilization under the reign of just and equal laws." What once I had occasion to say of the protection afforded by our government, I repeat: "It is certainly something in which a citizen of the United States may feel a generous pride that the government of his country extends protection to all persons within its jurisdiction, and that every blow aimed at any of them, however humble, come from what quarter it may, 'is caught upon the broad shield of our blessed constitution and our equal laws.'" Ah Kow v. Nunan, 5 Sawy. 552–563.

I utterly dissent from, and reject, the doctrine expressed in the opinion of the majority, that "congress, under the power to exclude or expel aliens, might have directed any Chinese laborer found in the United States without a certificate of residence to be removed out of the country by executive officers, without judicial trial or examination, just as it might have authorized such officers absolutely to prevent his entrance into the country."[4] An arrest in that way, for that purpose, would not be a reasonable seizure of the person, within the meaning of the fourth article of the amendments of the constitution. It would be brutal and *756 oppressive. The*existence of the power thus stated is only consistent with the admission that the government is one of unlimited and despotic power, so far as aliens domiciled in the country are concerned. According to this theory, congress might have ordered executive officers to take the Chinese laborers to the ocean, and put them into a boat, and set them adrift, or to take them to the borders of Mexico, and turn them loose there, and in both cases without any means of support. Indeed, it might have sanctioned towards these laborers the most

[4]These words were originally printed in the opinion of the court, to which the dissent was directed. In the revision of the opinion the phraseology is changed.

shocking brutality conceivable. I utterly repudiate all such notions, and reply that brutality, inhumanity, and cruelty cannot be made elements in any procedure for the enforcement of the laws of the United States.

The majority of the court have, in their opinion, made numerous citations from the courts and the utterances of individuals upon the power of the government of an independent nation to exclude foreigners from entering its limits, but none, beyond a few loose observations, as to its power to expel and deport from the country those who are domiciled therein by its consent. The citation from the opinion in the recent case of Nishimura Ekiu v. U. S., (the Japanese Case,) 142 U. S. 651, 12 Sup. Ct. Rep. 336; the citation from the opinion in Chae Chan Ping v. U. S., (the Chinese Exclusion Case,) 130 U. S. 603, 604, 606, 9 Sup. Ct. Rep. 623; the citation in the case before the judiciary committee of the privy council,—all have reference to the exclusion of foreigners from entering the country. They do not touch upon the question of deporting them from the country after they have been domiciled within it by the consent of its government, which is the real question in the case. The citation from Vattel is only as to the power of exclusion; that is, from coming into the country. The citation from Phillimore is to the same effect. As there stated, the government allowing the introduction of aliens may prescribe the conditions on which they shall be allowed to remain, the conditions being imposed whenever they enter the country. There is no dispute about the power of congress to prevent the landing of aliens in the country. The question is as to the power *757 of congress to deport them, *without regard to the guaranties of the constitution. The statement that in England the power to expel aliens has always been recognized, and often exercised, and the only question that has ever been as to this power is whether it could be exercised by the king without the consent of parliament, is, I think, not strictly accurate. The citations given by Mr. Choate in his brief show conclusively, it seems to me, that deportation from the realm has not been exercised in England since Magna Charta, except in punishment for crime, or as a measure in view of existing or anticipated hostilities. But, even if that power were exercised by every government of Europe, it would have no bearing in these cases. It may be admitted that the power has been exercised by the various governments of Europe. Spain expelled the Moors; England, in the reign of Edward I., banished 15,000 Jews;[1] and Louis XIV., in 1685, by revoking the edict of Nantes, which gave religious liberty to Protestants in France, drove out the Huguenots. Nor does such severity of European governments belong only to the distant past. Within three years, Russia has banished many thousands of Jews, and apparently intends the expulsion of the whole race,—an act of barbarity which has aroused the indignation of all Christendom. Such was the feeling in this country that, friendly as our relations with Russia had always been, President Harrison felt compelled to call the attention of congress to it in his message in 1891, as a fit subject for national remonstrance. Indeed, all the instances mentioned have been condemned for their barbarity and cruelty, and no power to perpetrate such barbarity is to be implied from the nature of our government, and certainly is not found in any delegated powers under the constitution.

The government of the United States is one of limited and delegated powers. It takes nothing from the usages or the former action of European governments, nor does it take any power by any supposed inherent sovereignty. There is a great deal of confusion in the use of the word "sov- *758 eignty" *by law writers. Sovereignty or supreme power is in this country vested in the people, and only in the people. By them certain sovereign powers have been delegated to the government of the United States, and other sovereign powers reserved to the states or to themselves. This is not a matter of inference and argument, but is the express declaration of the tenth amendment to the constitution, passed to avoid any misinterpretation of the powers of the general government. That amendment declares that "the powers not delegated to the United States by the constitution, nor prohibited by it to the states, are reserved to the states, respectively, or to the people." When, therefore, power is exercised by congress, authority for it must be found in express terms in the constitution, or in the means necessary or proper for the execution of the power expressed. If it cannot be thus found, it does not exist.

It will be seen by its provisions that the sixth section recognizes the right of certain Chinese laborers to remain in the United States, but to render null that right it declares that if, within one year after the passage of the act, any Chinese laborer shall have neglected, failed, or refused to comply with the provisions of the act to obtain a certificate of residence, or shall be found within the jurisdiction of the United States without a certificate of residence, he shall be deemed and adjudged to be unlawfully within the United States, and may be arrested by any United States customs official, collector of internal revenue or his deputies, United States marshal or his deputies, and taken before a United States judge, whose duty it shall be to order that he be deported from the United States, unless he

[1] The Jews during his reign were cruelly despoiled, and in 1290 ordered, under penalty of death, to quit England forever, before a certain day. 6 Amer. & Eng. Enc. Law, p. 434.

shall establish clearly, to the satisfaction of the judge, that by reason of accident, sickness, or other unavoidable cause, he has been unable to secure his certificate, and to the satisfaction of the judge, by at least one credible white witness, that he was a resident of the United States at the time of the passage of the act. His deportation is thus imposed for neglect to obtain a certificate of residence, from which he can only escape by showing his inability to secure it from
*759 one of the causes named. That is the punishment*for his neglect, and that, being of an infamous character, can only be imposed after indictment, trial, and conviction. If applied to a citizen, none of the justices of this court would hesitate a moment to pronounce it illegal. Had the punishment been a fine, or anything else than of an infamous character, it might have been imposed without indictment; but not so now, unless we hold that a foreigner from a country at peace with us, though domiciled by the consent of our government, is withdrawn from all the guaranties of due process of law prescribed by the constitution, when charged with an offense to which the grave punishment designated is affixed.

The punishment is beyond all reason in its severity. It is out of all proportion to the alleged offense. It is cruel and unusual. As to its cruelty, nothing can exceed a forcible deportation from a country of one's residence, and the breaking up of all the relations of friendship, family, and business there contracted. The laborer may be seized at a distance from his home, his family, and his business, and taken before the judge for his condemnation, without permission to visit his home, see his family, or complete any unfinished business. Mr. Madison well pictures its character in his powerful denunciation of the alien law of 1798, in his celebrated report upon the resolutions, from which we have cited, and concludes, as we have seen, that if a banishment of the sort described be not a punishment, and among the severest of punishments, it will be difficult to imagine a doom to which the name can be applied.

Again, when taken before a United States judge, he is required, in order to avoid the doom declared, to establish clearly, to the satisfaction of the judge, that by reason of accident, sickness, or other unavoidable cause he was unable to secure his certificate, and that he was a resident of the United States at the time, by at least one credible white witness. Here the government undertakes to exact of the party arrested the testimony of a witness of a particular color, though conclusive and incontestable testimony from
*760 others may be adduced. The law might as well have said that unless the laborer*should also present a particular person as a witness, who could not be produced, from sickness, absence, or other cause, such as the archbishop of the state, to establish the fact of residence, he should be held to be unlawfully within the United States.

There are numerous other objections to the provisions of the act under consideration. Every step in the procedure provided, as truly said by counsel, tramples upon some constitutional right. Grossly it violates the fourth amendment, which declares that "the right of the people to be secure in their persons * * * against unreasonable searches and seizures shall not be violated, and no warrant shall issue but upon probable cause, supported by oath or affirmation, and particularly describing the * * * persons * * * to be seized."

The act provides for the seizure of the person without oath or affirmation or warrant, and without showing any probable cause by the officials mentioned. The arrest, as observed by counsel, involves a search of his person for the certificate which he is required to have always with him. Who will have the hardihood and effrontery to say that this is not an "unreasonable search and seizure of the person?" Until now it has never been asserted by any court or judge of high authority that foreigners domiciled in this country by the consent of our government could be deprived of the securities of this amendment; that their persons could be subjected to unreasonable searches and seizures, and that they could be arrested without warrant upon probable cause, supported by oath or affirmation.

I will not pursue the subject further. The decision of the court, and the sanction it would give to legislation depriving resident aliens of the guaranties of the constitution, fill me with apprehensions. Those guaranties are of priceless value to every one resident in the country, whether citizen or alien. I cannot but regard the decision as a blow against constitutional liberty, when it declares that congress has the right to disregard the guaranties of the constitution intended for the protection of all men domiciled in the country with the consent of the government, in their rights of person and *761
property.* How far will its legislation go? The unnaturalized resident feels it to-day, but if congress can disregard the guaranties with respect to any one domiciled in the country with its consent, it may disregard the guaranties with respect to naturalized citizens. What assurance have we that it may not declare that naturalized citizens of a particular country cannot remain in the United States after a certain day, unless they have in their possession a certificate that they are of good moral character, and attached to the principles of our constitution, which certificate they must obtain from a collector of internal revenue upon the testimony of at least one competent witness of a class or nationality to be designated by the government?

What answer could the naturalized citizen in that case make to his arrest for deportation, which cannot be urged in behalf of the Chinese laborers of to-day?

I am of the opinion that the orders of the court below should be reversed, and the petitioners should be discharged.

Mr. Chief Justice FULLER, dissenting.

I also dissent from the opinion and judgment of the court in these cases.

If the protection of the constitution extends to Chinese laborers who are lawfully within, and entitled to remain in, the United States, under previous treaties and laws, then the question whether this act of congress, so far as it relates to them, is in conflict with that instrument, is a judicial question, and its determination belongs to the judicial department.

However reluctant courts may be to pass upon the constitutionality of legislative acts, it is of the very essence of judicial duty to do so, when the discharge of that duty is properly invoked.

I entertain no doubt that the provisions of the fifth and fourteenth amendments, which forbid that any person shall be deprived of life, liberty, or property without due process of law, are, in the language of Mr. Justice Matthews, already quoted by my Brother Brewer, "universal in their application to all persons within the territorial jurisdiction, without *762 regard to any differences of race, of color, or of nationality;" and although in Yick Wo's Case, 118 U. S. 356, 6 Sup. Ct. Rep. 1064, only the validity of a municipal ordinance was involved, the rule laid down as much applies to congress, under the fifth amendment, as to the states, under the fourteenth. The right to remain in the United States, in the enjoyment of all the rights, privileges, immunities, and exemptions accorded to the citizens and subjects of the most favored nation, is a valuable right, and certainly a right which cannot be taken away without taking away the liberty of its possessor. This cannot be done by mere legislation.

The argument is that friendly aliens, who have lawfully acquired a domicile in this country, are entitled to avail themselves of the safeguards of the constitution only while permitted to remain, and that the power to expel them, and the manner of its exercise, are unaffected by that instrument. It is difficult to see how this can be so, in view of the operation of the power upon the existing rights of individuals; and to say that the residence of the alien, when invited and secured by treaties and laws, is held in subordination to the exertion against him, as an alien, of the absolute and unqualified power asserted, is to import a condition not recognized by the fundamental law. Conceding that the exercise of the power to exclude is committed to the political department, and that the denial of entrance is not necessarily the subject of judicial cognizance, the exercise of the power to expel, the manner in which the right to remain may be terminated, rests on different ground, since limitations exist or are imposed upon the deprivation of that which has been lawfully acquired. And while the general government is invested, in respect of foreign countries and their subjects or citizens, with the powers necessary to the maintenance of its absolute independence and security throughout its entire territory, it cannot, in virtue of any delegated power, or power implied therefrom, or of a supposed inherent sovereignty, arbitrarily deal with persons lawfully within the peace of its dominion. But the act before us is not an act to abrogate or repeal treaties or laws in respect of Chinese laborers entitled to remain in the United States, or *763 to expel them from the country, and no such intent can be imputed to congress. As to them, registration for the purpose of identification is required, and the deportation denounced for failure to do so is by way of punishment to coerce compliance with that requisition. No euphuism can disguise the character of the act in this regard. It directs the performance of a judicial function in a particular way, and inflicts punishment without a judicial trial. It is, in effect, a legislative sentence of banishment, and, as such, absolutely void. Moreover, it contains within it the germs of the assertion of an unlimited and arbitrary power, in general, incompatible with the immutable principles of justice, inconsistent with the nature of our government, and in conflict with the written constitution by which that government was created, and those principles secured.

Immigration Policies: Messages of Exclusion to African Americans

Bill Ong Hing*

Introduction

To many African Americans, the Coast Guard's interdiction and forced reversal of boats carrying Haitian refugees is a racist act directed at people of African descent.[1] Until recently, this perception of racism in the application of U.S. immigration policies was reinforced by the contrasting news of planeloads and boatloads of Cuban refugees being warmly received just miles down the coast in Florida.[2] The allegation of racism is further bolstered by the fact that compared to Asians, Latinos, and Irish persons, few African immigrants are admitted into the United States.

On one hand, African Americans are concerned about racism in the government's decisions about whom to admit to this country. However, many are also concerned that too many immigrants of all backgrounds are being admitted. Among many African Americans, there is concern that immigrants are taking away jobs, depressing their wages, or taking away business opportunities in their communities.

> In the hotel industry and in some of the other industries where you used to see predominantly blacks, you don't see blacks now; . . . [for example] waiting tables; [and] many of the construction jobs have

* Associate Professor of Law, Stanford Law School. Many thanks to my research assistants, Adisa Abudu and Dawn Chirwa, and to the immigration symposium editors of the *Howard Law Journal*.

1. *See, e.g.*, Malissia Lennox, Note, *Refugees, Racism, and Reparations: A Critique of the United States' Haitian Immigration Policy*, 45 Stan. L. Rev. 687 (1993).

2. *See Cuban Athletes Race to Defect; 39 at Puerto Rico Games Seek Asylum*, S.F. Examiner, Nov. 30, 1993, at A2; *13 Escape Cuba in Stolen Biplane*, St. Petersburg Times, Nov. 16, 1993, at A1; *America Accused of Double Standard on Aerial Hijackers; Haitians Prosecuted, Cubans Celebrated*, Ariz. Republic, Sept. 19, 1993, at A6; *Cubans Seeking Asylum Freed After Probe Into Miami Flight*, L.A. Times, Dec. 31, 1992, at A20. *But see* Walter Pincus & Roberto Suro, *Ripple in Florida Straits Overturned U.S. Policy; Reno and Chiles were Catalysts in Getting Administration to Slam the Door on Cuban Rafters*, Wash. Post, Sept. 1, 1994, at A31.

> been taken by the Hispanics and the Asians who have come into the area.[3]
>
> The immigrants over here . . . work at a very low scale. So that's what's hurting different businesses [that are black-owned]. . . . Where I'm making a labor scale of twenty-two dollars an hour and an immigrant would take on a job for maybe five or six dollars an hour. . . . That kind of cuts my throat.[4]
>
> African-Americans who burned and looted Korean businesses said they did so because they were not respected when they shopped there. They also complained that Koreans will not hire blacks to work in their businesses. "They (the Koreans) raped us," said one south-central resident. "They come in here and open up a liquor store on every corner."[5]

These excepts are examples of the feelings that many African Americans experience because of jobs which were once available but now are frequently given to immigrants.

In short, U.S. immigration policies send messages of exclusion to many African Americans. The message is not only one of physical exclusion from the country, but exclusion from deliberations on matters that appear to affect their community. The concerns of the African American community seem to have been brushed aside in favor of a policy that appears to allow an increasing number of immigrants into the United States. As the total number of immigrants increases, however, admission of people of color into the United States continues to be restricted.[6]

---

3. Interview with Joyce Punch, retired African American school teacher, on *America Becoming* (PBS television broadcast, Sept. 16, 1991).

4. Interview with Steve Holt, African American business owner, on *America Becoming* (PBS television broadcast, Sept. 16, 1991).

5. Earnest L. Perry, *Violence Is No Answer, But Understanding Can Be a Key*, Hous. Chron., May 10, 1992, Outlook Section, at 1.

6. This article focuses on the effect of the new immigration policy on African Americans, not because they are the only people of color in this country who are victimized by subordination, nor the only people which may be adversely affected by the new policy, but because they are the paradigmatic measure of how adverse the consequences of the policy may be. Not only are African-Americans the most voluminous group of people of color in this country, but they also fall behind most other racial groups in many measures of success, whether educational, social, or economic. For example, nearly one-half of all African-American children live in poverty, while one-third of all Hispanic children live in poverty. Debbie Goldberg, *Staying in School*, Wash. Post, Apr. 8, 1990, at R13. African-Americans also experience more residential segregation than Asians, Mexicans, or Cubans at similar income levels. Peter Schuck, *America and the New Immigrant Experience*, L.A. Times, May 5, 1991, at M1. Finally, in 1992, 33.3 percent of all African-Americans and 29.3 percent of Latino-Americans lived in poverty. *America's Poor Showing*, Newsweek, Oct. 18, 1993, at 44. This article does not seek to establish a "hierarchy of oppression." The article, however, is intended to analyze an immigration policy which results in the continued disregard of the need for African American educational achievement. Special emphasis will be placed on the extent to which the policy may ultimately affect

Whether or not these popular images are true, the implementation of our immigration laws and policies do little to dispel them. This article represents an initial foray into the bases of these concerns and provides preliminary proposals to respond to them that include African Americans in the formulation of immigration policies. Indeed if policy makers and immigrants' rights advocates care about the African American community, there is much that we can change in our immigration laws that could address at least some of their concerns. Furthermore, this can be done without negatively affecting the families of non-African immigrant groups who seek to be reunited under U.S. immigration laws.

This article proposes that the United States government should do more to understand the nature of African migration and provide extra visa numbers to African nationals to compensate for the historical underrepresentation of Africans among immigrant groups. This proposal mirrors the increases we have recently provided to Irish nationals. In addition, this article proposes that the United States government seriously consider eliminating visas for skilled workers, or substantially reduce their number and assess a fee on employers who hire the remainder. The fees collected would fund education and job training for underskilled American workers. These proposals would at least begin to address the important concerns described above.

Part I of this article is devoted to the issue of African immigration—its background, its relation to other countries, its response to current provisions, and possible changes to create uniformity. Part II provides a background on visas for skilled workers, an assessment of the socioeconomic needs of the African American community, and an argument supporting the need for changes in the policies on immigration of skilled workers.

---

subordinated people, including both unskilled American citizens and unskilled immigrants, as a whole.

I. African Immigration

A. *Numbers and the Current System*

Throughout the immigration history of the United States, Africans have been underrepresented as a voluntary immigrant group.[7] Before 1965, Africans represented less than one percent of the total immigrant population;[8] today Africans constitute only 2.3 percent of all immigrants.[9] Before the latter half of the twentieth century, the low numbers may have been attributable to the fact that most of the African continent was controlled by Europe; immigration simply was not an option for more than a handful of elite Africans. But even after most African countries attained independence in the 1950s and 1960s, African immigration to this country remained low.

In real numbers, this means that in the 1920s, for example, of 4.1 million immigrants, 6286 were from Africa; in the 1950s, of 2.5 million immigrants, 14,092 were African.[10] In the 1970s, Africans numbered 80,779 of the total 4.5 million, and in a typical year such as 1987, they were 15,730 of the total 601,516 immigrants.[11]

When one compares the number of African immigrants to those from other world regions, only Oceania (which includes Australia, Fiji, New Zealand, Tonga, and Western Samoa) is less represented, ranging from 0.4 percent to 0.8 percent of the total over the last thirty-five years. European immigrant representation has declined from over fifty-two percent in the 1950s to 10.4 percent in the 1980s.[12] Asian immigration has increased from six percent of the total in the 1950s to over thirty-seven percent in the 1980s when 2.7 million Asians immigrated.[13] Mexicans alone were 22.6 percent of the total in the 1980s,[14] a figure that was distorted by the large number of Mexicans granted amnesty under the 1986 legalization program.

Under current immigration laws, the primary means of immigration is through a family reunification category or an employment cate-

---

7. *See generally* John Hope Franklin, From Slavery to Freedom: A History of Negro Americans 59 (3d ed. 1967) (stating that before the Civil War, over 9.5 million Africans were captured and brought to the western hemisphere as slaves).

8. U.S. Immigr. and Naturalization Service, Stat. Y.B. of the Immigr. and Naturalization Service, 1989, at xvii (1990) [hereinafter 1990 INS Stat. Y.B.].

9. U.S. Immigr. and Naturalization Service, Stat. Y.B. of the Immigr. and Naturalization Service, 1990, at 52 tbl. 3 (1991) [hereinafter 1991 INS Stat. Y.B.] (Part of the reason was because many Mexicans attained permanent immigrant status that year under the 1986 legalization provisions.).

10. 1991 INS Stat. Y.B., *supra* note 9, at 49 tbl. 2.

11. 1990 INS Stat. Y.B., *supra* note 8, at 4 tbl. 2.

12. 1991 INS Stat. Y.B., *supra* note 9, at 49-50 tbl. 2.

13. 1991 INS Stat. Y.B., *supra* note 9, at 49-50 tbl. 2.

14. 1991 INS Stat. Y.B., *supra* note 9, at 49-50 tbl. 2.

gory. The family categories require a prospective immigrant to have a U.S. citizen or lawful permanent resident spouse, parent, or sibling. Most employment visas are set aside for professionals, corporate executives, outstanding scientists, educators, athletes, and other skilled persons. Additional visas are available for investors and so-called diversity immigrants, a category that will be explained later.[15] Individuals fleeing persecution may enter as refugees if they are from one of the countries named on the State Department list. In addition, others already in the United States, even those who entered surreptitiously who have a well-founded fear of persecution back home, may apply for asylum regardless of the country they flee.[16]

An unlimited number of visas are available those who qualify as "immediate relatives" of U.S. citizens. "Immediate relatives" includes spouses, children, and parents of adults.[17] Through complicated formulas, each country also gets about 25,000 visas for other relatives such as siblings, adult and married sons and daughters, spouses and children of resident aliens, up to a worldwide ceiling of 226,000.[18]

Under 1990 legislation, the number of visas set aside for employment categories was substantially increased from 54,000 to 140,000.[19] Each year, the number of visas set aside for refugees varies depending on the recommendations of the State Department.[20] For example, the number in 1980 was 234,000, but for 1994 the figure is 121,000.[21] In theory, the number of persons who can be granted asylum is unlimited. In practice, however, asylum is granted to no more than about 5000 individuals each year.[22]

Notwithstanding the increase in employment visas in 1990, the unlimited number of immediate relatives and the large number of visas available for other relatives indicate that family reunification remains the primary policy goal of U.S. immigration laws. Africans, however, have not been able to take advantage of this policy to the

---

15. *See infra* text accompanying notes 120-22.
16. BILL ONG HING, HANDLING IMMIGRATION CASES §§ 6.1-6.3, 6.31-6.32 (1985); Immigration and Nationality Act, 8 U.S.C. §§ 1157, 1158 (1994).
17. HING, *supra* note 16, §§ 4.2-4.6.
18. HING, *supra* note 16, §§ 4.34-4.35.
19. HING, *supra* note 16, §§ 4.34-4.35.
20. Immigration and Nationality Act, 8 U.S.C. § 1157 (1994).
21. *Clinton Administration to Admit 121,000 Refugees This Year*, 70 INTERPRETER RELEASES 1325, 1331 (1993).
22. *See, e.g.*, 1991 INS STAT. Y.B., *supra* note 9, 105 tbls. 19, 30.

same extent as some Asians and Latinos. In 1990, 9316 Africans immigrated in the immediate relative category—about three percent of the total. Spouses accounted for seventy-seven percent of African immediate relatives, followed by parents at 13.4 percent, and children at 9.6 percent.

Of the 7614 Africans who immigrated in 1990 in other relative and occupational categories subject to quotas, thirty percent entered in an occupational preference.[23] By way of comparison, the total quota immigrants from Mexico, 19,986, mainland China, 19,795, the Philippines, 19,588, India, 19,157, Korea, 18,624, and Vietnam, 8829, each outnumbered the sum of quota immigrants from the entire continent of Africa. The highest proportion of occupational visas in any of those countries was 17.1 percent for Korea. Mexico was under ten percent.[24]

In short, while African countries each had 20,000 visas available to them under the numerically restricted family and occupational categories in 1990, no African country came close to its ceiling prior to the increases of the 1990 legislation The Immigration and Naturalization Service (hereinafter "INS") reported these figures: Egypt 1768; South Africa 1046; Nigeria 628; Cape Verde 583; Kenya 532; Ghana 486; Ethiopia 381; Tanzania 336; Morocco 246; Uganda 220; Sierra Leone 198; and Liberia 151.[25] Botswana, Central African Republic, the Congo, Gabon, Guinea-Bissau, and Namibia had fewer than ten such immigrants in 1990, according to the INS.[26] Even when the unlimited immediate relative figures are included, the totals for individual countries are still relatively low: Egypt 2833; South Africa 1960; Nigeria 1896; Ghana 1159; Ethiopia 977; Cape Verde 788; Kenya 592; Morocco 465; Liberia 426; Uganda 343; Tanzania 339; Sierra Leone 267; and Senegal 106.[27] What the figures do reveal, however, is that Africans are interested in immigrating to the United States.

Of the 121,000 refugee visas set aside for 1994, 7000 are designated for Africa, up from 3000 for 1993. This compares with 45,000 for East Asia, 55,000 for Eastern Europe and the former Soviet Union, 4000 for Latin America and the Caribbean, and 6000 for the Near East and South Asia.[28] Until 1994, the number of refugee visas

---

23. 1991 INS Stat. Y.B., *supra* note 9, at 58-59 tbl. 5.
24. 1991 INS Stat. Y.B., *supra* note 9, at 58-59 tbl. 5.
25. 1991 INS Stat. Y.B., *supra* note 9, at 58-59 tbl. 5.
26. Letter from the Demographics Statistics Branch, Immigration and Naturalization Service, Washington, D.C. to Bill Hing (Mar. 2, 1993) (on file with author).
27. 1991 INS Stat. Y.B., *supra* note 9, 52-53 tbl. 3.
28. *Clinton Administration to Admit 121,000 Refugees This Year*, 70 Interpreter Releases 1325, 1331 (1993).

designated for Africa remained stable in spite of fluctuations in refugee slots. For example in 1985, when the total refugee pool was only 70,000, the refugee quota for Africa remained at 3000, while the total for the Soviet Union and Eastern Europe was down to 9000.[29]

The number of nonimmigrants from Africa is also relevant to the number of Africans who become immigrants. Nonimmigrants are individuals who enter the United States for a temporary purpose and period, such as tourists, students, media representatives, business representatives, and government personnel.[30] The number of nonimmigrant visas is unlimited. This factor is pertinent to immigrant numbers because tourists and students often become permanent immigrants. While nonimmigrants who wish to become immigrants must still meet the requirements of one of the family or employment categories, marriage to a U.S. citizen or contacts with prospective employers while here in a nonimmigrant status is much easier. For example, foreign students commonly obtain a practical training permit to work for a year after graduation.[31] After the practical training period, an employer can help the person change status to a temporary worker visa. Many of these workers later qualify for a permanent occupational category based upon the training and education received while in the United States as a nonimmigrant.

Two related nonimmigrant categories facilitate permanent relocation for many aliens. Treaty traders and treaty investors who are either engaging in trade between the United States and their home countries or purchasing businesses in the United States can remain in this country indefinitely.[32]

As long as annual reporting requirements are satisfied, extensions of the treaty status are granted. But in order to qualify for such visas, the United States must have a qualifying trade or commerce treaty with the particular country.[33] Only six African nations enjoy such treaty status with the United States compared to sixteen European

29. Hing, *supra* note 16, § 6.31.
30. Immigration and Nationality Act, 8 U.S.C. § 1101(a)(15) (1994).
31. Hing, *supra* note 16, § 3.6.
32. Immigration and Nationality Act, 8 U.S.C. § 1101 (a) (15) (e) (1994); *see also* Hing, *supra* note 16 §§ 3.38-3.40 (examining the current law, procedures, forms and strategies in the area of immigration).
33. Immigration and Nationality Act, 8 U.S.C. § 1101(a)(15)(e).

and fifteen Asian countries.[34] Those six countries—Cameroon, Ethiopia, Liberia, Senegal, Togo, and Zaire—accounted for 0.5 percent of all treaty traders and investors entering the United States in 1989; Europeans and Asians represented 36.6 percent and 60.2 percent respectively.

As in the case of immigrants, Africans comprise a small percentage of the nonimmigrants who enter the United States. Almost half of all nonimmigrants originate from one of four countries: Japan, the United Kingdom, Mexico, or Germany.[35] In 1992, 206,535 nonimmigrants were from Africa, but this was only one percent of the total. One hundred sixteen thousand four hundred forty-two were tourists, and about 7374 were students.[36]

B. *Some Explanations for Low African Immigration Rates*

1. *A History of Exclusion*

Scholars generally trace the beginning of racially restrictive U.S. immigration policies to laws directed at various Asian immigrant groups, national origin quotas that disfavored southern and eastern Europeans, and policies that alternatively attracted and deported Mexican migrants depending on our nation's economic needs.[37] However, since African Americans were slaves for the first one hundred years of this nation's history, they were unable to take advantage of immigration rights anyway. By the time they were freed, slavery had destroyed African ancestral family and cultural structure. Familial and social ties with Africa had long been severed so reunification with African relatives was impossible. Moreover, what African in his or her right mind would have wanted to come to the United States dur-

---

34. HING, *supra* note 16, § 3.35.

35. U.S. IMMIGR. AND NATURALIZATION SERVICE, STAT. Y.B. OF THE IMMIGR. AND NATURALIZATION SERVICE, 1992, at 128 tbl.39 (1993).

36. *Id.*

37. *See generally* BILL ONG HING, MAKING AND REMAKING ASIAN AMERICA THROUGH IMMIGRATION POLICY, 1850-1990 (1993) (analyzing the development of American immigration policies and the impact such policies have on the demographics and culture of Asian-American communities); ALFREDO MIRANDE, GRINGO JUSTICE (1987) (arguing that since the end of the war between the United States and Mexico, Mexicans have been subjected to prejudicial and discriminatory treatment); Gerald P. López, *Undocumented Mexican Migration: In Search of a Just Immigration Law and Policy*, 28 UCLA L. REV. 615, 641 (1981) (proposing that the United States has a moral obligation to Mexican workers because the workers make positive contributions to the community); JOHN HIGHAM, STRANGERS IN THE LAND: PATTERNS OF AMERICAN NATIVISM, 1860-1925 (1963) (examining the history of American anti-foreigner spirit).

ing or immediately after slavery?[38] Thus, an African exclusion law was not necessary to discourage African immigration.

Prior to 1870, the subordination of people of African descent was further underscored by the fact that people from Africa could not become U.S. citizens through naturalization. The Nationality Act of 1790 limited naturalization to "free white persons" and specifically excluded African Americans and Native Americans.[39] However, in 1870, Congress extended naturalization rights to outlanders of African descent.[40] Still, African immigrants were not widely accepted by the United States.

In 1912, Congress began consideration of legislation that would eventually result in a literacy requirement for immigrants. In the midst of this debate, a Mississippi senator offered an amendment to exclude persons of African descent, which was narrowly rejected, twenty-eight to twenty-five, with forty-two not voting.[41] But two years later, a Missouri senator succeeded in persuading the Senate to pass immigration legislation that included a provision requiring the exclusion of "all members of the African or black race." However, the Senate was forced to recede from the African exclusion when the legislation went to conference committee.[42]

The reactionary and isolationist political climate following World War I manifested itself in the "Red Scare" of 1919-20 and led to exclusionist demands that went beyond attacks directed at Asians. By the early 1920s, the pressure resulted in the overwhelming passage of the

38. Even after the abolition of slavery, every African American born in the United States was considered three-fifths of a person. U.S. CONST. art. I, § 2 *amended by* U.S. CONST. amend XIV, § 2.

39. Years later, when Japanese immigrants unsuccessfully sought naturalization as "free white persons," the Supreme Court reaffirmed that the 1790 naturalization statute intended that "Negroes and Indians" were to be denied naturalization. Ozawa v. United States, 260 U.S. 178, 196 (1922).

40. Act of July 14, 1870, 16 Stat. 254 (1870). In the same legislation, Congress deliberately denied Chinese persons that right because of their "undesirable qualities." *Id. See also* HING, *supra* note 37, at 22-23 (noting that a strong anti-Chinese sentiment arose in California and other western states); EDWARD P. HUTCHINSON, LEGISLATIVE HISTORY OF AMERICAN IMMIGRATION POLICY 1798-1965, 57-60 (1981) (documenting the legislative history of immigration laws and policy). The Senate defeated one amendment that would have read: "That the naturalization laws are hereby extended to aliens of African nativity, and to persons of African descent, and to persons born in the Chinese empire." *Id.* at 58.

41. Act of July 14, 1870, 16 Stat. 254 (1870). *See also* HUTCHINSON, *supra* note 40, at 152 (tracking the legislative history of early immigration legislation).

42. Act of July 14, 1870, 16 Stat. 254 (1870).

landmark Immigration Act of 1924.[43] This Act, opposed by only six Senators, took aim at southern and eastern Europeans, who the Protestant majority in the United States viewed with dogmatic disapproval.[44] The arguments advanced in support of the bill stressed several recurring themes: the racial superiority of Anglo-Saxons, the belief that immigrants would cause the lowering of wages, the unassimilability of foreigners, and more generally, the threats to the nation's social unity and order posed by immigration.[45]

The 1924 Act cleverly, yet transparently, restructured criteria for admission to respond to nativist demands, and embodied a general selection policy that remained in place until 1952. The Act provided that the immigration of any particular nationality would be limited to two percent of the 1890 population of that nationality. Thus, the racial composition of the country in 1890 served as the model for years to come.

The law had the greatest impact upon Jews, Italians, Slavs, and Greeks, who had immigrated in great numbers after 1890, and who would be most disfavored by a quota system based on their numbers in the United States in 1890.[46] "Racial homogeneity" was to be maintained by allocating national quotas to conform to the estimated composition by national origin of the population of the United States in 1920. In actuality, only the portion of the population of direct European origin was to be considered in the computation. Excluded from the 1920 base population were immigrants from the Western Hemisphere and their descendants, "aliens ineligible to citizenship or their descendants, the descendants of slave immigrants, or the descendants of American aborigines."[47]

This national origins quota system was finally repealed in 1965.[48] In its place, Congress installed a system that favored family reunification categories and provided annual per-country quotas of 20,000 for countries of the Eastern Hemisphere, including Asia, Europe, and Africa.[49] A similar system was established for countries of the Western Hemisphere, such as Canada and Mexico, in 1978. Beginning in 1952,

---

43. Immigration Act of 1924, ch. 190, 43 Stat. 153.

44. CHARLES GORDON & STANLEY MAILMAN, IMMIGRATION LAW AND PROCEDURE § 1.2(c) (1992); ABBA P. SCHWARTZ, THE OPEN SOCIETY 107 (1968).

45. *See* JOHN F. KENNEDY, A NATION OF IMMIGRANTS 74-76 (1964) and HUTCHINSON, *supra* note 40, at 483-85.

46. HIGHAM, *supra* note 37, at 318. Except for Filipinos, the 1924 law completely closed off immigration possibilities for Asian immigrants. HING, *supra* note 37, at 32-33.

47. Immigration Act of 1924, ch. 190, 43 Stat. 153.

48. Immigration and Nationality Act of 1952, ch. 477, 66 stat. 163.

49. *Id.*

immediate relatives of United States citizens could enter in unlimited numbers.[50]

A combination of the current immigration laws, their implementation, and events in Africa provide a starting point in understanding African immigration rates to the United States.

2. *International Migration Within Africa*[51]

One factor that may explain the underrepresented status of Africans as immigrants and nonimmigrant visitors to the United States is international migration within Africa. Intracontinental migration is a common phenomenon throughout Africa. In fact, Africa has one of the most mobile populations in the world.[52] Of the approximately 383 million Africans living on the continent, an estimated thirty-five million people, or ten percent, live outside their own country for some period of time.[53] Migrants account for ten percent of the total population in Sub-Saharan Africa,[54] and West Africa is recognized as one of the few areas in the world where relatively large numbers of people continue to move freely across international borders.[55]

Contemporary international migration in West Africa is the result of three interrelated migrational stages: (1) those contemporary movements that have continued from the pre-colonial period; (2) those movements that originated in the colonial period; and (3) those movements that are essentially a product of the independence period.[56] Since pre-colonial times, West African tribes and smaller kin-

50. Act of June 27, 1952, 66 Stat. 163.

51. This article will discuss African immigration on three geographic levels—continental, regional (West Africa), and country-specific (primarily Ghana).

52. SERGIO RICCA, INTERNATIONAL MIGRATION IN AFRICA: LEGAL AND ADMINISTRATIVE ASPECTS v (1989).

53. *Id.* at 3.

54. *Id.* at v.

55. K.C. ZACHARIAH & JULIEN CONDE, MIGRATION IN WEST AFRICA: DEMOGRAPHIC ASPECTS 3 (1981). This section of the article focuses primarily on West Africa.

56. W.T.S. Gould, African Population Mobility Project Working Paper No. 16 (1974) (on file with the University of Liverpool Department of Geography). *See also* Elliot P. Skinner, *Labor Migration and National Development in Africa*, *in* AFRICAN MIGRATION AND NATIONAL DEVELOPMENT 18 (Beverly Lindsay ed., 1985) (arguing that modern labor migration in Africa is the direct result of Western expansion and "the fitful but progressive incorporation of Africa into the emergent global capitalist economy." Skinner believes that it all started with the involuntary migration of millions of Africans to America, which led to the growth of a Euro-American capitalist base. This created the opportunity to colonize Africa, which in turn led to the need for African labor.)

ship groups have migrated for a variety of reasons, including trade, marriage, war, and economic prosperity. The colonial boundaries arbitrarily drawn by European colonizers in the late 1800s, which cut across many traditional tribal territories, were maintained as state borders when the African colonies became independent countries. As a result, contemporary movement within traditional tribal areas often resulted in "international" migration, at least in the technical sense of the word.[57] Although exact statistics about informal border crossings are not available, it is estimated that hundreds of thousands of Africans cross international borders daily without going through any formalities.[58]

West African migration expanded significantly during the colonial period when the European colonizers created artificial economic centers of activity in certain colonies such as Ghana and Togo. At the same time, they left other areas such as Burkina Faso, and Mali virtually undeveloped. Because prosperity of the colonial economic centers depended on a continual supply of labor, the European colonizers "resorted to all sorts of subterfuges"[59] to force Africans to migrate from less favored areas to the colonial centers and work as soldiers or as laborers in the railroad, mining, and agricultural industries.[60]

The impact of West African migration during the colonial period cannot be underestimated. Entire villages moved to Ghana in the early 1900s as a result of forced labor, military recruitment, tribal uprisings and famine that was often caused by the colonists' favoritism toward certain tribes.[61] In 1922, Upper Volta was forced to send 6000 workers to build railroads in Sudan[62] after attempts to import labor from Cuba and China failed.[63] Similarly, in the late 1930s, over 20,000 Upper Voltans became forced laborers in the Ivory Coast, and that number swelled to over 150,000 during World War II.[64] Even after forced labor was abolished in the 1930s, household or "hut" taxes were used to force large numbers of Africans to take paid jobs in the colonial centers.[65]

---

57. One example of this is the Yoruba farmers who live on the Dahomey/Nigeria border and cultivate crops in both countries. Gould, *supra* note 56, at 13-14.
58. Ricca, *supra* note 52, at 61.
59. Ricca, *supra* note 52, at 61.
60. Skinner, *supra* note 56, at 25.
61. Zachariah & Conde, *supra* note 55, at 31.
62. Skinner, *supra* note 56, at 24.
63. Skinner, *supra* note 56, at 24.
64. Zachariah & Conde, *supra* note 55, at 6.
65. Ricca, *supra* note 52, at 12.

International migration in West Africa still bears the "indelible imprint" of colonization[66] despite the fact that West African countries gained independence over thirty years ago. New forms of political organization were introduced during the colonial period that "fundamentally reordered the spatial organization of society and economy" and facilitated the movement of Africans within the continent.[67]

The development of certain areas of West Africa during the colonial period, to the exclusion of others, created migratory flows from the less developed rural areas to the urban centers. With few exceptions, this trend continues today "regardless of whether there is an international border between point of departure and destination."[68] Many West Africans have come to depend on the economic remuneration they receive as migrant laborers.

Colonization-created migration to West African urban centers increased shortly after African independence when some West African countries became prosperous as they secured control of resources previously monopolized by the colonizers. For instance, until the mid-1970s, Ghana and the Ivory Coast were the main recipients of migrant laborers from neighboring countries, with thirty-five percent of the Ivory Coast's labor force consisting of non-Ivorians.[69] Economic prosperity, however, often proved to be a temporary or inconsistent phenomenon. As a result of economic decline and other factors, some West African countries found themselves unable to absorb the steady flow of migrants to their urban areas, "so that for the first time labor supply out-paced demand and there was increasing overt unemployment."[70]

A prime example of this phenomenon is Ghana. After gaining independence in 1957, Ghana became one of the most prosperous countries in Africa due to its status as the world's largest cocoa exporter. Migrants from Upper Volta, Togo, and Nigeria flocked to Ghana, expecting to make a better living there. In 1960, 828,000 foreign nationals, twelve percent of the country's total population, resided in Ghana.[71] Ten years later, the number had decreased to

66. RICCA, *supra* note 52, at 12.
67. Gould, *supra* note 56, at 31.
68. RICCA, *supra* note 52, at 13.
69. Gould, *supra* note 56, at 2-3.
70. Gould, *supra* note 56, at 25.
71. ZACHARIAH & CONDE, *supra* note 55, at 37.

562,000; due to Ghana's economic downturn, better opportunities elsewhere in the region—most notably the Ivory Coast—and enforcement of the Alien Compliance Order of 1969. This Ghanaian decree required all foreign workers without valid residence permits to leave the country immediately.[72]

Thirty years after independence, economic development in West Africa still varies from country to country. Although a few countries, such as the Ivory Coast, Nigeria, and Gambia are relatively prosperous, most West African countries remain underindustrialized and poor; indeed some countries have experienced a net deterioration in economic productivity.[73] As a result, unemployment, underemployment, and low income continue to be among the main factors influencing international migration in West Africa.[74] The economic and political instability prevalent throughout the continent have created significant refugee populations that have added to international migration in the region. In 1986, there were six million refugees in Africa, as many as the rest of the world combined.[75]

Contemporary West African migration comes in many forms, including economic refugees who have left their home country because of hunger or poverty, pilgrims and other temporary visitors, public or private sector employees who have been transferred abroad, and students studying abroad. But the largest group of migrants leaves their country of origin due to unemployment, underemployment, and low income.[76] Though labor migrants can be broken down into "two distinct occupational groups,"[77] namely unskilled or semi-skilled labor migrants and highly skilled migrants, most tend to be relatively uneducated, and thus, they occupy low-status jobs in their countries of destination.[78]

The response of West African governments to international migration has varied. Most have not limited migrant entry into their country or restricted migration of their citizens elsewhere. Those governments that have tried either to prohibit or restrict international migration across their borders generally have not been successful for any length of time. As a result, "most migrations continue and the pattern

72. Zachariah & Conde, *supra* note 55, at 38.
73. Kofi K. Apraku, African Emigres in the United States: A Missing Link in Africa's Social and Economic Development at xix (1991).
74. Ricca, *supra* note 52, at 3.
75. Ricca, *supra* note 52, at v.
76. "Migrants in the pre-colonial, colonial and independent periods have sought alternative economic opportunities and have been prepared to migrate from one 'state,' however defined, to another to avail themselves of these." Gould, *supra* note 56, at 33.
77. Ricca, *supra* note 52, at 4.
78. Zachariah & Conde, *supra* note 55, at 49.

of international movement established in the colonial period has been carried into independence."[79]

Statistics indicate that international migration in West Africa is on the rise.[80] In the Ivory Coast and Gabon, foreigners account for twenty-five and twenty percent of the populations respectively.[81] Many migrants choose to remain in their host countries for longer periods of time, sometimes permanently. The implications of widespread and increasing international migration in West Africa for immigration from the region to the United States are potentially limitless. Intra-continental immigration and migration serve the important purpose of stimulating economic development[82] by providing jobs for the many unskilled and semi-skilled labor migrants who do not stand a chance of immigrating to the United States. As intra-continental migration stimulates development of certain West African centers, a more diversified economy may evolve, creating higher paying and more skilled jobs. If so, economic incentives to remain in Africa, as opposed to migration to the United States, may follow.

3. *Europe as an Immigration Option*

After African independence, many West Africans initially looked to Europe as an immigration haven, due partly to the cultural and economic ties that remained between the colonizers and their former subjects.[83] This was especially true for French-speaking Africans.[84]

---

79. Gould, *supra* note 56, at 21.

80. *See also* RICCA, *supra* note 52, at 4 (suggesting that there are six or seven times more international African migrant workers today than in 1958).

81. RICCA, *supra* note 52, at 4.

82. "International migration has been both the cause and effect of economic development in Africa." Gould, *supra* note 56, at 34.

83. "The decolonization of Africa . . . did not end the economic dependence of the emergent nation states on a still largely European-controlled global economic system. Independent Africans are now part of a much expanded and more universal labor migratory system still geared largely to European interests." Skinner, *supra* note 56, at 19.

84. The relative closeness of French colonies to France has much to do with the two different styles of colonization employed by the French and the British—direct versus indirect rule. The French practiced "direct rule" in which they moved to their colonies in significant numbers, established ties with local business, became actively involved in local government, and enforced a closer relationship to their subjects. France still participates in the parliamentary decision-making of its former African colonies and has formal influence over many of its former colonies. In contrast, Britain took a much more "hands off" policy in governing its colonies. Under "indirect rule," Britain maintained a colonial governmental presence in its colonies, but that was the extent of interaction between Britain and its colonial subjects. Skinner, *supra* note 56, at 19-20.

A second factor that played a significant role in the decision to immigrate to Europe, as opposed to America, was the ease with which West Africans were able to travel in and out of Europe. Because West African nations had been former colonies of France and Great Britain, their citizens enjoyed Commonwealth [of Nations] citizenship and did not need a visa to enter France or Great Britain, which were "free port" nations.[85]

As a result of these influences and incentives, many Africans, particularly French-speakers, immigrated to France and Great Britain in the first twenty years after independence. The largest group of unskilled or semi-skilled laborers came from the francophone countries of West and North Africa and immigrated to France and other European countries.[86] However, a parallel movement by immigrants from former British colonies in Africa did not occur to the same extent for two reasons. First, "the jobs taken by African workers in France are occupied in Britain by West Indians, Indians, and Pakistanis."[87] Second, although Ghanaians and other English-speaking Africans enjoyed commonwealth status, Africans perceived the door closing in their faces. As one Ghanian interviewed noted: "[Y]ou could be excluded for arbitrary and capricious reasons. [Prime Minister Margaret] Thatcher made it particularly difficult for commonwealth citizens to enter, especially those who are dark, because she thought they were an economic burden."[88]

However, immigration opportunities in Europe may be diminishing. Great Britain and France have been restricting access to their lands for several years, and neither country is still a "free port" nation. Great Britain began restricting African immigration after the Africans boycotted the Commonwealth Games in 1986 because South Africa had been invited to participate.[89] While the boycott may have precipitated some tightening, Great Britain and France have continued the exclusionist trend on broader grounds, including "right-wing hysteria, high unemployment . . ., the confusion left by the breakup of old empires,"[90] and racism.[91]

85. Interview with Patricia, Ghanaian foreign student, in Oakland, Cal. (Mar. 12, 1992).
86. Gould, *supra* note 56, at 10.
87. Gould, *supra* note 56, at 10.
88. Interview with Charles, lawful permanent resident from Ghana, in San Francisco, Cal. (Oct. 15, 1992).
89. Interview with Patricia, Ghanaian foreign student, in Oakland, Cal. (Mar. 12, 1992); *see also IOC Chief Says Boycotting Countries Should Be Punished*, Reuters North European Service, AM cycle, July 25, 1986 (arguing that boycotting is disruptive to the purpose of such an event).
90. Stan Sloan, *Europe Slams the Door on Immigrants of Color*, Emerge Magazine, Apr. 1992, at 14.

The precise impact of French and British immigration restrictions on the desire of Africans to immigrate to the United States is difficult to measure. Until now, France and Great Britain have been workable alternatives to immigration into the United States for thousands of Africans. Today, France is home to thousands of primarily French-speaking, unskilled African laborers who work as street sweepers, factory workers, and the like.[92] Recently, however, many Ghanaians and Nigerians have begun moving into Saudi Arabia and, to a lesser extent, Libya, due to the economic opportunities there and Europe closing its door on immigrants.[93] However, the United States is indubitably perceived to be a viable alternative.

4. *Cultural, Economic, and Institutional Barriers*

All prospective immigrants are likely to face cultural, economic, and institutional barriers to immigration to the United States. Africans' perception of racism in the United States is likely to be a deterrent as well. Beyond those critical, unmeasured factors, however, there does not seem to be unique intrinsic or indigenous cultural factors among Africans that result in a decreased desire to immigrate to the United States. For instance, although Ghanaians have been referred to as "nonmigratory" in nature,[94] INS statistics show Ghanaian nationals to be the fifth largest African immigrant group in the United States today, outnumbered only by Egyptians, Ethiopians, Nigerians, and South Africans. The following statement is typical of views expressed in interviews with Ghanaian immigrants:

> 91. Now it's hell getting a visa to get into England because they say that most Africans enter England for economic reasons, not education or other legitimate purposes. They use economic reasons to keep the British Isles lily white. It's pitiful when you get to Gatwick and see how they've separated out the Africans and Caribbeans. [But] if you wave an American passport, no questions asked.

Interview with Charles, lawful permanent resident from Ghana, in San Francisco, Cal. (Oct. 15, 1992).

92. Skinner, *supra* note 56, at 36.

> 93. This brain drain is mostly nurses and doctors. Everybody's going because of the money. Most people going are Christian not Muslim, so they are sacrificing their enjoyment for the money. The money is so good that they can solve all their financial problems by working there for a year or two.

Interview with Patricia, Ghanaian foreign student, in Oakland, Cal. (Mar. 12, 1992).

94. "The Ghanaians were, and still are, nonmigratory; very few have migrated to other countries in the region for economic reasons, though some live in neighboring countries for family reasons, such as marriage to a foreign national." ZACHARIAH & CONDE, *supra* note 55, at 6.

To come to America is the ideal of most Ghanaians. Most everybody wants to go to America, but it's really hard to get a visa, so people settle for less. If it was as easy to get a visa, however, people would come much more. They would save up their money like immigrants from other countries, and simply immigrate.[95]

Immigration is also expensive. Relocation expenses faced by all immigrants can be a practical obstacle for Africans. All immigrants, as opposed to refugees, must demonstrate that they have a job offer, independent wealth, or someone in the United States who can support them.[96] But for many Africans, even the most basic part of relocation—travel funds—may be difficult to accumulate. As one Ghanaian noted:

A plane ticket from Ghana to this country costs somewhere around a year's unspent salary. But what are you going to be able to save, when everything you make goes to pay for food and transportation? The ordinary person can't just buy a ticket without the help of a relative who lives outside the country. Most Ghanaians [who want to travel abroad] have relatives living outside the country who send in the hard currency, because the exchange rate is so bad.[97]

Yet the economic rewards of immigration can be profound. A Ghanian who earns a few hundred Cedis (less than $1.00 on the black market) a month working as a civil servant in one of the government ministries can make around $50 to $200 a day driving a cab in the United States. In Washington, D.C., many Africans drive cabs and wash dishes, and are content to be earning a living rather than struggling to survive in their home countries.[98]

As was the case for many Africans during periods of concentrated Asian and Latino immigration, "going to America" does not necessarily mean immigrating permanently. However, "[m]ost people do not want to stay forever. They just want to stay for awhile and get a comfortable life before going back. Why get your education here and rush back [to no job opportunity]?"[99]

African women, however, may face a unique problem. If the applicant lives in a country where women are discouraged from working,

---

95. Interview with Susan, Ghanaian American who has spent a significant amount of time in Ghana, in Oakland, Cal. (Sept. 6, 1992).

96. HING, *supra* note 15, §§ 4.46, 4.47 (discussing the public charge ground of exclusion in the immigration laws).

97. Interview with Patricia, Ghanaian foreign student, in Oakland, Cal. (Mar. 12, 1992).

98. Interview with Charles, lawful permanent resident from Ghana, in San Francisco, Cal. (Oct. 15, 1992).

99. Interview with Patricia, Ghanain foreign student, in Oakland, Cal. (Mar. 12, 1992). A study of 250 "African emigres" found that "nearly all . . . plan to resettle in their home countries sometime in the future." APRAKU, *supra* note 73, at 16.

she may have a difficult time establishing that she will be able to find work. The economic development problem faced by women in the Third World can affect their ability to immigrate—certainly independently—because of the need to establish the desire and ability to work in the United States.

The biggest impediment to African immigration and nonimmigrant visitation, according to many Africans, is at the consular level. A visa is needed to enter the United States, and with few exceptions, nonimmigrant and immigrant visas to the United States are issued by consular officials located at U.S. embassies. Consular officers play a critical role in visa issuance since they make the decisions on eligibility. Eligibility depends upon whether a visa is available, whether the person meets visa requirements, whether the person is excludable, and in some cases, whether a waiver of excludability is merited.[100]

Because consular decisions are not subject to judicial review and many instances of abuse have been documented, the question of nonreviewability of consular decisions has been a common focus of inquiry on the part of many practitioners.[101] An extensive study of consular-level decisions concluded that most consuls did not follow the guidelines for visa adjudication set forth in the State Department's Foreign Affairs Manual. It also found that nearly two-thirds of visa officers had less than two years experience in making visa decisions, and that visa refusals were not documented.[102] Each year, American consuls deny over 550,000 visas.[103]

In the area of nonimmigrant visas, for example, the degree of difficulty in obtaining a visa is related to the perceived level of local eco-

---

100. Edward B. Quist, *Justice for the Alien: The Adequacy of the Consular Visa Issuance System*, 7 OKLA. CITY U. L. REV. 461 (1982).

101. *See, e.g.*, Sam Bernsen, *Consular Absolutism in Visa Cases*, *in* REVIEWABILITY OF CONSULAR DECISIONS (American Immigration Lawyers Association Conference, Jan. 19, 1987) (on file in the Stanford Law School Library, 26 International Relations Pamphlets). *See also* Ulrich v. Kellogg, 30 F.2d 984 (D.C. Cir. 1929), *cert. denied*, 279 U.S. 868 (1929) (establishing the tradition against judicial review of consular visa decisions); Ventura-Escamilla v. INS, 647 F.2d 28 (9th Cir. 1981) (upholding the *Ulrich* tradition); HING, *supra* note 16, § 10.10 (analyzing the role of consular officials in visa adjudication and arguing against over-specialization in the field of immigration).

102. Kim R. Anderson & David A. Gifford, *Consular Discretion in the Immigrant Visa-Issuing Process*, 16 SAN DIEGO L. REV. 87 (1978) (summarizing 360 questionnaires mailed to consular officials world-wide, intensive personal interviews with visa-issuing officers, and observations of visa interviews).

103. Bernsen, *supra* note 101, at 3.

nomic pressure to immigrate to the United States. If such pressures are perceived as negligible, as in much of Europe and Japan, the process of applying for a visa is likely to be a pro forma matter handled by mail or through travel agencies. Conversely, if the demand for immigrant visas is high, most nonimmigrant visa applicants must personally visit the consulate to establish their qualifications for a nonimmigrant visa and lack of intention to immigrate.[104]

Consular-level policies, attitudes, and actions, protected by the shield of nonreviewability, may be the greatest obstacle facing potential African immigrants and nonimmigrant visitors. Although the INS claims that "tourists . . . are encouraged to visit as a boon to the U.S. economy,"[105] Africans do not receive such hospitality from consular officials. Instead, they experience inconsistent and nebulous standards.[106]

5. *Limitations of the Current Immigration Laws*

In theory, Africans who are numerically exempt and qualify as immediate relative immigrants are eligible to enter this country. However, the underrepresentation of immediate relative African immigrants suggests that either there simply are not many eligible Africans (i.e., the pool of Africans already in the United States who can petition for relatives in Africa is small), or that intrinsic or extrinsic considerations (some already mentioned) are at play.

The exact number of immediate African relatives of U.S. citizens is impossible to calculate, and annual statistics on the number of Africans who unsuccessfully petition for immigrant status but are rejected are unavailable. But the low number of Africans entering in the immediate relative category annually obviously is related to a small pool of U.S. citizens who have spouses, children, or parents in Africa who are not U.S. citizens. Two unique, reinforcing historical barriers to African immigration—colonization and slavery—are particularly relevant here.

European colonization of the African continent, which started in the late eighteenth century and continued until the recent past or present (depending on one's perspective),[107] severely restricted the

---

104. Hing, *supra* note 16, § 3.13.

105. 1990 INS Stat. Y.B., *supra* note 8, at xxxii.

106. "They'll look at your hair, and if it's too kinky, then they won't like you . . . . It depends on which side of the bed [the official] got up on that particular day." Interview with Charles, lawful permanent resident from Ghana, in San Francisco, Cal. (Oct. 15, 1992).

107. Colonization is widely noted as having ended in the 1970s, with Angola gaining independence in 1974. Arguably, colonization is still alive and well today in southern Africa—Namibia only recently gained independence on March 21, 1990, and although some might distin-

movement of most Africans outside of the continent until the last thirty years. Under the European colonial regime, African nations were not independent and Africans were not given European citizenship. From 1920 to 1965, the national origins quota system governed, which effectively barred African immigration.[108] And even after 1965, colonies were limited to annual quotas of two hundred.[109]

The impact of slavery was manifold. Family ties were brutally severed and generally lost. Indigenous African economic and social life was also destroyed. In addition, slavery clearly created a serious disincentive for voluntary African immigration during the nineteenth century, and perhaps the early twentieth century. Because the system is primarily family based, and African countries have historically sent only a few nationals to the United States as either immigrants or nonimmigrants, only a small number of U.S. citizens or lawful permanent residents have spouses, children, parents, or siblings in Africa.

The legal barriers for nonimmigrant visas are also problematic. Student F-1 visas are a prime example of the way in which nonimmigrant visa requirements can be barriers to obtaining a visa. Most foreign students enter the United States with F-1 visas. In order to obtain an F-1 visa, the student must prove that she (a) will be engaged in fulltime study; (b) speaks English proficiently or plans to learn English upon her arrival to this country; (c) intends to leave the United States after completing her studies; and (d) has sufficient funds for her studies.[110] Specifically, the student must provide the relevant consulate office with documentation that she has funds at her disposal to cover the expense of her first year of study, and that barring an unforeseen change in circumstances, funds will be available for each school year in the same amount and from either the same source or other equally reliable sources.[111]

The financial requirement proves the most troubling for many African students. Most students are not in the position to personally fund a U.S. college education due to unfavorable currency exchange

guish South Africa from other colonial situations because whites actually settled there in large numbers, this distinction is of dubious merit. Paul Taylor, *No News is Bad News in Namibia*, WASH. POST, Aug. 22, 1993, at A23.

108. HUTCHINSON, *supra* note 40, at 485. *See supra* notes 37-50 and accompanying text.

109. HING, *supra* note 37, at 41.

110. HING, *supra* note 16, §§ 3.2-3.3.

111. Hing, *supra* note 16, §§ 3.2-3.3.

rates or because of lack of funds. In addition, African students are ineligible to receive U.S. federal student loans because they are only available to U.S. citizens. Even those students who do have the funding often encounter daunting or insurmountable opposition from immigration officers at the consular or post-entry level who are predisposed to disbelieve their educational intentions.[112]

Furthermore, in some cases, higher education in the United States is more cost prohibitive and less desirable than European alternatives. Indeed, some European countries offer either free education or substantial scholarships to "Third World" students,[113] either as payback for having colonized a particular African country or for humanitarian reasons. For French-speaking Africans, France is often a more attractive educational alternative to the United States for obvious cultural and language reasons.

The Immigration Reform and Control Act of 1986 created the Visa Waiver Pilot Project (hereinafter "VWPP") designed to increase business and tourism investment in the United States. VWPP allows certain nonimmigrants, more specifically visitors for pleasure and business representatives from certain countries, to enter temporarily without holding a visa.[114] The United Kingdom and Japan were the first two countries to participate in the program. The program has since been expanded to include France, Germany, Switzerland, Italy, Sweden, and the Netherlands.[115] In 1989, 2,348,959 tourists and 99,665 business visitors took advantage of the VWPP.[116]

In 1992, the INS implemented a new program entitled "INSPASS" to expedite the entry of frequent business travelers.[117] The program is being initially tested at Newark International Airport and New York's JFK International Airport. Participation in the INSPASS program is voluntary and open to aliens who travel to the United States on business three or more times a year. They can apply for an INSPASS document that will allow them to proceed quickly through

112. Interview with Charles, lawful permanent resident from Ghana, in San Francisco, Cal. (Oct. 15, 1992).

113. One Ghanaian sought his post-graduate education in Norway, despite the fact that he had never visited that country, did not speak the language (and subsequently had to spend his entire first year in Norway taking intensive Norwegian classes), and positively hated cold weather. After years of trying unsuccessfully to raise enough money to pay for a U.S. education and obtain a student visa to study here, he gave up and pursued a free education in Norway instead. Interview with Edward, lawful permanent resident from Ghana, in Oakland, Cal. (Mar. 12, 1992).

114. 1990 INS STAT. Y.B., supra note 7, at xxxii.

115. 1990 INS STAT. Y.B., *supra* note 8, at xxxii.

116. 1990 INS STAT. Y.B., *supra* note 8, at xxxii.

117. *See INS Announces New Expedited Inspections for Frequent Business Travelers*, 69 INTERPRETER RELEASES 1040 (1992).

INS inspections by means of biometric identification technology (handprint). Presently, however, the program is limited to citizens of Andorra, Austria, Belgium, Bermuda, Canada, Denmark, Finland, France, Germany, Iceland, Italy, Japan, Liechtenstein, Luxembourg, Monaco, the Netherlands, New Zealand, Norway, San Marino, Spain, Sweden, Switzerland, and the United Kingdom.[118]

Apparently, any country that qualifies can be accepted into the program.[119] However, the economic crisis that permeates most of the African continent makes it unlikely that any African country will qualify for this program in the near future. As long as Africa remains one of the poorest world regions, blanket nonimmigrant admissions programs like the VWPP will be unavailable, and African visitors will continue to be subject to heightened scrutiny in the issuance of nonimmigrant visas.

Legislation in 1990 does offer the potential for Africans to enter and establish the seeds or a pipeline to take advantage of the family categories. Beginning October 1, 1994, 55,000 "diversity" visas will be made available annually in a lottery-type program to nationals of countries from which immigration was lower than 50,000 over the preceding five years.[120] Under a complicated formula that weights countries and regions of the world and uses relative populations as well, the State Department will determine the distribution of visas. A single state can never get more than seven percent (3,850) of the total diversity visas available each year. In one projection of the formula, every African nation will qualify, and the first year's regional distribution would look like this: Africa 20,900, Asia 6,380, North America 0, Europe 24,310, Latin America 2,530, and Oceania 880.[121]

On its face, the new diversity program represents a step in the right direction as far as visa availability for Africans. The justification for special treatment for Irish nationals who have wanted to immi-

118. *Id.*

119. For example, according to INS, "qualified" means "approved countries which offer a reciprocal waiver of visas to U.S. citizens." 1990 INS STAT. Y.B., *supra* note 7, at xxxii. *See also* Christine Davidson, *A Comparison of Alien Admissions Before and After IRCA*, INS IMMIGR. ISSUES 6 (Jan. 1991) (arguing that although IRCA altered migration in the United States, it was not responsible for all of the changes that occurred between 1984-1989. "The long term effects of IRCA will not be known until well into the future"). *Id.*

120. Immigration and Nationality Act, 8 U.S.C. § 1153(c) (1990).

121. *The Immigration Act of 1990 Analyzed: Part 8—Diversity Immigrants*, 68 INTERPRETER RELEASES 137, 139 (1991).

grate the past few years was the recognition that under the primarily family-based immigration system, a special seed or pipeline category needed to be established in order to get a significant number of people in who could then take advantage of the family categories. This had not occurred in the case of the Irish nationals. Although Asians were not given that kind of advantage and had to slowly build up their pool beginning after the 1965 amendments,[122] Africans certainly deserve the extra push just as much as the Irish.

The notion that the 1965 amendments were not intended to go out of the way to help Asian immigration is supported by the fact that those who would benefit most from a family-based system would be those who were here in greater numbers in 1965, i.e., those of European ancestry.[123] However, European immigration actually declined. This should have been predicted because family reunification for those nationalities had been completed.

Instead, the number of Asian immigrants began to increase in the early 1970s, then exploded in the late 1970s and 1980s, when Asian and Latin immigrants represented over eighty percent of the total.[124] Therefore, although the country's population was still overwhelmingly white, Congress, in 1986, added a provision to the Immigration Reform and Control Act to help thirty-six countries that had been "adversely affected" by the 1965 changes. This was the beginning of the special push categories. However, these categories did not benefit African immigrants.

To be considered "adversely affected," a country must have been issued fewer visas after 1965 than before. Thus, the list did not include any African countries.[125] The 1986 law provided an extra 5000 such visas a year for 1987 and 1988, but the number increased to 15,000 per year for 1989 and 1990 through additional legislation.[126]

In 1988, Congress set aside an extra 20,000 visas to increase immigration diversity over a period of two years. This time, the lottery for the visas was available to nationals of countries that were "under-represented," namely a foreign state that used less than twenty-five percent of its 20,000 preference visas in 1988.[127] As a result, all but thirteen countries in the world were eligible, including all of Africa. Over 3.2 million applications were received for the 20,000 visas—2121

122. HING, *supra* note 37, at 39-40.
123. HING, *supra* note 37, at 117-18.
124. HING, *supra* note 37, at 1.
125. BILL ONG HING, HANDLING IMMIGRATION CASES § 4.22B (Supp. 1994).
126. Immigration Amendments of 1988, Pub. L. No. 100-658, § 2, 102 Stat. 3908 (1988).
127. *Id.* § 3.

were granted to Africans. Some individual figures were: Algeria 11, Angola 1, Cameron 21, Ghana 155, Ivory Coast 15, South Africa 140, Uganda 48, Nigeria 254, Liberia 20, Gambia 15, Chad 1, Congo 1, Tanzania 62, Egypt 790, Libya 3, Malawi 4, Guinea 6, Togo 1, Mali 10, Niger 1, Somalia 82, Sudan 21, Cape Verde 32, Tunisia 17, Mozambique 1, Senegal 96, Sierra Leone 11, Zaire 5, Zambia 35, Zimbabwe 29, Morocco 580, and Kenya 138.[128] These small but diverse numbers are further evidence of the level of desire on the part of Africans to immigrate to the United States.

While Africans may be able to benefit under the permanent diversity provisions of the 1990 law that will go into effect in October 1994, Africans are excluded from the transitional diversity program currently in effect. During the transitional period, 40,000 visas are available each year, but again, only to those "adversely affected" by the 1965 amendments again, and forty percent are reserved for Irish nationals.[129]

While some Africans were able to benefit from the 1988 diversity program, we cannot assume that they will benefit in the same manner under the program effective October 1994. In order to take part in the lottery program in 1988, an applicant needed only to submit a letter with family and other biographical data.[130] Under the 1994 diversity program, however, the applicant must have a high school education (or its equivalent), or within five years of application, the applicant must have at least two years of work experience in an occupation that requires at least two years of training or experience.[131] Thus, the amount of assistance the program actually provides to African immigrants ultimately will depend upon the socioeconomic characteristics of various African nations.

C. *Recommendations*

If we are indeed serious about facilitating the immigration of Africans to the United States, the program that is about to go into effect in October 1994 should be modified. The history of Asian im-

128. *State Department Reports Country Figures for OP-1 Program*, 66 Interpreter Releases 659, 669-70 (1989).
129. Hing, *supra* note 125, § 4.22C.
130. Hing, *supra* note 125, § 4.22B.
131. Hing, *supra* note 125, § 4.22C.

migration to the United States after the 1965 amendments—particularly from Korea and India—indicates that a small critical mass needed to be developed in the United States before sufficient numbers of Asians could benefit from the family immigration categories.[132] Policymakers have recognized this need in their attempts to establish a pool of Irish nationals who could take advantage of the family categories. The diversity lotteries in 1986 and 1988, and particularly the current transition program for countries "adversely affected" by the 1965 amendments were especially intended to help Irish nationals. Forty percent of the transition visas are set aside for Irish nationals. That type of set-aside is necessary for African immigrants if we are to earnestly increase their numbers.

The requirement of a high school diploma or work experience in a non-entry-level job is also troubling when compared to the absence of similar requirements for Irish nationals under current diversity programs. Since policymakers have determined that the imposition of standard grounds of exclusion (e.g., public charge and criminal exclusions) were sufficient for earlier diversity programs, are they not adequate for the October 1994 program under which Africans stand to benefit? Policymakers ought to reconsider these requirements; otherwise, they risk continuing a message of exclusion to Africans and African Americans.

II. Immigration Categories for Skilled Workers and Their Messages to African Americans

A. *Background*

It is difficult to respond to African American concerns that immigrants take away jobs or lower their wages. This article does not intend to comprehensively address these issues. However, generally speaking, while most economic studies show that the presence of large numbers of immigrants in certain geographic areas keeps wages low in certain service jobs, the presence of immigrants actually stimulates the economy and creates new jobs.[133] Based on these studies, many commentators have concluded that immigrants do not hurt the U.S. econ-

132. Hing, *supra* note 37, at 117-18.

133. *See generally* Julian L. Simon, The Economic Consequences of Immigration (1989) (examining the economic impact immigration has on Americans and concluding that in most ways, immigration is economically beneficial to United States citizens); George J. Borjas, Friends or Strangers: The Impact of Immigrants on the U.S. Economy (1990) (weighing the economic benefits and detriments immigration has on American workers and the economy).

omy.[134] The problem is, these types of studies do little to change the perceptions of African American and other communities with large segments that are poor and unemployed. The results of these studies do little to change the perceptions of those who see Latino or Asian immigrants working in their neighborhoods in jobs where African Americans worked before, and those who believe that immigrant workers are favored.[135] Such studies simply do not answer the concerns that are specific to African American communities.

Among other things, future studies should focus on whether the impact of immigrant workers varies between industries, and how wages are set in each industry; whether immigrants are taking jobs that native workers, including African Americans, are no longer interested in; and if immigrants help the economy, whether they have helped the African American community, and if so, how.

While most of these questions remain for separate consideration, this article does consider one aspect of the United States immigration selection system—occupational visas. This particular category of visas sends an economic message to African Americans that is unnecessary and must be reconsidered.

As surprising as it may be for the unemployed, according to some, the United States has a massive problem on its hands—an ever-shrinking labor force. The problem is not necessarily a physical shrinkage in size even though empirically, the labor force will be numerically smaller in the coming years;[136] rather, the problem is a "capacity" shrinkage. The future labor force is in danger of being too poorly equipped with the skills necessary to handle the technological changes that are fast transforming the working world.

---

134. *See* López, *supra* note 35; Peter H. Schuck, *The Emerging Political Consensus on Immigration Law*, 5 GEO. IMMIG. L.J. 1 (1990); Frank Sharry, *Why Immigrants Are Good for America*, ORLANDO SENTINEL, Sept. 22, 1991, at G1.

135. *See, e.g.*, Jack Miles, *Blacks vs. Browns*, ATLANTIC MONTHLY, Oct. 1992, 41 (citing the replacement of African Americans in janitorial jobs by Latinos); Stephanie Chavez, *Racial Tensions over South L.A. Jobs Grow*, L.A. TIMES, July 22, 1992, at B1, B3 (describing African-Americans' threat to shut down construction sites where Latinos were hired but African Americans were not).

136. Presently, the "Baby-boomers" are aging, and they have not reproduced with the same abundance as their parents. This has led to a decrease in the numerical size of the future labor force. Howard Fullerton, Jr., *Labor Force Projections: The Baby Boom Moves On*, 114 MONTHLY LAB. REV. 31 (1991). Additionally, the labor force, which rose by 2.9% per year during the 1970s, will only rise by 1% per year during the 1990s. Jack Whelan, *Training for the Millennium: The Role of American Business in Education*, 6 PAC. N.W. EXECUTIVE 21 (1990).

Recent studies have assessed the composition of the labor market in the upcoming millennium.[137] These studies reveal that not only will the appearance of the labor force change, but also that if the quality of education and training of the U.S. population continues at the current pace, the amount and quality of skills now possessed by the labor force will have degenerated by the next century.[138] This failure to instruct our next generation of workers is especially apparent among people of color, particularly African Americans, who historically have been undereducated by the system of education in the United States and are therefore ill-prepared to handle structural changes in the economy.

The various pronouncements of the projected decline in the capacity of U.S. workers to perform future jobs adequately has created a profound sense of impending doom among government leaders and captains of industry alike.[139] Thus, the question becomes, how are we handling this problem? There are two major solutions that have been proposed and implemented.

The first, and most obvious solution to some, is to finally begin paying some attention to our long-neglected educational system. The current attention being paid to education is too little. It may soon be too late if more is not done. The Clinton administration is attempting to squeeze extra dollars from our financially strapped federal government to upgrade the nation's schools,[140] but state systems are less inclined to "pour" more money into education than the federal government.[141] In addition, although some American businesses have

---

137. *See, e.g.*, WILLIAM B. JOHNSTON, WORKFORCE 2000: WORK AND WORKERS FOR THE TWENTY-FIRST CENTURY (1987) [hereinafter WORKFORCE 2000]. For a discussion of Workforce 2000, *see infra* notes 164-171 and accompanying text.

138. *See, e.g.*, WORKFORCE 2000 *supra* note 138. *See also* William Miller, *Firms Lag Behind in Coping with Massive Workforce Changes, Charges Workforce 2000 Update*, INDUS. WK., Aug. 20, 1990, at 34 (noting that many businesses are beginning to feel the effects of a deteriorating workforce).

139. *See infra* text accompanying notes 164-72.

140. During the Bush administration, the only response to the increasingly poor educational system was an escalation in the privatization of education, accompanied by an increase in the number of standardized tests given to students. Money was continuously said to "not [be] the answer." REUTERS, Jan. 28, 1992, AM cycle (President Bush's "choice" plan would give parents the choice to send their children to public or private schools and would allocate a greater amount of federal monies to private schools); Dele Olojede, *A New Agenda for Black Caucus*, NEWSDAY, Sept. 15, 1991, at 17 (Bush's education agenda, America 2000, emphasizes expanded standardized testing). Although President Clinton's proposed budget eliminated thirty-three education programs ranging from reading instruction to bilingual vocational courses, the money saved will be used to expand aid to poorer school districts. An estimated $600 million to $700 million in new funds will be provided to create new national academic and skill standards for elementary and secondary schools. James Risen, *'95 Budget Pares Funds Requested by Pentagon*, L.A. TIMES, Dec. 23, 1993, at A1.

141. *See generally* JONATHAN KOZOL, SAVAGE INEQUALITIES (1991) (discussing the abysmal state of education in many of America's public schools).

taken it upon themselves to reach out to communities and provide the additional resources needed to educate and train young people, these businesses are still rare enough to be legitimately termed "enlightened." U.S. corporations devote only $30 billion annually to education and on-the-job training programs, just 1.4 percent of the nation's payroll. This contrasts sharply with the approximately $300 billion per year businesses spend on plants and equipment.[142] Meanwhile, the education of the poor, especially poor people of color, is frighteningly abysmal. Unless our priorities change immediately, this situation will not reverse.[143]

A second solution to this prospect of a future work force lacking essential skills—one increasingly being looked to with enthusiasm by big business interests and government powerbrokers alike—is immigration. That is, the "right" kind of immigration. These leaders see the immigration of ever-larger numbers of highly skilled immigrants as an immediate quick-fix to a thorny situation.[144]

The Immigration Act of 1990[145] was a step toward this solution—one that had long been sought after and lobbied for by American businesses. The Act was a significant, some say revolutionary, revision of the focus of U.S. immigration law.[146] After passage of the Act,

142. John Katzenberger et al., *Illiteracy Could Cost Hoosier Businesses as Much as $1 Billion*, INDIANAPOLIS BUS. J., Sept. 3, 1990, at A1. Workforce 2000 discovered that only five percent of companies spend more than $5,000 per year in classroom and on-the-job training. Miller, *supra* note 138, at 34.

143. According to Arnold Packer, co-director of the Workforce 2000 study, "If we don't take an across-the-board corporate interest in the quality of our work force, then . . . when the "[baby] boomers" retire . . . [we will be] turning over the reins to an untrained, unskilled and undermotivated work force." When asked if such a rein-turning would definitely take place, he replied, "[I]t's a 50-50 proposition." *Tom Brown, Workforce 2000: The Question Is: Will It Work?*, INDUS. WK., Aug. 6, 1990, at 19.

144. *See* Lisa Keefe, *Chicago's New Immigrants*, CRAIN'S CHI. BUS., Feb. 24, 1992, at 1 (asserting that the increasingly poor level of skills possessed by job applicants "gives businesses two choices: improve the public schools, or import talent from abroad. Last year, big business made it clear that it will pursue the second course, perhaps not as the only solution, but certainly as a leading one."). Barry Chiswick, Dean of Economics and Business Administration at the University of Illinois, makes a consistent point, arguing that the entire immigration selection system ought to be shifted from a family-based to a skills-based immigration policy. *See Hearings Before the Subcommittee on Economic Resources, Competitiveness, and Security Economics of the Joint Economic Committee of Congress*, May 29, 1986, at 235 (testimony of Barry Chiswick); Barry Chiswick, *The Economic Progress of Immigrants: Some Apparently Universal Patterns*, *in* THE GATEWAY: U.S. IMMIGRATION ISSUES AND POLICIES 119, 123-30; 141-45 (1982).

145. Immigration Act of 1990, Pub. L. No. 101-649, 1043 Stat. 4978 (1990).

146. *See* Alice E. M. Aragones, Note, *The Immigration Act of 1990: Changes in Employment-Based Immigration*, 5 GEO. IMMIGR. L.J. 109, 109 (1991) and Eustace T. Francis, *Taking Care of*

although the main thrust of immigration law continues to be family reunification,[147] highly-skilled immigrants will be deliberately encouraged to resettle in this country more than ever before.

Under the new law, occupational visas are divided into five categories:[148] (1) 40,000 visas for priority workers who possess extraordinary ability in the arts, sciences, education, business, or athletics,[149] or outstanding professors and researchers,[150] and certain multinational executives;[151] (2) 40,000 visas for professionals holding advanced degrees or aliens of exceptional ability;[152] (3) 40,000 visas for skilled workers, professionals with baccalaureate degrees, and unskilled workers for jobs for which qualified American workers are not available (only 10,000 visas can be issued for unskilled workers);[153] (4) 10,000 visas for special immigrants,[154] and (5) 10,000 visas for employment creation immigrants or investors whose investments create at least ten new jobs.[155]

This emerging policy favoring the "importation" of highly-skilled immigrants to satisfy a growing vacuum in the labor pool creates the danger of providing a way for our nation's leaders to escape the dilemma caused by the undereducation of Americans. Furthermore, it sends a message to African Americans that the United States government would rather bring in immigrants to fill skilled jobs than to train available workers to fill them, and avoid a solution that would have helped African American economic progress.

Business: The Potential Impact of Immigration Reform on Corporate Strategic Planning, 5 GEO. IMMIGR. L.J. 79 n.2 (1991).

147. *See supra* text accompanying notes 15-18.

148. HING, *supra* note 125, §§ 4.34, 5.1, 5.1A, 5.1B.

149. Immigration and Nationality Act, 8 U.S.C. § 1153(b)(1)(A) (1990). To show extraordinary ability an alien must demonstrate: (1) sustained national or international acclaim; (2) achievements in the field recognized through extensive documentation; (3) intent to work in that particular area of extraordinary ability; and (4) that the alien should "substantially benefit prospectively" the United States. Immigration and Nationality Act, 8 U.S.C. § 1153(b)(1)(A)(i)-(iii) (1990).

150. Immigration Act of 1990, 8 U.S.C. § 1101 (1990) (creating new subsection of the Immigration and Nationality Act, 8 U.S.C. § 1153(b)(1)(B) (1990)). Aliens under this provision must demonstrate (1) international recognition in a specific area; (2) at least three years of experience in teaching or research in that field; and (3) seek to enter the United States for a tenured position or tenure-track position, or comparable research position. *Id.*

151. Multinational executives must have been employed by a sponsor firm or corporation for at least one year in the three years preceding the time of the alien's application for admission into the U.S. and must have the intent to continue to render services to that employer, or its subsidiary or affiliate in the United States. *Id.*

152. Immigration Act of 1990, 8 U.S.C. § 1101 (1990) (creating new subsection of the Immigration and Nationality Act, 8 U.S.C. § 1153(b)(2) (1990)).

153. Immigration and Nationality Act, 8 U.S.C. § 1153(b)(3) (1990).

154. Immigration and Nationality Act, 8 U.S.C. § 1153(b)(4) (1990). This category includes religious workers and ministers among others.

155. Immigration and Nationality Act, 8 U.S.C. § 1153(b)(5) (1990).

Most of the available immigrant visas are still reserved for relatives of U.S. citizens and lawful permanent residents. However, the near tripling of occupational visas from 54,000 to 140,000 in the 1990 legislation signals the beginning of a shift in the focus of U.S. immigration law from concern with family reunification toward a policy of importing skilled workers.[156] In fact, before 1990, up to one-half of the 54,000 occupational visas could be used for unskilled workers who were not displacing American workers; but the figure for unskilled workers now has been reduced to 10,000.[157]

An additional note on the investors or employment-creation visa is in order. This new category, not to be confused with the nonimmigrant treaty trader,[158] is designed for immigrants seeking to engage in a new commercial enterprise. The law requires that the investment be at least $1 million, but the amount is reduced to $500,000 if the enterprise is located in a targeted employment area, such as a rural area or any area that has experienced high unemployment of at least 150 percent of the national average rate.[159] No fewer than 3,000 of the 10,000 visas are reserved for applicants whose enterprises will be located in targeted employment areas.[160]

In the first year of availability, fewer than two hundred applications were made for the 10,000 available investors visas. Immigration practitioners who worked with prospective investors offer several explanations. First, under U.S. tax laws, when a person becomes a lawful permanent resident of the United States, the person is taxed on her worldwide income. Most prospective investors have substantial income abroad from countries with very favorable tax laws, and they are reluctant to give that up for permanent resident status in the United States. Second, many potential investors balk at a regulatory requirement that their sources of income for the past ten years be disclosed. While this requirement was imposed to ferret out investors whose source of income was illegal drug trafficking, many legitimate inves-

156. Senator Alan Simpson (R-Wyo.) referred to the Act as the final result of a decade-long effort to "close the back door" [of undocumented immigration] while we open the front door wider to skilled immigrants of a more diverse range of nationalities." 136 CONG. REC. S17,109 (daily ed. Oct. 26, 1990) (statement of Sen. Simpson).

157. HING, *supra* note 16, §§ 4.12, 4.34.

158. *See supra* text accompanying notes 32-34.

159. HING, *supra* note 125, § 5.1(B).

160. HING, *supra* note 125, § 5.1(B).

tors prefer that the sources of their wealth remain private.[161] Finally, many who would qualify for the investor category also qualify for less onerous categories such as an intra-company transferee working for a multinational corporation.[162] As a result, few takers for the new job-creation category have stepped forward, and areas of high unemployment have not been helped. An opportunity to send a positive message to the African American community has thus been thwarted.

B. *Purpose of 1990 Revisions*

In the latter half of the 1980s, business leaders became concerned with reports that the native work force would be shrinking, both in size and in the capacity to meet the rapidly changing demands of our technological era. The reports suggested that these trends would inevitably result in an even greater decline in America's economic competitiveness. One such report that not only revealed much of the Congressional debate on immigration policy, but also transformed the popular discussion on the changing American labor market was Workforce 2000.[163]

Workforce 2000 concluded that the emerging labor market in America contained a decreasing number of white males and an increasing number of females, both white and of color, and males of color.[164] The report also concluded that the future labor force would not only be comprised of a very different racial mix, but also would be comprised of a larger number of low-skilled or unskilled persons.[165] This is dire news at a time when higher-skilled jobs have been, and will continue to be, the fastest-growing category.[166]

---

161. Interview with Angeli Cheng, attorney with the law firm of Fallon, Bixby & Cheng in San Francisco, Cal. (June 2, 1992); interview with Lance Nagel, attorney with the law firm of Graham & James in San Francisco, Cal. (Sept. 13, 1992).

162. HING, *supra* note 16, § 5.20.

163. WORKFORCE 2000, *supra* note 137.

164. White males will become a decreasing factor in labor force participation according to the report, adding only a net increase of fifteen percent to the work force; the fastest growing labor force group will be black females. This represents, in demographic terms, almost a 180-degree change in the racial composition of the work force. Nicholas Hadgis, *Moving Beyond Affirmative Action into Equal Opportunity: Restaurant Industry*, 24 NATION'S RESTAURANT NEWS NEWSPAPER 44 (1990).

165. John Iwaniec, *Immigration Law Impacts Housing: Immigration Act of 1990*, 33 NAT'L REAL EST. INVESTOR 30 (1991) (stating this conclusion of Workforce 2000 and adding, "The report provides evidence to support the case that the labor shortages which threaten the 1990s can be relieved by immigration.").

166. Since 1983, eighty percent of new jobs have been in higher-skilled job categories, leading to a "growing skills gap in an economy creating more jobs than our education system can fill." Warren Brookes, *New Studies Show Why It Pays to Be Well-Educated*, S.F. CHRON., July 3, 1990, at C3.

Business leaders took Workforce 2000 and similar reports to heart and began to lobby Congress for an immigration-based solution.[167] The legislative history of the Immigation Act of 1990 reveals that Congress appeared to acquiesce with the general perception that the lower level of skills among workers, rather than any actual decrease in the size of the work force, presented the greatest threat to the economy.[168] In addition, both Congress and business leaders had cause to be alarmed over the current state of the nation's work force, let alone its future state. One study of the quality of America's domestic work force states that four million adults in the United States are unable to recognize letters and numbers, and twenty-seven million have basic skills below the fifth-grade level.[169]

Furthermore, Congress must have believed that, although the number of workers in this country may be adequate to fulfill employers' needs, their abilities fall below par. It is hard to imagine any other reason why Congress would have so readily acceded to employers' claims of increasing difficulty in locating qualified workers at a time of such high unemployment.

---

167. In fact, business interests underwent a "10-year lobbying blitz" to achieve passage of the Act. Keefe, *supra* note 145. Much of the reporting surrounding passage of the Act highlighted the hardship undergone by corporations and other business associations in attempting to find qualified U.S. workers to fill job vacancies.

A sample of the perceptions held by employers about the quality of U.S. workers includes the following statement made by the corporate counsel to Amoco in Chicago: "If you want to compete and hire good employees with good backgrounds, you have to hire foreign nationals. The labor pool for civil and geotechnical engineering . . . [is comprised] predominantly [of] foreign-born people . . . ." Todd White, *Employers to Be Affected by New Immigration Act*, 12 L.A. Bus. J. 30 (1990).

Of the respondents surveyed by "Workforce 2000," seventy percent reported difficulty in recruiting technical personnel, sixty-three percent reported difficulty in recruiting professionals, and fifty-five percent reported a problem hiring secretarial and clerical workers. *Much of Workforce 2000 Is Here, Witnesses Tell Senate Task Force*, Bureau of National Affairs, Inc. DLR No. 140, A12 (July 20, 1990).

In fact, employer associations perceived their need for foreign workers to be so severe that a possible penalty of a one-year bar from applying for immigrant visas for foreign workers that would be issued against employers for underpaying foreign workers was termed the "death penalty." *President Signs Immigration Bill Easing Temporary Work Visa Rules*, BNA Wash. Insider, Dec. 16, 1991.

168. H.R. Rep. No. 101-723(I), 101st Cong., 2d Sess. (1991), *reprinted in* 1990 U.S.C.C.A.N. 6710, 6726 (citing the Workforce 2000 report and stating that "the so-called 'de-skilling of America' is considered by many to be the biggest single problem facing the U.S. workplace in the near and distant future.").

169. Katzenberger et al., *supra* note 142, at A1.

Eventually, the Immigration Act of 1990 was the answer given to those business interests.[170] Congress made it clear that the Act was the beginning of a much more liberal approach to the immigration of skilled foreign workers. The inclusion of a three-year review period in the legislation provides further evidence that the Act is only the beginning of a new focus on importing skilled workers. Congress has the option, at the end of this three-year period, to increase the number of visas allocated to the occupational preference categories if it decides that the need for qualified workers continues.[171]

Predictably, the bill met with a great deal of resistance from labor organizations. They were mollified, however, by the fact that the labor certification requirements remained essentially the same as they were prior to passage of the Act,[172] and were not streamlined as in the House version of the bill.[173]

Labor groups' attempts to prevent any perceived displacement of U.S. workers by the new influx of skilled foreign workers may or may not work.[174] In any event, any protections that are won will only be of use to those American workers who possess the qualifications required by a particular employer. These organizations provide no protection for domestic workers whose abilities to acquire necessary qualifications are severely limited or non-existent. For them, a measure that only provides protection for domestic workers who are equally as "qualified" or "skilled" as alien workers is an empty gesture.

African Americans comprise one sector of society whose access to the institutions that bestow the skills necessary to compete in to-

---

170. Keefe notes that the Act is the first piece of immigration legislation to reflect the interests of business lobbies. Keefe, *supra* note 145, at 1.

171. When asked if the U.S. should open its doors to even more employment-based immigration, Gene McNary, former Commissioner of the Immigration and Naturalization Service, replied that if the new immigration law proves successful and increased numbers would be beneficial, then the numbers would be increased in three to four years. Jefferson Morley, *America and the New Immigrant Experience; Los Angles Times Interview of Gene McNary*, L.A. Times, May 5, 1991, at M1.

172. Under current labor certification procedures, an employer often hires the foreign worker first, then establishes through advertising and recruitment efforts that no native worker is available at the prevailing wage. State employment offices, as well as the U.S. Department of Labor, have to be satisfied. *See* Hing, *supra* note 37, at §§ 5.2 - 5.10.

173. *See* Aragones, *supra* note 147, at 120. The House bill, H.R. 4300, suggested a labor "attestation" instead of a labor "certification" requirement that would have made employers prove that they had attempted to locate qualified American workers before they turned to the foreign work force. H.R. Rep. No. 101-723(I), 101st Cong., 2d Sess. (1991) *reprinted in* 1990 U.S.C.C.A.N. 6710, 6741-46.

174. One of the other stated purposes for streamlining the labor certification requirements of immigration law was to provide protection to the domestic labor force by modifying the existing requirements which were considered too unweildy and too difficult to enforce. H.R. Rep. No. 101-723(I), 101st Cong., 2d Sess. (1991) *reprinted in* 1990 U.S.C.C.A.N. 6710, 6741-46.

day's job market is becoming increasingly restricted. As a consequence they are found in disproportionate numbers among the "unskilled" or "underskilled." Therefore, from their vantage point, any pronouncements that employers are having an ever more difficult time locating qualified workers are viewed suspiciously.

This is not to say that the employers' claims of hardship are not accurate, at least to some extent. It is also true, however, that this nation, through its sometimes benign but often deliberate neglect, has undereducated and undertrained the poor in general, and African Americans in particular, thus creating and perpetuating the current situation. This underutilization of already existing domestic potential must be addressed before even larger numbers of skilled workers are imported to fill an even greater perceived shortage in the labor pool.

C. *Underutilizing African American Potential*

The situation of African Americans, three decades after the beginning of civil rights era advances, is still woefully behind white Americans in virtually all measures of economic and educational success. The percentage of African American families earning an average income has steadily decreased.[175] African Americans are disproportionately poverty-stricken.[176] Furthermore, African Americans are represented in disproportionate numbers among the prison population,[177] high-school dropouts,[178] and teen homicides.[179]

175. This percentage dropped from 56.3% of black families earning an average income in 1970 to 46.7% by 1988. David H. Swinton, *The Economic Status of Black Americans During the 1980's: A Decade of Limited Progress*, *in* NATIONAL URBAN LEAGUE, THE STATE OF BLACK AMERICA 29 (1990).

176. The percentage of black families living below the poverty level has risen from 33% in 1970 to 39.2% in 1988. *Id.* at 29. Thus, another generation of black children has been born into poverty. The number of black children under eighteen born in poverty is now 44.1%, up from 41.5% in 1970. *Id.* at 34. Those born in poverty today will also more than likely remain in poverty because of changes in the economy which make it more difficult than ever for poor Americans to become upwardly mobile.

177. One in four black men in their twenties is in jail, on probation or out on parole, while only one in five are in college. Dirk Johnson, *Milwaukee Creating 2 Schools Just for Black Boys*, N.Y. TIMES, Sept. 30, 1990, § 1, at 1.

178. At least 30 percent of black youth are not graduating from high school, and more than half of all suspensions, expulsions and dropouts in metropolitan school districts are black youth. Antoine Garibaldi, *Black School Pushouts and Dropouts: Strategies for Reduction*, *in* BLACK EDUCATION: A QUEST FOR EQUITY AND EXCELLENCE 227, 227-29 (Willy D. Smith and Eva W. Chunn eds., 1989). *See also* Laura J. Young and Susan L. Melnick, *Forsaken Lives, Abandoned Dreams: What Will Compel Us to Act?*, 58 HARV. EDUC. REV. 389 (1988) (reviewing WILLIAM J.

Many of these problems are due to unemployment among African Americans, which is reaching epidemic proportions. In 1989, the unemployment rate for African American men over the age of twenty was 9.8 percent compared with 3.9 percent for white men. The unemployment rate for African American youth aged 16 to 19 was 32.7 percent versus 12.6 percent for white youth in that age range.[180] These unemployment figures do not include the people who have dropped out of the labor market altogether by ceasing to look for employment.[181]

Many of these somber figures are due to overt racism manifested in discrimination in employment and in schools as well as discrimination in the manner in which criminal "justice" is implemented. On the other hand, many of the figures are due to institutional racism that is much more insidious. This section addresses the institutional neglect, caused by racism, within the educational system; neglect that leads to African American unemployment, underemployment, and dissappearance from the labor market.

1. *Underemployment among African Americans*

John Work conducted a study on the underemployment problem among African Americans. He concluded that when African American workers are employed, especially in white collar positions, they are clustered in the lower echelon, lower paying occupations.[182] When African American workers do reach management level posi-

---

JOHNSON, THE TRULY DISADVANTAGED: THE INNER CITY, THE UNDERCLASS, AND PUBLIC POLICY (1987), which analyzes social dislocation and poverty in America's inner cities).

179. In 1986, Blacks constituted 44 percent of the murder victims although they comprised only 12 to 15 percent of the U.S. population. Carl C. Bell, *Preventing Black Homicide*, *in* NATIONAL URBAN LEAGUE, THE STATE OF BLACK AMERICA 143 (1990).

180. Swinton, *supra* note 175, at 25, 47.

181. Kevin Phillips has shown that the official employment figures tabulated by the government are often "underflated" by not reporting the numbers of people who have ceased looking for work, reporting underemployed workers along with fully employed workers, and reclassifying some people who are still looking for work as "unemployable" in order to exclude them from the statistical tables. KEVIN PHILLIPS, POLITICS OF RICH AND POOR 20 (1990).

Therefore, the low unemployment figures (around 5.3 percent) during the years of the "Reagan Miracle" were misleading because of the tactic of excluding certain definitions from "unemployment." For example, in 1988 in New York City,

> 45.3% of the city's residents over the age of 16 could not be counted as labor force participants because of poverty, lack of skills, drug use, apathy or other problems.... [The number] of uncountables for the nation as a whole was 34.5%. Thus the paradox: *millions of jobs might be going begging, but huge numbers of Americans remained either unemployed or unemployable.*

Id. (emphasis in original).

182. JOHN W. WORK, RACE, ECONOMICS, AND CORPORATE AMERICA 89-97 (1984). The lower-level white collar occupations include positions in the mailroom, secretarial and low-level clerical positions. Although this finding may not seem profound, it is crucial. In these days of widespread sentiment that black Americans have made "enough progress, and should just stop complaining," empirical proof to the contrary is significant. Work also discovered that corporate

tions, they are often found in departments from which much less upward mobility is available, such as personnel and public relations departments.[183]

Work attributed much of the underemployment of African Americans in white collar positions to racial discrimination and racist assumptions among employers. He concluded that many employers will discount an African American applicant's qualifications, even if they are facially equivalent to a competing white applicant's qualifications.[184] He also found that even with affirmative action programs, the current social disapproval of overt racism, and the prevalent ideal of an interviewing and selection process based solely on merit, employers continue to wield a large amount of arbitrary discretion in hiring decisions. This militates against African American job applicants even when their qualifications are superior to those of white applicants.[185]

The experiences of African American workers are debilitating and humiliating. These workers endure the daily frustration of being underutilized at work, and they are often overlooked for promotions. However, these African Americans are the lucky ones. Many do not even get a foot in the door of a company. This lack of access to even the opportunity to be underemployed is primarily due to sustained segregation and inadequate education.[186]

2. *Unemployment among African Americans*

Extrême segregation flowing from racial antipathy is a contributing factor to the high unemployment level among African Americans. Studies have shown that desegregation is favorably linked to occupational advancement among African Americans.[187] More than forty years after *Brown v. Board of Education*[188] outlawed de jure segregation, many African Americans still experience almost complete resi-

---

America created a "caste" system that perpetuated the notion of certain jobs being "Negro jobs." *Id.* at 47-49.

183. *Id.* at 90-91.

184. *Id.* at 112-17.

185. *Id.*

186. Work, *supra* note 182, at 3, 25.

187. *See, e.g.*, Robert Crain & Jack Strauss, School Desegregation and Black Occupational Attainments: Results From a Long-Term Experiment 1-2 (1985).

188. Brown v. Board of Educ., 347 U.S. 483 (1954).

dential and educational segregation.[189] The link between segregation and unemployment has become more evident over the past years, as many jobs have moved out of urban areas to the suburbs. Because so many members of the African American community are isolated in urban areas, and therefore do not live near many desirable jobs, they often do not have the social contacts that would enable them to hear of job vacancies. In addition, they find it difficult to get to jobs when they do hear of them.[190]

However, not all unemployment or even underemployment of African American workers can be traced to racism in hiring. It remains an inescapable truth that the dearth of skills in the African American community is greater than that among almost any other racial group. Growing numbers of African American youth are not acquiring the skills needed to enter the job market and compete for high paying, prestigious positions. This problem begins and ends with education. In the United States, the education of many African American children, especially poor African American children, is almost completely neglected. This neglect is traceable to racism and this society's almost mystical ability to avoid making the endemic problems of the country's citizens of color a priority.

Jonathan Kozol wrote an extraordinary chronicle depicting the state of education for the nation's poor children of color. He found that for these children, and for African American children especially, education is not only being neglected, but in some cases, as with dropouts, is almost non-existent.[191] Although the book is not an empirical study, it presents an extremely detailed and precise observation of the vast differences between poor inner city and relatively wealthy suburban schools, covering a number of major metropolitan areas throughout the United States. A few of the situations described in Kozol's book in which poor African American children are living and being educated follow. These conditions can only be described as horrific.

---

189. *See generally* KOZOL, *supra* note 141 (evaluating the disparities in education between poor and wealthy children).

190. For example, in Chicago, the "el," the most extensive public transportation network in the city, bypasses almost all of the black neighborhoods which are most in need of public transportation. *America's Blacks — A World Apart*, ECONOMIST, Mar. 30, 1991, at 17, 21. In Washington, D.C., the Metro system almost completely bypasses most predominantly poor or working- class black neighborhoods. This non-provision of the public transportion service needed in these communities can only be described as deliberate. Interview with Dawn Chirwa, associate with Arnold & Porter in Washington, D.C. (Dec. 13, 1993).

191. *See generally* KOZOL, *supra* note 141 (intending to include all poor children regardless of race, but focussing instead on poor children of color whose educational prospects are most dim).

a. East St. Louis

In East St. Louis, the school system is ninety-eight percent African American.[192] The schools do not have enough books for the children, and those they do have are often outdated.[193] The science labs in the high schools have no running water and almost no lab equipment. One high school has no VCRs, a crucial teaching tool these days.[194] One of the main high schools is often shut down due to an antiquated sewer system that sometimes backs up, flooding the school with raw sewage.[195]

The children of East St. Louis who must struggle to learn in these schools, live in neighborhoods that are highly contaminated with toxic waste and dangerous levels of lead from nearby chemical plants.[196] Some of their "playgrounds" are nothing more than stews of industrial waste.[197] As if toxic poisoning has not wrought enough damage, the children also live with complications due to poor nutrition, insufficient health care, and indequate immunization, not to mention the daily threat of physical violence that surrounds them.[198] In spite of this misery that rivals that of many third world nations, some pupils do manage to achieve.[199] However, only fifty-five percent of the students at East St. Louis High School, for example, will graduate. Of this number, only twenty percent are projected to go on to four-year colleges, while only another ten to twenty percent may receive some other form of higher education. The majority of students join the large ranks of the East St. Louis unemployed and unemployable.[200]

192. KOZOL, *supra* note 141, at 7.
193. KOZOL, *supra* note 141, at 28.
194. KOZOL, *supra* note 141, at 27-29.
195. KOZOL, *supra* note 141, at 23-24.
196. KOZOL, *supra* note 141, at 9-11.
197. KOZOL, *supra* note 141, at 10.
198. KOZOL, *supra* note 141, at 20-22.
199. A science teacher at East St. Louis High School has had two students in twenty-three years graduate from MIT. This may be a faint consolation, but given the circumstances, it is an amazing achievement. KOZOL, *supra* note 141, at 28.
200. KOZOL, *supra* note 141, at 30. Many of these young people will join the seventy-five percent of East St. Louis' population that is on welfare. KOZOL, *supra* note 141, at 7.

b. Chicago

In Chicago, one of the most racially segregated cities in the United States,[201] nearly half of the 595 public schools are 95 to 100 percent African American.[202] On Mondays and Fridays during springtime in Chicago, 18,000 children will have no teacher due to the lack of substitutes within the system.[203] Goudy Elementary School, for example, has no playground, no science labs, and no art or music teachers. Soap, paper towels, and toilet paper are scarce. A history class is taught from textbooks in which Richard Nixon is still president. According to Kozol, this is far from the worst school in Chicago's public school system.[204]

Only about $5,500 is spent per pupil in the Chicago school system. This level of spending and the concomitant condition of the schools contrasts sharply with New Trier High School located in the Chicago suburb of New Trier, which spends $9,000 per pupil.[205] New Trier boasts seven gyms, a fencing room, courses in music, art, drama, and six foreign languages, as well as the best a school can offer in all the "basics."

While at New Trier High the ratio of students to counselors is twenty-five to one, it is not uncommon for the ratio in Chicago's urban schools to be four hundred to one.[206] The library at a suburban school, such as New Trier, may contain 60,000 volumes, while a Chicago school's library "is lucky to have 13,000 volumes."[207] Ninety-three percent of New Trier's seniors go on to four-year colleges, many to Ivy League universities.[208] Not surprisingly, in Chicago's high schools, the graduation rate is dismally low. For instance, at North Lawndale's high school, located in an all-African American Chicago city neighborhood, the graduation rate is thirty-eight percent,[209] while the *non-completion* rate at Chicago's community colleges, largely attended by students of color who have come from the city's public schools, is ninety-seven percent.[210] In light of this disparity between Chicago's city and suburban schools, many urban Chicago residents cannot be convinced that "more money isn't the answer" to their

201. *America's Blacks — A World Apart*, *supra* note 190, at 20-21.
202. Jean L. Griffin, *Suburbs School City Blacks In Success*, Chi. Trib., Sept. 19, 1985, at C1.
203. Kozol, *supra* note 141, at 53-54.
204. Kozol, *supra* note 141, at 63.
205. Kozol, *supra* note 141, at 54.
206. Kozol, *supra* note 141, at 76-77.
207. Kozol, *supra* note 141, at 77.
208. Kozol, *supra* note 141, at 65.
209. Kozol, *supra* note 141, at 45.
210. Kozol, *supra* note 141, at 59.

problems, especially when they have none to buy needed books or such simple things as toilet paper.[211]

c. New York

In New York, whose poor residents and whose public schools are heavily African American and Latino, schoolchildren are overcrowded into almost prison-like conditions. In one New York school district, an old roller-skating rink has been converted into a primary school. Classrooms have no windows or sufficient ventilation, let alone air-conditioning, and the heating frequently breaks down in the winter. At Public School 261 in District 10, 1,300 children are crowded into a building that has a capacity for 900. Because of the overcrowding, children must share scarce textbooks and even scarcer counselors. There are 930 students to each student counselor. This school is ninety percent African American and Latino.[212]

As in the former cities mentioned, New York City's schools lack basic supplies. Some schools, such as Morris High School in the South Bronx, are literally disintegrating around the students' ears.[213] As in Chicago and East St. Louis schools, the drop out rate is "impressive." Fifty percent of New York City's children do not graduate from high school.[214] Clearly, many of these desperate conditions in New York City's schools can be traced to the disparity in educational spending that ranges from $5,500 per pupil in New York's public schools to $11,000 per pupil in the surrounding suburban public school districts.[215]

These stories are by no means atypical. Such conditions exist in public schools which are attended by poor children of color, particularly African American children, across the nation. It does not require a genious to understand that children cannot learn under these conditions. Given the abysmal and worsening conditions in the public

211. KOZOL, *supra* note 141, at 77-80.
212. KOZOL, *supra* note 141, at 85-88.
213. KOZOL, *supra* note 141, at 99-107.
214. KOZOL, *supra* note 141, at 112. Kozol suggests this 50% drop-out rate in most school districts is often actually higher than official statistics estimate, since these statistics only report the drop-out rate at the high-school level. He found that there are large numbers of children who drop-out or are expelled in junior high school or even sooner, long before high school. The official drop-out rate does not include these children. KOZOL, *supra* note 141, at 112.
215. KOZOL, *supra* note 141, at 83.

schools attended by poor African American children, African Americans understandably question the wisdom and intent of policymakers who would respond to the alleged dearth of qualified workers with increased immigration rather than with better education and job training.

If Congress is seriously worried about the education of the future American work force, it should address the disparity in funding between schools that leads to such a cavernous gap in quality. Huge financial resources are needed to bring the schools in urban areas up to the quality level of the relatively well-funded suburban schools. However, better to pay this price now than in the future, when the cost of the problems will be much higher. Still, the question remains as to the link, if any, between the critical conditions of the educational prospects of poor African American children and the "importation" of skilled workers.

D. *Is the "Importation" of Skilled Foreign Workers Another Obstacle to African American Economic Progress?*

In spite of its economic woes, most would agree that the United States continues to have tremendous human resources potential. Yet, Congress has taken the position that increasing numbers of skilled immigrants are needed to fill skilled jobs that could be filled by American workers. Given the presence of significant numbers of unemployed and under-skilled native workers, Congress' actions seem to constitute a severe disregard of the nation's human potential. One wonders whether this nation's leaders have taken reasonable steps to utilize this potential instead of looking to immigration to solve the projected shortage of skilled workers. One also wonders whether expanding employment-based visas, without paying concurrent meaningful attention to the education and training of native residents, will only further damage the economic prospects of African Americans and create more structural economic problems in the long run.

Whatever Congress' motivation, those who are skeptical about this nation's commitment to addressing the plight of unemployed African Americans could look at the Immigration Act of 1990 and conclude that the nation is now beginning to use immigration to avoid improving the situation of native unemployed and under-skilled workers. A reasonable conclusion is that United States' leaders are either deliberately or subconsciously searching for a way to continue to avoid repairing the desperate situation of people whom they and the rest of the power structure have abandoned.[216]

To the skeptics, this argument is not far-fetched. The nation's leaders continue to allow countless children to be undereducated or to live in miserable conditions. These leaders appear to be unwilling to do what is necessary to correct the nightmare in which so many people in our county live and in which children are expected to learn. Given that background, it is not hard to imagine that these leaders would take a relatively easy road to solving the inevitable consequences of societal neglect.[217]

The skeptics' view is also supported by Congress' rejection of a provision in the House version of the bill that was to become the Immigration Act of 1990.[218] This provision would have levied a fee against employers for every skilled foreign worker they hired. The fee would have been funneled into an Educational Assistance and Training Fund[219] to educate and retrain the unemployed and underemployed.[220] The fee was set at $1,000 per alien hired, with a lower amount to be levied against small businesses or charitable organizations.[221] The money raised would have been allocated to states who have both a high ratio of alien to U.S. employees and high levels of unemployment.[222]

The fee would have generated only approximately $50 million a year.[223] However, the provision at least acknowledged that the "importation" of skilled foreign workers may have an adverse impact on

216. These people include poor domestic citizens as well as poor, unskilled immigrants. Both groups consist overwhelmingly of people of color.

217. *See supra* notes 137-44 and accompanying text for a synopsis of other examples of a government and business community which is unwilling to pay the price to increase the quality of the skills of domestic workers.

218. H.R. 4300, 101st Cong., 2d Sess. (1990).

219. Job Training Reform Amendments of 1992, 29 U.S.C. § 1506 (1993).

220. Act of Nov. 29, 1990, Pub. L. No. 101-649, 1990 U.S.C.C.A.N. (104 Stat.) 6710, 6725-26.

221. *Id.* at 6766.

222. *Id.* at 6752-53.

223. *Id.* at 6768. There are currently approximately ten million people unemployed in this country, corresponding to a 7.3% unemployment rate. *See* Daily Lab. Rep. (BNA) No. 66 at D-1 (Apr. 6, 1992). Thus, the $50 million generated would have meant only an extra $5 per person per year for training purposes, almost a cruel joke to those people whose problems the provision was meant to address.

These funds also would only cover people listed on unemployment rolls, meaning those who are currently actively looking for work. The people who have given up and are no longer looking for work would not be included in the fund's benefits.

the nation's incentive to educate and train our own domestic workers. Unfortunately, when the House bill reached the Committee on Ways and Means, the employer fee provision received short-shrift. It was struck from the House bill by the Committee on the grounds that it was a "tax", an impermissible appropriation under such a bill, and not a "user fee" that presumably would have been permissible.[224] Currently, the Educational Assistance and Training Fund is still written into the immigration laws, but no method of funding it was ever adopted. Thus, this nominal attempt to provide for the people who may be harmed by the new immigration policy was stillborn.

Our nation's legislators claim that the skilled immigrant provisions of the new immigration act are needed for our economy. In the short-run, this is more than likely true. However, our country simply cannot continue ignoring a growing, uneducated, unemployed and unemployable "underclass" if we want to achieve real and sustainable economic growth. We must realize that our most valuable resources are our human resources, and this includes everyone. Ignoring the plight of the poor in general, and African Americans in particular, is shortsighted. These underserved members of the community will continue to exist, and our society will have to provide for their needs somehow.

It would be much more advisable, ultimately less costly, and more productive for the United States to begin making the education, training, and employment of every person, particularly the most subordinated members of our society, a high priority. It is not too late for our nation to realign its priorities and to attack the root of the problem of our decreasingly skilled labor force, instead of only applying a "band-aid" to the symptom. If we make education and retraining national priorities, we will be able to build a stronger economy that will satisfy everyone, including big business.

The policy behind the Immigration Act of 1990 must be revised. It seems that neither compassion nor long-term planning has been able to motivate government or businesses to invest in the nation's school system to the extent needed. Perhaps more drastic measures are necessary. If immigration of skilled foreign workers was completely eliminated or dramatically curtailed, even for only a few years

224. The Committee members proclaimed that the fee imposed on employers for hiring foreign skilled workers was a tax, not a permissible user fee because it was

> not designed solely to compensate the Federal Government for the provision of services to the particular person (i.e., the petitioning employer) that is required to pay the fee. Rather the funds generated from the fee would be used primarily to provide training to U.S. persons in fields which have a shortage of employees.

H.R. Rep. No. 101-723(II), 101st Cong., 2d Sess., pt. 2, at 6781-82 (1990).

in niches that cannot be easily moved offshore, more corporations would become so desperate for workers that they would have to invest in our urban schools. We have the resources to improve our schools; this nation is still extremely wealthy. However, we must do something soon before more generations become mired in hopelessness.

CLOSING

> Why are we more shocked when a dozen people are killed in Vilnius than [by] a massacre in Burundi? Because they are white people. That's who we are. That's where America comes from.[225]
>
> Patrick Buchanan
> Presidential Candidate
> January 1992

During his failed run for the presidency in 1992, Patrick Buchanan repeatedly expressed, through his statements on immigration, crime, multiculturalism, and economic policies, that his vision of the United States is a Euro-centric one that excludes all people of color. While most policymakers are quick to disassociate themselves from his type of blatant racism, our current immigration policies send a similar message to African Americans. The message is one of exclusion rather than inclusion.

Is it difficult to understand why the messages of U.S. immigration policies are offensive to African Americans? In addition to the U.S. government's highly-publicized exclusionary policies toward Haitian refugees, racism is suggested by the small numbers of African immigrants who are admitted each year. While many African èmigrès look within the continent and to Europe instead, evidence that Africans wish to immigrate to the United States is clear. Yet we have done little by way of affirmative measures to increase the numbers of African immigrants. In addition, the increase in visas for skilled immigrants sends a message to African Americans that their critical need for better education and job training is not a high priority.

We can do better. If we are serious about helping nationals of countries that have been underrepresented, let's consider underrepresentation not just since 1965, but also underrepresentation re-

---

225. *A Nasty Campaign of "Us" vs. "Them," Buchanan's Appeal to Voters' Fears*, L.A. TIMES, Mar. 5, 1992, at B6.

sulting from affirmative exclusion policies and national origin tests. Since African immigration is grossly underrepresented under any test, then just as extra numbers have been specifically set aside for Irish nationals without special educational or job experience requirements, let us set aside extra numbers for Africans on the same terms. Finally, we must place pressure on the business community to look to our own marginalized people as resources. Business ingenuity and know-how can be put to good use by retraining the unemployed and making education a high priority for future generations.

My examination of these questions is very preliminary. Much more should be considered and developed, including: a better understanding of the cultural, economic, and institutional barriers to African immigration; differences that exist between various African countries; and the impact a larger class of African immigrants would have on the African American community. For every study that tends to show a generally positive impact that large numbers of immigrants would have on the economy, we need more detail on specific job niches and geographic locations. We should consider the impact of so-called fair trade agreements on the domestic workforce; especially when even their staunchest supporters concede that as many as 150,000 U.S. jobs will be lost in the agreement with Mexico alone.[226] As we look for other responses beyond the few simple ones I offer, we must not underestimate the complexity of the economic issues. Depending on what we find after hard analysis, policymakers and immigrants' rights advocates may have to make some difficult choices when the social and economic well-being of other subordinated groups is found to be at stake.

Only through these kinds of measures and efforts can we finally start sending positive messages to African Americans through our immigration policies—messages of inclusion, rather than continued exclusion.

---

226. *See, e.g.*, Keith Bradsher, *Trade Pact Could Cost up to 150,000 U.S. Jobs*, N.Y. Times, Sept. 11, 1992, at D1 (quoting Bush Administration Labor Secretary Lynn Martin).

FEDERAL REGULATION OF ALIENS AND THE CONSTITUTION

*By T. Alexander Aleinikoff**

In two cases decided in the afterglow of the centennial of the U.S. Constitution, the Supreme Court said nearly everything the modern lawyer needs to know about the source and extent of Congress's power to regulate immigration. The cases sustained the constitutionality of the first significant congressional restrictions on immigration—the shameful Chinese exclusion laws. Constitutional analysis of the immigration power has yet to shed the original sins of the Court's ratification of Congress's ignominious legislation.

I. THE IMMIGRATION POWER AT THE CENTENNIAL

The *Chinese Exclusion Case*[1] confirmed congressional authority to prevent aliens from entering (the *exclusion* power); and *Fong Yue Ting v. United States*[2] ratified congressional authority to send resident aliens home (the *deportation* power). Both cases denied virtually any authority for the judiciary to review substantive decisions as to which classes of aliens should be entitled to enter or remain in the country. The decisions also established that deportation is not "punishment" in a constitutional sense and thus rendered inapplicable constitutional protections afforded in criminal proceedings, as well as prohibitions on cruel and unusual punishment, ex post facto laws and bills of attainder.

[62] Michael, *supra* note 60, at 831:

> If one lesson from *Hardwick* is that the Supreme Court is changing again, another might be that comparative law could be of assistance there (despite the attitude expressed by one attorney as "we already have 51 jurisdictions to compare, and that's enough"). With US law being argued to Strasbourg institutions, and the House of Lords referring to both US and Strasbourg law, perhaps it is now the Americans who have assumed the attitude once ascribed by Sir James Fawcett QC to the British: when told how things are done in another country they simply say "How funny."

Compare the remarks of Justice Scalia, *supra* note 47, with the approach taken by Justice Gubbay in *State v. Ncube*, *supra* text at notes 10–16. *See also* Kirby, *The Role of the Judge in Advancing Human Rights by Reference to International Human Rights Norms*, 62 AUSTL. L.J. 514 (1988).

[63] Lester, *supra* note 3, at 560–61.

* Professor of Law, University of Michigan Law School.

[1] Chae Chan Ping v. United States, 130 U.S. 581 (1889).

[2] 149 U.S. 698 (1893).

The cases further suggested that Congress may authorize executive branch officials to exclude or remove aliens without judicial trial or examination, and that exercise of this authority may be insulated from judicial review. In the *Japanese Immigrant Case*,[3] the Court ignored the broadest reading of its earlier dicta and held that the Due Process Clause applied to *deportation* proceedings. This decision, coming 10 years after *Fong Yue Ting*, ought to be viewed as part of the *fin de siècle* understanding of the scope of the immigration power.

In analyzing the constitutional issues in these cases, the Court did not start with the text or structure of the Constitution and ask how a power to regulate immigration might be inferred. Rather, it approached the question of congressional power from the perspective of the conduct of foreign affairs. Just as the United States must possess the power to protect itself against invading armies, so, too, it must be able to defend itself against "vast hordes of [a foreign] people crowding in upon us."[4] The power to exclude aliens, the Court said in the *Chinese Exclusion Case*, "is an incident of every independent nation. . . . If it could not exclude aliens it would be to that extent subject to the control of another power."[5] The deportation power was established in similar dire terms. The Court announced in *Fong Yue Ting* that the right to deport foreigners is "an inherent and inalienable right of every sovereign and independent nation, essential to its safety, its independence and its welfare."[6]

To find that regulation of immigration implicates the foreign relations of the United States does not entail a holding that international law provides the source of constitutional rules. Certainly, the Constitution may supply constraints on the foreign affairs power quite distinct from those recognized under international law. But the Court viewed the issues as all of a piece, and it deemed international law as relevant to, if not the basis of, constitutional norms. Before announcing its conclusion in *Fong Yue Ting* that Congress possessed the power to deport aliens, the Court quoted extensively from Vattel, Ortolan, Phillimore and Bar.

Seeing immigration regulation through an international law lens is understandable. Aliens, after all, are citizens of foreign nations, and the treatment of foreign nationals certainly invokes traditional concerns of international law. The Court recognized early on that *state* regulation of aliens could embroil the federal Government in "disastrous quarrels with other nations."[7] Furthermore, the Court could sensibly reason that those who drafted and ratified the Constitution fully expected the United States to take its place among nation-states as an equal. Defining the immigration power without consideration of how other nations regulated immigration ran the risk of leaving the United States in a position of comparative weakness.

[3] [Yamataya v. Fisher,] 189 U.S. 86 (1903). [4] *Chinese Exclusion*, 130 U.S. at 606.
[5] *Id.* at 603–04. [6] 149 U.S. at 711.
[7] *See, e.g.*, Chy Lung v. Freeman, 92 U.S. 275, 280 (1875) (invalidating California statute providing for inspection of arriving aliens).

The foreign affairs/international law model had important—and lasting—implications for constitutional analysis of the immigration power. First, according to the international law commentators of the 19th century, "reception [of foreigners] into the territory is a matter of pure permission, of simple tolerance, and creates no obligation."[8] Constitutional law has repeatedly reflected this view, noting that an alien can have no "right" to enter the United States. Admission constitutes an act of beneficence on the part of the sovereign.[9]

Second, seeing the immigration power as an aspect of international relations suggested a very limited—or nonexistent—role for the courts.[10] It is a nice question whether locating the immigration power elsewhere—for example, as a part of the power to regulate commerce with foreign nations—would have persuaded the Court to exercise somewhat stricter scrutiny of immigration laws; but it certainly would have forced apart the issues of judicial role and congressional authority.[11]

Focusing only on the immigration cases would paint a false picture of the status of aliens in the United States in the late 19th century. The Supreme Court recognized important rights on behalf of aliens residing in the United States. Three years before *Chinese Exclusion*, the Court held, in *Yick Wo v. Hopkins*, that the Fourteenth Amendment protected aliens from unfriendly state regulation.[12] And in *Wong Wing v. United States*, it invalidated a section of the 1892 immigration statute—whose deportation provisions it had approved in *Fong Yue Ting*—providing for "the imprisonment at hard labor" of any Chinese national judged by executive authorities to be in the United States illegally. The Court reaffirmed its holding that deportation of such aliens was constitutionally permissible; but it found the section under scrutiny to be plainly punitive, and such "infamous punishment" could not be imposed without "a judicial trial to establish the guilt of the accused."[13]

A full description of the constitutional status of aliens at the end of the 19th century would therefore identify two norms: (1) as to matters relating to admission and expulsion, Congress possessed "plenary" power virtually unfettered by the Constitution;[14] and (2) resident aliens could invoke con-

[8] *Fong Yue Ting*, 149 U.S. at 708 (quoting 1 T. ORTOLAN, RÈGLES INTERNATIONALES ET DIPLOMATIE DE LA MER 297 (4th ed. 1864)).

[9] For the most recent statement, see Landon v. Plasencia, 459 U.S. 21, 32 (1982). *See generally* Legomsky, *Immigration Law and the Principle of Plenary Congressional Power*, 1984 SUP. CT. REV. 255, 269–70.

[10] *E.g.*, *Chinese Exclusion*, 130 U.S. at 606.

[11] It should be made clear that an international law perspective did not compel either plenary congressional power or judicial abdication. The international authorities recognized that the power of a sovereign to exclude and deport aliens under international law "may be subjected, doubtless, to certain forms by the domestic laws of each country." *Fong Yue Ting*, 149 U.S. at 708 (quoting 1 T. ORTOLAN, *supra* note 8, at 297). But there can be little doubt that associating the immigration power with the exercise of the foreign affairs power suggested a rather limited judicial role.

[12] Yick Wo v. Hopkins, 118 U.S. 356 (1886).

[13] Wong Wing v. United States, 163 U.S. 228, 237 (1896).

[14] The exception to this norm was The Japanese Immigrant Case, 189 U.S. 86 (1903), which subjected deportation procedures to (albeit modest) constitutional scrutiny.

stitutional protections in situations where citizens would be entitled to such protections (for example, criminal prosecutions). Both aspects of the Court's constitutional jurisprudence have been strengthened in the 20th century.

II. The Immigration Power at the Bicentennial

The "plenary power" cases—harsh in their implications as they are—have been reaffirmed and even extended in the Constitution's second hundred years. Immigration law has remained blissfully untouched by the virtual revolution in constitutional law since World War II, impervious to developments in due process, equal protection and criminal procedure. News of the death of the "rights/privilege" distinction has not reached immigration law: initial entry into the United States is still described as a "privilege,"[15] and Congress is apparently free to make whatever distinctions it desires in the distribution of government benefits to aliens.[16] Substantive regulations defining what classes of aliens may enter and remain in the United States are still regarded as largely beyond judicial scrutiny.[17]

Again, this somber picture is not a full description of the constitutional status of aliens. The legacy of *Yick Wo* has also survived and flourished in the 20th century. Outside the immigration process, aliens receive most of the constitutional protections afforded citizens.[18] Furthermore, beginning with *Graham v. Richardson*[19] in 1971, the courts have invalidated scores of state laws that excluded aliens from public programs and economic opportunities in the private sphere.[20] While states may still deny aliens political rights and employment identified with the exercise of sovereign power,[21] the Supreme Court's decisions mean that little distinguishes the day-to-day lives of citizens from those of permanent resident aliens. Indeed, in a dramatic application of the Equal Protection Clause, the Court held that Texas could not exclude undocumented alien children from its public schools—at least absent a clear federal policy authorizing such exclusion.[22]

Most important, despite its recognized authority to regulate immigration as it deems fit, Congress has hardly reduced resident aliens to a subordinate class in the United States. Permanent resident aliens, refugees and asylees are entitled to nearly all federal benefits on equal terms with citizens and are

[15] Landon v. Plasencia, 459 U.S. 21, 32 (1982).

[16] Mathews v. Diaz, 426 U.S. 67 (1976).

[17] The Court nominally applies a low level of scrutiny. But the standard is satisfied upon the merest of showings. *See, e.g.*, Fiallo v. Bell, 430 U.S. 787 (1977); Kleindienst v. Mandel, 408 U.S. 753 (1972).

[18] *E.g.*, Almeida-Sanchez v. United States, 413 U.S. 266 (1973) (Fourth Amendment); Bridges v. Wixon, 326 U.S. 135 (1945) (First Amendment).

[19] 403 U.S. 365 (1971).

[20] *E.g.*, Examining Bd. of Eng'rs v. Otero, 426 U.S. 572 (1976); *In re* Griffiths, 413 U.S. 717 (1973) (bar admission); Sugarman v. Dougall, 413 U.S. 634 (1973) (state civil service employment).

[21] The Supreme Court has upheld state laws excluding aliens from employment as parole officers, Cabell v. Chavez-Salido, 454 U.S. 432 (1982); teachers, Ambach v. Norwick, 441 U.S. 68 (1979); and police officers, Foley v. Connelie, 435 U.S. 291 (1978).

[22] Plyler v. Doe, 457 U.S. 202 (1982).

generally allowed to work at whatever jobs they choose. Under federal law, aliens in immigration proceedings are entitled to be represented by a lawyer, confront witnesses and present evidence. Most decisions by immigration officials are reviewable in federal court. U.S. citizenship is rather easily attainable for aliens who have been admitted as permanent residents. While there are exceptions to each of these generalizations (exceptions that strike many as unjust and unreasonable[23]), it is fair to conclude that Congress, in keeping with the American tradition of trying to assimilate those aliens we choose to admit, treats most resident aliens with a considerable degree of concern and respect.

We have therefore ended the second century of the Constitution as we ended its first. The descendants of *Yick Wo* and *Wong Wing* extend constitutional protections to aliens; yet they stand side by side with the progeny of *Chinese Exclusion* and *Fong Yue Ting*, which leave federal regulation of immigration virtually free from constitutional restrictions.

III. An Agenda for the Third Century of the Constitution

The analysis and implications of *Chinese Exclusion* and *Fong Yue Ting* are so patently inconsistent with modern constitutional principles that a thoroughgoing reconceptualization of the immigration power is in order.[24] The first step is to end appeals to "ancient principles of the international law of nation-states"[25] in determining the source and scope of the immigration power. The power to regulate the admission and residence of aliens may be securely located in the commerce power or implied from a structural analysis of the Constitution.[26] One may call the immigration power "plenary," just as the commerce power has been labeled "plenary." But to say that Congress has all the immigration power it could be granted is not to say that the exercise of the power is free of limits established elsewhere in the Constitution. Eliminating the talk of sovereignty and inherent power ought to help decision makers recognize that the immigration power does not stand above or before the Constitution. It is an ordinary, if powerful, constitutionally delegated prerogative. Accordingly, immigration regulations ought to be subject to the judicial scrutiny accorded other exercises of federal power.

[23] For example, there is no judicial review of visa denials by consular officers (*see* T. Aleinikoff & D. Martin, Immigration: Process and Policy 205–12 (1985)); aliens are not entitled to government-provided counsel in deportation proceedings, despite the complexity of the law and the interests at stake (*see id.* at 414–22); the exclusionary rule does not apply in deportation proceedings, INS v. Lopez-Mendoza, 468 U.S. 1032 (1984); and homosexuality is a bar to naturalization, *In re* Longstaff, 716 F.2d 1439 (5th Cir. 1983), *cert. denied*, 467 U.S. 1219 (1984).

[24] Professor Henkin's recent review minces no words: "*Chinese Exclusion*—its very name an embarrassment—must go." Henkin, *The Constitution and United States Sovereignty: A Century of* Chinese Exclusion *and Its Progeny*, 100 Harv. L. Rev. 853, 863 (1987).

[25] Kleindienst v. Mandel, 408 U.S. 753, 765 (1972).

[26] For a structural analysis, see T. Aleinikoff & D. Martin, *supra* note 23, at 16–18.

Once the immigration power is brought within the fold of other constitutional powers, we can begin to chart a course for its development in the next century. I will suggest three areas of reform: (1) elimination of the border/interior distinction in procedural due process analysis; (2) more serious scrutiny of substantive regulations of immigration; and (3) reconceptualization of equal protection analysis regarding federal regulation of aliens.

Due Process and the Border / Interior Distinction

Under current doctrine, aliens seeking to enter the United States for the first time are entitled only to those procedures Congress decides to provide; resident aliens whom the Government seeks to deport, however, are accorded fairly substantial due process protections. In *Landon v. Plasencia*, the Supreme Court appeared to undermine the border/interior distinction by holding that a resident alien who had left the country and sought reentry was entitled to due process at the border. Yet the case did not scrap the distinction; it simply reoriented it by distinguishing between initial entrants and returning resident aliens. Henceforth, resident aliens—whether in exclusion or deportation proceedings—are entitled to due process. However, "an alien seeking initial admission" remains at the mercy of Congress; he "requests a privilege and has no constitutional rights regarding his application."[27]

The border/interior distinction, as many have noted, cannot withstand analysis.[28] To take just a single example, it seems peculiar, to say the least, that an alien who seeks entry on an immigrant visa based on marriage to a U.S. citizen lacks constitutional rights when stopped at the border, but an alien who is apprehended after entering the country surreptitiously enjoys the full panoply of constitutional protections. Some may argue that what distinguishes the initial entrant from the alien who has entered and taken up residence is the very trigger of the Due Process Clause: the alien at the border has no "property" or "liberty" interest that is being taken; the alien who has entered has developed "the ties that go with permanent residence."[29] But this distinction will not work. Aliens who present immigrant visas at the border have already established "ties" with the United States (either a close family relative or a job waiting for them). Furthermore, it is hardly obvious why due process requires that we acknowledge the "ties" established by an alien who has entered the country without inspection. Rather, it seems sensible to conclude that governmental action that seeks to remove an alien from the territory of the United States implicates a "liberty" interest of the targeted individual.[30]

[27] Landon v. Plasencia, 459 U.S. 21, 32 (1982).

[28] *E.g.*, Martin, *Due Process and Membership in the National Community: Political Asylum and Beyond*, 44 U. PITT. L. REV. 165, 173–75 (1983); Hart, *The Power of Congress to Limit the Jurisdiction of Federal Courts: An Exercise in Dialectic*, 66 HARV. L. REV. 1362, 1386–96 (1953).

[29] Landon v. Plasencia, 459 U.S. at 32; *see Fong Yue Ting*, 149 U.S. at 762 (Fuller, C.J., dissenting) (deportation constitutes "deprivation of that which has been lawfully acquired").

[30] For another way to the same result—which, I believe, has received too little attention—see Van Alstyne, *Cracks in "The New Property": Adjudicative Due Process in the Administrative State*,

To say that due process applies is not to answer the question of what protections are mandated by the Constitution. Under the Court's current flexible approach,[31] reasonable arguments can be made for varying degrees of process depending on a host of factors.[32] But elimination of the border/ interior distinction will ensure that Congress and the courts analyze these factors in crafting procedures for the border.

Scrutiny of Substantive Immigration Decisions

Despite the best efforts of generations of commentators, deportation grounds remain virtually immune to constitutional challenge.[33] I would like to focus here on what may seem the most counterintuitive aspect of the current doctrine: that aliens may apparently be deported for speech or association that is protected under the First Amendment.

Under prevailing First Amendment doctrine, a statute making advocacy of communism or mere membership in the Communist Party criminal would not be sustained.[34] And this protection clearly extends to aliens resident in the United States, as well as to citizens.[35] What, then, does one make of provisions in the immigration code that mandate the deportation of aliens who "advocate or teach . . . opposition to all organized government"; or who "are members of . . . the Communist Party of the United States"; or who "knowingly circulate . . . or display . . . any written or printed matter . . . teaching . . . the economic, international, and governmental doctrines of world communism"?[36]

The legacies of *Fong Yue Ting* and *Wong Wing* seem to have produced a curious paradox in constitutional law: an alien can be deported for conduct for which, by dint of the First Amendment, he or she could not be imprisoned. The impact of this anomaly is not restricted to the immigration context. Because deportation is usually seen as a drastic penalty (at times, perhaps, more harsh than imprisonment), aliens are likely to forgo speech condemned by the deportation provisions. The immigration statute therefore chills the exercise of First Amendment rights outside the immigration context.

The case that is usually credited with burying First Amendment rights in the deportation context is *Harisiades v. Shaughnessy*,[37] a decision that has not

---

62 CORNELL L. REV. 445 (1977) (suggesting that Due Process Clause includes within it a "liberty" interest in freedom from nonarbitrary adjudicative procedures).

[31] *See* Mathews v. Eldridge, 424 U.S. 319 (1976).

[32] *See, e.g.*, Martin, *supra* note 28, at 208–34.

[33] *See, e.g.*, Maslow, *Recasting Our Deportation Law: Proposals for Reform*, 56 COLUM. L. REV. 309 (1956); Hesse, *The Constitutional Status of the Lawfully Admitted Permanent Resident Alien: The Inherent Limits of the Power to Expel*, 69 YALE L.J. 262 (1959); Note, *Immigration and the First Amendment*, 73 CAL. L. REV. 1889 (1985).

[34] *See, e.g.*, Brandenburg v. Ohio, 395 U.S. 444 (1969) (per curiam).

[35] Bridges v. Wixon, 326 U.S. 135, 148 (1945); Schneider v. State, 308 U.S. 147, 160 & n.8 (1939).

[36] 8 U.S.C. §1251(a)(6)(B), (C) and (G) (1982).

[37] 342 U.S. 580 (1952).

lost luster even with the passing of the Cold War era of which it is surely a product.[38]

But read carefully, *Harisiades* does not support the claim that the First Amendment does not limit the deportation power. Nowhere does Justice Jackson's majority opinion suggest that the deportation of an alien for membership in the Communist Party raises no First Amendment issue. Rather, the opinion sustains the statutory provision by relying, in summary fashion, upon the Court's decision a year before in *Dennis v. United States.*[39] "Different [constitutional] formulae have been applied in different situations," wrote Jackson, "and the test applicable to the Communist Party has been stated too recently to make further discussion at this time profitable. [Here a footnote cites *Dennis.*] We think the First Amendment does not prevent the deportation of these aliens."[40] *Dennis* was a case about domestic activity; it upheld the convictions, under the Smith Act, of Communist organizers. Thus, read properly, *Harisiades* establishes that deportation grounds are to be judged by the same standard applied to other burdens on First Amendment rights.

Because First Amendment doctrine has grown considerably more protective since *Dennis,*[41] deportation grounds based on speech should face some fairly tough scrutiny. In the bright light of modern free speech jurisprudence, the archaic deportation grounds appear fatally overbroad.[42]

Equal Protection: Regulation of Immigration and of Aliens

Equal protection challenges to immigration rules have failed in the face of strong judicial statements about the broad scope of Congress's power to exclude and remove aliens.[43] If we can get beyond the blinding power of the word "plenary," we can perceive at least two kinds of circumstances in which modern notions of equality and fairness ought to apply to congressional regulation of aliens.

First, the courts have wrongly assumed that every federal regulation based on *alienage* is necessarily sustainable as an exercise of the *immigration* power.[44] It should be apparent that some statutes burdening aliens are based

[38] *See, e.g.*, Fiallo v. Bell, 430 U.S. 787, 792 (1977) (citing *Harisiades* favorably).

[39] 341 U.S. 494 (1951).

[40] 342 U.S. at 592.

[41] *See, e.g.*, Hess v. Indiana, 414 U.S. 105 (1973) (per curiam); Brandenburg v. Ohio, 395 U.S. 444 (1969) (per curiam).

[42] It is fitting that in the centennial year of *Chinese Exclusion*, a federal court has concluded, for the first time, that the deportation grounds mentioned in the text "are substantially overbroad and violate the First Amendment." American-Arab Anti-Discrimination Comm. v. Meese, 714 F.Supp. 1060 (C.D. Cal. 1989).

[43] *E.g.*, Fiallo v. Bell, 430 U.S. 787 (1977). For a rare case invalidating an immigration regulation as a violation of equal protection, see Francis v. INS, 532 F.2d 268 (2d Cir. 1976) (striking down, not a statutory provision, but rather an administrative interpretation of the statute).

[44] For a similar suggestion, see Rosberg, *The Protection of Aliens from Discriminatory Treatment by the National Government*, 1977 SUP. CT. REV. 275.

on considerations other than a policy judgment regarding the number and classes of aliens who may enter or remain in the United States. (Exclusion of aliens from the federal civil service provides a good example.) Rather than invoking "the immigration power" as a barrier to constitutional analysis, courts ought to examine the justifications offered on behalf of federal regulations based on alienage to see if they meet traditional constitutional standards of permissibility.

A second set of equal protection norms that ought to be brought to bear on immigration law are those that demand substantial justification for governmental regulations that burden fundamental constitutional rights. For instance, the Supreme Court has extended special constitutional protection to marriage and family, subjecting laws that significantly interfere with such relationships to " 'critical examination' of the state interests advanced in support of the classification."[45] Yet the immigration laws bristle with severe burdens on family relationships. To focus on one example: Citizens and permanent resident aliens do not enjoy similar opportunities for reuniting with immediate family members located outside the United States. The entry of the spouse or minor child of a resident alien is subject to a numerical limitation (if the relationship was formed after the alien entered the country), while the entry of the spouse or minor child of a U.S. citizen is not. The impact of the statutory scheme is dramatic because the numerical limits have created a waiting list of tens of thousands of aliens. Thus, if a permanent resident alien marries a Mexican national, the alien spouse may have to wait as long as 10 years to be eligible for entry. However, if a U.S. citizen marries a Mexican national, the alien spouse may enter as soon as immigration authorities process the paper work.

This difference in treatment would likely trigger close judicial scrutiny outside the immigration context. Yet the preferential reunification rights granted citizens have been upheld without careful examination of the Government's justifications for such a policy.[46] Perhaps Congress could offer sound and substantial justifications; the mere fact that the policy burdens aliens should not be grounds for denying that Congress has an obligation to do so.

* * * *

The proposals suggested here stem from the same, rather straightforward, idea: immigration law must be brought into the fold of modern constitutional law.[47] Other congressional powers have flourished in the 20th

[45] Zablocki v. Redhail, 434 U.S. 374, 383, 388 (1978) (requiring statute that interferes with right to marry to be justified by "sufficiently important state interests and . . . closely tailored to effectuate only those interests"). *See* Turner v. Safley, 107 S.Ct. 2254 (1987) (burden on right to marry); Moore v. City of E. Cleveland, 431 U.S. 494 (1977) (interference with decisions on family living arrangements).

[46] Noel v. Chapman, 508 F.2d 1023 (2d Cir.), *cert. denied*, 423 U.S. 824 (1975).

[47] *See* Henkin, *The Constitution as Compact and as Conscience: Individual Rights Abroad and at Our Gates*, 27 WM. & MARY L. REV. 11 (1985).

century despite the judiciary's active and creative development of constitutional rights on behalf of individuals and groups. There is no reason that immigration law cannot thrive within the constitutional boundaries established for other delegated and implied federal powers. Indeed, as *Yick Wo* and *Wong Wing* demonstrate, the nation has well been able to tolerate the extension of constitutional rights to aliens on terms equal to those enjoyed by citizens.

We ought to honor the bicentennial of the Constitution by effacing *Chinese Exclusion* and *Fong Yue Ting*—cases that stained its centennial. Repudiation of ill-begotten tradition is itself a venerable constitutional tradition.

SEGREGATION'S LAST STRONGHOLD: RACE DISCRIMINATION AND THE CONSTITUTIONAL LAW OF IMMIGRATION

Gabriel J. Chin*

For over a century, the Supreme Court has granted federal immigration laws a unique immunity from judicial review. Relying on the so-called plenary power doctrine, the Court has said that "over no conceivable subject" is federal power greater than it is over immigration; even modern federal cases, for example, state that Congress may freely discriminate on the basis of race in this context. In recent years, the Court has defended this approach primarily on stare decisis grounds. In this Article, Professor Gabriel Chin suggests that the plenary power doctrine should be reexamined because the foundational cases are unsound. Chae Chan Ping v. United States *and* Fong Yue Ting v. United States, *the cases that established the plenary power doctrine and remain the leading cases, upheld federal discrimination against immigrants based on race. The plenary power cases were decided by the same Court that decided* Plessy v. Ferguson; *like* Plessy, *they reflect principles that have been emphatically rejected by the Court since* Brown v. Board of Education.

---

* Associate Professor, University of Cincinnati College of Law. LL.M., Yale Law School; J.D., Michigan Law School; B.A., Wesleyan University. E-Mail: gchin@aya.yale.edu.

This Article won the 1998 Association of American Law Schools Scholarly Paper Competition and was presented at the 1998 AALS Annual Meeting in San Francisco.

Thanks to Owen Fiss for his suggestions on an earlier draft and to Bruce Ackerman, Alex Aleinikoff, Len Baynes, Bob Chang, Roger Daniels, Richard Delgado, Don Dunn, Suzanne Emam, Paul Finkelman, George Fishman, Jim Gardner, Anne Goldstein, Leora Harpaz, Chris Iijima, Kevin Johnson, David Kelly, Joan Mahoney, Stanley Mailman, Earl Maltz, Jay Mootz, Hiroshi Motomura, Jim Nafziger, Sheldon Novick, Vic Romero, Theodore Ruthizer, Koroush Salehi, Michael Scaperlanda, Peter Schuck, Mark Tushnet, and Frank Wu for their comments and other assistance. The 1994–95 Yale Law School Graduate Fellows, Gerald Neuman and the other participants in the 1996 AALS Immigration Law Workshop at the University of Colorado School of Law, and the participants in the 1996 Asian Pacific American Law Faculty Conference at UCLA School of Law offered useful comments. Finally, the University of Cincinnati College of Law and Western New England College School of Law generously provided summer research support for this project.

This Article is dedicated to the memories of my grandfather, Frank Chew Chin Sr., whose illegal entry into this country as a "paper son" in defiance of the Chinese Exclusion Act made it possible for me to be here, and my great-great-great-great-great uncle, John Percival Jones, an ardent supporter of Chinese exclusion during his representation of Nevada in the United States Senate from 1873 to 1903, *see, e.g., infra* notes 155, 166–167, 173 and accompanying text, in the hope that, in retrospect, at least, he would recognize the error of his position.

Just as the Jim Crow laws were designed to exclude those of African descent from American society, the laws excluding Asian immigrants upheld in Chae Chan Ping *and* Fong Yue Ting *betray a belief in racial separation. This vision of the United States as a nation reserved for caucasians had a substantial effect on the groups it targeted. Other legal premises that were central to the decisions in the plenary power cases no longer pertain. The Court derived the plenary power doctrine from a particular conception of international law, but contemporary international law does not grant nations unlimited rights to discriminate against immigrants. Moreover, the Supreme Court has recently recognized the availability of some limited judicial review within the plenary power doctrine. Racial classifications would not survive that review. This Article concludes that the plenary power cases reflect values deeply at odds with those of contemporary society, and maintains that such traditional notions of the doctrine on their own terms have been undermined. Under these circumstances,* Chae Chan Ping *and* Fong Yue Ting *should no longer constitute the foundation of such an important branch of public law.*

INTRODUCTION

Brown v. Board of Education[1] marked the beginning of the modern era of American constitutional law by introducing the presumption against racial discrimination. Because of the revolution it started, *Brown* has been called "the most important political, social and legal event in America's twentieth-century history."[2] *Brown* led the Supreme Court to invalidate racial discrimination in employment, public benefits, voting, jury selection, criminal justice, marriage and the family—that is, in virtually every area of American law and life.[3] However, there is one major exception to judicial suspicion of racial classifications: immigration law.

In immigration law alone, racial classifications are still routinely permitted. In recent decades, courts in the District of Columbia[4] and the First,[5] Second,[6] Fifth,[7]

1. 347 U.S. 483 (1954).

2. J. HARVIE WILKINSON III, FROM BROWN TO BAKKE: THE SUPREME COURT AND SCHOOL INTEGRATION, 1954–1978, at 6 (1979).

3. *See* 3 RONALD D. ROTUNDA & JOHN E. NOWAK, TREATISE ON CONSTITUTIONAL LAW: SUBSTANCE AND PROCEDURE ch. 18 (2d ed. 1992) (citing numerous cases).

4. *See* Unification Church v. INS, 547 F. Supp. 623, 628 (D.D.C. 1982) (assuming that Congress could authorize racial discrimination). Similarly, in *Olsen v. Albright*, 990 F. Supp. 31 (D.D.C. 1997), the court explained:

> Although the Constitution clearly prohibits discrimination on the basis of race, ethnicity or national origin, the Supreme Court declared that "the power to expel or exclude aliens [is] a fundamental sovereign attribute exercised by the Government's political departments largely immune from judicial control." During most of its history, the United States openly discriminated against individuals on the basis of race and national origin in its immigration laws.

Id. at 37 (citations omitted).

5. *See* Ramos v. United States Civil Serv. Comm'n, 376 F. Supp. 361, 366 n.8 (D.P.R. 1974) (three-judge court) ("Congress may discriminate on racial . . . lines in deciding whom to admit [or deport]."), *aff'd in part, dismissed in part*, 426 U.S. 916 (1976).

6. *See* Vigile v. Sava, 535 F. Supp. 1002, 1016 (S.D.N.Y.) ("Congress may employ race or national origin as criteria in determining which aliens to exclude . . . ."), *rev'd on other grounds sub nom.* Bertrand v. Sava, 684 F.2d 204, 211–12 (2d Cir. 1982) (agreeing that an alien may be denied entry on ground of race); Hitai v. INS, 343 F.2d 466, 467 (2d Cir. 1965) (rejecting an equal protection challenge to a racial classification limiting Asian immigration); Fiallo v. Levi, 406 F. Supp. 162, 165 (E.D.N.Y. 1975) (three-judge court) (stating that an alien may be denied entry on the basis of race), *aff'd sub nom.* Fiallo v. Bell, 430 U.S. 787 (1977).

7. *See* Pierre v. United States, 547 F.2d 1281, 1290 (5th Cir.) ("[A]liens may be denied entrance on grounds [of] race . . . ."), *vacated on other grounds*, 434 U.S. 962 (1977). Similarly, in *In re Longstaff*, 716 F.2d 1439 (5th Cir. 1983), the court explained:

> Congress has unbounded power to exclude aliens from admission to the United States. Our national immigration policy was for many years based on national origin quotas that reflected racial and ethnic prejudice. Congress can bar aliens . . . for discriminatory and arbitrary reasons, even those that might be condemned as a denial of equal protection or due process . . . .

Id. at 1442 (footnote omitted).

Seventh,[8] Eighth,[9] Ninth,[10] Tenth,[11] and Eleventh[12] Circuits have said not only that aliens may be excluded or deported[13] on the basis of race without strict scrutiny, but also that such racial classifications are lawful per se. Apparently no court has even hinted to the contrary.[14] These circuits merely honor an unbroken line of Supreme Court decisions holding that "Congress may exclude aliens of a particular race from the United States"[15] and, more broadly, that an alien seeking admission "has no constitutional rights regarding his application."[16]

8. *See* Achacoso-Sanchez v. INS, 779 F.2d 1260, 1264 (7th Cir. 1985) (Easterbrook, J.) ("Congress has the power to make distinctions on account of race [in the immigration context].").

9. *See* Kiev v. Glickman, 991 F. Supp. 1090, 1097 (D. Minn. 1998) ("'No limits can be put by the courts upon the power of Congress to protect . . . the country from the advent of aliens whose race or habits render them undesirable as citizens, or to expel such if they have already found their way into our land, and unlawfully remain therein.'") (quoting with approval Wong Wing v. United States, 163 U.S. 228, 237 (1896)); *see also* Montgomery v. Ffrench, 299 F.2d 730, 733–35 (8th Cir. 1962) (noting limits to judicial review and citing cases authorizing racial discrimination).

10. *See* Dunn v. INS, 499 F.2d 856, 858 (9th Cir. 1974) ("Congress has the power to exclude aliens of a particular race . . . .").

11. *See* Lopez v. INS, 758 F.2d 1390, 1393 (10th Cir. 1985) ("[Aliens] may be deported for considerations of race . . . . "); Rodriguez-Fernandez v. Wilkinson, 654 F.2d 1382, 1386 (10th Cir. 1981) (same).

12. *See* Louis v. Nelson, 544 F. Supp. 973, 998 (S.D. Fla. 1982) (assuming racial discrimination was permissible), *aff'd in part, rev'd in part sub nom.* Jean v. Nelson, 711 F.2d 1455 (11th Cir. 1983), *aff'd in part, rev'd in part on other grounds en banc*, 727 F.2d 957, 970 (11th Cir. 1984) ("Haitian plaintiffs . . . cannot claim equal protection rights under the fifth amendment . . . ."), *aff'd*, 472 U.S. 846 (1985); *see also* Cuban Am. Bar Ass'n v. Christopher, 43 F.3d 1412, 1427 (11th Cir. 1995) ("Aliens may be excluded . . . on grounds that might be 'suspect in the context of domestic legislation . . . .'" (citation omitted)).

13. "Exclusion" traditionally referred to prohibiting an alien from entering the United States; "deportation" referred to removing an alien already here. Current law now refers to both acts as "removal."

14. Cases decided in the circuits that have not addressed race directly nevertheless recognize the deference due to congressional decisions in this area. *See, e.g.*, Almario v. Attorney General, 872 F.2d 147, 150 (6th Cir. 1989) (noting limits to judicial review and citing cases authorizing racial discrimination).

15. Yamataya v. Fisher, 189 U.S. 86, 97 (1903); *see also* Harisiades v. Shaughnessy, 342 U.S. 580, 597 (1952) (Frankfurter, J., concurring) (arguing that the Court must defer to Congress even when immigration policy rests on "discredited racial theories[,] anti-Semitism or anti-Catholicism"); Oceanic Steam Navigation Co. v. Stranahan, 214 U.S. 320, 336 (1909); United States v. Ju Toy, 198 U.S. 253, 261 (1905); United States *ex rel.* Turner v. Williams, 194 U.S. 279, 291 (1904); Wong Wing v. United States, 163 U.S. 228, 237 (1896); *cf.* Reno v. Flores, 507 U.S. 292, 305–06 (1993) ("[I]n the exercise of its broad power over immigration and naturalization, Congress regularly makes rules that would be unacceptable if applied to citizens." (internal quotation marks omitted) (citations omitted)); Petitioner's Oral Argument 1997 WL 699809, at *16, Miller v. Albright, 118 S. Ct. 1428 (1998) (No. 96-1060) (question from unidentifed Justice) ("Now, Congress, however, under its immigration policies, can certainly discriminate on the basis of race, can't it, all the time? . . . It's done this from the beginning.").

16. Landon v. Plasencia, 459 U.S. 21, 32 (1982).

The Supreme Court's immigration jurisprudence represents the last vestige of an antique period of American law. The Supreme Court created constitutional immigration law under Chief Justice Melville Fuller in the waning years of the nineteenth century. In Fuller's first frenetic decade on the Court, he and his colleagues gave their imprimatur to racial segregation in *Plessy v. Ferguson*,[17] solidified the legality of statutory exclusion of women from public life in *In re Lockwood*,[18] and approved of the criminal punishment of members of an unpopular religion who tried to vote in the face of a statute disenfranchising them in *Davis v. Beason*.[19] The Fuller Court's immigration cases bear the same indicia of bigotry. In *Chae Chan Ping v. United States* (the *Chinese Exclusion Case*),[20] the Court held that resident aliens returning from overseas visits could be excluded on the basis of their race. In *Fong Yue Ting v. United States*,[21] the Court concluded that aliens lawfully living in the United States could be deported if Congress determined that their race was undesirable. The Congress that passed the laws at issue,[22] the Justice Department that defended them,[23] and the Supreme Court that ultimately held them constitutional[24] explicitly relied on the racial characteristics of the Chinese as justification for their action.

Fong Yue Ting and *Chae Chan Ping* are the foundation for what has come to be known as the plenary power doctrine, the rule that "the power of Congress over the admission of aliens to this country is absolute."[25] The Court is fond of saying that "'[o]ver no conceivable subject is the legislative power of Congress more complete than it is over' the admission of aliens,"[26] a sobering claim in a regime in which some powers of Congress are simply beyond judicial review.[27]

---

17. 163 U.S. 537 (1896), *overruled by* Brown v. Board of Educ., 347 U.S. 483 (1954).

18. 154 U.S. 116 (1894) (stating that a female member of the Supreme Court and District of Columbia bars could be denied admission to the Virginia bar). Here, the Court built on *Bradwell v. Illinois*, 83 U.S. (16 Wall.) 130, 137–39 (1872), which upheld the exclusion of a woman from admission to a state bar, and *Minor v. Happersett*, 88 U.S. (21 Wall.) 162, 178 (1874), which held that female citizens had no constitutional right to vote.

19. 133 U.S. 333 (1890) (upholding the conviction of a Mormon who tried to vote), *overruled by* Romer v. Evans, 517 U.S. 620, 634 (1996).

20. 130 U.S. 581 (1889).

21. 149 U.S. 698 (1893).

22. *See infra* notes 153–192 and accompanying text.

23. *See infra* notes 109–116 and accompanying text.

24. *See infra* notes 117–131 and accompanying text.

25. 4 ROTUNDA & NOWAK, *supra* note 3, § 22.2, at 598.

26. Fiallo v. Bell, 430 U.S. 787, 792 (1977) (quoting Oceanic Steam Navigation Co. v. Stranahan, 214 U.S. 320, 339 (1909)); *accord* Reno v. Flores, 507 U.S. 292, 305 (1993).

27. *See, e.g.*, Nixon v. United States, 506 U.S. 224, 226 (1993) (stating that the Senate has unreviewable discretion over impeachment proceedings).

Plessy, *Lockwood*, *Davis*, and other disgraceful cases of that era are not just dead but dishonored, usually discussed if at all as evidence of a lamentable history of bigotry in American law.[28] The cases that created the plenary power doctrine, by contrast, not only continue to be cited but, in the words of one distinguished authority, "said nearly everything the modern lawyer needs to know about the source and extent of Congress's power to regulate immigration."[29] Four decades after *Brown*, the plenary power doctrine is said to make racial discrimination in the immigration context lawful per se.[30]

Congress has not hesitated to use the plenary power recognized by the Supreme Court to discriminate. Many groups have been identified as undesirable—homosexuals,[31] Mormons,[32] the mentally retarded,[33] Southern and

28. *See, e.g.*, OWEN M. FISS, TROUBLED BEGINNINGS OF THE MODERN STATE, 1888–1910, ch. 12 (1993) (discussing, in a chapter entitled "Plessy, Alas," the errors of the Court); Carol Sanger, *Curriculum Vitae (Feminae): Biography and Early American Women Lawyers*, 46 STAN. L. REV. 1245 (1994) (discussing *Bradwell* and *Lockwood*).

Other cases notable for their intolerance may also be passing into legal history, leaving the plenary power cases even more anomalous. The notorious *Bowers v. Hardwick*, 478 U.S. 186 (1986), held that a state could, consistent with the Constitution, criminalize homosexual sodomy. Justice Scalia and Ronald Dworkin seem to believe that the Court's decision in *Romer v. Evans*, 517 U.S. 620 (1996), invalidating a state constitutional amendment that repealed laws prohibiting discrimination against homosexuals, begins the process of overruling *Bowers*. *See id.* at 631–33 (Scalia, J., dissenting) (arguing that the result in *Romer* is inconsistent with *Bowers*); Ronald Dworkin, *Sex, Death, and the Courts*, N.Y. REV. BOOKS, Aug. 8, 1996, at 44, 50. The Japanese-American incarceration cases, *Korematsu v. United States*, 323 U.S. 214 (1944), and *Hirabayashi v. United States*, 320 U.S. 81 (1943), likewise, recently received authoritative criticism in *Adarand Constructors, Inc. v. Pena*, 515 U.S. 200 (1995), although in a context not entirely congenial to many supporters of affirmative action. *See* Reggie Oh & Frank Wu, *The Evolution of Race in the Law: The Supreme Court Moves from Approving Internment of Japanese Americans to Disapproving Affirmative Action for African Americans*, 1 MICH. J. RACE & L. 165, 178 (1996); *see also infra* note 261 and accompanying text (noting the argument of Justice Stewart that internment cases were decided before the recognition of equal protection obligation on the national government).

29. T. Alexander Aleinikoff, *Federal Regulation of Aliens and the Constitution*, 83 AM. J. INT'L L. 862, 862 (1989).

30. *See supra* notes 4–12 and accompanying text.

31. *See* HOUSE JUDICIARY COMM., 100TH CONG., GROUNDS FOR EXCLUSION OF ALIENS UNDER THE IMMIGRATION AND NATIONALITY ACT: HISTORICAL BACKGROUND AND ANALYSIS 80–83 (Comm. Print 1988) [hereinafter GROUNDS FOR EXCLUSION]; *cf.* Larry Cata Backer, *Constructing a "Homosexual" for Constitutional Theory: Sodomy Narrative, Jurisprudence, and Antipathy in United States and British Courts*, 71 TUL. L. REV. 529 (1996) (discussing judicial construction of a negative image of homosexuals).

32. *See* GROUNDS FOR EXCLUSION, *supra* note 31, at 113 (describing historical exclusion of those who practice or "believe in" polygamy).

33. *See id.* at 77.

Eastern Europeans,[34] persons of African descent,[35] and Mexicans,[36] to name but a few. Indeed, as recently as April 1998, the Court refused to strike down a statute regulating the citizenship of children of American citizens born abroad that discriminated on the basis of sex and "illegitimacy."[37] However, the power to exclude aliens on the basis of race was used against only a single group—Asians. Through a series of laws in effect from 1882 to 1965, most people of Asian ancestry were barred from immigrating to the United States.[38]

The influence of the plenary power doctrine extends beyond racial and religious discrimination. For example, the Court has upheld lifetime detention of an alien on security grounds based on secret information without judicial or administrative review,[39] deportation proceedings conducted in English against a non-English-speaking alien who had the benefit of neither counsel nor translation,[40] as well as deportation of a long-term resident alien for lawful political activity.[41]

Immigration scholars love to hate the plenary power doctrine; they have argued forcefully for years that it should be reexamined or abolished, in

34. *See* Gabriel J. Chin, *The Civil Rights Revolution Comes to Immigration Law: A New Look at the Immigration and Nationality Act of 1965*, 75 N.C. L. REV. 273, 279 (1996) (describing the national origins quota system and its aim of reducing southern and eastern European immigration).

35. *See id.* at 280 (describing various techniques designed to reduce African immigration).

36. *See* Gilbert Paul Carrasco, *Latinos in the United States*, *in* IMMIGRANTS OUT! THE NEW NATIVISM AND THE ANTI-IMMIGRANT IMPULSE IN THE UNITED STATES 190, 190–204 (Juan F. Perea ed., 1997) (describing brutal and unfair tactics sometimes used to deport Mexicans) [hereinafter IMMIGRANTS OUT!]; Kevin R. Johnson, *The New Nativism: Something Old, Something New, Something Borrowed and Something Blue*, *in* IMMIGRANTS OUT!, *supra*, at 171–73; *see also* Rachel F. Moran, *Foreword—Demography and Distrust: The Latino Challenge to Civil Rights and Immigration Policy in the 1990s and Beyond*, 8 LA RAZA L.J. 1 (1995) (discussing contemporary nativism).

37. *See* Miller v. Albright, 118 S. Ct. 1428, 1432 (1998). For a further discussion of *Miller v. Albright*, see *infra* note 254. *See also* Fiallo v. Bell, 430 U.S. 787, 799–800 (1977) (upholding a statute discriminating on the basis of sex and illegitimacy in the granting of immigration privileges). *See generally* Joan Fitzpatrick, *The Gender Dimension of U.S. Immigration Policy*, 9 YALE J.L. & FEMINISM 23 (1997) (discussing how American immigration law disproportionately burdens women).

38. *See infra* notes 78–92 and accompanying text.

39. *See* Shaughnessy v. United States *ex rel.* Mezei, 345 U.S. 206, 214–15 (1953). For background about this case, see Charles D. Weisselberg, *The Exclusion and Detention of Aliens: Lessons from the Lives of Ellen Knauff and Ignatz Mezei*, 143 U. PA. L. REV. 933, 964–84 (1995).

40. *See* Yamayata v. Fisher, 189 U.S. 86, 101–02 (1903).

41. *See* Harisiades v. Shaughnessy, 342 U.S. 580 (1952) (finding that long-term resident aliens could be deported because of past membership in the Communist Party, even though membership was not a ground for deportation when the aliens were members). *See generally* Kevin R. Johnson, *The Antiterrorism Act, the Immigration Reform Act, and Ideological Regulation in the Immigration Laws: Important Lessons for Citizens and Noncitizens*, 28 ST. MARY'S L.J. 833 (1997) (describing historical and present ideological restrictions on immigration).

general[42] and in particular contexts.[43] While the plenary power doctrine is, as others have pointed out, "aberrational,"[44] "a maverick, a wild card,"[45] "a constitutional fossil,"[46] an "oddity,"[47] theoretically unsatisfying,[48] and inconsistent with modern international law,[49] the implications of its disreputable

42. *See, e.g.*, STEPHEN H. LEGOMSKY, IMMIGRATION AND THE JUDICIARY: LAW AND POLITICS IN BRITAIN AND AMERICA (1987); Louis Henkin, *The Constitution and United States Sovereignty: A Century of* Chinese *Exclusion and Its Progeny*, 100 HARV. L. REV. 853, 885–86 (1987); Stephen H. Legomsky, *Ten More Years of Plenary Power: Immigration, Congress, and the Courts*, 22 HASTINGS CONST. L.Q. 925, 937 (1996); Stephen H. Legomsky, *Immigration Law and the Principle of Plenary Congressional Power*, 1984 SUP. CT. REV. 255, 306–07 [hereinafter Legomsky, *Plenary Congressional Power*]; Hiroshi Motomura, *Immigration Law After a Century of Plenary Power: Phantom Constitutional Norms and Statutory Interpretation*, 100 YALE L.J. 545, 607–13 (1990) [hereinafter, Motomura, *Phantom Norms*]; Hiroshi Motomura, *The Curious Evolution of Immigration Law: Procedural Surrogates for Substantive Constitutional Rights*, 92 COLUM. L. REV. 1625, 1699–1704 (1992); James A.R. Nafziger, *The General Admission of Aliens Under International Law*, 77 AM. J. INT'L L. 804, 845–47 (1983); Victor C. Romero, *The Congruence Principle Applied: Rethinking Equal Protection Review of Federal Alienage Classifications After* Adarand Constructors, Inc. v. Peña, 76 OR. L. REV. 425 (1997); Michael Scaperlanda, *Polishing the Tarnished Golden Door*, 1993 WISC. L. REV. 965, 1022–31; Peter H. Schuck, *The Transformation of Immigration Law*, 84 COLUM. L. REV. 1, 73–90 (1984); Samuel M. Silvers, *The Exclusion and Expulsion of Homosexual Aliens*, 15 COLUM. HUM. RTS. L. REV. 295, 322 (1984); Frank H. Wu, *The Limits of Borders: A Moderate Proposal for Immigration Reform*, 7 STAN. L. & POL'Y REV. 35, 48 (1996); *see also* Denise M. Fabiano, Note, *Immigration Law—*Flores v. Meese*: A Lost Opportunity to Reconsider the Plenary Power Doctrine in Immigration Decisions*, 14 W. NEW ENG. L. REV. 257, 293 (1992); Richard F. Hahn, Note, *Constitutional Limits on the Power to Exclude Aliens*, 82 COLUM. L. REV. 957, 997 (1982).

43. *See, e.g.*, Linda Kelly, *Preserving the Fundamental Right to Family Unity: Championing Notions of Social Contract and Community Ties in the Battle of Plenary Power Versus Aliens' Rights*, 41 VILL. L. REV. 725, 771–83 (1996); Kikuyo Matsumoto-Power, *Aliens, Resident Aliens, and U.S. Citizens in the Never-Never Land of the Immigration and Nationality Act*, 15 U. HAW. L. REV. 61, 64–80 (1993) (arguing that sham marriage investigation procedures are unfair); James A.R. Nafziger, *Review of Visa Denials by Consular Officers*, 66 WASH. L. REV. 1, 38–46 (1991) (stat-ing that visa denials should be reviewable notwithstanding the existence of the plenary power doctrine); Jim Rosenfeld, *Deportation Proceedings and Due Process of Law*, 26 COLUM. HUM. RTS. L. REV. 713, 739–41 (1995) (criticizing the judicial understanding of due process in deportation); Margaret H. Taylor, *Detained Aliens Challenging Conditions of Confinement and the Porous Border of the Plenary Power Doctrine*, 22 HASTINGS CONST. L.Q. 1087, 1156–58 (1995) (arguing that the plenary power doctrine should not preclude challenges by aliens to conditions of confinement); Andrés Snaider, Note, *The Politics and Tension in Delegating Plenary Power: The Need to Revive Nondelegation Principles in the Field of Immigration*, 6 GEO. IMMIGR. L.J. 107, 126–27 (1992) (arguing that the plenary power doctrine should be mediated by the nondelegation doctrine)

44. Motomura, *Phantom Norms*, *supra* note 42, at 549.

45. Schuck, *supra* note 42, at 1.

46. Henkin, *supra* note 42, at 862.

47. Legomsky, *Plenary Congressional Power*, *supra* note 42, at 255.

48. *See, e.g.*, Hahn, *supra* note 42.

49. *See* Berta Esperanza Hernández-Truyol, *Natives, Newcomers and Nativism: A Human Rights Model for the Twenty-First Century*, 23 FORDHAM URB. L.J. 1075 (1996); Nafziger, *supra* note 42, at 816–23; Scaperlanda, *supra* note 42, 1022–31.

birth in cases authorizing racial discrimination—and sympathetic to that discrimination—have not been fully explored.[50]

Reading the plenary power cases as race cases offers several useful perspectives. First, the racial composition of the immigration stream is once again a significant public policy issue. Anti-immigration activists such as Peter Brimelow, who support reintroduction of racial criteria to America's immigration laws,[51] receive respectful hearings before congressional committees and influence public opinion on immigration law and policy.[52] Reflecting the potential seriousness of the issue, a leading immigration law casebook asks students to evaluate the constitutionality of a hypothetical racial restriction.[53]

Congress recently granted statutory authority to engage in race discrimination in processing visa applications. Although the Immigration and Nationality Act (INA) generally prohibits discrimination on the basis of "race, sex, nationality, place of birth, or place of residence,"[54] the Illegal Immigration Reform and Immigrant Responsibility Act of 1996[55] created a substantial exception, authorizing the State Department to use race (as well as religion, sex, and other factors) in establishing visa application procedures and locations. Thus, a future secretary of state could permit would-be immigrants of African or Asian descent (or, for that matter, who are Jews, Muslims, or Scientologists) to apply only to the consulate in Berne or Sao Paolo, or impose procedural requirements that would make it practically impossible for them to receive visas. Although it seems unlikely that any secretary of state would do so, the INA itself currently presents the issue of whether the plenary power doctrine continues to permit classification of

50. This is not to say that the racial origins of the plenary power cases have gone unnoticed, but generally scholars have pointed out aspects of the discriminatory history in the course of projects analyzing other features of the immigration laws. *See, e.g.*, Nancy Morawetz, *Rethinking Retroactive Deportation Laws and the Due Process Clause*, 73 N.Y.U. L. REV. 97, 98 n.9 (1998) (noting near-contemporaneity of *Plessy* and plenary power cases); Motomura, *Phantom Norms*, *supra* note 42, at 554–55; Michael Scaperlanda, *Partial Membership: Aliens and the Constitutional Community*, 81 IOWA L. REV. 707, 725 (1996) ("Plenary immigration power was founded on the back of blatant racial discrimination.").

51. *See* PETER BRIMELOW, ALIEN NATION: COMMON SENSE ABOUT AMERICA'S IMMIGRATION DISASTER 78 (1995); *see also* LAWRENCE AUSTER, THE PATH TO NATIONAL SUICIDE: AN ESSAY ON MULTICULTURALISM AND IMMIGRATION (1990).

52. *See Legal Immigration Reform Proposals: Hearing Before the Subcomm. on Immigration and Claims of the Judiciary Comm., House of Representatives*, 104th Cong., 28–42 (1995) (testimony of Peter Brimelow).

53. *See* STEPHEN H. LEGOMSKY, IMMIGRATION AND REFUGEE LAW AND POLICY 94–96 (2d ed. 1997).

54. Immigration and Nationality Act § 202(a)(1)(A), 8 U.S.C. § 1152(a)(1)(A).

55. Pub. L. No. 104-208, subtit. c, § 633, 110 Stat. 3009-701 (codified at Immigration and Nationality Act § 202(a)(1)(B), 8 U.S.C. § 1152(a)(1)(B) (Supp. 1998)).

aliens by race. Perhaps more importantly, treating the immigration cases as race cases offers a new critique of the plenary power doctrine in all of its applications. While the Court has shown little inclination to abandon the plenary power doctrine, underscoring the racial origins of the doctrine ought to give the Court pause.

Part I of this Article discusses the origins of the plenary power doctrine.[56] The Supreme Court has said that stare decisis is a central rationale for retention of an analytical approach at odds with all other areas of judicial review.[57] The precedents, however, are tainted; the cases that created the plenary power doctrine suggest racial bias on the part of the Justice Department that prosecuted the cases and the Justices of the Supreme Court who decided them.[58] The plenary power doctrine cases may have been decided as they were because of the race of the aliens involved. Thus, we cannot be sure how a Supreme Court unburdened by invidious prejudice would have approached the question.

Part I proposes that the statutes approved in the plenary power cases were neither less invidious nor more justifiable than the discrimination approved in *Plessy*. When debating the Chinese Exclusion Act, many legislators candidly appealed to Chinese racial inferiority and the importance of protecting white culture. Because they were motivated by racism rather than economic protectionism or other legitimate reasons, the exercises in racial classification justified by the plenary power doctrine were similar to the domestic racial classifications that are widely recognized as illegitimate.[59]

Part I also suggests that Asian exclusion had effects similar to domestic racial discrimination, by disadvantaging a group of nonwhite Americans. Just as the Jim Crow laws approved by *Plessy* had the purpose and effect of excluding African Americans from the benefits of citizenship, the Asian Exclusion Laws were intended to minimize Asian-American population and influence, and they succeeded, ensuring that Asian Americans would remain, as they are today, a tiny minority.[60] In sum, the race-based laws upheld under the plenary power doctrine had the same purpose and effect as domestic racial discrimination, namely, promoting white supremacy.[61] The

56. *See infra* notes 68–95 and accompanying text.
57. *See infra* notes 96–99 and accompanying text.
58. *See infra* notes 100–131 and accompanying text.
59. *See infra* notes 132–183 and accompanying text.
60. *See infra* notes 184–240 and accompanying text.
61. The term "white supremacy" is used here advisedly because it may have a somewhat hyperbolic ring today. Yet, court decisions and political actors discussing segregation and other race-conscious laws used this term. *See, e.g.*, Loving v. Virginia, 388 U.S. 1, 11 (1967) (noting that miscegenation laws were "designed to maintain White Supremacy"); Louisiana v. United

similarities between racial classifications in immigration law and in domestic contexts suggest that there is no reason not to apply comparable legal standards to both areas.

Part II puts aside the racial motivation of those who created the plenary power doctrine and evaluates the doctrine on its own terms. As a technical matter, the Court's reliance on precedent to sustain the doctrine is tenuous because the plenary power doctrine has not been consistently or coherently applied. Indeed, many of the outcomes in the old immigration cases have been rejected even as the Court sustained the general principle that Congress has plenary power over immigration.[62]

Moreover, the internal logic of the plenary power cases themselves suggests that the plenary power doctrine should now change, at least to the extent that it would no longer justify discrimination on suspect grounds. It would be most satisfying to conclude that the recognition of Fifth Amendment equal protection review in *Bolling v. Sharpe*[63] casts doubt on the old cases, but so far the Court has refused to apply domestic equal protection principles to immigration law. Thus, without questioning the general validity of the plenary power doctrine, it is not possible to argue that principles like strict scrutiny of suspect classifications should apply directly in this context. However, the Court itself has suggested that international law supplies the content of plenary power.[64] Limits on congressional authority logically may be found there. Part II suggests that international law now disfavors racial discrimination.[65]

Further, in recent decades, the Supreme Court has recognized that some level of judicial review over immigration decisions is permitted even under the plenary power doctrine; even immigration decisions will be subjected to judicial review at the rational basis level.[66] Part II also argues that racial discrimination would not now be permitted under the rational basis standard.[67]

States, 380 U.S. 145, 149 (1965) (noting that after *Brown*, state legislature and "Citizen's Councils" cooperated "to preserve white supremacy"); Gaston Co. v. United States, 288 F. Supp. 678, 684 n.11 (D.D.C. 1968) (three-judge court) (quoting a 1900 *Raleigh News & Observer* article announcing "White Supremacy Made Permanent" after North Carolina adopted a literacy test for voting), *aff'd*, 395 U.S. 285 (1969); Elmore v. Rice, 72 F. Supp. 516, 520 (E.D.S.C.) (quoting a remark of South Carolina's governor that "we South Carolinians will use the necessary methods to retain white supremacy in our primaries and to safeguard the homes and happiness of our people"), *aff'd*, 165 F.2d 387 (4th Cir. 1947).

62. *See infra* notes 253–276 and accompanying text.
63. 347 U.S. 497 (1954).
64. *See infra* notes 277–286 and accompanying text.
65. *See infra* notes 287–298 and accompanying text.
66. *See infra* notes 299–323 and accompanying text.
67. *See infra* notes 324–361 and accompanying text.

This Article concludes that the plenary power doctrine is not just old, it is anachronistic. It has been so thoroughly undermined by its creation to serve white supremacy, changes in international law, and changes in the Court's understanding of judicial review, that there is virtually nothing left of the foundational cases. Under these circumstances, it is time for the Court to consider revisiting the doctrine.

I. THE RACIAL ORIGINS OF THE PLENARY POWER DOCTRINE

In *Chae Chan Ping v. United States*,[68] the Court held that Congress had a sovereign power to regulate immigration, not enumerated in the Constitution, that could be exercised to exclude aliens of a particular race, and was not subject to judicial review.[69] The circumstances of the case illustrate the potency of the plenary power the Court recognized. Chae Chan Ping was a legal alien returning from an overseas visit who had lived in San Francisco for over a decade. The Chinese Exclusion Act prohibited Chinese people, like Chae Chan Ping, from entering the United States, but he "left . . . having in his possession a certificate . . . entitling him to return to the United States . . . issued to him by the collector of customs of the port of San Francisco" pursuant to statute.[70] Chae Chan Ping was sailing home but had not reached port when Congress voided the reentry certificates with immediate effect. Congress made no exception for those like Chae Chan Ping, who left the country in reliance on the government's assurance that they could return.[71]

Although the Court "conceded that [the statute at issue] is in contravention of express stipulations" of two treaties between the United States and China,[72] it unanimously concluded that the action was within the power of Congress, and thus that Chae Chan Ping could be excluded from the United States.[73] Implicitly, the Court upheld the constitutionality of the Chinese Exclusion Act itself, which prohibited the immigration of virtually all persons of Chinese ancestry.

Four years later, in *Fong Yue Ting v. United States*,[74] the Court upheld a requirement that Chinese residents of the United States, and they alone,

68. 130 U.S. 581 (1889).
69. *See id.* at 606.
70. *Id.* at 582.
71. *See id.*
72. *Id.* at 600.
73. *See id.* at 609.
74. 149 U.S. 698 (1893).

register with the federal government upon pain of deportation.[75] In the course of its analysis, the Court determined that aliens could be deported solely because of their race.

In both cases, the Court seemed to say that it simply had no power of judicial review over matters involving immigration.[76] Since then, the Court has repeatedly held that the plenary power includes the right to exclude aliens on the basis of their race.[77]

Congress has frequently drawn lines on the basis of race in immigration and naturalization law. The first naturalization act, in 1790, permitted only free white persons to become naturalized citizens;[78] persons of African nativity and descent were added in 1870.[79] When persons of "races indigenous to the Western Hemisphere" were added in 1940,[80] only members of Asian races could not naturalize.

Congress prohibited Asian immigration from Asian and non-Asian countries in a series of statutes and treaties referred to here as the Asian Exclusion Laws. The Chinese Exclusion Act of 1882,[81] upheld in *Chae Chan Ping*, was the first express racial restriction. The beginning of Japanese immigration around 1900 resulted in the Gentlemen's Agreement of 1907–08 under which Japan promised not to issue travel documents to those planning to come to America.[82] In 1917, after Indians began

75. *See id.* at 727–28. A general requirement of alien registration was not part of American law until 1940. *See* Alien Registration Act of 1940, ch. 439, 54 Stat. 670.

76. *See Fong Yue Ting*, 149 U.S. at 712 (declaring that "final decision" over immigration "has been committed by the Constitution to the other departments of the government"). The Court explained:

> The question whether, and upon what conditions, these aliens shall be permitted to remain within the United States being one to be determined by the political departments of the government, the judicial department cannot properly express an opinion upon the wisdom, the policy, or the justice of the measures enacted by Congress in the exercise of the powers confided to it by the Constitution over this subject.

Id. at 731; *see also Chae Chan Ping*, 130 U.S. at 606 (stating that congressional "determination is conclusive upon the judiciary").

77. *See* cases cited *supra* note 15.

78. *See* Act of Mar. 26, 1790, ch. 3, 1 Stat. 103 (repealed 1795).

79. *See* Act of July 14, 1870, ch. 254, § 7, 16 Stat. 254, 256.

80. Nationality Act of 1940, § 303, 54 Stat. 1137, 1140. Mexicans in territory ceded to the United States after the Mexican war were offered citizenship under the Treaty of Guadalupe-Hidalgo, Feb. 2, 1848, U.S.-Mex., arts. 8–9, 9 Stat. 922, 929–30 (1848), and Indians were naturalized as a group by statute in 1924, *see* Act of June 2, 1924, ch. 233, 43 Stat. 253.

81. Act of May 6, 1882, ch. 126, 22 Stat. 58, *repealed by* Act of Dec. 17, 1943, ch. 344, 57 Stat. 600.

82. *See Memorandum by the Division of Far Eastern Affairs*, *in* 2 U.S. DEPT. OF STATE, PAPERS RELATING TO THE FOREIGN RELATIONS OF THE UNITED STATES 1924, at 339 (1939) (attaching portions of the Gentlemen's Agreement as an appendix), *reprinted in part in* BILL ONG HING, MAKING AND REMAKING ASIAN AMERICA THROUGH IMMIGRATION POLICY, 1850–1990, at 207–12 (1993).

immigrating in quantity, Congress adopted a more comprehensive solution, prohibiting immigration of all races indigenous to the so-called Asiatic Barred Zone.[83] This process culminated in the Immigration Act of 1924,[84] which tied immigration to the right to naturalization. "Aliens ineligible to citizenship" were excluded entirely under the 1924 law. This phrase was a euphemism for Asians because the act defined an alien "ineligible to citizenship" as one covered by the Chinese Exclusion Act or racially ineligible to naturalize.[85] After 1924, then, most Asians could not immigrate,[86] and almost all who were already here were barred from becoming naturalized citizens. The Asian Exclusion Laws operated on the basis of race, not nationality or place of birth. Thus, a person of Japanese ancestry born and raised in Brazil would be treated as Japanese, not Brazilian.[87]

In 1943, Congress allowed a handful of Chinese to immigrate, and permitted those here to naturalize;[88] in 1946, Filipinos and Indians received these privileges.[89] In 1952, Congress lifted the remaining absolute bars to immigration and naturalization of other Asians.[90] Asians, however, continued to be disadvantaged with low quotas—105 Chinese per year, for example—and rules charging persons of Asian ancestry to the small Asian

83. *See* Act of Feb. 5, 1917, ch. 29, § 3, 39 Stat. 874, 876, created the Asiatic Barred Zone, prohibiting immigration of natives or descendants of natives of many continental Asian countries. "The barred zone includes the East Indies, western China, French Indochina, Siam, Burma, India, Bhutan, Nepal, eastern Afghanistan, Turkestan, the Kirghiz Steppe, and the southeastern portion of the Arabian Peninsula. This bar does not apply to white persons born in the zone . . . ." SENATE JUDICIARY COMM., THE IMMIGRATION AND NATURALIZATION SYSTEMS OF THE UNITED STATES, S. REP. NO. 81-1515, at 67 (1950). Chinese were already barred by the Chinese Exclusion Act; Japanese immigrants were covered by the Gentlemen's Agreement, which, however, was not particularly effective in eliminating Japanese immigration. *See* ROGER DANIELS, THE POLITICS OF PREJUDICE 44–45 (2d ed. 1977).

84. Immigration Act of 1924, ch. 190, 43 Stat. 153.

85. *Id.* § 28(c), 43 Stat. at 168.

86. Asian ministers, teachers, tourists, and students could enter, and lawful residents could return from trips abroad. *See id.* §§ 4 & 13(a), 43 Stat. at 155, 161–62. These and other modest exceptions—as well as illegal immigration—meant that the influx of Asian immigrants was never entirely shut off.

87. *See* Hitai v. INS, 343 F.2d 466, 467 (2d Cir. 1965).

88. *See* Act of Dec. 17, 1943, ch. 344, §§ 2–3, 57 Stat. 600–01 (establishing an annual quota of 105 persons of Chinese ancestry and making Chinese ineligible to naturalize).

89. *See* Act of July 2, 1946, ch. 534, 60 Stat. 416.

90. *See* Immigration and Nationality Act of 1952, ch. 477, § 201(a), 66 Stat. 163, 175 (establishing quotas for Asian countries); *id.* § 311, 66 Stat. at 239 ("The right of a person to become a naturalized citizen of the United States shall not be denied or abridged because of race or sex . . . .").

quotas regardless of their citizenship or nativity.[91] In 1965, Congress eliminated these last forms of anti-Asian racial policy.[92]

Under current law, no races are explicitly favored in the awarding of immigrant or nonimmigrant visas, and many believe that no particular nations are advantaged or disadvantaged as an indirect means of racial preference.[93] Yet, the power to select immigrants on the basis of race is said to remain at the ready. *Chae Chan Ping* and *Fong Yue Ting* continue to be cited in modern decisions of the Supreme Court;[94] because all constitutional immigration law flows from these cases, even decisions that do not cite them must rely on cases that do.[95]

Like baseball's antitrust exemption,[96] the plenary power doctrine's longevity is attributable in part to "the enormous force . . . of almost a century of precedent."[97] In *Galvan v. Press*,[98] Justice Frankfurter, writing for the Court, recognized that the plenary power doctrine was inconsistent with

---

91. For an example of the operation of the racial attribution rules for Asian immigrants, see *Wagio Kong Tjauw Wong v. Esperdy*, 214 F. Supp. 264, 266–69 (S.D.N.Y. 1963).

92. *See* Immigration and Nationality Act Amendments of 1965, Pub. L. No. 89-236, 79 Stat. 911 (codified as amended in scattered sections of 8 U.S.C.); Chin, *supra* note 34.

93. There is controversy over whether the current laws are nondiscriminatory because even the landmark 1965 amendments retained per-country limits. Now, for example, no country may receive more than 7% of the quota numbers. *See* Immigration and Nationality Act § 202(a)(2), 8 U.S.C. § 1152(a)(2). As a result, would-be immigrants from high immigration demand countries, such as Mexico and the Philippines, may have to wait years for their number to come up, while an otherwise similarly situated native of another country who is eligible for the same immigration status might be able to enter immediately. *See* Howard F. Chang, *Immigration Policy, Liberal Principles, and the Republican Tradition*, 85 GEO. L.J. 2105, 2108 (1997); Jan C. Ting, *"Other Than a Chinaman": How U.S. Immigration Law Resulted from and Still Reflects a Policy of Excluding and Restricting Asian Immigrantion*, 4 TEMP. POL. & CIV. RTS. L. REV. 301, 308 (1995) (noting that the limit is imposed "regardless of the size of its population, and regardless of the number of qualified immigrants waiting for visas").

94. *See* Browning-Ferris Indus. v. Kelco Disposal, Inc., 492 U.S. 257, 263 (1989) (citing *Fong Yue Ting*); INS v. Lopez-Mendoza, 468 U.S. 1032, 1038 (1984) (citing *Fong Yue Ting*); Fiallo v. Bell, 430 U.S. 787, 792 (1977) (citing *Fong Yue Ting* and *Chae Chan Ping*). As Professor Henkin explained: "The Supreme Court has never reexamined" *Chae Chan Ping*, and "it has shown no disposition to do so." Henkin, *supra* note 42, at 854.

95. For example, in *Reno v. Flores*, 507 U.S. 292 (1993), all five of the cases the Court cited for the scope of judicial review of immigration statutes in turn cited to *Fong Yue Ting* or *Chae Chan Ping*, or other cases that did. *See id.* at 305–06 (citing *Fiallo*, 430 U.S. at 792; Mathews v. Diaz, 426 U.S. 67, 81 n.17 (1976); Carlson v. Landon, 342 U.S. 524, 534 n.18, 537 n.27 (1952); Oceanic Steam Navigation Co. v. Stranahan, 214 U.S. 320, 336 (1909); Wong Wing v. United States, 163 U.S. 228, 231, 236 (1896)).

96. *See* Flood v. Kuhn, 407 U.S. 258, 283–84 (1972) (reaffirming, on stare decisis grounds, *Toolson v. N.Y. Yankees, Inc.*, 346 U.S. 356 (1953), and *Federal Baseball Club Inc. v. National League of Prof'l Baseball Clubs*, 259 U.S. 200 (1922), which held that baseball was not subject to antitrust laws because it was not commerce).

97. Legomsky, *Plenary Congressional Power*, *supra* note 42, at 291.

98. 347 U.S. 522 (1954).

modern constitutional law, but concluded it was too firmly entrenched to be reconsidered.

> In light of the expansion of the concept of substantive due process as a limitation upon all powers of Congress, even the war power, much could be said for the view, were we writing on a clean slate, that the Due Process Clause qualifies the scope of political discretion heretofore recognized as belonging to Congress in regulating the entry and deportation of aliens. . . .
>
> But the slate is not clean. As to the extent of the power of Congress under review, there is not merely "a page of history," but a whole volume. Policies pertaining to the entry of aliens and their right to remain here are peculiarly concerned with the political conduct of government. . . . [T]hat the formulation of these policies is entrusted exclusively to Congress has become about as firmly embedded in the legislative and judicial tissues of our body politic as any aspect of our government.[99]

Stare decisis is an established and important prudential principle. However, the plenary power doctrine appears to have been motivated by racism, which history has shown to be a sound basis for departing from precedent.

A. Racial Bias of the Justice Department and the Court in the Plenary Power Cases

At the oral argument in *Brown*, John W. Davis, Topeka's lawyer, urged the Court to refuse to reconsider *Plessy*; he explained, "somewhere, sometime to every principle comes a moment of repose when it has been so often announced, so confidently relied upon, so long continued, that it passes the limits of judicial discretion and disturbance."[100] Davis insisted that racial segregation was such a rule; of course, Davis was wrong.

---

99. *Id.* at 530–31 (citations omitted) (quoting N.Y. Trust Co. v. Eisner, 256 U.S. 345, 349 (1921)); *see also* United States v. Valenzuela-Bernal, 458 U.S. 858, 875 n.1 (1982) (O'Connor, J., concurring) (citing *Galvan*, 347 U.S. at 531); *Fiallo*, 430 U.S. at 792 n.4 ("We are [not] inclined to reconsider this line of cases.") (citing *Galvan*, 347 U.S. at 530–32); Kleindienst v. Mandel, 408 U.S. 753, 767 (1972) ("We are not inclined in the present context to reconsider this line of cases.") (citing *Galvan*, 347 U.S. at 531–32); Marcello v. Bonds, 349 U.S. 302, 314 (1955) ("We perceive no special reasons . . . for overturning our precedents on this matter . . . .") (citing *Galvan*, 347 U.S. at 522). As is developed below, the Court has not really followed its precedents with fetishistic consistency. *See infra* notes 253–276 and accompanying text.

100. Oral Argument, John W. Davis, *in* 49A LANDMARK BRIEFS AND ARGUMENTS OF THE SUPREME COURT OF THE UNITED STATES: CONSTITUTIONAL LAW 490 (Philip B. Kurland & Gerhard Casper eds., 1975) [hereinafter LANDMARK BRIEFS].

Cases authorizing racial discrimination have proved to be the leading example of precedent vulnerable to overruling. The Court that bowed to precedent in *Galvan* had reached the opposite result in *Brown* and *Bolling*[101] just seven days earlier. Racial segregation was well entrenched doctrinally and socially; *Brown* repudiated a long line of cases,[102] as did *Bolling*.[103] In *Galvan*, the Court modestly declared that "[w]e are not prepared to deem ourselves wiser or more sensitive to human rights than our predecessors."[104] But in the segregation decisions, they correctly found that the Fuller Court's wisdom and sensitivity to human rights were deeply flawed.

Hunter v. Underwood[105] is another example of judicial suspicion of legal rule making potentially distorted by racism. In *Hunter*, the Court held unconstitutional a rule that was facially neutral and that accomplished a substantively permissible end because it had become law as a result of invidious prejudice.[106] The law disenfranchised people convicted of certain crimes; it was designed to keep African Americans from voting.[107] The Court declined to speculate what the state would have done without the impermissible motive; if the state chose to reenact the law, it would have to do so without the distorting influence of racial prejudice.[108]

---

101. 347 U.S. 497 (1954) (holding school segregation in the District of Columbia violated the equal protection component of the Due Process Clause of the Fifth Amendment).

102. *See* McLaurin v. Oklahoma, 339 U.S. 637, 642 (1950) (declaring separate educational opportunity not equal in the facts of the case); Sweatt v. Painter, 339 U.S. 629, 636 (1950) (same); Missouri *ex rel.* Gaines v. Canada, 305 U.S. 337, 349–52 (1938) (same); Gong Lum v. Rice, 275 U.S. 78, 85–87 (1927) (holding that a Chinese-American student was not entitled to attend a white school); McCabe v. Atchison, Topeka & Santa Fe Ry. Co., 235 U.S. 151, 160 (1914) (upholding separate but equal transportation facilities); Berea College v. Kentucky, 211 U.S. 45 (1908) (stating that a school could be prohibited from offering integrated programs); Cumming v. County Bd. of Educ., 175 U.S. 528, 545 (1899) (Harlan, J.); Plessy v. Ferguson, 163 U.S. 537, 550–51 (1896). Of course, *Plessy*'s repudiation did not come from nowhere; it had been attacked over time by the methodical litigation campaign described in RICHARD KLUGER, SIMPLE JUSTICE (1976).

103. Mantra-like, the Court had repeated in a series of cases that the "Fifth Amendment has no equal protection clause." La Belle Iron Works v. United States, 256 U.S. 377, 392 (1921); *see also* Hirabayashi v. United States, 320 U.S. 81, 100 (1943); Detroit Bank v. United States, 317 U.S. 329, 337–38 (1943); Helvering v. Lerner Stores Corp., 314 U.S. 463, 468 (1941); Sunshine Anthracite Coal Co. v. Adkins, 310 U.S. 381, 401 (1940); Paramino Lumber Co. v. Marshall, 309 U.S. 370, 379–80 (1940); Currin v. Wallace, 306 U.S. 1, 14 (1939); United States v. Carolene Products Co., 304 U.S. 144, 151 (1938); Steward Mach. Co. v. Davis, 301 U.S. 548, 584 (1937); Heiner v. Donnan, 285 U.S. 312, 338 (1932) (Stone, J., dissenting).

104. *Galvan*, 347 U.S. at 531.

105. 471 U.S. 222 (1985).

106. *See id.* at 226–29.

107. *See id.* at 229.

108. *See id.* at 232–33. *See also* McLaughlin v. Florida, 379 U.S. 184, 188 (1964) (striking down racial classification and rejecting state court's reliance on *Pace v. Alabama*, 106 U.S. 583 (1883)) ("*Pace* represents a limited view of the Equal Protection Clause which has not withstood analysis in the subsequent decisions of this Court.").

The plenary power cases betray the same kinds of flaws in formation as did the provision at issue in *Hunter*; it is by no means inevitable that the cases would have come out as they did if the aliens involved had not been people of color.[109] At one level, the solicitor general's brief in *Fong Yue Ting* advanced an ordinary doctrinal argument for the color-blind principle that the United States had the power to expel aliens for any reason.[110] At the same time, the brief reflects the Justice Department's candid reliance on racial justifications for legal action; in particular, that nations could justly prefer one race over another. The solicitor exhaustively demonstrated that racial expulsions were an accepted part of domestic policy around the world, noting that the English had expelled Egyptians and Turks,[111] suggesting that "[a]nalogies might also be drawn from the forcible removal by the United States of Indians [sic],"[112] and reminding the Court that "Russia is now engaged in expelling the Jews, and laws looking to a limitation upon their rights and regulating their status have been in recent years a source of party strife in the German Empire."[113] While pogrom, racism, brutality toward Native Americans, and "regulating" the status of Jews in Germany were apparently reassuring precedents to a majority of the Court in 1893, they are now a telling, even chilling, impeachment of the solicitor's position.

The Justice Department also argued that the actions of Congress should be upheld because of the racial characteristics of the people who were the objects of the law. The solicitor insisted that Chinese were excluded from citizenship not because of racial prejudice, but because their race made it impossible for them to be citizens: "They cannot, by reason of their habits, become citizens in thought, and they are, therefore, forbidden by law to become such in fact."[114] The solicitor also reminded the Court that the Chinese are

> a people not suited to our institutions, remaining a separate and distinct race, incapable of assimilation, having habits often of the most

109. *Fong Yue Ting*, for example, reflects Justice Brewer's evident belief that it was a powerful critique of the majority's position that the deportation power might extend beyond Chinese; "[i]t is true this statute is directed only against the obnoxious Chinese; but if the power exists, who shall say that it will not be exercised to-morrow against other classes and other people?" Fong Yue Ting v. United States, 149 U.S. 698, 743 (1893) (Brewer, J., dissenting).

110. As did the government's brief in *Chae Chan Ping*, which was a short doctrinal treatment. *See* Brief for the United States, Chae Chan Ping v. United States, 130 U.S. 581 (1889) (No. 88-1446), *reprinted in* 9 LANDMARK BRIEFS, *supra* note 100, at 255.

111. *See* Brief for the Respondents at 28–29, *Fong Yue Ting*, 149 U.S. at 698 (Nos. 92-1345, -1346, -1347).

112. *Id.* at 30.

113. *Id.* at 54.

114. *Id.* at 39.

> pernicious character, working at wages that debase our own laboring classes, not bound by any considerations of the sanctity of an oath, given to evasions of other laws of Congress, and by that body, . . . declared to be a people of such a character and so inimical to our interests as to require that their coming shall be prohibited . . . . To such a people, so described both by the learned members of this court and by the Congress of the United States, this law is to be applied.[115]

The solicitor also pointed out that any rule had to be drawn with the nature of the litigants in mind:

> In view of past experiences of civilized communities with hordes of barbarians, it does not seem possible that a rule never yet ingrafted upon this principle of self-preservation will have its origin in this court.
>
> Again, it is now generally conceded that the most insidious and dangerous enemies to the State are not the armed foes who invade our territory, but those alien races who are incapable of assimilation, and come among us to debase our labor and poison the health and morals of the communities in which they locate.[116]

The solicitor had reason to expect that his arguments would be sympathetically received by some members of the Court.[117] One member of the *Chae Chan Ping* Court, Lucius Q.C. Lamar, supported the Chinese Exclusion Act in 1882 as one of Mississippi's senators,[118] suggesting that he thought Chinese exclusion was wise as well as constitutional.[119] Justice Howell E. Jackson, part of the *Fong Yue Ting* majority, also supported

115. *Id.* at 12.

116. *Id.* at 55.

117. The *Chae Chan Ping* Court was shorthanded; Justice Stanley Matthews had died and David Brewer had not yet taken his seat. This circumstance may have deprived posterity of a dissent; Matthews, the author of *Yick Wo v. Hopkins*, 118 U.S. 356 (1886), opposed an earlier version of the Chinese Exclusion Act while in the Senate. *See, e.g.*, 8 CONG. REC. 1400 (1879) (reflecting Matthews's vote against the bill). Brewer was the most faithful judicial friend of Asian immigrants in the United States. *See* Gabriel J. Chin, *The Plessy Myth: Justice Harlan and the Chinese Cases*, 82 IOWA L. REV. 151, 165 n.95 (1996).

118. *See* 13 CONG. REC. 3412 (1882) (Lamar was not present for the vote, but made his position known by "pairing"—agreeing in advance not to show up and vote in person—with a senator who opposed the bill. A colleague announced the positions of each. Through the pair device, the two senators were relieved of having to appear in person, avoided having their absences affect the outcome, and set out their positions on the record).

119. Lamar, a slave owner, left Congress to campaign for secession; during the Civil War, he served in military and political positions in the Confederate government. *See* Arnold M. Paul, *Lucius Quintus Cincinnatus Lamar*, *in* 2 THE JUSTICES OF THE UNITED STATES SUPREME COURT, THEIR LIVES AND MAJOR OPINIONS, at 694, 696–99 (Leon Friedman & Fred L. Israel eds., 1997) [hereinafter 2 JUSTICES].

the Chinese Exclusion Act during his Senate service.[120] Chief Justice Fuller's biographer noted that "[r]acist assumptions permeated both popular attitudes and scientific thought in late nineteenth century America . . . . There is little doubt that Fuller shared the prevalent racial outlook."[121] Justice Harlan, as his *Plessy* dissent reflects, believed that the Chinese were racially unsuited to live in the United States.[122] Justice Field, author of *Chae Chan Ping*, supported exclusion of the Chinese as a matter of policy.[123] Similarly, *Plessy* author Henry Billings Brown's belief that blacks "were inherently inferior to whites"[124] may have made it difficult for him to be objective about cases involving members of other nonwhite races.

Finally, the opinions themselves reflect that racial considerations were in the minds of the Justices. In *Chae Chan Ping*, the Court explained that there was "a well-founded apprehension—from the experience of years—that a limitation to the immigration of certain classes from China was essential to the peace of the community on the Pacific coast, and possibly to the preservation of our civilization there."[125] The Chinese, reasoned the Court, "remained strangers in the land, residing apart by themselves, and

---

120. *See* 13 CONG. REC. 1749 (1882) (reflecting a vote in favor of an amendment explicitly prohibiting the naturalization of Chinese); *id.* at 1753 (reflecting a vote in favor of a bill excluding Chinese); *id.* at 3412 (reflecting support for a bill excluding Chinese through "pair" with a senator opposed to bill).

121. JAMES W. ELY, JR., THE CHIEF JUSTICESHIP OF MELVILLE W. FULLER, 1888–1910, at 155 (1995). Fuller's voting record was not unmixed; he dissented in *Fong Yue Ting*, yet he dissented in *United States v. Wong Kim Ark*, 169 U.S. 649 (1898)—disagreeing with the majority's holding that persons of Chinese ancestry born in the United States were citizens.

122. In *Plessy*, he wrote: "There is a race so different from our own that we do not permit those belonging to it to become citizens of the United States. Persons belonging to it are, with few exceptions, absolutely excluded from our country. I allude to the Chinese race." Plessy v. Ferguson, 163 U.S. 537, 561 (1896) (Harlan, J., dissenting). Justice Harlan voted with the majority in *Chae Chan Ping*, but did not participate in *Fong Yue Ting*. *See* 149 U.S. iii, n.1 (noting that Justice Harlan was absent for the portion of the Term in which *Fong Yue Ting* was decided). There is little doubt, however, that Harlan supported the reasoning and result of the case. *See* Chin, *supra* note 117, at 161 & n.71 (listing cases in which Justice Harlan relied on *Fong Yue Ting*).

123. As Charles McCurdy wrote:

> "You know I belong to the class who repudiate the doctrine that this country was made for the people of all races," Field wrote to a California friend in 1882. "On the contrary, I think it is for our race—the Caucasian race." The Chinese ought to be excluded from the United States, he said, for "[t]he manners, habits, mode of living, and everything connected with the Chinese prevent the possibility of their ever assimilating with our people. They are a different race, and, even if they could assimilate, assimilation would not be desirable."

Charles W. McCurdy, *Stephen J. Field and the American Judicial Tradition*, *in* THE FIELDS AND THE LAW 5, 17 (1986). Nonetheless, Field dissented in *Fong Yue Ting*.

124. Joel Goldfarb, *Henry Billings Brown*, *in* 2 JUSTICES, *supra* note 119, at 771.

125. Chae Chan Ping v. United States, 130 U.S. 581, 594 (1889).

adhering to the customs and usages of their own country. It seemed impossible for them to assimilate with our people or to make any changes in their habits or modes of living."[126] The point having been made, Justice Horace Gray's decision in *Fong Yue Ting* likewise contains an abbreviated discussion of the racial justifications for Chinese exclusion.[127]

A century later, in *Planned Parenthood v. Casey*,[128] the Court identified two cases, *Lochner v. New York*[129] and *Plessy*, as examples of constitutional decisions appropriately overruled because the Court that decided them had failed to understand the facts of economic and social life, and the implications of the rules it announced; it is no coincidence that both cases were legacies of the Fuller Court.[130] Scorned by Justices Marshall and Brennan as "not the strongest precedents in the United States Reports,"[131] *Fong Yue Ting* and *Chae Chan Ping* were unsound for precisely the same reasons as *Plessy*: None of the decisions reflected the now-axiomatic disfavor of racial discrimination. As in *Hunter*, the possibility that race infected the decision-making process undermines the argument that stare decisis should insulate *Chae Chan Ping* and *Fong Yue Ting* from reconsideration.

---

126. *Id.* at 595.

127. *See* Fong Yue Ting v. United States, 149 U.S. 698, 717 (1893).

128. 505 U.S. 833 (1992).

129. 198 U.S. 45 (1905).

130. *Planned Parenthood*, 505 U.S. at 861–64. *Planned Parenthood v. Casey* supports the idea that racial precedents are uniquely vulnerable. In *Casey*, the Supreme Court explained that it traditionally considered overruling cases when "related principles of law have . . . left the old rule no more than a remnant of abandoned doctrine," and when "facts have so changed, or come to be seen so differently, as to have robbed the old rule of significant application or justification." *Id.* at 855 (citations omitted). The Court refused to overrule *Roe v. Wade* because "[n]o development of constitutional law since the case was decided has implicitly or explicitly left *Roe* behind as a mere survivor of obsolete constitutional thinking." *Id.* at 857. All of these criteria apply with full force to the old cases approving racial discrimination. *See* Anne E. Pettit, Note, *"One Manner of Law": The Supreme Court, Stare Decisis and the Immigration Law Plenary Power Doctrine*, 24 FORDHAM URB. L.J. 165, 202–18 (1996). In Justice Brandeis's oft-cited words,

> in cases involving the Federal Constitution, where correction through legislative action is practically impossible, this Court has often overruled its earlier decisions. The Court bows to the lessons of experience and the force of better reasoning, recognizing that the process of trial and error, so fruitful in the physical sciences, is appropriate also in the judicial function.

Burnet v. Coronado Oil & Gas Co., 285 U.S. 393 , 406–08 (1932) (Brandeis, J., dissenting), *quoted in* United States v. Scott, 437 U.S. 82, 101 (1978). *See generally* William O. Douglas, *Stare Decisis*, 49 COLUM. L. REV. 735 (1949); Earl M. Maltz, *Some Thoughts on the Death of Stare Decisis in Constitutional Law*, 1980 WIS. L. REV. 467.

131. Kleindienst v. Mandel, 408 U.S. 753, 781 (1972) (Marshall, J., joined by Brennan, J., dissenting).

B. Congress's Racial Motives for Asian Exclusion

The evident biases of the Justices and the Justice Department impeach the decisions the Fuller Court ultimately handed down. However, that the decisions were racist does not necessarily mean that they were consequential, that the Court's failures of perception and judgment had any impact beyond the U.S. Reports, or have any effects today. To understand the full import of the decisions requires examination of what kind of statutes they upheld, what kind of future conduct the Court said was within the constitutional power of Congress and the executive, and what subsequently happened in the United States.

This is the analysis the Court employed in *Brown*. There, the Court did not focus exclusively on the *Plessy decision*. It did not, for example, rest its conclusion on the Fuller Court's faulty legal reasoning or misanalysis of precedent. Instead, the *Brown* Court rejected *segregation* because that doctrine was motivated by a desire to uphold white supremacy and because of its pernicious effects on innocent whites, African Americans, and others who were separated by law.[132] It is not a priori true that racial classifications in immigration law are analytically identical to domestic ones, that *Fong Yue Ting* should be thought of as *Plessy's* cousin, because the nature, justifications, and consequences of discrimination in immigration, theoretically, could be different.[133] As it turns out, the nature and effects of the racial discrimination authorized by *Fong Yue Ting* and *Chae Chan Ping* were not materially different from the racial discrimination authorized by *Plessy*.

The legislative history of the statutes approved by the Court in the plenary power cases indicates that they were not primarily motivated by a desire to influence foreign policy or international affairs, or even to protect American labor, but instead to foster white supremacy by defending white civilization against an undesirable race. The Asian Exclusion Laws thereby accomplished their goals, keeping the Asian-American population small and ensuring that group's cultural, political, and economic marginalization. Because these policies had effects in America on Americans of all races, they give rise to the same concerns as domestic racial classifications.

132. *See Casey*, 505 U.S. at 863 ("[T]he *Plessy* Court's explanation for its decision was so clearly at odds with the facts apparent to the Court in 1954 that the decision to reexamine *Plessy* was on this ground alone not only justified but required.").

133. *Cf.* Adarand Constructors, Inc. v. Pena, 515 U.S. 200, 237 (1995) ("[W]e wish to dispel the notion that strict scrutiny is 'strict in theory, but fatal in fact.'" (citation omitted)).

1. Asian Americans and African Americans in the Regime of White Supremacy

In some respects, the experiences of those of Asian and African descent in the United States could not have been less similar. The times and circumstances of their arrival, the places they lived, the work they did, their languages, customs, and beliefs, as well as their numerical representation in this country, were all quite different.[134] Some commentators seem to believe that Asian Americans and African Americans occupy the opposite ends of a certain spectrum; Asian Americans are called the "model minority" in pointed contrast to African Americans, while African Americans are regarded as the paradigmatic case of a racially subjugated group. When considering historical racial discrimination, some commentators seem either unaware of Asian exclusion or conclude, explicitly or implicitly, that Asian exclusion was less malevolent, less consequential, or both, than domestic discrimination. Further, some authorities and, it might be reasonable to speculate, more members of the general public, understand Asian exclusion as an aspect of legitimate economic protectionism, minimizing any racial motivation.[135]

Concretely, some skeptics of race-conscious programs argue that race-conscious remedies are unwarranted *even for* African Americans who, these commentators admit, experienced terrible injustices.[136] Meanwhile, some

---

134. There is a growing number of excellent historical works on Chinese and other Asians in the United States. *See, e.g.*, SUCHENG CHAN, THIS BITTERSWEET SOIL: THE CHINESE IN CALIFORNIA AGRICULTURE, 1860–1910 (1986); ROGER DANIELS, ASIAN AMERICA: CHINESE AND JAPANESE IN THE UNITED STATES SINCE 1850 (1988); CHARLES J. MCCLAIN, IN SEARCH OF EQUALITY: THE CHINESE STRUGGLE AGAINST DISCRIMINATION IN NINETEENTH-CENTURY AMERICA (1994); LUCY E. SALYER, LAWS HARSH AS TIGERS: CHINESE IMMIGRANTS AND THE SHAPING OF MODERN IMMIGRATION LAW (1995); ALEXANDER SAXTON, THE INDISPENSABLE ENEMY: LABOR AND THE ANTI-CHINESE MOVEMENT IN CALIFORNIA (1971); RONALD TAKAKI, STRANGERS FROM A DIFFERENT SHORE: A HISTORY OF ASIAN AMERICANS (1989).

135. *See* GROUNDS FOR EXCLUSION, *supra* note 31, at 8 ("The Chinese Exclusion Act of May 6, 1882 . . . was enacted out of a concern for protecting domestic labor from foreign competition, combined with racial prejudice." (citation omitted)); JUDITH BENTLEY, AMERICAN IMMIGRATION TODAY: PRESSURES, PROBLEMS, POLICIES 62 (1981) ("Orientals were excluded so long because Americans feared their potentially large numbers, competition for jobs, and the vast difference in cultures."); *cf.* 1 JOHN H. WIGMORE, EVIDENCE IN TRIALS AT COMMON LAW § 516, at 645 (1st ed. 1904) (attributing discrimination against Chinese to "[t]he condition of public feeling in [California] against the economic encroachments of Chinese laborers"). One acquaintance defended the exclusion laws on the ground that "we were at war," evidently confusing Asian exclusion with the incarceration of Japanese Americans during World War II.

136. Paul Carrington wrote that "there might be a principled basis for distinguishing groups if one were to limit the compensation to genuinely severe disadvantage imposed by law, or by the government of the United States. On this basis, one could reasonably compensate those victimized by slavery without attempting to compensate for lesser wrongs." Paul D. Carrington, *Diver-*

defenders of race-conscious policies, perhaps engaging in triage in difficult times, suggest that African Americans may be the only group with a history of discrimination clearly worthy of redress.[137] Thurgood Marshall wrote: "The

---

sity!, 1992 UTAH L. REV. 1105, 1156. Charles Fried also contended that African Americans have been uniquely harmed:

> Grounding a theory of benign governmental discrimination in a particular historical injustice . . . would make the equal protection principle of the fourteenth amendment applicable especially or only to the descendants of former slaves. To generalize it beyond African-Americans to Asian-Americans, Native Americans, Spanish-surnamed persons, and, as sometimes happens, Eskimos and Aleuts requires a theory of deprivation that is quite indeterminate and infinitely manipulable.

Charles Fried, Metro Broadcasting, Inc. v. FCC: *Two Concepts of Equality*, 104 HARV. L. REV. 107, 125 (1990) (footnote omitted); *see also, e.g.*, City of Richmond v. J.A. Croson Co., 488 U.S. 469, 527 (1989) (Scalia, J., concurring in the judgment) ("It is plainly true that in our society blacks have suffered discrimination immeasurably greater than any directed at other racial groups."); G. Turnbull, *Reflections on Affirmative Action in Academia*, *in* AFFIRMATIVE ACTION AND THE UNIVERSITY: A PHILOSOPHICAL INQUIRY 135 (Steven M. Cahn ed., 1992) [hereinafter AFFIRMATIVE ACTION AND THE UNIVERSITY] ("Since American history includes the moral blot of black slavery and its continuing aftermath, the addition of African Americans into academia is a special problem.").

137. For example, Stephen Steinberg argues:

> Only in the most general sense can it be claimed that Asians, West Indians and African Americans are all "races" that have been victims of racial stereotyping and discrimination. Although true, this proposition obscures the unique oppression that blacks have endured throughout American history, beginning with two centuries of slavery and another century of official segregation, reinforced by the lynch mob and systematically unequal treatment in all major institutions.

STEPHEN STEINBERG, THE ETHNIC MYTH: RACE, ETHNICITY, AND CLASS IN AMERICA 269 (1989). Similarly, Alex Johnson wrote:

> [O]ther ethnic, racial and religious groups . . . were not subject to slavery and the intense institutional racism that has been directed at blacks by American society. Simply put, these groups by and large have not suffered the effects of racism to the same degree as blacks because they have not been subjected to the same extent and intensity of racism.

Alex M. Johnson, Jr., *Defending the Use of Quotas in Affirmative Action: Attacking Racism in the Nineties*, 1992 U. ILL. L. REV. 1043, 1072; *see also, e.g.*, MICHAEL HARRINGTON, THE OTHER AMERICA: POVERTY IN THE UNITED STATES 75 (rev. ed. 1981) ("[T]he Negro is colored, and no other group in the United States has ever faced such a problem . . . ."); ANDREW KULL, THE COLOR-BLIND CONSTITUTION 223 (1992) ("No other racial or ethnic group among America's immigrants has a comparable claim to special treatment, and the moral awkwardness of asking black Americans to be content with nondiscrimination should not stop us from giving that answer to everyone else."); Guido Calabresi, Bakke *as Pseudo-Tragedy*, 28 CATH. U. L. REV. 427, 432–33 (1979) (suggesting that remedies for past discrimination should be "limited to those groups having suffered such special ill-treatment that they were made the object of a special constitutional status or adjustment" and that "application of this test would limit remedies to blacks and perhaps to American Indians"); K.G. Jan Pillai, *Affirmative Action: In Search of a National Policy*, 2 TEMP. POL. CIV. RTS. L. REV. 1, 25 (1992) ("Among the various minority groups in the United States, the history of suffering endured by the black population is unique."); Adolphus Levi Williams, Jr., *A Critical Analysis of the* Bakke *Case*, 16 S.U. L. REV. 129, 130 (1989) ("Although many groups in the United States . . . have been, and are, discriminated against [including Asians], Afro-Americans, perhaps for a variety of reasons, have been and continue to be particularly vulnerable to racism, prejudice and discrimination."); *cf.* Kenneth L. Karst, *Paths to Belonging: The Constitution and Cultural Identity*, 64 N.C. L. REV. 303, 321 (1986) (arguing that

experience of Negroes in America has been different in kind, not just in degree, from that of other ethnic groups. It is not merely the history of slavery alone but also that a whole people were marked as inferior by the law."[138] Marshall helped make the growth of the Asian-American community possible because his work as a lawyer in *Brown* and other cases paved the way for civil rights legislation in the 1960s, including the Immigration and Nationality Act Amendments of 1965. As a Justice, Marshall championed the rights of immigrants and aliens of all races.[139] Yet, his assertion that only African Americans were "marked as inferior by the law" necessarily implies that Asian Americans and other racial groups were not. That this great man could make such a claim means either, as some have argued, Asian Americans are appropriately identified as a group that did not suffer particularly serious discrimination in the past,[140] or (as this Article suggests) their experiences are so little known and understood that they are virtually invisible.[141]

There is a limited but genuine similarity between treatment of Asians and African Americans in the late nineteenth century. It was no coincidence that, virtually contemporaneously, the Fuller Court approved segre-

"[d]iscrimination against other ethnic groups typically has not reached the level of intensity that discrimination against blacks reached in the South" but discussing various forms of discrimination against other groups).

138. Regents of the Univ. of Cal. v. Bakke, 438 U.S. 265, 400 (1978) (Marshall, J., concurring in the judgment in part and dissenting in part).

139. *See* Michael Scaperlanda, *Justice Thurgood Marshall and the Legacy of Dissent in Federal Alienage Cases*, 8 GEO. IMMIGR. L.J. 1 (1994).

140. *See, e.g.*, Kent Greenawalt, *Judicial Scrutiny of "Benign" Racial Preference in Law School Admissions*, 75 COLUM. L. REV. 559, 598–99 (1975) (supporting the discretion of the university to exclude Japanese and Chinese from its affirmative action program); Andrew Oldenquist, Remarks on Affirmative Action, *in* AFFIRMATIVE ACTION AND THE UNIVERSITY, *supra* note 136, at 190 ("No one needs to worry that Jews or Chinese Americans are underrepresented on university faculties . . . ."); Robert M. O'Neil, *Racial Preference and Higher Education: The Larger Context*, 60 VA. L. REV. 925, 942 (1974) ("Thus if the University of Washington determined that Asian applicants were less in need of preferential consideration than other minorities, that would appear to be a constitutionally permissible choice."); Richard A. Posner, *The* Bakke *Case and the Future of "Affirmative Action,"* 67 CAL. L. REV. 171, 176 (1979) (contending that Asians, like Jews, Catholics, and Mormons, "have overcome any handicaps imposed by discrimination"); Robert L. Simon, *Affirmative Action and the University: Faculty Appointment and Preferential Treatment*, *in* AFFIRMATIVE ACTION AND THE UNIVERSITY, *supra* note 136, at 48, 62 ("While [remedial] principle clearly supports extending preferential treatment for black Americans, it arguably would exclude Asians, and perhaps women.").

141. Pat Chew observes that "[d]ialogue that defensively focuses on which minority group is worse off, and thus most deserving of attention and help, is divisive." Pat K. Chew, *Asian Americans: The "Reticent" Minority and Their Paradoxes*, 36 WM. & MARY L. REV. 1, 88 (1994). This Article's argument is not meant to minimize or deny the harms inflicted on other racial groups, but merely to suggest that for Asian Americans, too, there is a difference between today's situation and the situation that would exist but for invidious discrimination.

gation of African Americans in *Plessy* and exclusion of Asians in *Chae Chan Ping* and *Fong Yue Ting*. Indeed, after the Civil War, the fortunes of the two groups waxed and waned virtually in lockstep. July 28, 1868 was a momentous date for both the Chinese and the freedmen who hoped for opportunity in America. On that day, Secretary of State William H. Seward signed the Burlingame Treaty with China, which promised that Chinese persons could immigrate on the same basis as all other foreigners,[142] and also issued a proclamation that the Fourteenth Amendment had become part of the Constitution.[143]

That promising start was short-lived, of course. The year 1877 was a nervous one for both groups as federal troops were withdrawn from the South after the House resolved the Tilden-Hayes election; Congress then turned its attention towards the Chinese.[144] By the mid-1890s, with *Plessy* in 1896 and *Fong Yue Ting* in 1893, the Supreme Court made clear that the Constitution would not intervene if the state and national legislatures wished to exclude nonwhites from the benefits of American society.[145] As Oscar Handlin explained: "By the end of the century the pattern of racist practices and ideas seemed fully developed: the Orientals were to be totally excluded; the Negroes were to live in a segregated enclave; the Indians were to be confined to reservations as permanent wards of the nation . . . ."[146]

142. The treaty recognized "the inherent and inalienable right of man to change his home" and assured Chinese aliens enjoyment of "the same privileges, immunities, or exemptions in respect to travel or residence" that the United States afforded to "citizens . . . of the most favored nation." Burlingame Treaty, July 28, 1868, U.S.-China, 16 Stat. 739–40.

143. *See* Proclamation, 15 Stat. 708 (July 28, 1868).

144. *See* Shirley Hune, The Issue of Chinese Immigration in the Federal Government, 1875–1882 (1979) (unpublished Ph.D. dissertation, George Washington University) (on file with author) (discussing history of Chinese Exclusion legislation); Hudson N. Janisch, *The Chinese Exclusion Laws: Congress and the Politics of Unbridled Passion*, *in* ASIAN AMERICANS AND CONGRESS 71, 78 (Hyung-Chan Kim ed., 1996).

145. *Yick Wo v. Hopkins*, 118 U.S. 356 (1886), which invalidated selective enforcement of municipal ordinance against the Chinese, was an unusual judicial victory. Professor Joo argues that "*Yick Wo* was possible only because the interests of Chinese aliens in fighting state discrimination converged with the interests of the federal judiciary in extending the Fourteenth Amendment to protect economic interests from state interference." Thomas Wuil Joo, *New "Conspiracy Theory" of the Fourteenth Amendment: Nineteenth Century Chinese Civil Rights Cases and the Development of Substantive Due Process Jurisprudence*, 29 U.S.F. L. REV. 353, 355 (1995).

146. OSCAR HANDLIN, RACE AND NATIONALITY IN AMERICAN LIFE 48 (1957). Similarly, a major theme of Alexander Saxton's study is that America had experienced "three great racial confrontations: with the Indian, with the African , and with the Oriental. . . . What happened to Orientals in America, while similar in many ways to what happened to other immigrants, is generally more like what happened to blacks . . . ." SAXTON, *supra* note 134, at 1–2. Justice David J. Brewer had made a similar point decades before, noting that

> [w]ithin our borders by virtue of an earlier right, through our compulsion, or at our solicitation, are to-day the three despised races of America [namely, Native Americans,

Even the reasoning the Court employed to justify discriminatory treatment of African Americans and Asians was virtually identical: Forcible separation of people of color from whites, the Court said, simply did not imply inferiority. In *Plessy*, for example, the Court explained:

> We consider the underlying fallacy of the plaintiff's argument to consist in the assumption that the enforced separation of the two races stamps the colored race with a badge of inferiority. If this be so, it is not by reason of anything found in the act, but solely because the colored race chooses to put that construction upon it.[147]

The same reasoning was applied to Asians denied citizenship on the basis of race. In *Ozawa v. United States*,[148] the Supreme Court held that Japanese people could not become naturalized citizens, but emphasized that "[o]f course there is not implied—either in the legislation or in our interpretation of it—any suggestion of individual unworthiness or racial inferiority. These considerations are in no manner involved."[149]

By the early 1950s, thanks in part to the need to respond to Communist propaganda, the national government concluded that it was inappro-

---

> African Americans and Chinese Americans]—races whom to wrong is for so many a habit, whom to plunder and oppress is justified by our greed and their weakness.

MICHAEL J. BRODHEAD, DAVID J. BREWER: THE LIFE OF A SUPREME COURT JUSTICE, 1837–1910, at 121–22 (1994) (quoting DAVID J. BREWER, JUBILEE ANNIVERSARY: AN ADDRESS DELIVERED AT THE FIFTIETH ANNUAL MEETING OF THE AMERICAN MISSIONARY ASSOCIATION, OCTOBER 21, 1896 (n.p., n.d.)). Supporters of racial exclusion also candidly acknowledged the relationship between Asian immigrants and African Americans. Woodrow Wilson, for example, stated:

> In the matter of Chinese and Japanese coolie immigration . . . I stand for the national policy of exclusion. We cannot make a homogeneous population out of a people who do not blend with the Caucasian race. . . . Oriental coolieism will give us another race problem to solve and surely we have had our lesson.

DANIELS, *supra* note 83, at 55 (quoting letter from Woodrow Wilson (May 3, 1912)); *see also* MICHAEL LIND, THE NEXT AMERICAN NATION: THE NEW NATIONALISM AND THE FOURTH AMERICAN REVOLUTION 95 (1995) ("The structure of race relations in Euro-America . . . was built on the interlocking framework of disenfranchisement of blacks and Asians, social segregation, a segregated labor market and restrictions on the immigration and naturalization of nonwhites."); ELIOT GRINNELL MEARS, RESIDENT ORIENTALS ON THE AMERICAN PACIFIC COAST 1 (1928) ("'Injustice has been the only American way of meeting a race problem. We dealt unjustly by the Indian, and he died. We dealt unjustly with the Negro, and he submits. If the Japanese ever come in sufficient numbers to constitute a race problem, we shall deal unjustly with them . . . . '" (quoting CHESTER ROWELL, 26 WORLD'S WORK (1913)).

147. Plessy v. Ferguson, 163 U.S. 537, 551 (1896).

148. 260 U.S. 178 (1922).

149. *Id.* at 198. The Court was equally solicitous of the feelings of Hindus in holding that they, too, could never become naturalized citizens of the United States: "It is very far from our thought to suggest the slightest question of racial superiority or inferiority. What we suggest is merely racial difference, and it is of such character and extent that the great body of our people instinctively recognize it and reject the thought of assimilation." United States v. Thind, 261 U.S. 204, 215 (1923).

priate to continue exclusion of Asians[150] and segregation of African Americans.[151] The beginnings of more substantial legal change came a decade later in the form of civil rights legislation advanced by the Kennedy and Johnson administrations.[152]

The parallels between the African-American and Asian-American legal experience in America are not coincidental. Rather, the similarity exists because the presence of Asians, like African Americans, threatened white supremacy. Consequently, the Asian threat, like the African-American threat, was eliminated by laws that were, in turn, blessed by the Supreme Court.

The invidious nature of American treatment of Africans and African Americans is virtually axiomatic. The evidence of racist motives for Congress's treatment of Chinese and other Asians is far less well known, so this Article deals with it in some detail.

2. Racial, Not Economic, Protection

The Chinese Exclusion Act of 1882 was the first Asian Exclusion Law, and the one that generated the most contemporary controversy. Its proponents denied that protection of jobs alone justified the measures; that would have been a difficult claim in a period in which immigration was numerically unlimited. Instead, the factor that made Chinese immigration undesirable was the immigrants' race and culture. Senator Saulsbury stated that he

> would not close the gates against the immigration of the Chinese to this country simply upon the ground that they enter into competition with labor, but I put it upon another ground, which is that it introduces a distinct race of people with a different civilization from that to which we are accustomed, wholly incapable of assimilation

150. *See* Immigration and Nationality Act of 1952, ch. 477, § 201(a), 66 Stat. 163, 175 (awarding all quota areas at least 100 visas per year); *cf.* Chin, *supra* note 34, at 287–88 (describing the congressional concern about Communist propaganda as one motivating factor for immigration reform).

151. *See* Brown v. Board of Educ., 347 U.S. 483 (1954); Mary L. Dudziak, *Desegregation as a Cold War Imperative*, 41 STAN. L. REV. 61 (1988) (describing the Cold War motivation of *Brown*).

152. The Civil Rights Act of 1964, Pub. L. No. 88-352, 78 Stat. 241 (codified as amended in 42 U.S.C. §§ 1971–1975), prohibited racial discrimination in various areas of American life. The Voting Rights Act of 1965, Pub. L. No. 89-110, 79 Stat. 437 (codified as amended at 42 U.S.C. §§ 1971, 1973 to 1973bb-1), eliminated many of the artifices certain states used to disenfranchise voters on the basis of race. The Immigration and Nationality Act Amendments of 1965, Pub. L. No. 89-236, 79 Stat. 911 (codified as amended in scattered sections of 8 U.S.C.), eliminated the last forms of racial discrimination from the Immigration and Nationality Act (INA).

> with our people. In my opinion, their presence among us will not improve their condition morally or in any other respect . . . while it may lead to discord and possible collision with the Anglo-Saxon race among whom they live.[153]

Senator Miller agreed that the key was excluding Asians from American society rather than economic effects per se. Chinese immigration "is a substitution of Chinese and their civilization for white men and Anglo-Saxon civilization. This involves considerations higher than mere economic theories."[154] Other legislators agreed that labor concerns per se were not paramount.[155]

Mass immigration also was not the problem; Chinese represented a fraction of total immigration. Moreover, labor competition from white aliens was not criticized. Thus, Representative George urged foreign nations to "[g]ive us of the people of our own race."[156] However, "[e]very Chinaman

153. 13 CONG. REC. 1584 (1882) (statement of Sen. Saulsbury). When discussing the Chinese Exclusion Act, this Article uses without distinction legislative history from two separate exclusion bills that were considered in 1882. The first was passed by the House and Senate in March, but vetoed by President Chester A. Arthur on April 4, 1882. The second bill, introduced on April 6, 1882, was passed by the House and Senate in April and signed by the president on May 6, 1882. The major difference between the bills was the period of suspension of Chinese immigration—20 years under the vetoed bill, 10 under the bill that became law. *See* E.P. HUTCHINSON, LEGISLATIVE HISTORY OF AMERICAN IMMIGRATION POLICY, 1789–1965, at 81–82 (1981). The legislative history of the first bill is relevant to the second, given that the same people were considering the same issues at virtually the same time; there had been no intervening congressional election, the bills were similar in substance, and they were both passed within a two-month period. Accordingly, this Article assumes that views expressed with respect to the March bill were still held in April.

154. *Id.* at 1484 (statement of Sen. Miller).

155. Representative Butterworth did not oppose the immigration to protect labor alone. He stated:

> I oppose the immigration of these Chinese not more on account of the influence of their labor than for the moral degradation which does and the political degradation which may attend it. Why may we not feel concerned touching the social and moral influences exerted by contact with this servile, corrupt people?

Id. at 2132 (statement of Rep. Butterworth); *see also id.* at 3406 (statement of Sen. Call). According to Senator Grover,

> [t]he question of encouragement or restraint of Chinese immigration presents a deeper problem than that of mere effect on our present well-being. The abiding characteristics of the Chinese as a people are so variant from our own, and so opposed to the purposes of our civilization, that the encouragement or restriction of this class of immigration presents the alternative of choosing for the Pacific States and Territories between the institutions of America and the institutions of China.

Id. at 1545 (statement of Sen. Grover). According to Senator Jones, "a deterioration by a very moderate percentage of the character and force of a population by intermixing with it an inferior race may involve a greater national loss of wealth, and that of the most reliable and permanent form, than all which can be gained by employers from a reduction of wages." *Id.* at 1743 (statement of Sen. Jones).

156. *Id.* at 2165 (statement of Rep. George).

in this country today is a substitute for and has kept out one honest German, Irishman, Welshman, Englishman, or other Caucasian."[157] Representative Deuster pointed out that

> [n]o voice is raised against the immigration from across the Atlantic because we feel ourselves akin to the people who thus come to our shores. They are the offspring of the nations from which the founders of the Republic, its defenders, and heroes sprang. They are not only of our race, but allied in moral and religious sentiments, in the degree of culture attained.[158]

By contrast, "[t]he Chinaman is neither socially, morally, nor politically fit to assimilate with us."[159]

Desire for white immigrants was no mere esthetic preference. Senator Jones contended that

> [f]ree institutions are only possible with the favored races. It is not because they are a monopoly of the favored races, but because no other race is capable of creating them; no other race is capable of perpetuating them; no other race is capable of treading freedom's heights with firm and unwavering step.[160]

Accordingly, admission of Asians could destroy American civilization. Representative Berry agreed: "[S]hould they be ever clothed with the right of American citizenship—which God forbid—it will be but a question of time when the light of Christianity will be obscured in the gloom of heathen darkness."[161] "[I]f the Republic endures it must be a homogeneous population."[162]

157. *Id.* at 2043 (statement of Rep. Brumm).

158. *Id.* at 2031 (statement of Rep. Deuster).

159. *Id.* Some legislators were less explicit about the inferiority of the Chinese, but nevertheless did not want them to be part of the American family. Senator Pendleton, for example, argued: "They are all men, they are all human beings; I believe they are all working out, in one form and another, their progress and advancement and improvement. [But] Mr. President, you cannot put the spot upon the camelia and expect it to retain its unsullied purity and whiteness." *Id.* at 3409 (statement of Sen. Pendleton); *see also id.* at 2206 (statement of Rep. Sherwin).

160. *Id.* at 1742 (statement of Sen Jones).

161. *Id.* at 2035 (statement of Rep. Berry). Senator Miller, likewise, believed that allowing the races to exist side by side would create "a mongrel race, half Chinese and half Caucasian, [producing] a civilization half pagan, half Christian, semi-oriental, altogether mixed and very bad." *Id.* at 1483 (statement of Sen. Miller).

162. *Id.* at 2223 (statement of Rep. Cannon). Opponents of the bill denied that the campaign of hostility towards the Chinese and their legal ineligibility for naturalization created the appropriate environment for a fair test of their eagerness to assimilate, or not. Senator Brown said that if the government would treat "the Chinaman . . . as you now treat the African . . . you will find him assimilate much more readily than he does with . . . no encouragement offered to him, but every possible discouragement. No, the Chinaman does not assimilate. We do not permit him to assimilate." *Id.* at 1643 (statement of Sen. Brown); *see also id.* at 1706 (statement of Sen. Platt); *id.* at 2185 (statement of Rep. Joyce); *id.* at 1981 (statement of Rep. Taylor).

"We religiously believe that unrestricted Mongolian immigration means ultimate destruction."[163]

Legislators of this era candidly explained the white supremist bases of their arguments. Senator Teller, for example, explained that "[t]he Caucasian race has a right, considering its superiority of intellectual force and mental vigor, to look down upon every other branch of the human family. . . . We are the superior race today. We are superior to the Chinese . . . ."[164] Senator Miller explicitly appealed to notions of racial inferiority.

> It is not numbers that are needed; quality is of more importance than quantity. One complete man, the product of free institutions and a high civilization, is worth more to the world than hundreds of barbarians. Upon what other theory can we justify the almost complete extermination of the Indian, the original possessor of all these States? I believe that one such man as Washington, or Newton, or Franklin, or Lincoln, glorifies the Creator of the world and benefits mankind more than all the Chinese who have lived, struggled and died on the banks of the Hoang Ho [sic].[165]

---

163. *Id.* at 1978 (statement of Rep. Cassidy). Many legislators were not simply arguing that birds of a feather should flock together; in addition to being different, the Chinese were bad. Representative Calkins, for example, argued that

> [w]herever they come they spread mildew and rot throughout the entire community. And yet, upon a mere notion of humanitarianism, you throw open the doors because you say the sentiment of the people of this great Republic is to invite the people of other nations to these shores. By so doing you plant a cancer in your own country that will eat out its life and destroy it from among the nations of the earth.

Id. at 1904 (statement of Rep. Calkins). One legislator urged exclusion to avoid the prospect that the Chinese, "in time of excitement [would] pillage and burn property, and . . . murder your citizens." *Id.* at 1985 (statement of Rep. Blount). Senator Vest described the Chinese as "parasites, like those insects which fasten themselves upon vegetables or upon animals and feed and feed until satiety causes them to release their hold." *Id.* at 3358 (statement of Sen. Vest); *see also id.* at 2164 (statement of Rep. George) ("[T]here is no recognized principle that the ignorant, the savage, the barbarous, the brutal, the servile, the riffraff, the dangerous, or the refuse element of any foreign nation are entitled of right to enter our body-politic . . . .").

164. *Id.* at 1645 (statement of Sen. Teller). Perhaps embarrassed by the immodesty of his remarks, the senator tried to clarify them: "I did not mean to claim that the Anglo-Saxon was in all respects superior, . . . I meant to say that there were other branches of the human family that stood lower than ours in intellectual strength, in mental vigor, and in moral worth, and a man is blind to history who does not know that that is a fact." *Id.* at 1712–13.

165. *Id.* at 1487 (statement of Sen. Miller). Senator Miller added:

> The two civilizations which have here met are of diverse elements and character, both the result of evolution under different conditions, radically antagonistic, and as impossible of amalgamation as are the two great races who have produced them. The attempt to merge them must result, as both reason and experience teaches, in the displacement of one or the other. Like the mixing of oil and water, neither will absorb the other.

Id. at 1483.

> Forty centuries of Chinese life have made the Chinaman what he is. An eternity of years cannot make him such a man as the Anglo-Saxon. It is as impossible to bring the

The Chinese were said to have a peculiar resistance to assimilation with westerners. Senator Jones explained that

> [t]he main objection to the incoming of the Chinese cannot be brought against Europeans of our own race and nearly related to us. They soon assimilate with us, they are bone of our bone and flesh of our flesh, and have hopes, fears and traditions identical with our own. They readily adapt or manners, customs, and language, and soon become an undistinguishable part of our body-politic.[166]

However, Chinese were different: "The presence of an inferior race either inevitably expels the superior race or reduces it to the ways, modes of life, and all the other conditions of the new-comers."[167]

> Chinaman up to the American standard as it is cruel and wicked to risk, by any experiment, the degradation of the American laborer to the Chinese standard.

Id. at 1485. This was an objection peculiar to the Chinese. Senator Miller stated:

> But it has been said that all the arguments we make against the admission of the Chinese cooly into direct competition with our free American laborer apply with equal force to the immigrants who come from Europe . . . particularly the Irish. . . . [I]t is not true. European immigrants are men of the like mental and physical characteristics as the American laborer. They are of the same or a kindred race, trained under a like civilization, with similar aspirations, hopes and tendencies.

Id. at 1485; *see also id.* at 2034 (statement of Rep. Berry).

166. *Id.* at 1741 (statement of Sen. Jones).

167. *Id.* at 1744. Similarly, Representative Page explained that he was

> opposed to Chinese immigration, first, because the Chinese are a people eminently unfit to become desirable citizens in a country the people of which cherish the genius of free institutions . . . . [T]his people because of their habits and characteristics are unfit to become not alone citizens of the United States but even inhabitants thereof.

Id. at 1932 (statement of Rep. Page). Representative McClure concurred:

> Alien in manners, servile in labor, pagan in religion, they are fundamentally un-American. Our civilization, which is the most potent in the world in blotting out race distinctions and amalgamating nationalities, is utterly powerless to efface in a single aspect the primeval national characteristics of the Chinaman. He is literally iron-clad to the genius of our institutions. He is the same unadulterated Mongolian on the banks of the Sacramento river as he is on the Hoang Ho [sic].

Id. at 2126 (statement of Rep. McClure). Representive Butterworth added:

> That these people do not assimilate with us is certain; that they could not is equally clear in my mind. My honored colleague says our laws forbid it. No, sir; they cannot assimilate with us, not because of any law traced upon our statute-book, but because of a law traced by the finger of the Deity, a law which this House cannot repeal nor provide against.

Id. at 2131 (statement of Rep. Butterworth). Many others agreed. *See, e.g., id.* at 1674 (statement of Sen. Edmunds); *id.* at 2030–31 (statement of Rep. Deuster); *id.* at 2043–44 (statement of Rep. Brumm); *id.* at 2162 (statement of Rep. Washburn); *id.* at 2164–65 (statement of Rep. George); *id.* at 2211 (statement of Rep. Pacheco); *id.* at 2220 (statement of Rep. Curtin); *id.* at 2223 (statement of Rep. Cannon).

3. The African Analogy

At the time, many legislators on both sides of the issue insisted that the African-American situation was no different from the Asian-American situation. Opponents of restriction argued that the Chinese could become part of the fabric of American society with the same degree of success as African Americans. Restrictionists agreed, but concluded that the African-American experiment had been a failure.

Early on, opponents of racial restriction invoked the "all men are created equal" line from the Declaration of Independence. To support their position, as had Chief Justice Taney in *Dred Scott v. Sandford*,[168] supporters of exclusion candidly explained the racial boundaries of the principle. Senator Grover pointed out that the Founding Fathers'

> conduct towards the aborigines of this country, whom they found in lawful occupancy here when they came, must be construed as fixing a limit to the meaning of their public declarations upon the rights of man. When they declared that all men were created equal, and were endowed with the inalienable right of life, liberty, and the pursuit of happiness, they undoubtedly meant all men like themselves, and in like manner joined in the bonds of civil society. And when they dedicated this country to be an asylum for the oppressed of all nations they undoubtedly meant all nations whence they came, peoples of character and conditions similar to their own.[169]

In short, "[l]iberty and equal rights are liberty and equal rights of men, not of slaves, not of coolies."[170]

Senator Platt, an opponent of restriction, argued that the bill was "race legislation; it is avowed to be race legislation, and all the old arguments that we heard about the danger of social equality between the negro and the white man are resuscitated and rehabilitated for the occasion."[171] Proponents of the

168. 60 U.S. (19 How.) 393, 410 (1857) (concluding that all men are *not* created equal, notwithstanding the language of the Declaration, because "it is too clear for dispute, that the enslaved African race were not intended to be included"). *See generally* PAUL FINKELMAN, DRED SCOTT V. SANDFORD: A BRIEF HISTORY WITH DOCUMENTS (1997).

169. 13 CONG. REC. at 1546 (statement of Sen. Grover); *see also id.* at 1583 (statement of Sen. Maxey) ("[N]o one who framed the Constitution or voted for its adoption ever dreamed of incorporating people of the Mongolian race into our civilization.... [T]he Mongolian race are not of our household; they are not of our people."); *id.* at 2128 (statement of Rep. Bayne) (arguing that fundamental principles of American government do not include the Chinese because "[w]hen this continent was first settled, and during the whole period of its history from that time until about twenty-five years ago, it was never dreamed that the people of Asia would be likely to immigrate into it").

170. *Id.* at 2030 (statement of Rep. Flower); *see also id.* at 1978 (statement of Rep. Cassidy).

171. *Id.* at 1705 (statement of Sen. Platt). Senator Hoar also made an explicit comparison with slavery. "[I]f this legislation be not wrong, as Abraham Lincoln said of slavery, nothing is

bill responded to the claim that Chinese could be assimilated into American society like African Americans by arguing that making the freed slaves citizens was a dreadful mistake. Senator George, for example, suggested presciently that the position of African Americans was no more secure than that of the Chinese immigrants.

> [I]t is not so clear that our Constitution, with all its great strength and vigor, after embracing the African, can take in with safety even the more enlightened countrymen of Confucius.
>
> . . . .
>
> The Constitution was ordained and established by white men, as they themselves declared in its preamble, "to secure the blessings of liberty to themselves (ourselves) and their (our) posterity," and I cannot doubt that this great pledge thus solemnly given will be as fully redeemed in favor of the white people of the South, should occasion for action arise, as I intend on my part and on their behalf to redeem it this day in favor of the white people of the Pacific States, by my vote to protect them against a degrading and destructive association with the inferior race now threatening to overrun them.[172]

Other members of Congress, though perhaps less explicitly contemptuous of African Americans, nevertheless concluded that their presence in this country was beneficial to neither black nor white and therefore should not be repeated with the Chinese. Senator Jones argued:

> Does anybody pretend to tell me that it is a blessing to this country that [African Americans] are here? It is no fault of ours that they

---

wrong. If it be not a violation of the moral law, there is no moral law as it has been pronounced to us." *Id.* at 1709 (statement of Sen. Hoar). Representative Taylor agreed:

> Mr. Speaker, in this city men, women, and children have gone from the auction block under the Stars and Stripes of this Government; under the same cry we hear to-day as against the Chinese they have been sold within the sound of my voice under the hammer. The defense was that they were an inferior race—the only defense that ever could be given. The old slogan is heard again now.

Id. at 1983 (statement of Rep. Taylor); *see also id.* at 2174 (statement of Rep. Hawk).

172. *Id.* at 1637–38 (statement of Sen. George). Senator Vest stated:

> [The] experiment in the Southern States of putting the negro above the white man is a miserable failure, and it stands to-day a monument to the fact that the intellect of the Caucasian race, bayoneted to the earth, plunged into blood up to the chin, will yet rise to the full height of the destiny which God has placed before it.

Id. at 3358 (statement of Sen. Vest). Senator Teller also contended that the presence of Africans in this country was a mistake. As to making Africans citizens, the senator argued that "[i]t was not a desirable thing to do . . . . There was not an intelligent American who looked upon that problem at that time who did not fear that the result would be that the antagonism of the two races would eventually swamp this country." *Id.* at 1713 (statement of Sen. Teller). He also suggested that the same fate would meet African immigrants, if they came over in mass, as Congress proposed for Chinese. *See id.* at 1645.

> are here; it is no fault of theirs; it is the fault of a past generation; but their presence here is a great misfortune to us to-day, and the question of the adjustment of the relations between the two races socially and politically is no nearer a settlement now than it was the day Sumter was fired upon. . . . [The Chinaman's] race is socially more incongruous to ours and less capable of assimilation with us than is the negro race. . . . What encouragement do we find in the history of our dealings with the negro race or in our dealings with the Indian race to induce us to permit another race-struggle in our midst?[173]

Many members of the House shared this attitude.[174]

Proponents of Asian exclusion considered themselves fortunate because they dealt with the problem early, before Asians established a substantial presence in the United States. Unlike the African problem, the Chinese issue had been caught in time: "[I]n California to-day we find the people subjected to exactly the same condition of affairs that this country suffered from when the mother country permitted African slavery to be introduced, and we find a question to-day in California equally demanding our attention."[175]

Efforts to explain why African Americans had nothing to fear from opening this particular Pandora's Box ring flat, for many who praised African Americans for their Americanism did so while supporting the principle of racial hierarchy. Senator Slater, for instance, argued that "the negro is thoroughly American, if he is not Caucasian; the Chinese are neither Caucasian nor American, but are alien to our race, customs, religion and civilization."[176] According to Representative Hayne, American laborers "believe America is for the Caucasian race, and accepting the results of the rebellion recognize the African as a co-laborer and citizen. Not so with the

---

173. *Id.* at 1744–45 (statement of Sen. Jones).

174. Representative Page proposed "that there is not a member upon this floor, of either party, who believes that the coming of the African race to this country originally was a blessing to us or to the African himself." *Id.* at 1932 (statement of Rep. Page). Representative Buckner agreed:

> And now, sir, when we have had but a slight foretaste of the fallacy, if not the criminality, of the practical enforcement of the dogma that all men are entitled to equal rights in this Government of ours, we are called upon to vote for a proposition that not only excludes millions of men from citizenship and suffrage, but which imposes what I hope will be a perpetual prohibition upon their polluting our soil even with the soles of their feet. How true it is that experience ever is contradicting the theories of men.

Id. at 2138 (statement of Rep. Buckner); *see also id.* at 2221–22 (statement of Rep. Curtin).

175. *Id.* at 2169 (statement of Rep. McLane); *see also id.* at 1977 (statement of Rep. Cassidy) ("If this undesirable element is ever to be checked now is the time to do it."); *id.* at 2162–63 (statement of Rep. Washburn) ("[T]here is no better time than the present . . . [to] deal definitely and effectively with this question of Chinese immigration and cheap labor . . . labor of a lower grade, if possible, than the slave labor of the South . . . .").

176. *Id.* at 1636 (statement of Sen. Slater).

Mongolian; for him they have no affiliation nor welcome . . . ."[177] Many legislators argued that African Americans were superior to the Chinese;[178] indeed, some legislators justified exclusion on the ground that it would protect African Americans from labor competition[179] or moral corruption.[180] However, admitting the principle that any race could be subordinated put every nonwhite race at risk. As Professor Konvitz wrote more than five decades ago,

> [a]fter 1876 the Negro problem and the Chinese question were linked when it came to voting in Congress on anti-Chinese measures . . . . The South, it has been said, "was quite willing to join with the Pacific Coast in fitting the Chinese into a caste system which, in many respects, closely resembled that which prevailed throughout the former slave belt."[181]

4. Later Asian Exclusion Legislation

Subsequent laws excluding Chinese and other Asians have comparatively abbreviated legislative histories, but they reflect similar sentiments. In 1902, for example, when the Chinese Exclusion Act was made permanent, legislators were again concerned that "[t]he Chinaman in

177. *Id.* at 2129 (statement of Rep. Hayne). Representative Cassidy argued that potentially worrisome principles of the bill were inapplicable to African Americans because of differences between them and the Chinese; the interests of African Americans "are in no sense invaded or threatened by . . . this bill . . . . It is aimed at a non-voting, non-progressive, and non-assimilative people, who have no attachments for this country . . . . None of these things will lie against the citizen of African descent any more than against the man of Caucasian blood." *Id.* at 1979 (statement of Rep. Cassidy).

178. Representative Speer explained that African Americans were superior to Chinese: "A typical negro is infinitely superior to a typical Chinaman. . . . Sir, you will never see the spectacle of the entire South rising up against the negro as you see the entire West rising up against the Chinaman." *Id.* at 2029 (statement of Rep. Speer). Representative Brumm agreed: "You have the negro; he is here; you cannot put him below the Chinaman; he is a citizen and a voter; he is the superior of the Chinaman in almost every respect." *Id.* at 2043 (statement of Rep. Brumm). Senator Farley stated: "[T]he African people . . . are a superior people to the Chinese in many respects. . . . " *Id.* at 1582 (statement of Sen. Farley).

179. *See id.* at 2177 (statement of Rep. Richardson).

180. Senator Maxey, for example, argued:

> The colored man is naturally superstitious, and how far he might be carried away from Christian civilization after the Joss god of the Chinaman no one can tell. . . . Do not let those colored men be brought in contact with these Chinese people. It will not do. Let them be alone with us.

Id. at 1584 (statement of Sen. Maxey).

181. MILTON R. KONVITZ, THE ALIEN AND THE ASIATIC IN AMERICAN LAW 12 (1946) (quoting CAREY MCWILLIAMS, BROTHERS UNDER THE SKIN 83 (1943)).

America is forever and always an alien,"[182] that "the Mongolian race [is] not a desirable addition to our population,"[183] and recognized that "[t]he Chinese problem is to the Pacific coast what the negro problem is to the Southern States."[184]

In short, the Asian Exclusion Laws bear the same indicia of belief in racial hierarchy that other forms of discrimination did. They were motivated by the idea that the United States was particularly reserved for white

---

182. 35 CONG. REC. 3807 (1902) (statement of Rep. Mondell); *id.* at 3685 (statement of Rep. Clark); *id.* at 3693 (statement of Rep. Kahn); *id.* at 3695–96 (statement of Rep. Naphen); *id.* at 3731 (statement of Rep. Otjen).

183. *Id.* at 3781 (statement of Rep. Lacey). The Representative continued: "[W]ith the world to choose from, we should select the best instead of the least desirable of the world's races." *Id.* Repsentative Naphen stated:

> With one exception, the tests imposed or suggested against those seeking the hospitality of our shores, have been standards of character, education and property, not racial. We have drawn the race line only against one nationality. . . . Obviously, there must exist special reasons for the exception made in this particular class. The characteristics of this race justify our action with overwhelming force.

Id. at 3695 (statement of Rep. Naphen). Representative Brick argued:

> They would not only supplant American workmen and degrade American labor, but they would come to us in multiplied thousands, devoid of all racial and political homogeneity, bringing with them social vices and national habits that would surely contaminate the clear stream of Christian civilization and American institutions, that must inevitably lower the standard of American citizenship and eventually undermine the Republic.

Id. at 3738 (statement of Rep. Brick); *see also id.* at 3685 (statement of Rep. Clark) ("It is largely a racial question, and it raises the paramount issue, 'Shall the white man continue to dominate the Western Hemisphere, or shall he be placed in the process of ultimate extinction and be supplanted by the yellow man?'"); *id.* at 3794 (statement of Rep. Sulzer) ("It is largely a racial question of supremacy and essentially for the protection of labor."); *id.* at 3718 (statement of Sen. Fairbanks) ("There is nothing immoral in our exclusion of those who do not tend to elevate our civilization.").

184. *Id.* at 3681 (statement of Rep. Clark). Representative Clark continued:

> I believe, moreover, that the white people of the South are the most capable of dealing with their race question, just as the white people of the Pacific coast are most competent to deal with their Chinese race question. [Applause.] Upon these race questions I unhesitatingly take my position with the white people of the South and the white people of the Pacific coast.

Id. Representative Adams observed that the African-American question "does not concern an evil which can be prevented, as does the measure involving the Chinese question. In the one case the evil is an accomplished fact; the other, I am thankful to say, involves an evil which it is in our power to avert." *Id.* at 3687 (statement of Rep. Adams). Representative Kern stated:

> We have a race problem in this country now crying aloud for adjustment. . . . I refer to the negro problem. It is idle to contend that the civil war solved that problem. The civil war only presented it for the future generations to deal with. . . . Let us honestly try in an impartial spirit and with patriotic resolve to deal with the race question which the fathers have forced upon us, prompted by considerations of humanity and equal justice and love of our country and of the unfortunate race that is in our midst, but let us in wisdom refrain from taking upon ourselves the burden of bringing into this country another race problem to menace American institutions.

Id. at 3741–42 (statement of Rep. Kern).

people. Inevitably, the task of calculating the relative worth of one race, the Chinese, opened the question of the merits of other races, such as Americans of African descent.

C. Where Are the Rest of Us? The Present Effects of the Asian Exclusion Laws

The effects of the laws upheld in *Fong Yue Ting* and *Chae Chan Ping* are with us today. The consequences of racial discrimination in immigration are similar to domestic legal manifestations of white supremacy. Proponents of the Chinese Exclusion Act spoke of the Chinese "question":[185] The question, specifically, was whether there should be any Chinese in the United States. One proponent of Chinese exclusion explained:

> Now, sir, the friends of this measure have no disposition to disguise its purposes. It means to suspend and practically extinguish the future coming of Chinese coolies to the United States.... Let the future coming of Chinese laborers be suspended under the operations of this act, and let those who are now here, unless time and experience shall prove them worthy of citizenship, quietly and without violence melt away.[186]

When Asian exclusion was made comprehensive in 1924, Congress echoed this sentiment: "All must agree that nothing can be gained by permitting to be built up in the United States colonies of those who cannot, under the law, become naturalized citizens...."[187]

The structure of the laws suggested that the ideal number of Chinese Americans was zero. When debating the Chinese Exclusion Act, some senators and representatives who opposed a blanket exclusion argued that any ban should be limited to "undesirable" immigrants, such as unskilled laborers, those without families, immigrants who could not read and write

185. *See* 13 CONG. REC. 1751 (1882) (statement of Sen. Jones) ("We want this question settled."); *id.* at 2126 (statement of Rep. McClure) ("Now is the time to settle this question.").

186. *Id.* at 2209–10 (remarks of Rep. Hazelton). Others on both sides of the issue agreed that the question was practical elimination of the Chinese presence: "You call it a suspension, but no use of words can disguise the fact that it is absolute prohibition for twenty years." *Id.* at 1702 (statement of Sen. Platt). "[I]t is about as near an absolute prohibition as an ordinary lease of nine hundred and ninety-nine years is to a sale." *Id.* at 1938 (statement of Rep. Tyler). "Twenty years may not be too long for our national welfare to close our ports against Mongolian pauper labor; half a century, one hundred years, forever, may ultimately become our wisest policy." *Id.* at 2129 (statement of Rep. Scranton).

187. H.R. REP. NO. 68-350, at 6 (1924).

English, prostitutes, criminals, and diseased persons.[188] Nevertheless, the House rejected several amendments that would have provided for exclusion of only these classes,[189] as well as an amendment that would have dealt with the putative "invasion" problem by imposing a ceiling of 10,000 Chinese immigrants per year.[190] Accordingly, the evidence gives credence to Senator Hoar's claim that this legislation was not aimed at undesirable Chinese, but at all Chinese.

> This legislation says: "If your skin be yellow, if your father and mother were Mongolians, you may be as pure, as good, as law-abiding, as useful, as honorable, as honest as any man who lives, but because of that into which no moral quality possibly can enter, the color of your skin, you shall be made an exile or a criminal, or sent to the penitentiary, if you stay."[191]

In sum, the majority that voted for the bill presumably were persuaded by Representative Cassidy's argument that "[a]ll of them are bad."[192] Subsequent statutes regulating Asian immigration also imposed blanket bans rather than selective restriction.

Numbers mean power, or at least, potential power. As Eldridge Cleaver wrote in 1968:

> There are 20,000,000 of these blacks in America, probably more. Today they repeat, in awe, this magic number to themselves: there

---

188. Senator Sherman argued that "[i]f there are intelligent Chinamen, skilled laborers, able to read and write, with wives and families, who choose to come here and make their homes among us and help us develop this country, there is no danger from that kind of immigration . . . . I think the bill has gone too far." 13 CONG. REC. at 1749 (statement of Sen. Sherman). Representative Williams invited his colleagues to impose

> any reasonable restrictions . . . ; provide for the exclusion of the vicious, the criminal, the profligate, the dissolute; go further, and provide exceptional remedies for exceptional emergencies; require that Chinamen shall learn our language, adopt our dress, attend our schools, conform to our manners and customs, become freeholders, and live in families, . . . or anything within the bounds of reason, and I will vote for the bill, however harsh these conditions may seem.

Id. at 2039 (statement of Rep. Williams). Senator Platt said he would support a bill that excluded immigrants who posed the dangers identified by the bill's advocates: "coolies," criminals, prostitutes, and diseased persons. *Id.* at 1703 (statement of Sen. Platt). Because the law imposed a blanket ban, he concluded that "its only aim and result [is] the extirpation and the exclusion of the Chinaman from this country. . . . The true intent . . . is to declare that henceforth, excepting only the Chinese now here and the colored people now here, no man shall work in the United States except he be a white man." *Id.* at 1705; *see also, e.g.*, *id.* at 1938 (statement of Rep. Tyler); *id.* at 1981 (statement of Rep. Taylor); *id.* at 2035 (statement of Rep. Moore); *id.* at 2037 (statement of Rep. Carpenter); *id.* at 2177 (statement of Rep. Browne); *id.* at 2183 (statement of Rep. Dawes); *id.* at 2185 (statement of Rep. Joyce); *id.* at 2189 (statement of Rep. Orth).

189. *See id.* at 2226.

190. *See id.*

191. *Id.* at 1709 (statement of Sen. Hoar); *see also id.* at 2174 (statement of Rep. Lord).

192. *Id.* at 1979 (statement of Rep. Cassidy).

> are 20,000,000 of us! They shout this to each other in humiliated astonishment. No one need tell them that there is vast power latent in their mass. They know that 20,000,000 of anything is enough to get some recognition and consideration.[193]

In the United States today, there are more Jews than there are in Israel,[194] more persons of Irish descent than there are in Ireland,[195] and more than half as many people of English and German descent as there are residents of England and Germany.[196] America is one of the largest Latino countries in the world,[197] and the American population of African ancestry is larger than the population of twenty-nine of the thirty-five African nations.[198] But, though half the world is Asian, in this "nation of immigrants," there are fewer Asian Americans than there are residents of a single large Asian city such as Bombay or Jakarta.[199]

No one knows what the Asian-American population would be today if Asians had been allowed to immigrate on the same basis as members of other racial groups. The effect of Asian exclusion was no doubt significant. Asians have been eager to come to the United States since the immigration laws were liberalized;[200] over a third of the post-1965 immigration has come

193. ELDRIDGE CLEAVER, SOUL ON ICE 136 (1968).

194. *See* BUREAU OF THE CENSUS, U.S. DEPT. OF COMMERCE, STATISTICAL ABSTRACT OF THE UNITED STATES 1996, at 70 tbl.89 (116th ed. 1996) (estimating the Jewish population of the United States to be 5,880,000) [hereinafter ABSTRACT]; INFORMATION PLEASE ALMANAC 128 (48th ed. 1995) (estimating Israel's population to be 5,400,000) [hereinafter INFORMATION PLEASE].

195. *See* ABSTRACT, *supra* note 194, at 53 tbl.56 (estimating the American population of Irish descent to be 38,736,000); INFORMATION PLEASE, *supra* note 194, at 128 (estimating the population of Ireland to be 3,600,000).

196. *See* ABSTRACT, *supra* note 194, at 53 tbl.56 (estimating the American population of German ancestry to be 57,947,000 and English ancestry to be 32,652,000); INFORMATION PLEASE, *supra* note 194, at 128–29 (estimating the population of Germany to be 81,200,000 and the United Kingdom to be 58,400,000).

197. *See* ABSTRACT, *supra* note 194, at 14 tbl.12 (estimating the Hispanic origin population of the United States as of 1995 to be 26,994,000).

198. *See* CENTRAL INTELLIGENCE AGENCY, THE WORLD FACTBOOK 1995 (1995) (listing populations of all countries); ABSTRACT, *supra* note 194, at 14 tbl.12 (estimating the African-American population as of 1995 to be 33,141,000).

199. *See* ABSTRACT, *supra* note 194, at 14 tbl.12 (estimating the Asian-American population as of 1995 to be 9,287,000); INFORMATION PLEASE, *supra* note 194, at 130 (estimating the population of Bombay to be over 12 million and Jakarta to be over 10 million).

200. *See* Brinley Thomas, *From the Other Side: A European View*, 367 ANNALS AM. ACAD. POL. SOC. SCI. 63, 67–68 (1966). Thomas noted that a 1962 law allowing immigration of scientists and engineers resulted in a 23% increase in 1963 of European scientists and engineers, but a 182% increase of Asians.

> This indicates that skilled Asian immigration has a remarkably high elasticity with respect to a moderate liberalizing of the restrictions. If the relatively minor supplement to the law in October 1962 led to an almost threefold increase in the influx of highly

from Asia.[201] Even during the exclusion period, illegal Asian immigration to the United States was common.[202] Moreover, throughout the exclusion period, some large Asian countries, such as China and Japan, have had "surplus" population and have had substantial emigration.[203]

Politically, Asians are not interchangeable with members of other races.[204] For example, over half of the nearly ten million Asian Americans live outside of Hawaii and California, but all of the five Asian members of Congress and Senators are from those two states. Asian Americans are disadvantaged because the bloc from which they may expect support has been kept artificially small; as Justice (then-judge) Stephen Breyer explained in a Voting Rights Act case, writing about Boston but with wider implications, "the Asian population . . . is so small that its 'submergence' is inevitable."[205]

Their relatively small population in the United States harms Asians in the same way it does any other small group; politicians have less reason to be responsive to them. As Bruce Ackerman explained:

> Consider, for example, an American constituency that includes 12 percent blacks and .5 percent Jehovah's Witnesses among its popula-

---

> qualified Asians, one wonders what the basic alteration in the principles of selection enacted in 1965 is likely to do.

Id.

201. *See* ABSTRACT, *supra* note 194, at 11 tbl.8 (providing statistics from 1971–94); U.S. DEPARTMENT OF JUSTICE, 1970 ANNUAL REPORT, IMMIGRATION AND NATURALIZATION SERVICE 64 tbl.14 (providing statistics for 1965–70).

202. *See* HING, *supra* note 82, at 73–75 (discussing illegal immigration).

203. Since 1965, Asians have been able to immigrate on the same basis as other groups and, as a result, have substantially increased their population. But it was not designed as a remedy; there was no preference for Asians to permit them to make up for lost ground. It does not put them in the position they would have enjoyed absent discrimination; had there been no Asian exclusion, they would have had 80 years of nondiscriminatory immigration *plus* what they enjoyed following the 1965 law.

204. Judges deciding Voting Rights Act cases often decline to aggregate Asian Americans with other minority groups for purposes of determining whether the minority population is large enough to warrant creation of a majority-minority district. *See, e.g.*, Debaca v. County of San Diego, No. 92-55661, 1993 WL 379838, at *2 (9th Cir. Sept. 27, 1993) (affirming the rejection of a Voting Rights Act claim premised on aggregation of the population of blacks, Hispanics, and Asian Americans because there was "no showing" of "political cohesiveness among the multi-ethnic class as a whole"); Brewer v. Ham, 876 F.2d 448, 453–54 (5th Cir. 1989) (affirming a finding of noncohesiveness between blacks, Asian Americans, and Hispanics); Latino Political Action Comm. v. City of Boston, 609 F. Supp. 739, 746 (D. Mass. 1985) (finding no evidence of cohesion between blacks and Latinos, or Asian Americans and blacks), *aff'd*, 784 F.2d 409 (1st Cir. 1986) (Breyer, J.).

205. *Latino Political Action Comm.*, 784 F.2d at 415; *see also, e.g.*, Barnett v. Daley, 32 F.3d 1196, 1200 (7th Cir. 1994) (noting that Asian Americans and Native Americans are "too small and scattered to elect anyone"); *cf.* Jeffrey Goldberg, *The Overachievers*, N.Y. MAG., Apr. 10, 1995, at 43, 46 (describing the "idea of a Korean-American running for mayor of New York" as "sheer fantasy"). *But see* Frank H. Wu, *The Making of an Asian American President*, A. MAG.: INSIDE ASIAN AMERICA, Oct./Nov. 1996, at 34.

> tion. Doubtless, both groups will be encouraged by the pluralist vision of democracy, since it suggests that neither group will inexorably be excluded from the pluralist bazaar. Nonetheless, it should be plain that these two groups have absolutely no reason to find the prospect of pluralist bargaining *equally* gratifying. To the contrary, the fact that blacks greatly outnumber Witnesses is bound to play an important role in any plausible bargaining theory. Thus, even if the two groups could somehow be compensated for their . . . disadvantages, the Witnesses could not reasonably expect to win substantive victories nearly as often as the blacks.[206]

The Civil Rights Act of 1991 offers an illuminating example of the fact that Asian Americans cannot count on other groups treating their interests as highly as their own. Section 402(b) of the act provides: "Notwithstanding any other provision of this Act, nothing in this Act shall apply to any disparate impact case for which a complaint was filed before March 1, 1975, and for which an initial decision was rendered after October 30, 1983."[207] This odd restriction applied to a single lawsuit: *Wards Cove Packing Co. v. Atonio*,[208] an employment discrimination suit against salmon canneries involving an Asian Pacific American and Alaskan Native plaintiff class.[209] Exclusion of this particular case from the law's coverage was necessary to garner support for the controversial bill.

It was not arbitrary that a suit involving Asian Americans, rather than members of some other group, was used as a bargaining chip. It is hard to imagine the Congressional Black Caucus agreeing to give up an important lawsuit, for example, the litigation over systematic discrimination against African Americans at the Denny's restaurant chain, in order to get the bill passed. If the Congressional Asian Pacific American Caucus were a large group, it would not have voluntarily sacrificed a case so prominent in the Asian-American community. Section 402(b) became law because Asian

206. Bruce A. Ackerman, *Beyond* Carolene Products, 98 HARV. L. REV. 713, 721 (1985). It is for this reason that the size of the homosexual population is sometimes hotly debated. *See, e.g.*, Frank Bruni, *Activists Slam Study That Puts Gays at 1%*, DETROIT FREE PRESS, Apr. 16, 1993, at 5A (noting the reaction to a survey putting the gay population at 1% and stating that "[g]ay activists predicted that opponents of gay rights will use the new figure to argue that gays represent a tiny, irrelevant minority"); Thaddeus Herrick, *Groups at Odds over Study on Numbers of Gays*, ROCKY MTN. NEWS, Apr. 16, 1993, at 4A.

207. Civil Rights Act of 1991, Pub. L. No. 102-166, § 402(b), 105 Stat. 1071, 1099.

208. 490 U.S. 642 (1989).

209. *Id.* at 663 n.4 (Stevens, J., dissenting). Chinese, Japanese, Filipino, and Mexican workers had a decades-long tradition of employment in the canneries, where "the season was short, the hardship great and the pay low." Hudson N. Janisch, The Chinese, the Courts, and the Constitution: A Study of the Legal Issues Raised by Chinese Immigration to the United States, 1850–1902, at 924–30 (1971) (unpublished J.S.D. dissertation, University of Chicago) (on file with author).

Americans did not have the power to stop it, and Asian Americans did not have the power to stop it because they were too few in number to elect many representatives to Congress.

Congressional reaction to the incarceration of Japanese Americans might have been quite different if there had been a powerful bloc of Asian-American lawmakers. Even Chinese- or Indian-American legislators might have been suspicious of this action, even though it was directed against persons of a different national ancestry. Similarly, the perception that Asian-American applicants to colleges and graduate schools are treated less advantageously than whites[210] might engender a different reaction if Asian Americans, rather than coming hat in hand, had the political power to force schools to take their interests seriously. Indeed, in 1882, one representative argued: "If the one hundred and five thousand Chinamen now in this country were armed with the freeman's great weapon of defense, the ballot, you would never have heard of [the Chinese Exclusion Act]."[211]

Population has economic effects as well. Because Asians are a small group, as a matter of simple arithmetic it is clear that they must either spread thin or congregate in particular occupations or geographical areas. In either event, it may be more difficult for Asian Americans who choose police work rather than engineering or live in Indianapolis, Indiana rather than Sacramento, California. There will be fewer relatives and friends with advice about how to get into a particular line of work, fewer fellow Asians who have paved the way, and fewer potential contacts and clients.

In cultural areas, too, members of a smaller group are at a natural disadvantage. A writer of Asian-American history or literature has a smaller built-in market. Even leaving aside political considerations, in a situation of scarcity, it may seem less urgent for a university to have an Asian-American studies department because the group is so much smaller than some other minorities. As a result, Asian-American scholars studying Asian-American issues may find it more difficult to find work.

---

210. *See* GABRIEL CHIN, SUMI CHO, JERRY KANG & FRANK WU, BEYOND SELF-INTEREST: ASIAN PACIFIC AMERICANS TOWARD A COMMUNITY OF JUSTICE, A POLICY ANALYSIS OF AFFIRMATIVE ACTION 23 & n.149 (1996), *reprinted in* UCLA AS. PAC. AM. L.J. (forthcoming); John H. Bunzel & Jeffrey K.D. Au, *Diversity or Discrimination? Asian Americans in College*, PUB. INTEREST, Spring 1987, at 49; Grace W. Tsuang, Note, *Assuring Equal Access of Asian Americans to Highly Selective Universities*, 98 YALE L.J. 659, 660–65 (1989).

211. 13 CONG. REC. 2184 (1882) (statement of Rep. Joyce). Representative Taylor stated:

> I want to know now if you can protect one class of laborers upon any principle of exclusion of other laborers. Are you doing it when you keep them away? But you say: "Oh, that is very different, because the other people are white men." That is, if you go clear to the bottom, "because the other people's friends have ballots." When you go there you strike the bottom rock, although it might not be popular to say that.

Id. at 1982 (statement of Rep. Taylor).

Asian immigrants share the disadvantages of native born citizens; they find themselves to be members of a smaller and less powerful group. There are fewer people who speak their language, making their initial adjustment more difficult or requiring them to stay close to cities with higher Asian populations. Asian immigrants have been damaged in another way: Some recent arrivals would already be Americans had their forebears been allowed to immigrate. Their position would be easier because their parents or grandparents would have already established themselves in the United States by, for example, learning English and unfamiliar customs.[212]

Discrimination in immigration also sheds light on the domestic discrimination Asian Americans experienced. The Asian Americans who made it to the United States in previous generations experienced familiar forms of racial discrimination at the hands of the law.[213] They were subject to electoral disenfranchisement,[214] exclusion from desirable neighborhoods through restrictive covenants,[215] testimonial disqualification,[216] school segre-

212. Thus, the argument that immigrants "have no conceivable claim upon our nation's conscience" is debatable. *See* Mark Krikorian, *Affirmative Action and Immigration*, *in* DEBATING AFFIRMATIVE ACTION: RACE, GENDER, EQUALITY, AND THE POLITICS OF INCLUSION 301 (Nicolaus Mills ed., 1994).

213. *See generally* 1–4 ASIAN AMERICANS AND THE LAW (Charles McClain ed., 1994); CHIN ET AL., *supra* note 210; HING, *supra* note 82.

214. *See* CAL. CONST. art. II, § 1 (1879) (repealed 1926) (making a "native of China" ineligible to vote). Disenfranchising Chinese aliens was not merely symbolic in 1879; at the time, aliens could vote in many states. *See* Jamin B. Raskin, *Legal Aliens, Local Citizens: The Historical, Constitutional and Theoretical Meanings of Alien Suffrage*, 141 U. PA. L. REV. 1391 (1993); Gerald M. Rosberg, *Aliens and Equal Protection: Why Not the Right to Vote?*, 75 MICH. L. REV. 1092 (1977). California adopted a literacy test designed in part to prevent Chinese Americans from voting. *See* Castro v. State, 466 P.2d 244, 248 n.11 (Cal. 1970).

215. In *Amer v. Superior Court* and *Yin Kim v. Superior Court*, the petitioners were

> United States citizens and decorated overseas veterans of the Armed forces. Petitioner Amer [was] of Chinese ancestry; petitioner Kim [was] of Korean ancestry. [They were] threatened with state court injunctive orders directing them to vacate their homes because their properties [were] covered by restrictive covenants which provide[d] that persons who are not of the Caucasian race cannot occupy the premises except in the capacity of servants.

Consolidated Petitions for Writs of Certiorari at 4, Amer v. Superior Court, 334 U.S. 813 (1948) (No. 47-429) & Yin Kim v. Superior Court, 334 U.S. 813 (1948) (No. 47-430). The Supreme Court vacated the injunctions and directed the California Supreme Court to reexamine the cases in light of the just-decided *Shelley v. Kraemer*, 334 U.S. 1 (1948), which held racially restrictive covenants judicially unenforceable. *See also* Gandolfo v. Hartman, 49 F. 181 (C.C.S.D. Cal. 1892) (invalidating an anti-Chinese restrictive covenant). *See generally* CLEMENT E. VOSE, CAUCASIANS ONLY: THE SUPREME COURT, THE NAACP, AND THE RESTRICTIVE COVENANT CASES 212 (1959).

216. *See* People v. Hall, 4 Cal. 399, 404–05 (1854) (finding it reversible error to admit testimony of Chinese against a white defendant in a murder case); People v. Washington; 36 Cal. 658, 670–72 (1869) (holding that Chinese witnesses could not testify against an African American); *see also* State v. Mah Jim, 10 P. 306, 306–07 (Or. 1886) (stating that "[e]xperience convinces every one that the testimony of Chinese witnesses is very unreliable" and reversing

gation,[217] prohibitions on property ownership,[218] racial violence,[219] prohibitions on marriage to whites,[220] unequal enforcement of facially neutral laws,[221] and disqualification from many businesses and professions.[222]

The legal history of Asian Americans, including the discrimination they experienced, is far from fully developed. It is clear, however, that certain whites have made sustained efforts to keep Asians out of the social and economic life of the nation. Nevertheless, if considered in isolation from Asian exclusion, the legal disabilities that Asian Americans suffered would be a less important factor for them than for, say, African Americans. Part of this is a consequence of demographics. Many African-American adults alive today personally experienced de jure racial discrimination; the parents or grandparents of most African Americans almost certainly did. By contrast, more than 85% of the Asian-American population was born or immigrated after 1960, so relatively few lived under overtly discriminatory American laws or had parents who did so.[223] The Asian-American subgroups who have been in the United States the longest—and therefore most likely to have experienced Jim Crow type discrimination—are

murder conviction); 14 THE ENCYCLOPAEDIA OF EVIDENCE 159 (Edgar W. Camp ed., 1909) ("Chinese and other oriental races, are recognized as having little regard for the obligation of an oath, but their testimony may not be arbitrarily rejected." (footnotes omitted)).

217. *See* Gong Lum v. Rice, 275 U.S. 78, 86–87 (1927) (holding that a Chinese-American student was not entitled to attend schools with whites); CHARLES M. WOLLENBERG, ALL DELIBERATE SPEED: SEGREGATION AND EXCLUSION IN CALIFORNIA SCHOOLS, 1855–1975, at 8–107 (1976) (describing segregated schools for Chinese, Japanese, Native Americans, and African Americans in California).

218. *See, e.g.*, Dudley O. McGovney, *The Anti-Japanese Land Laws of California and Ten Other States*, 35 CAL. L. REV. 7 (1947).

219. *See* CHIN ET AL., *supra* note 210, at 22–23.

220. *See, e.g.*, *In re* Shun T. Takahashi's Estate, 129 P.2d 217 (Mont. 1942) (ruling that the marriage between a Japanese and a Caucasian is void); Naim v. Naim, 87 S.E.2d 749, 756 (Va.) (declaring the marriage between a Chinese and a Caucasian void), *vacated*, 350 U.S. 891 (1955), *adhered to on remand*, 90 S.E.2d 849, 850 (Va.), *motion to recall mandate denied*, 350 U.S. 985 (1956).

221. *See* Yick Wo v. Hopkins, 118 U.S. 356, 374 (1886) (holding unconstitutional a laundry licensing provision discriminatorily applied against Chinese).

222. *See In re* Hong Yen Chang, 24 P. 156, 157 (Cal. 1890) (holding that because a Chinese American was racially ineligible to naturalize, he could not be admitted to the bar); *In re* Takuji Yamashita, 70 P. 482, 483 (Wash. 1902) (same for a Japanese American).

223. In 1990, there were 7,458,000 Asian and Pacific Islanders in the United States. *See* ABSTRACT, *supra* note 194, at 14 tbl.12. Of those, almost two-thirds were themselves foreign born, *see id.* at 52 tbl.55, and others were undoubtedly young children of immigrants. In 1960, there were just over a million Asian and Pacific Islanders in the United States. *See* 1 BUREAU OF THE CENSUS, U.S. DEP'T OF COMMERCE, CENSUS OF POPULATION: 1960, pt. 1, at 144, tbl.44 [hereinafter 1 CENSUS]. *See generally* Robert S. Chang & Keith Aoki, *Centering the Immigrant in the Inter/National Imagination*, 85 CAL. L. REV. 1395 (1997), 10 LA RAZA L.J. 309 (1998).

often better off economically than the newer arrivals[224] and, on average, have more income and years of education than African Americans.[225] In sum, if the Asian-American experience of discrimination is viewed as simply a variation of the African-American situation, the impact of such discrimination might well be seen as having been less intense and consequential.

This analysis, however, ignores significant attitude-shaping effects of the Asian Exclusion Laws. A central harm of discrimination is that it reinforces notions of group inferiority that may continue long after the discrimination stops.[226] Justice O'Connor's observation that de jure discrimination may harm persons of the affected race by denigrating "their very worth as citizens"[227] is particularly compelling when the classification denies them membership as citizens in the American community itself. Prohibiting immigration of aliens of a particular race stigmatizes Americans of the race deemed undesirable by their nation's laws. Laws justified only "as measures designed to maintain White Supremacy"[228] are painful and insulting to those disadvantaged, regardless of whether a particular law directly affects a particular person on a particular day.[229]

---

224. *See, e.g.*, Robert S. Chang, *Toward an Asian American Legal Scholarship*, 81 CAL. L. REV. 1241, 1261 & n.83 (1993) (noting that Laotians, Hmong, Cambodians, and Vietnamese have poverty rates from 3.5 to 7 times the national average, and only Japanese and Filipinos have below average rates). *See generally* HERBERT BARRINGER ET AL., ASIANS AND PACIFIC ISLANDERS IN THE UNITED STATES ch. 8 (1993) (discussing relative incomes of various Asian American groups).

225. *See* Chang, *supra* note 224, at 1261–62. *See generally* ANDREW HACKER, TWO NATIONS: BLACK AND WHITE, SEPARATE, HOSTILE, UNEQUAL (1992) (describing racial differences in income, education, and employment).

226. This, of course, was one of the reasons advanced by the Court in *Brown*: "To separate [African American students] from others of similar age and qualifications solely because of their race generates a feeling of inferiority as to their status in the community . . . in a way unlikely ever to be undone." Brown v. Board of Educ., 347 U.S. 483, 494 (1954); *see also* Nadine Strossen, *Blaming the Victim: A Critique of Attacks on Affirmative Action*, 77 CORNELL L. REV. 974, 979 (1992) ("It . . . seems a matter of common sense that a group of people who know they are to be systematically denied certain opportunities would become discouraged, and such discouragement would breed a lack of will, a lack of motivation, a lack of incentive.").

227. Metro Broad., Inc. v. FCC, 497 U.S. 547, 604 (1990) (O'Connor, J., dissenting).

228. Loving v. Virginia, 388 U.S. 1, 11 (1967). When the Kennedy and Johnson administration proposals to end race discrimination in the INA were before Congress, Secretary of State Dean Rusk pointed out the domestic effect of race discrimination in immigration law: "[W]e do not wish to be in the position of turning to one of our own fine citizens . . . and of saying to him, 'Well, you were lucky. You just barely squeaked through. We are glad to have you, we hope you do well, but we don't want many more like you.'" *Immigration: Hearings on H.R. 2580 Before Subcomm. No. 1 of the House Comm. on the Judiciary*, 89th Cong. 93 (1965). Rusk's point seems inescapable.

229. *See, e.g.*, Allen v. Wright, 468 U.S. 737, 755 (1984) ("[T]he stigmatizing injury often caused by racial discrimination . . . is one of the most serious consequences of discriminatory government action and is sufficient in some circumstances to support standing."); Norwood v.

The stigma of discrimination is not a one-way street. An old saw proposes "that if you can think about something that is related to something else without thinking about the thing to which it is related, then you have the legal mind."[230] Even the keenest jurisprudential intellect, however, might find it difficult to discriminate against aliens on the basis of their race while treating American citizens of that race with fairness and respect. Americans of disfavored races would be justifiably suspicious that judges or legislators who found discrimination against members of one race to be desirable or even just legal would be unenthusiastic about preventing domestic discrimination against citizens of that race. Exclusion of aliens of particular races who were "in other respects fully qualified" constituted "an assertion of their inferiority, and a stimulant to that race prejudice which is an impediment to securing to individuals of the race that equal justice which the law aims to secure to all others."[231]

The Asian Exclusion Laws enshrined the idea that Asians were ineradicably foreign and un-American, that they were dangerous and inferior. Some of these attitudes persist,[232] even as some of the racial ideology

---

Harrison, 413 U.S. 455, 466 (1973) ("A State may not grant the type of tangible financial aid here involved if that aid has a significant tendency to facilitate, reinforce and support private discrimination."); Reitman v. Mulkey, 387 U.S. 369, 381 (1967) (invalidating a state constitutional amendment on the ground that it would "significantly encourage and involve the State in private discriminations"). Professor Karst puts it this way:

> When a city segregates the races on a public beach, the chief harm to the segregated minority is not that those people are denied access to a few hundred yards of surf. Jim Crow was not just a collection of legal disabilities; it was an officially organized degradation ceremony, repeated day after day in a hundred ways, in the life of every black person within the system's reach.

KENNETH L. KARST, BELONGING TO AMERICA: EQUAL CITIZENSHIP AND THE CONSTITUTION 4 (1989).

230. LON L. FULLER, THE MORALITY OF LAW 4 (rev. ed. 1969) (attributing the statement to Thomas Reed Powell).

231. Strauder v. West Virginia, 100 U.S. 303, 308 (1880).

232. *See, e.g.*, *Anti-Asian Violence: Oversight Hearing Before the Subcomm. on Civil and Constitutional Rights, of the Comm. on the Judiciary of the House of Representatives*, 100th Cong. 4 (1987) (Testimony of Rep. Mineta) ("To many people, then and now, Americans of Asian ancestry are more Asian than they are American . . . ."); U.S. COMMISSION ON CIVIL RIGHTS, CIVIL RIGHTS ISSUES FACING ASIAN AMERICANS IN THE 1990S, at 20 (1992) ("The perception of Asian Americans as foreigners may also impede their acceptance in all areas of their lives and contribute to subtle as well as overt forms of discrimination against them in education, employment and other arenas."); U.S. COMMISSION ON CIVIL RIGHTS, RECENT ACTIVITIES AGAINST CITIZENS AND RESIDENTS OF ASIAN DESCENT 31 (1986) ("Asian American groups perceive that some in our society still believe that all Asian Americans are foreigners."); Sumi K. Cho, *Converging Stereotypes in Racialized Sexual Harassment: Where the Model Minority Meets Suzie Wong*, 1 J. GENDER, RACE & JUST. 177 (1997); Chris K. Iijima, *The Era of We-Construction: Reclaiming the Politics of Asian Pacific American Identity and Reflections on the Critique of the Black-White Paradigm*, 29 COLUM. HUM. RTS. L. REV. 47 (1997); Natsu Taylor Saito, *Alien and Non-Alien Alike: Citizenship, "Foreignness," and Racial Hierarchy in American Law*, 76 OR. L. REV. 261 (1997); Murray

underlying Jim Crow and slavery continues to affect African Americans. For example, the scandal now raging regarding contributions by Asians and Asian Americans to the Democratic National Committee for the 1996 presidential election has taken on an extremely racialized tone.[233] Similarly, the controversy over whether Asians should represent such a substantial proportion of the immigration stream[234] implies that an alien from Asia has less potential for becoming a satisfactory American than an alien from Europe.

Asian exclusion therefore made stereotypes about Asians more difficult to eradicate. Lived experience is a critical method of ameliorating prejudice and stereotypes; it is easier to hold strangers in contempt than fellow citizens, neighbors, and friends. For example, seeing a competent professional of a surprising color or gender may dispel the notion that some tasks are beyond the capabilities of particular groups.[235] Exposure of this kind is to a large degree a numbers game. Assume, for example, that Asian

---

Polner, *Asian Americans Say They Are Treated Like Foreigners*, N.Y. TIMES, Mar. 7, 1993, § 13 (Long Island ed.), at 1.

233. *See* Phil Tajitsu Nash & Frank Wu, *Asian-Americans Under Glass*, NATION, Mar. 31, 1997, at 15. *See generally* Bruce D. Brown, *Alien Donors: The Participation of Non-Citizens in the U.S. Campaign Finance System*, 15 YALE L. & POL'Y REV. 503 (1997).

234. Massachusetts Congressman Brian Donnelly supported the so-called diversity visa program by arguing that "our Nation must reintroduce into the immigrant stream . . . the peoples who have traditionally made up our immigrant stock." *Legal Immigration Legislation: Hearings Before the Subcommittee on Immigration, Refugees and International Law of the House Committee on the Judiciary*, 100th Cong. 7 (1988). Designed to favor Ireland and other Northern European countries, it became the Immigration Amendments of 1988, Pub. L. No. 100-658, 102 Stat. 3908 (1988), and, as modified, is now a permanent part of the INA. *See* Immigration and Nationality Act § 203(c); 8 U.S.C. § 1153(c). *See generally* Stephen H. Legomsky, *Immigration, Equality and Diversity*, 31 COLUM. J. TRANSNAT'L L. 319 (1993) (criticizing diversity program). Others have expressed concern about the increase of Asian and other nonwhite immigration. *See, e.g.*, BRIMELOW, *supra* note 51; John Berthelsen, *Politics of Exclusion: Anti-Immigrant Lobbyists Target Asians Again*, FAR E. ECON. REV., Nov. 18, 1993, at 27; Carl M. Cannon, *Focus of Immigration Policy Might Change; Europeans Now Feel Alienated*, MIAMI HERALD, Oct. 22, 1989, at 12A ("'We hope Congress takes into account the contributions which have been made to American culture by the Irish and other Europeans,' Donald Martin, of the Irish Immigration Reform Movement implored Congress recently."). The Irish immigration reform movement is driven in part by the notion that countries adversely affected by the 1965 immigration reform have been, somehow, discriminated against or treated unfairly. *See, e.g.*, Marvine Howe, *Working to Help Irish Immigrants Stay, Legally*, N.Y. TIMES, Nov. 27, 1988, at A52.

235. *See, e.g.*, Wygant v. Jackson Bd. of Educ., 476 U.S. 267, 315 (1986) (Stevens, J., dissenting) ("It is one thing for a white child to be taught by a white teacher that color, like beauty, is only 'skin deep'; it is far more convincing to experience that truth on a day-to-day basis . . . ."); T. Alexander Aleinikoff, *A Case for Race-Consciousness*, 91 COLUM. L. REV. 1060, 1071 (1991) ("Continued white ignorance of blacks and lack of contact in daily life makes white understandings of race difficult to alter."); Kenneth S. Tollett, *Black Institutions of Higher Learning: Inadvertent Victims or Necessary Sacrifices?*, 3 BLACK L.J. 162, 166 (1973) ("Many whites need integration to develop respectful, non-stereotypical perceptions of non-whites.").

Americans are perceived by some not to be "natural" lawyers[236] or military officers[237] or executives.[238] Even if, in a relatively nondiscriminatory world, they may be able to get positions as lawyers or officers in accordance with their numbers, proportional representation of Asians in such professions may be insufficient to make Asians anything other than oddities.

When considered in conjunction with Asian exclusion, it is harder to dismiss domestic discrimination against Asians as merely Jim Crow "lite"—less intense because Asian Americans were less despised. Domestic discrimination against Asians targeted a group that the Asian Exclusion Laws ensured was too small to be consequential. In 1965, for example, Asian Americans were six-tenths of one percent of the United States population;[239] the African-American proportion was more than seventeen times larger.[240] It would have been a gross misallocation of resources to devote too much energy to oppressing the tiny group of Asians because no matter how many professional jobs they obtained, or houses they purchased, they could not present a serious challenge to the status quo. Their insignificant place in the national society had been resolved by the exclusion laws.[241]

The much larger body of African Americans, by contrast, was far more dangerous to white supremacy because of their much greater potential to become a political and economic force. Proving the old adage that an ounce of prevention is worth a pound of cure, the work that had been done efficiently and effectively with respect to Asians through exclusion had to be painstakingly accomplished through Jim Crow separation.

236. *Cf.* David Quan, *Asian Americans and the Law: Fighting the Myth of Success*, 38 J. LEGAL EDUC. 619 (1988) (discussing the underrepresentation of Asian Americans in the legal profession).

237. *Cf.* Art Pine, *Marines Pin Bars on Man They Dismissed*, L.A. TIMES, Mar. 19, 1994, at A4 (discussing Bruce Yamashita, a Japanese American discharged from Marine officer training who successfully proved that his dismissal resulted from racial bias, for example, being told "[w]e don't want your kind around here—go back to your country").

238. *Cf.* Pan Suk Kim & Gregory B. Lewis, *Asian Americans in the Public Service: Success, Diversity and Discrimination*, 54 PUB. ADMIN. REV. 285 (1994) (suggesting that Asian Americans in government service are channeled into nonsupervisory roles); Susan Katz Miller, *Asian-Americans Bump Against Glass Ceilings*, 258 SCIENCE 1224, 1225 (1992); *America's Asians: The Glass Ceiling*, ECONOMIST, June 3, 1989, at 23 (discussing the "notion, conveniently subscribed to by the white majority, that Asian-American talents flower better in the back room, the laboratory or the class-room than in senior management").

239. *See* 1 CENSUS, *supra* note 223, at 144 tbl.44 (stating that Japanese, Chinese, Filipino, and other Asian and Pacific races constituted 1,096,021 of a total population of 179,323,175).

240. *See id.* (estimating African Americans to be 10.5% of the population).

241. This is not to deny that at particular times and in particular areas, such as the West Coast, relatively more substantial Asian population gave them the potential for economic power, and they were subject to relatively more intense discrimination. However, in most of the country and for most of American history, Asian Americans have represented a small percentage of the population.

The aims of Asian exclusion and its substantial success qualifies the perception that, "viewed in broad historical and social perspectives, . . . the attitude toward blacks has been qualitatively different—more persistent, more manifest, and more resistant to eradication."[242] That may be true, but it contains the questionable implication that the white attitude towards Asians was more benevolent. The Asian Exclusion Laws ensured that significant communities of Asian Americans would not be established in the United States. The violence and hostility towards Asians that led to the Asian Exclusion Laws did not diminish because white Americans came to like and respect Asians and embrace the idea of having them live in the United States. Instead, the goals of the project were substantially and permanently achieved.

Constant vigilance was required to deal with the threat to white supremacy represented by millions of African Americans living in this country who wanted freedom and justice. No mechanism known to slavery or Jim Crow could permanently silence their demands for equality; segregation had to be practiced every day, otherwise it might cease to work. By contrast, the Chinese inspectors and other immigration authorities simply kept most Asians out; bureaucratic efficiency offered a nearly complete solution to the problem Congress perceived.[243]

D. Depriving America of the Benefits of Citizens of Asian Ancestry

Many Asian Americans, like people who trace their forebears to other parts of the world, pay taxes, serve in the armed forces, study and work hard, and have contributed parts of their intellectual, artistic, and culinary traditions to the nation's culture. But the other immigrants who arrived during the period of exclusion also made worthwhile contributions to the United States. A major form of injury to the United States resulted not simply from the absence of Asian Americans, but from their deliberate exclusion.

Unless we have a second Civil War, the United States has no choice but to accept its situation as a multiracial society. The details will be debated and worked out over time, including important and controversial issues like what should be done to ensure the full participation of people of color in society, the question of "Official English," and how many immigrants should be admitted annually and what qualifications they should

242. Geoffrey C. Hazard, Jr., *Permissive Affirmative Action for the Benefit of Blacks*, 1987 U. ILL. L. REV. 379, 385.

243. *See* SALYER, *supra* note 134 (discussing the treatment of Chinese immigrants at the hands of the Bureau of Immigration).

have. But in the Reconstruction of the 1860s and the second in the 1960s, the United States made a fundamental and, one hopes, irrevocable decision that there will be no second-class citizens defined by race. The people of the United States know that a house divided against itself cannot stand; therefore, the techniques of Johannesburg or Nuremberg have been definitively rejected.

If this is right, then a principal harm of legal reaction against equality and citizenship for nonwhites was fostering the illusion that this closed question was still open. C. Vann Woodward explained that, in the immediate post-Reconstruction period, race relations were to some degree in flux—some African Americans still voted, for example, and served in public office in the South until years after the compromise of 1877.[244] Accordingly, *Plessy* was a catalyst for segregation and racism, rather than simply bowing to the situation as it was. The Court's decision gave legal sanction to decades of brutality toward African Americans (and others), but it was ultimately futile; the Court could only delay, not prevent, the day when African Americans would be elected to Congress, attend every state university, and work in every profession. The victims of the Court's mistake were the people of color subjected to these laws, of course, but they were also the whites deceived into compromising their consciences for chimeras like white supremacy. The *Plessy* Court's sin was to permit the law to interfere with the process of allowing citizens of different races to recognize their common humanity.

The Asian Exclusion Laws and the decisions that approved them caused similar harms. Race-based immigration restriction only put off the day when Asian Americans would live here in large numbers. The United States is still committed to large-scale immigration and is still a desirable immigration destination. Half of the world's population (that is to say, half of the potential immigrant pool) lives in Asian countries. Accordingly, it seems inevitable that Asians will constitute a growing part of the American scene. As it turned out, then, instead of preventing problems, the Asian Exclusion Laws created them, by crystalizing and amplifying prejudices that otherwise might have been ameliorated by time.[245]

---

244. *See* C. VANN WOODWARD, THE STRANGE CAREER OF JIM CROW 33 (3d ed. 1974).

245. Justice Stanley Matthews made a similar point in a speech passionately opposing a version of the Chinese Exclusion Act. Then a senator from Ohio, Matthews argued that the problem was racism rather than race per se. "[W]e should be very careful," he warned,

> how we endanger the peace and security of our age by courting the essential conflict of antagonist races. The same cry made the Jew an exile for hundreds of years and crowded him into the Ghetto at Venice because he was alien in race and killed our Saviour. The same thing made a chattel slave of the poor African that had been torn from his home and brought here against his will. And we are reminded of the dire and bloody conflicts

These generalities illuminate real cases. In 1982, Detroit saw one of the most notorious acts of violence against an Asian American, Vincent Chin, who was murdered in a bar confrontation by two white men wielding bats, Michael Nitz and Ronald Ebens, autoworkers resentful of Japanese competition.[246] Who can know for sure. But if there had been no Asian exclusion and white Americans had been able to meet the spectral Asian-American immigrants who-might-have-been, as friends, neighbors, and coworkers who faced the effects of foreign competition just like all other Americans, perhaps Nitz and Ebens would have perceived Vincent Chin as the American citizen he was, not as a foreigner. Perhaps Nitz and Ebens would have been able to distinguish the people of Japanese ancestry whom they resented from the people of Chinese ancestry who were not selling cars in the United States, just as they could, presumably, distinguish an O'Reilly from a Schultz. Maybe Vincent Chin would be alive, and Nitz and Ebens would not have to live with the knowledge that they are killers.[247]

In retrospect, no matter how desirable or urgent racial subordination seemed at the time (it always does), it has inevitably turned out to be a dreadful mistake. The tradition of Justices confessing error on pivotal issues after it is too late[248] includes Henry Billings Brown's tacit acknowledgment

---

> which his presence inspired and carried on. . . . [W]as it the presence of the black man on this continent that brought in upon us all this woe, or was it his presence here as a slave? Was it rather not the presence of that invisible specter, the injustice with which we treated him, which came upon us with the retribution of a just Heaven and inflicted upon us the punishment for our joint and common sin?

8 CONG. REC. 1275 (1879).

246. *See* United States v. Ebens, 800 F.2d 1422 (6th Cir. 1986); Paula C. Johnson, *The Social Construction of Identity in Criminal Cases: Cinema Verité and the Pedagogy of Vincent Chin*, 1 MICH. J. RACE & L. 347 (1996).

247. Similarly, ex-Klansman Henry Alexander was apparently tortured by his conscience after participating in the 1957 murder of Willie Edwards, Jr., an African-American Winn-Dixie truck driver in 1957; "my life hasn't meant nothing," Alexander said decades later. Adam Rossiter, *Long Overdue Shame for Klan Killing*, SAN JOSE MERCURY NEWS, Sept. 4, 1993, at 1A. Alexander avoided responsibility for his crime because charges were twice dismissed on technical grounds before he died in 1992. *See id.*

248. Among the most notable, Justice Blackmun refused to "tinker with the machinery of death," Callins v. Collins, 510 U.S. 1141, 1145 (1994) (Blackmun, J., dissenting from denial of certiorari), after supporting the death penalty in many key cases. When discussing his vote to uphold a law criminalizing homosexual sodomy in *Bowers v. Hardwick*, 478 U.S. 186 (1986) (5–4 decision), Justice Powell admitted that he "probably made a mistake." Linda Greenhouse, *When Second Thoughts in Case Come Too Late*, N.Y. TIMES, Nov. 5, 1990, at A14. Similarly, Justice Douglas wrote in his autobiography: "I have always regretted that I bowed to my elders [in *Korematsu v. United States*, 323 U.S. 214 (1944)] . . . . The evacuation case . . . was ever on my conscience." WILLIAM O. DOUGLAS, THE COURT YEARS, 1939–1975: THE AUTOBIOGRAPHY OF WILLIAM O. DOUGLAS 280 (1980).

that the *Plessy* majority opinion, which he authored, was wrong.[249] America has apologized for the internment of Japanese Americans during World War II and regularly pays to compensate Native Americans for the brutality and dishonesty it displayed in its dealings with them. The plenary power doctrine is the product of a similar set of mistakes. Nothing in America's history or law warrants honoring cases as fallacious, costly, distracting, brutal, and destructive as *Fong Yue Ting* and *Chae Chan Ping*. The law should reflect that the fantasy of an America as white as it once was is just that, a fantasy.

II. LIMITING THE PLENARY POWER DOCTRINE ON ITS OWN TERMS

The plenary power doctrine's disreputable origins impeach it, but in an ironic way the history also sustains it. The extreme nature of the plenary power cases underscores for all three branches of government the power of this doctrine. If, as the cases say, Congress can discriminate on the basis of race, virtually anything it does is permissible no matter how inconsistent with cherished American values like due process or nondiscrimination. However, even taking the plenary power cases seriously, there is a strong argument that Congress can no longer discriminate on the basis of race.

In the past century, at least eight Justices have insisted that the Constitution prohibits deportation or exclusion on the basis of race.[250] Louis

249. *See* H. B. Brown, *The Dissenting Opinions of Mr. Justice Harlan*, 46 AM. L. REV. 321, 338 (1912) (noting that Harlan "assumed what is probably the fact, that the statute had its origin in the purpose, not so much to exclude white persons from railroad cars occupied by blacks, as to exclude colored people from coaches occupied or assigned to white persons"); Joel Goldfarb, *Henry Billings Brown*, *in* 2 JUSTICES, *supra* note 119, at 772–73.

250. Never five Justices at once, of course. Most recently, Justices Marshall and Brennan suggested that a policy of denying parole to undocumented Haitian aliens would be unconstitutional if it were based on racial prejudice. *See* Jean v. Nelson, 472 U.S. 846, 881 (1985) (Marshall, J., dissenting). Similarly, Justice Murphy was "unable to believe that the Constitution sanctions" deportation of aliens based on "[t]he color of their skin, their racial background or their religious faith." Bridges v. Wixon, 326 U.S. 135, 160–61 (1945) (Murphy, J., concurring). Justices Douglas and Black seemed to agree with Justice Brewer's dissent in *Fong Yue Ting* that Congress had no right to expel aliens on the basis of race. *See* Harisiades v. Shaughnessy, 342 U.S. 580, 599–600 (1952) (Douglas, J., dissenting) (quoting Fong Yue Ting v. United States, 149 U.S. 698, 737–38 (1893) (Brewer, J., dissenting)); *see also* Kleindienst v. Mandel, 408 U.S. 753, 770 & n.1 (1972) (Douglas, J., dissenting). In *Fong Yue Ting*, Justice Field also agreed with Brewer that Chinese could not be deported, arguing that as "men having our common humanity, they are protected by all the guaranties of the Constitution," 149 U.S. at 754 (Field, J., dissenting), and called historical expulsions of particular races by other nations acts of "barbarity and cruelty," insisting that power to commit such expulsions "certainly is not found in any delegated powers under the Constitution." *Id.* at 757. Chief Justice Fuller argued that the due process principle, which "as much applies to Congress under the Fifth Amendment as to the States under the

Henkin wrote that he did not believe "that the Court would hold today that the Constitution permits . . . exclusion on racial . . . grounds."[251] The *Restatement (Third) of the Foreign Relations Law of the United States* notes that "[i]t was long assumed that there are no constitutional limitations on the power of Congress to determine whom it will admit or exclude . . . . It is now open to question, however, whether Congress could exclude an alien . . . solely on account of race or religion."[252] Professor Henkin and the authors of the *Restatement* do not explain the basis for their doubts, but their instincts are justified by legal doctrine as well as enlightened attitudes about race.

The plenary power of today is different from the plenary power of the Gilded Age. As is described below, plenary power was derived from international law. It has changed, as has the Court's conception of judicial review within the regime of Plenary power.

A. The Court's Sporadic Adherence to Stare Decisis

In *Galvan*, the Supreme Court insisted that a century of consistent precedent insulated the plenary power doctrine from reconsideration;[253] this notion still has great influence over the Court in immigration cases.[254]

Fourteenth," prohibited deportation of lawfully resident Chinese. *Id.* at 762 (Fuller, C.J., dissenting).

251. Henkin, *supra* note 42, at 863 (footnote omitted).

252. 2 RESTATEMENT (THIRD) OF THE FOREIGN RELATIONS LAW OF THE UNITED STATES § 722 cmt. l (1987) [hereinafter 2 RESTATEMENT (THIRD)]; *see also id.* § 722(2) (stating that aliens "may not be denied the equal protection of the laws, but equal protection does not preclude reasonable distinctions between aliens and citizens, or between different categories of aliens"); Michael J. Perry, *Modern Equal Protection: A Conceptualization and Appraisal*, 79 COLUM. L. REV. 1023, 1062 n.201 (1979) (arguing that because race is a "*morally irrelevant* trait[,] . . . [e]qual protection should be understood to prohibit the federal government from discriminating against a person, who happens to be an alien, on the basis of race"); Hahn, *supra* note 42, at 997 ("[A]dministrative practices that single out aliens for discriminatory treatment based upon race or nationality . . . must be regarded as a denial of equal protection of the laws.").

253. *See supra* notes 96–99 and accompanying text.

254. *Miller v. Albright*, 118 S. Ct. 1428 (1998), is the most recent example of a decision upholding discrimination in the immigration context that would have been unlikely to survive in the domestic context. A majority of the Court refused to strike down statutes granting citizenship more readily to "illegitimate" children of female American citizens born abroad than to such children of male U.S. citizens. While there were strong dissents, none of the Justices directly questioned the principle of extreme deference to Congress in the immigration context.

Chief Justice Rehnquist joined Justice Stevens's opinion upholding the laws on the merits. Their opinion took pains to make clear that this was a citizenship case, not an immigration case, suggesting that the limited review in which they engaged would have been even more truncated had the case involved a would-be immigrant. *See id.* at 1436 ("It is of significance that the petitioner in this case . . . is not challenging the denial of an application for special status. . . . If she were to prevail, the judgment in her favor would confirm her pre-existing citizenship rather than

However, the *Galvan* Court's characterization of the clarity and uniformity of the precedents is magnificently overstated.

Even the basic conclusion that the plenary power cases obligate the courts to refrain from normal Fifth Amendment equal protection review is not obvious. As late as 1943, the Supreme Court emphasized that the "Fifth Amendment contains no equal protection clause."[255] The Court applied the equal protection principle to the national government through the Due Process Clause of the Fifth Amendment in *Bolling*,[256] more than six decades after it decided *Fong Yue Ting* and *Chae Chan Ping*. "The Court [in *Chae Chan Ping*] did not address the question of whether exclusion of Chinese nationals on racial grounds violated the Constitution;"[257] thus, while these cases were race cases, they were not equal protection cases. The Court could not have been expected to decide whether these racial classifications implicated a Fifth Amendment equal protection principle that did not yet exist.[258]

---

grant her any rights that she does not now possess."). Justices Thomas and Scalia argued that the judiciary had no power to grant citizenship other than as authorized by Congress, noting that "[j]udicial power over immigration and naturalization is extremely limited." *Id.* at 1447 (Scalia, J., concurring).

Justices Kennedy and O'Connor relied primarily on the alien's lack of standing to challenge the discrimination against her father. *See id.* at 1442–43 (O'Connor, J., concurring). They were somewhat more skeptical of the classification, questioning whether "any gender classifications based on stereotypes can survive heightened scrutiny." *Id.* at 1445–46. However, they left open the question of "whether an alien may assert constitutional objections when he or she is outside the territory of the United States." *Id.* at 1445 (citation omitted). Further, they stated that "under rational scrutiny, a statute may be defended based on generalized classifications unsupported by empirical evidence. This is particularly true when the classification is adopted with reference to immigration, an area where Congress frequently must base its decisions on generalizations about groups of people." *Id.* at 1446 (citation omitted).

Finally, Justices Breyer, Ginsburg, and Souter dissented in opinions written by Justices Breyer and Ginsburg. They believed the statutes were unconstitutional, but not on grounds that necessarily signal any dissatisfaction with the plenary power doctrine in general. They noted that because the challenged laws "confer the status of citizenship 'at birth,' they do not involve the transfer of loyalties that underlies the naturalization of aliens where precedent sets a more lenient standard of review." *Id.* at 1458 (Breyer, J., dissenting, joined by Souter and Ginsburg, JJ.) (citing Fiallo v. Bell, 430 U.S. 787 (1977)). However, they recognized that "[t]he Court has applied a deferential standard of review in cases involving aliens, not in cases in which only citizens' rights were at issue." *Id.* at 1460.

255. Hirabayashi v. United States, 320 U.S. 81, 100 (1943) (citing Detroit Bank v. United States, 317 U.S. 329, 337–38 (1943)).

256. Bolling v. Sharpe, 347 U.S. 497, 499 (1954).

257. Scaperlanda, *supra* note 42, at 976–77 (footnote omitted); *see also* Ibrahim J. Wani, *Truth, Strangers, and Fiction: The Illegitimate Uses of Legal Fiction in Immigration Law*, 11 CARDOZO L. REV. 51, 81–82 (1989) (questioning whether the Court engaged in any form of judicial review in *Chae Chan Ping*).

258. As the Court has recognized, "[i]t is . . . contrary to all traditions of our jurisprudence to consider the law on [a] point conclusively resolved by broad language in cases where the issue was not presented or even envisioned." R.A.V. v. City of St. Paul, 505 U.S. 377, 386 n.5 (1992).

However, all members of the Court seemed to agree that Fong Yue Ting enjoyed the protection of the Constitution. The majority explained that the "power to exclude or to expel aliens . . . is vested in the political departments of the government . . . except so far as the judicial department has been authorized by treaty or by statute, *or is required by the paramount law of the Constitution*, to intervene."[259] The dispute between the majority and the dissenters may boil down to whether, as the dissenters insisted, Fong Yue Ting had a Fifth Amendment liberty interest in remaining in the United States.[260] *Fong Yue Ting* and *Chae Chan Ping* may be understood as turning on the fact that "those cases were decided . . . in an era before the Court had held that the Due Process Clause of the Fifth Amendment imposes the same equal protection standard upon the Federal Government that the Fourteenth Amendment imposes upon the States."[261] Perhaps the Court should consider *Chae Chan Ping* and *Fong Yue Ting* to have been limited or overruled by *Bolling*.

In any event, judicial treatment of immigration has been fluid over time. *Chae Chan Ping* and *Fong Yue Ting* themselves represent a revisionist view of the nature and source of federal power over immigration. In the *Head Money Cases*,[262] decided only five years before *Chae Chan Ping*, the Supreme Court unanimously held that immigration was subject to federal regulation under the commerce power, not under the inherent sovereignty theory, which is the basis for the plenary power doctrine. The *Head Money* Court cited precedents dating back many decades that, it said, had taken the same approach.[263] Because there is no question that the commerce

259. Fong Yue Ting v. United States, 149 U.S. 698, 713 (1893) (emphasis added); *see id.* at 724 ("Chinese laborers . . . are entitled, so long as they are permitted by the government of the United States to remain in the country, to the safeguards of the constitution . . . ."); *id.* at 730 (concluding that Fong Yue Ting "has not . . . been deprived of life, liberty or property, without due process of law"); *id.* at 732 (affirming the legality of the deportation order is "the only conclusion which appears to us to be consistent with the principles of international law [and] with the Constitution and laws of the United States").

260. *See id.* at 733 (Brewer, J., dissenting) (arguing that the statute at issue "deprives them of liberty . . . without due process of law"); *id.* at 759 (Field, J., dissenting) (under the statute the "foreigner . . . is withdrawn from all the guaranties of due process of law prescribed by the Constitution"); *id.* at 762 (Fuller, C.J., dissenting) (arguing that due process principle "as much applies to Congress under the Fifth Amendment as to the States under the Fourteenth").

261. Fullilove v. Klutznick, 448 U.S. 448, 524 n.3 (1980) (Stewart, J., dissenting) (explaining why Japanese-American incarceration cases were no longer valid authority).

262. 112 U.S. 580 (1884).

263. *Id.* at 590–94; *see* Mary Sarah Bilder, *The Struggle over Immigration: Indentured Servants, Slaves, and Articles of Commerce*, 61 MO. L. REV. 743, 745 (1996) ("The . . . plenary power doctrine . . . was the second theory of immigration authority—not adopted until 1889. For the first one hundred years, the Court debated the question of immigration power under the Commerce Clause.").

authority is limited by the Bill of Rights, if the Court reverted to its original theory of immigration power, constitutional immigration law would be brought in to the mainstream.

Analytically, the most difficult immigration cases involve applicants for initial entry because the moral and legal arguments in their favor are less compelling than when dealing with resident aliens. Even here, though, the Supreme Court has been far from consistent. *Chae Chan Ping* and *Fong Yue Ting* suggested congressional power over immigration was unlimited; in *Nishimura Ekiu v. United States*,[264] the Court held that, with respect to applicants for initial entry, "the decisions of executive or administrative officers, acting within powers expressly conferred by congress, are due process of law."[265] Yet, in *Fiallo v. Bell*,[266] a case involving applicants for initial entry, the Court concluded that there was "a limited judicial responsibility under the Constitution even with respect to the power of Congress to regulate the admission and exclusion of aliens."[267]

Similarly, the Court has flip-flopped on whether aliens in deportation proceedings had the right to procedural due process under the Fifth Amendment. *Fong Yue Ting* could be understood as meaning that resident aliens being deported from the United States, like aliens at the border seeking initial entry, had no constitutional right to procedural due process.[268] In *Yamataya v. Fisher*,[269] and a series of more recent cases,[270] the Court reached the opposite conclusion. If *Fong Yue Ting* were decided today, the outcome should be in favor of the Chinese, who were deported under a statute that allowed the testimony only of whites to prove lawful presence in the United States;[271]

---

264. 142 U.S. 651 (1892).

265. *Id.* at 660.

266. 430 U.S. 787 (1977).

267. *Id.* at 793 n.5.

268. *See* Fong Yue Ting v. United States, 149 U.S. 698, 728–32 (1893); *see also* Wong Wing v. United States, 163 U.S. 228, 237 (1896).

269. 189 U.S. 86, 100–01 (1903).

270. *See* Landon v. Plasencia, 459 U.S. 21, 32–33 (1982) (citing, inter alia, Wong Yang Sung v. McGrath, 339 U.S. 33, 49–50 (1950); Bridges v. Wixon, 326 U.S. 135, 153–54 (1945); United States *ex rel.* Tisi v. Tod, 264 U.S. 131, 133, 134 (1924)); *see also* Aleinikoff, *supra* note 29, at 863.

271. Actually, of the three appellants in *Fong Yue Ting*, only Lee Joe raised the argument that the racial limitation on competency was unfair. At that time, the Court found the argument meritless:

> [T]he requirement of proof, "by at least one credible white witness, that he was a resident of the United States . . . ," is within the acknowledged power of every legislature to prescribe the evidence which shall be received . . . . The competency of all witnesses, without regard to their color, to testify . . . rests on acts of Congress, which Congress may at its discretion modify or repeal.

149 U.S. at 729 (citations omitted). Professor Wigmore eloquently observed:

> Of these statutes it can only be said that they do not come consistently from a Legislature which in this series of enactments was itself breaking solemn treaty-faith with the

surely exclusion of otherwise competent witnesses on the basis of race would not now be deemed consistent with due process.[272]

The Court has also changed its mind on whether a resident alien seeking to return to the United States from an overseas visit is entitled to procedural due process before being found inadmissible. In *Lem Moon Sing v. United States*,[273] the Court suggested that returning resident aliens, like applicants for initial entry, had no constitutional rights.[274] However, in *Kwong Hai Chew v. Colding*[275] and *Landon v. Plasencia*,[276] the Court recognized that returning resident aliens had a connection to the United States that could not be severed without procedural due process. Chae Chan Ping's case might well have come out the other way under this view, for he was a long-term resident alien who left with official assurance that he could return. It is hard to see how his exclusion from his home after breaking a promise that he could return complied with the principles of procedural due process that the Court now acknowledges apply. Given the fluidity of the precedents, when faced with applications of the plenary power doctrine that are inconsistent with due process and equal protection, the Court clearly has choices, consistent with faithfulness to prudential concerns like stare decisis.

B. The International Law Foundation of the Plenary Power Doctrine

The Court derived the plenary power doctrine from its understanding of international law rather than any specific provision of the Constitution.[277] In

> very nation whose members it thus condemned as oath-breakers; and that the supposed special danger of perjury by Chinese attempting to evade those statutes of exile was precisely what might be expected from the people of any country when a hostile measure is attempted to be enforced by the harshest means.

1 WIGMORE, *supra* note 135, § 516, at 646.

272. Courts in various circumstances have found arbitrary exclusion of a witness to violate due process. *See* 1 JOHN H. WIGMORE, EVIDENCE IN TRIALS AT COMMON LAW, § 7.1, at 502 & n.62 (Tillers rev. 1983) (listing cases discussing the due process right to present evidence). Moreover, the equal protection rights of witnesses would be implicated by a racial competency test; an alien might well have third-party standing to raise such a claim. *See* Powers v. Ohio, 499 U.S. 400, 410–16 (1991) (holding that injury to defendant, defendant's motivation to pursue a claim effectively, and the improbability of a juror pursuing the issue independently warranted allowing the defendant to raise a claim of discriminatory exclusion of juror).

273. 158 U.S. 538 (1895).

274. *See id.* at 546–47.

275. 344 U.S. 590, 600–02 (1953).

276. 459 U.S. 21, 33–34 (1982).

277. There is no enumerated federal authority over immigration, so it is hard to argue that the text limits ordinary forms of judicial review. The idea that the Constitution applies only in the territorial United States, even if true, *see* Reid v. Covert, 354 U.S. 1, 5–6 (1957) (plurality opinion), does not explain the plenary power doctrine. Many people subject to immigration law

United States v. Curtiss-Wright Export Corp.,[278] the Court explained, "the power to expel undesirable aliens . . . which is [not] expressly affirmed by the Constitution, nevertheless exist[s] as inherently inseparable from the conception of nationality. This the court recognized, and . . . found the warrant for its conclusions not in the provisions of the Constitution, but in the law of nations."[279] Similarly, in *Nishimura Ekiu v. United States*,[280] the Court stated that "[i]t is an accepted maxim of international law, that every sovereign nation has the power, as inherent in sovereignty, and essential to self-preservation, to forbid the entrance of foreigners within its dominions, or to admit them only in such cases . . . as it may see fit to prescribe."[281] Modern cases continue to locate the font of plenary power in international law.[282]

In the last century, the law of nations may have imposed few limits on the acts of nations in this context.[283] However, as Michael Scaperlanda has pointed out, "[s]overeignty is not a static concept."[284] Of course, the Supreme Court has not been eager to rely on international law to limit the authority of Congress,[285] but the argument is different in the plenary power

are in the territorial United States—either they are resident aliens like Fong Yue Ting, who lived in the United States, or they are arrivals at ports of entry, like Chae Chan Ping, who was on a ship in San Francisco Bay. It has sometimes been suggested that immigration is a privilege, not a right, but the right/privilege distinction today cannot justify discrimination that would otherwise violate equal protection. *See* 3 ROTUNDA & NOWAK, *supra* note 3, § 17.2, at 582–83.

278. 299 U.S. 304 (1936). Many commentators have explained why the sovereignty theory is unpersuasive. *See, e.g.*, Legomsky, *Plenary Congressional Power*, *supra* note 42, at 273–75; Fabiano, *supra* note 42; Hahn, *supra* note 42, at 966–74. However, because the Court has relied on it, this analysis assumes, arguendo, its continuing validity.

279. *Curtiss-Wright*, 299 U.S. at 318 (citation omitted). The Court continues to cite *United States v. Curtiss-Wright Export Corp.* in immigration cases. *See, e.g.*, Sale v. Haitian Ctrs. Council, Inc., 509 U.S. 155, 188 (1993).

280. 142 U.S. 651 (1892).

281. *Id.* at 659; *see also* Fong Yue Ting v. United States, 149 U.S. 698, 705 (1893).

282. *See, e.g.*, Kleindienst v. Mandel, 408 U.S. 753, 765 (1972).

283. At least the Court so found; James Nafziger argues persuasively that international law does not now and never did grant nations an unlimited degree of discretion in admission and exclusion of aliens. *See* Nafziger, *supra* note 42.

284. Scaperlanda, *supra* note 42, at 970.

285. As the *Restatement* explains, while "[i]nternational law and international agreements of the United States are law of the United States," 1 RESTATEMENT (THIRD) OF THE FOREIGN RELATIONS LAW OF THE UNITED STATES § 111(1) (1987), "[a]n act of Congress supersedes an earlier rule of international law or a provision of an international agreement as law of the United States." *Id.* § 115(1)(a); *see also, e.g.*, Harry A. Blackmun, *The Supreme Court and the Law of Nations*, 104 YALE L.J. 39 (1994) (criticizing the refusal of the Supreme Court to follow international legal norms); Joan Fitzpatrick & William McKay Bennett, *A Lion in the Path? The Influence of International Law on the Immigration Policy of the United States*, 70 WASH. L. REV. 589 (1995) (arguing that various aspects of current immigration law violate international law); *cf.* Gordon A. Christenson, *Using Human Rights Law to Inform Due Process and Equal Protection Analyses*, 52 U. CIN. L. REV. 3 (1983).

context. By defining congressional power over immigration by reference to an external standard, the Court itself has provided for the possibility that domestic law will change as international law changes. As Professor Scaperlanda explains: "[W]here the international law of sovereignty provides the background norm for constitutional decisionmaking, as it does in [immigration] cases, the Court ought to look at the current norm with its recent limitations."[286]

C. The Nondiscrimination Principle in International Law

If the Court meant to grant Congress an unlimited power over immigration by linking the immigration power to international law, it failed. As Professors Hernandez and Scaperlanda have written, international legal doctrine has changed dramatically since the Court created the plenary power doctrine.[287] If international law ever permitted, or was neutral, on the issue of governmental racial discrimination, it is no longer; modern international law generally prohibits racial discrimination. Accordingly, the notion that an inherent part of national sovereignty is the power to discriminate against aliens on the basis of race is no longer a part of international law.

The International Convention on the Elimination of All Forms of Racial Discrimination[288] is perhaps the broadest expression of the antidiscrimination principle in international law. Signatory nations, including the United States, commit "to engage in no act or practice of racial discrimination against persons, groups of persons or institutions and to ensure that all public authorities . . . shall act in conformity with this obligation."[289] The International Covenant on Civil and Political Rights,[290] to which the

286. Scaperlanda, *supra* note 42, at 1015; *see also, e.g.*, Aleinikoff, *supra* note 29, at 863 ("The Court . . . deemed international law as relevant to, if not the basis of, constitutional norms."); Hahn, *supra* note 42, at 968 (observing that the exclusion power, according to the Supreme Court, is "defined by reference to international law").

287. *See* Hernández-Truyol, *supra* note 49, at 1113–29 (discussing the development of international human rights law); Scaperlanda, *supra* note 42, at 1002–21 (same); *see also* Berta Esperanza Hernández-Truyol & Kimberly A. Johns, *Global Rights, Local Wrongs and Legal Fixes: An International Human Rights Critique of Immigration and Welfare "Reform,"* 71 S. CAL. L. REV. 547 (1998).

288. U.N. GAOR 3d Comm., 20th Sess., 1406th plen. mtg., Annex, Agenda Item 58, U.N. Doc. A/Res/2106 *adopted* Dec. 21, 1965, 660 U.N.T.S. 212, 218, 5 I.L.M. 352 (entered into force Jan. 4, 1969, and ratified by the United States June 24, 1994).

289. *Id.* at pt. I, art. 2, § (1)(a).

290. International Covenant on Civil and Political Rights, *adopted* Dec. 19 1966, 999 U.N.T.S. 171, 6 I.L.M. 368.

United States is a party, also prohibits racial discrimination.[291] Other international agreements also make the point that racial discrimination is disfavored.[292]

The prohibition against racial discrimination applies to treatment of aliens as well as citizens. For example, the Protocol Relating to the Status of Refugees,[293] which the United States has ratified, requires signers to offer benefits to refugees "without discrimination as to race, religion or country of origin."[294] Similarly, the Geneva Convention Relative to the Treatment of Prisoners of War[295] prohibits "any adverse distinction based on race, nationality, religious belief or political opinions, or any other distinction founded on similar criteria."[296] Other treaties also make the point.[297]

---

291. Article 26 provides:

> All persons are equal before the law and are entitled without any discrimination to the equal protection of the law. In this respect, the law shall prohibit any discrimination and guarantee to all persons equal and effective protection against discrimination on any ground such as race, colour, sex, language, religion, political or other opinion, national or social origin, property, birth or other status.

Id. art. 26, 999 U.N.T.S. at 179, 6 I.L.M. at 375.

292. *See, e.g.*, U. N. CHARTER art. 1, para. 3 (stating that the goal of the United Nations is to "encourag[e] respect for human rights and for fundamental freedoms for all without distinction as to race"); Universal Declaration of Human Rights art. 7, adopted Dec. 10 1948; G.A. Res. 217(III)A, U.N. GAOR U.N. Doc. A/810, at 71, 73 (1948) ("All are equal before the law and are entitled without any discrimination to equal protection of the law. All are entitled to equal protection against any discrimination in violation of this Declaration and against any incitement to such discrimination."); *id.* art. 2 (banning race discrimination).

293. Protocol Relating to the Status of Refugees, *opened for signature* Nov. 1, 1968, 19 U.S.T. 6223, 6225, 686 U.N.T.S. 267.

294. Convention Relating to the Status of Refugees, *opened for signature*, July 28, 1951, art. 3, 19 U.S.T. 6223, 6264, 189 U.N.T.S. 150–156; *see also* Declaration of Human Rights of Individuals Who Are Not Nationals of the Countries in Which They Live, G.A. Res. 40.144, art. 2, para. 1 (1985). The Declaration allows nations "to establish differences between nationals and aliens. However, such laws and regulations shall not be incompatible with the international legal obligations of that State, including those in the field of human rights."

295. Geneva Convention Relative to the Treatment of Prisoners of War, *entered into force*, Oct. 21, 1950, 6 U.S.T. 3316, 3330, 75 U.N.T.S. 135, 148.

296. *Id.* at pt. II art. 16; *see also* Geneva Convention for the Amelioration of the Condition of the Wounded and Sick in Armed Forces in the Field, *entered into force*, Oct. 21, 1950, 6 U.S.T. 3114, 3122, 75 U.N.T.S. 31, 38 (prohibiting "any adverse distinction founded on sex, race, nationality, religion, political opinions, or any other similar criteria" in treatment of sick or wounded members of the armed forces).

297. *See, e.g.*, International Convention on the Suppression and Punishment of the Crime of Apartheid, Nov. 30, 1973, art. II(c), 1015 U.N.T.S. 243, 245 (defining as apartheid "any legislative measures and other measures calculated to prevent a racial group or social group from participation in the political, social, economic and cultural life of the country"); Convention on the Prevention and Punishment of the Crime of Genocide, *entered into force*, Jan. 12, 1951, art. II 78 U.N.T.S. 277, 280, Hein's No. KAV 2303 (1948) (prohibiting destruction of "a national, ethnical, racial or religious group").

The *Restatement (Third) of the Foreign Relations Law of the United States* recognizes the effect of this body of law, concluding that "systematic racial discrimination" now violates customary international law.[298] Thus, while there is no doubt that international law permits nations to regulate and limit immigration, it does not authorize them to do so on the basis of race. Because an important support for the nation's right to discriminate on the basis of race in the immigration context has been removed, it seems logical that the right to do so should be as well.

D. The Quiet Expansion of Judicial Review: *Fiallo v. Bell*

The Court's understanding of the scope of judicial review of immigration statutes has undergone a subtle but meaningful shift that, like changes in international law, militates in favor of a different resolution to the problem the Court faced in *Chae Chan Ping* and *Fong Yue Ting*. While the Court remains deferential to substantive congressional determinations of who should be admitted and excluded, it is clear that it is no longer *absolutely* deferential.

At one level, decisions of the modern Court give little suggestion that the plenary power doctrine might be in jeopardy. In *Landon v. Plasencia*,[299] for example, decided in 1982, the Court stated that "an alien seeking initial admission to the United States requests a privilege and has no constitutional rights regarding his application, for the power to admit or exclude aliens is a sovereign prerogative."[300] Nevertheless, the plenary power of today is not entirely plenary. Comparison of the 1972 case of *Kleindienst v. Mandel*[301] and the 1977 case of *Fiallo* illustrates this point.

In *Mandel*, Ernest Mandel, a Belgian journalist and self-described "revolutionary Marxist" was denied a temporary visa based on sections of the INA barring aliens who wrote or advocated the doctrines of world communism or advocated the establishment of a totalitarian dictatorship in the United States. While the INA provided for a waiver upon the recommendation of the State Department and the approval of the attorney general, the attorney general refused to approve it. As a result, Mandel was forced to address an academic conference by telephone rather than in person. The plaintiffs, academics, claimed infringement of their First Amendment rights.[302]

298. *See* 2 RESTATEMENT (THIRD), *supra* note 252 § 702(f).
299. 459 U.S. 21 (1982).
300. *Id.* at 32 (citations omitted).
301. 408 U.S. 753 (1972).
302. *Id.* at 760–61.

The Court stated that Mandel himself had absolutely no First Amendment claim: "It is clear that Mandel personally, as an unadmitted and nonresident alien, had no constitutional right of entry to this country as a nonimmigrant or otherwise."[303] While the Court recognized that the First Amendment rights of Americans were implicated by Mandel's exclusion, because they were deprived of a face-to-face discussion, the Court stated that it was "not inclined in the present context"[304] to reconsider the cases granting Congress "plenary power to . . . exclude those who possess those characteristics which Congress has forbidden."[305] Thus, there was no real question in the case that Congress could have simply banned all persons in the class, and no one would have had any conceivable ground for legal complaint.

The analysis did not end there because the Court believed that Congress might have contemplated limited judicial review of the attorney general's denial of a waiver.[306] The Court held that when a "facially legitimate and bona fide reason" was given, "courts will neither look behind the exercise of that discretion, nor test it by balancing its justification against the First Amendment interests of those who seek personal communication with the applicant."[307] The Court found such a reason had been provided in that case.[308]

In *Fiallo*, the Court's superficially similar approach actually employed a very different standard. The appellants were "three sets of unwed natural fathers and their illegitimate offspring who sought, either as an alien father or an alien child, a special immigration preference by virtue of a relationship to a citizen or resident alien child or parent."[309] Under the form of analysis applied by the *Mandel* Court, *Fiallo* should have been an easier case than it was. The alien fathers or children have no claim because as "unadmitted and nonresident alien[s]," they simply have no right of entry. While under *Mandel*, the Court would recognize the associational interests of the American residents, in *Fiallo* there was no waiver provision or any other indication that the categories established by Congress were other than

303. *Id.* at 762 (citations omitted).
304. *Id.* at 767.
305. *Id.* at 766 (quoting Boutilier v. INS, 387 U.S. 118, 123 (1967)).
306. *See id.* at 769.
307. *Id.* at 770.
308. Namely, Mandel had several years before exceeded the terms of admission under a prior conditional visa. *See id.* at 758. While advancing a reason would be *sufficient* to sustain the attorney general's exercise of the discretion, the Court did not hold that a reason was *necessary*: "What First Amendment or other grounds may be available for attacking exercise of discretion for which no justification whatsoever is advanced is a question we neither address nor decide in this case." *Id.* at 770.
309. Fiallo v. Bell, 430 U.S. 787, 790 (1977).

absolute. Accordingly, when even with a waiver provision the *Mandel* Court found no occasion to balance the constitutional rights of the American residents against the government decision, there would be even less justification for doing so in *Fiallo*.

The Court did not take that approach. While the Court stated that, "[a]t the outset, it is important to underscore the limited scope of judicial inquiry into immigration legislation,"[310] it declined to apply the "no review whatsoever" standard suggested by prior law. The Court announced that it possessed the authority to evaluate the *substantive* choices made by Congress regarding the categories of admissible aliens:

> The [attorney general] argue[s] that the challenged sections of the Act, embodying as they do "a substantive policy regulating the admission of aliens into the United States, [are] not an 'appropriate subject for judicial review." Our cases reflect acceptance of a limited judicial responsibility under the Constitution even with respect to the power of Congress to regulate the admission and exclusion of aliens, and there is no occasion to consider in this case whether there may be actions of the Congress with respect to aliens that are so essentially political in character as to be nonjusticiable.[311]

The Court ultimately upheld the classifications, after an exceedingly deferential review. Nevertheless, it considered the claim of invalidity on its merits.[312] Because there was no question that the legislation at issue had been validly enacted or that the general subject matter was within the power of Congress, the nature of the claim had to be that the statute conflicted with the Constitution.[313] Thus, *Fiallo* stands for the proposition that

---

310. *Id.* at 792.

311. *Id.* at 793 n.5 (citation omitted).

312. In support of the classification, the Court stated: "The inherent difficulty of determining the paternity of an illegitimate child is compounded when it depends upon events that may have occurred in foreign countries many years earlier. Congress may well have given substantial weight . . . to these problems of proof and the potential for fraudulent visa applications . . . ." *Id.* at 799 n.8. The Court also noted that

> there are widely varying relationships and degrees of kinship, and it is appropriate for Congress to consider not only the nature of these relationships, but also problems of identification, administration and the potential for fraud. . . . Congress has determined that certain classes of aliens are more likely than others to satisfy national objectives without undue cost . . . .

Id. at 795 n.6.

313. Indeed, the briefs make clear that part of the argument the parties presented to the Court was whether the law created a classification violating the equal protection component of the Fifth Amendment. *See* Appellants' Brief at 23–31, *Fiallo* (No. 75-6297); Appellee's Brief at 34–47, *Fiallo* (No. 75-6297); Appellants' Reply Brief (arguing that "the challenged statutory classifications are utterly irrational").

substantive immigration classifications will be tested against the requirements of the Fifth Amendment.[314]

It is not crystal clear from the decision itself what degree of review the Court applied in *Fiallo*, but it seems to be the standard referred to in *Mandel*.[315] Thus, the *Fiallo* court did not require a facially legitimate and bona fide reason to ensure that executive discretion was exercised in accordance with what Congress may have intended,[316] but instead to scrutinize whether the actions of Congress itself satisfied an external, constitutional constraint.

There is some question about the content of the standard recognized in *Fiallo*. In *Reno v. Flores*,[317] the Supreme Court stated that immigration classifications "must . . . meet the (unexacting) standard of rationally advancing some legitimate governmental purpose."[318] Both before and after *Flores*, some lower court decisions have applied the *Mandel* test, scrutinizing immigration classifications under the "facially legitimate and bona fide" standard.[319] The prevailing judicial[320] and scholarly[321] view, though, is that rational basis review applies. This test is not entirely toothless; indeed, it

---

314. *But see* Hahn, *supra* note 42, at 989 (arguing that *Fiallo* did not decide that the Fifth Amendment had been satisfied).

315. The Court stated that it could "see no reason to review the broad congressional policy choice at issue here under a more exacting standard than was applied in *Kleindienst v. Mandel*." *Fiallo*, 430 U.S. at 795 (footnote omitted).

316. *See Mandel*, 408 U.S. at 767, 769.

317. 507 U.S. 292 (1993).

318. *Id.* at 306.

319. *See, e.g.*, Miller v. Christopher, 96 F.3d 1467, 1470–71 (D.C. Cir. 1996), *aff'd*, 118 S. Ct. 1428 (1998); Giusto v. INS, 9 F.3d 8, 9–10 (2d Cir. 1993) (per curiam); Garcia v. INS, 7 F.3d 1320, 1327 (7th Cir. 1993); Campos v. INS, 961 F.2d 309, 316 (1st Cir. 1992).

320. *See, e.g.*, Miller v. Christopher, 96 F.3d 1467, 1473 (D.C. Cir. 1996) (Wald, J., concurring), *aff'd*, 118 S. Ct. 1428 (1998); Hamama v. INS, 78 F.3d 233, 236 (6th Cir. 1996); Ablang v. Reno, 52 F.3d 801, 804 (9th Cir. 1995); Mendoza v. INS, 16 F.3d 335, 338 (9th Cir. 1994); Rodriguez-Padron v. INS, 13 F.3d 1455, 1459 (11th Cir. 1994); Wauchope v. United States Dep't of State, 985 F.2d 1407, 1414 n.3 (9th Cir. 1993); Azizi v. Thornburgh, 908 F.2d 1130, 1133 n.2 (2d Cir. 1990); Newton v. INS, 736 F.2d 336, 339 (6th Cir. 1984); Narenji v. Civiletti, 617 F.2d 745, 748 (D.C. Cir. 1979); Chan v. Reno, 916 F. Supp. 1289, 1304 (S.D.N.Y. 1996); Ademi v. Moyer, No. 88 C 7017, 1989 WL 56904, at *2 (N.D. Ill. May 23, 1989); Smith v. INS, 684 F. Supp. 1113, 1116 (D. Mass. 1988); Akbari v. Godshall, 524 F. Supp. 635, 642 (D. Colo. 1981). *But see* Anetekhai v. INS, 876 F.2d 1218, 1222 n.7 (5th Cir. 1989) ("The INS argues that the test for determining the constitutionality of an immigration provision is whether it is based upon a "facially legitimate and bona fide reason." . . . Because [the statute] easily passes a rational basis test, we need not decide whether some lesser standard might be appropriate.").

321. *See* Gerald L. Neuman, *The Lost Century of American Immigration Law (1776–1875)*, 93 COLUM. L. REV. 1833, 1839 n.31 (1993) (arguing that the *Fiallo* test "appears roughly equivalent to the rational basis test"); Michael S. Satow, Note, *A Journey Through the Fog: Due Process Analysis of I.N.A. Section 242(a)(2)*, 5 GEO. IMMIGR. L.J. 677, 699 n.159 (1991); Steven J. Burr, Comment, *Immigration and the First Amendment*, 73 CAL. L. REV. 1889, 1905–06 (1985) (suggesting that *Fiallo* applied the rational basis test).

has been used in a handful of cases[322] to invalidate congressional immigration classifications.[323]

E. Applying *Fiallo* to Racial Classifications: Is Race Discrimination "Irrational"?

Domestic racial classifications are tested under the strict scrutiny standard, but under *Fiallo* the most that can be claimed even for racial immigration classifications is rational basis review, the deferential test usually applied to economic legislation. Is that enough?

The Supreme Court's equal protection cases can be understood as holding that racial classifications are not only forbidden as a matter of policy, but are actually irrational and are thus invalid, even under rational basis review. The classical expression of the idea is that for a distinction to be valid, "it must appear [that it] is one based upon some reasonable ground—some difference which bears a just and proper relationship to the attempted classification—and is not a mere arbitrary selection."[324] Put another way, a statute must "classify the persons it affects in a manner rationally related to legitimate governmental objectives."[325]

The idea that racial distinctions are affirmatively irrational finds support at least as far back as 1900. In *American Sugar Refining Co. v. Louisiana*,[326] the Supreme Court rejected an equal protection challenge to a law

322. *See, e.g.*, Yeung v. INS, 76 F.3d 337, 341 (11th Cir. 1996) (holding irrational a statute that permitted discretionary relief to those seeking to enter the country but not to those already here); Wauchope v. United States Dep't of State, 985 F.2d 1407 (9th Cir. 1993) (holding a classification irrational that conferred citizenship status on the children of some male citizens but not on the children of similarly situated female citizens); Aguayo v. Christopher, 865 F. Supp. 479, 490–91 (N.D. Ill. 1994) (same).

323. Another line of authority ultimately does not conflict with the *Fiallo* test. In *Mathews v. Diaz*, 426 U.S. 67 (1976), resident aliens challenged a Medicare regulation limiting eligibility to those individuals who had held permanent resident status for at least five years. *See id.* at 69. The Court held that the national government could discriminate in favor of citizens against aliens and could draw reasonable distinctions between groups of aliens. *See id* at 77–84. The Court unanimously upheld the rules because they were "reasonable" and not "wholly irrational," *id.* at 83, suggesting that irrational classifications are subject to judicial invalidation. This seems to be the same as the standard set by *Fiallo*; thus, a classification that flunks or passes the *Fiallo* test will earn the same result under *Diaz*. While Professors Aleinikoff, Martin, and Motomura plausibly understand *Diaz* as suggesting that "the Court is not likely to demand that Congress provide a compelling justification for immigration laws that discriminate on the basis of nationality or alienage," T. ALEXANDER ALEINIKOFF ET AL., IMMIGRATION: PROCESS AND POLICY 20 (3d ed. 1995), they do not suggest that *Diaz* authorizes racial discrimination.

324. Gulf, Colo. & Santa Fé Ry. Co. v. Ellis, 165 U.S. 150, 165–66 (1897).

325. Schweiker v. Wilson, 450 U.S. 221 (1981); *accord* Heller v. Doe, 509 U.S. 312, 319–21 (1993).

326. 179 U.S. 89 (1900).

taxing some sugar manufacturers but exempting planters who produced their own sugar and molasses.[327] The Court explained that discrimination, "if founded on a reasonable distinction in principle, is valid."[328] However,

> if such discrimination were purely arbitrary, oppressive or capricious, and made to depend upon differences of color, race, religious opinions, political affiliations, or other considerations having no possible connection with the duties of citizens as taxpayers, such exemption would be pure favoritism, and a denial of equal protection of the laws to the less favored classes.[329]

Since then, the Court has applied the principle to several areas of life. In *Bolling*, the Court used rationality language to strike down school segregation in the District of Columbia, holding that "[s]egregation in public education is not reasonably related to any proper governmental objective."[330] In *Steele v. Louisville & Nashville Railroad Co.*,[331] the Court stated that "discriminations [in the terms and conditions of employment embodied in a collective bargaining agreement] based on race alone are obviously irrelevant and invidious."[332] In *Hernandez v. Texas*,[333] the Court held that the exclusion of Mexican Americans from grand jury service was not a "reasonable classification."[334] Most directly relevant here, Justice Frank Murphy contended that racial classifications were irrational even with respect to aliens:

> The Constitution of the United States, as I read it, embodies the highest political ideals of which man is capable. It insists that our government, whether state or federal, shall respect and observe the dignity of each individual, whatever may be the name of his race, the color of his

327. *See id.* at 95.
328. *Id.* at 92.
329. *Id.*
330. Bolling v. Sharpe, 347 U.S. 497, 500 (1954).
331. 323 U.S. 192 (1944).
332. *Id.* at 203; *see also* Oyler v. Boles, 368 U.S. 448, 456 (1962) (holding that selective exercise of prosecutorial discretion was permissible unless "deliberately based upon an unjustifiable standard such as race, religion, or other arbitrary classification"); Kotch v. Board of River Port Pilot Comm'rs, 330 U.S. 552, 556 (1947) (noting that it would be unconstitutional "to deny a person a right to earn a living or hold any job because of hostility to his particular race, religion, beliefs, or because of any other reason having no rational relation to the regulated activities"); Reynolds v. Board of Pub. Instruction, 148 F.2d 754, 757 (5th Cir. 1945) ("[W]hile the equal protection clause of the Fourteenth Amendment makes no mention of race or color, and permits reasonable classifications and the exercise of reasonable discretion . . . a distinction in awarding public employment or compensation on no better basis than mere race or color would generally be arbitrary and forbidden.").
333. 347 U.S. 475 (1954).
334. *Id.* at 478, 482.

> skin or the nature of his beliefs. It thus renders irrational, as a justification for discrimination, those factors which reflect racial animosity.[335]

Conceivably, this line of cases used what is now recognizable as the vocabulary of rational basis review but actually engaged in strict scrutiny. If so, a naked rationality test might be insufficient to void a racial classification. Professor Michael Klarman persuasively responds to this possibility, arguing that "the language and the holding in several of these decisions are inconsistent with a presumptive rule against racial classifications" and that, before the development of the concept of strict scrutiny, "racial discrimination was impermissible with regard to certain fundamental rights, rather than across the board."[336]

Contemporary decisions of the Court issued under the tiered regime of equal protection review support the idea that racial classifications are irrational. Justice Stevens is the major modern champion of the idea that racial distinctions are forbidden because they are usually irrational. In *City of Cleburne v. Cleburne Living Center*,[337] he explained:

> The rational basis test, properly understood, adequately explains why a law that deprives a person of the right to vote because his skin has a different pigmentation than that of other voters violates the Equal Protection Clause. It would be utterly irrational to limit the franchise on the basis of height or weight; it is equally invalid to limit it on the basis of skin color. None of these attributes has any bearing at all on the citizen's willingness or ability to exercise that civil right. We do not need to apply a special standard, or to apply "strict scrutiny," or even "heightened scrutiny," to decide such cases.[338]

---

335. Oyama v. California, 332 U.S. 633, 663 (1948) (Murphy, J., concurring, joined by Rutledge, J.).

336. Michael Klarman, *An Interpretive History of Modern Equal Protection*, 90 MICH. L. REV. 213, 235 (1991). Professor Klarman uses the example of the emphasis in *Brown* on the importance of public education.

> Warren intimated, in other words, that had the framers of the Fourteenth Amendment foreseen the role that public education would one day play in American society, they unhesitatingly would have included it in the list of rights deemed too fundamental to permit of abridgement by racial discrimination. . . . [This implied] that other less fundamental rights could be distributed unequally according to race without infringing the Constitution. *Brown* thus confirms that the Justices were not yet thinking in terms of a racial classification rule.

Id. at 239. He points out that *Hirabayashi v. United States*, 320 U.S. 81 (1943), though apparently denouncing racial classifications, used highly deferential language in its review of the challenged measure. *See* Klarman, *supra*, at 232 & n.83.

337. 473 U.S. 432 (1985).

338. *Id.* at 452–53 (Stevens, J., concurring). *See generally* Note, *Justice Stevens' Equal Protection Jurisprudence*, 100 HARV. L. REV. 1146 (1987).

The Court applied a notion similar to the Stevens formulation in *Plyler v. Doe*,[339] explaining that "[c]lassifications treated as suspect tend to be irrelevant to any proper legislative goal."[340] Taken seriously, the idea that race is ordinarily irrelevant to a legitimate or proper state interest suggests that racial classifications would not pass even rational basis review.

Nevertheless, it must be recognized that all decisions to exclude aliens can be said to serve an important national policy. The Court has "often recognized that a 'primary purpose in restricting immigration is to preserve jobs for American workers.'"[341] Excluding any alien or class of aliens, including, for example, all Lutherans, Mexicans, Koreans, or Libertarians, promotes that goal to some degree.[342] If this is sufficient, then any conceivable immigration restriction would be rational. Moreover, in a world in which people and groups are not identical, if the Court applies the most deferential form of rational basis review, there would always be a rational, though spurious, basis for excluding any particular group.[343] For example, the Welsh could be excluded on the ground of preventing the spread of a disease one percent more prevalent in Great Britain than in other parts of the world or because of the high or low educational status or relative youth or maturity of the Welsh population.[344] Given the expansiveness of the kinds of rationales the Court has held to be rational, this kind of one-size-fits-all justification could also be applied to every other

---

339. 457 U.S. 202 (1982).

340. *Id.* at 216 n.14; *see also* Wygant v. Jackson Bd. of Educ., 476 U.S. 267, 301 (1986) (Marshall, J., dissenting) ("[R]acial distinctions are irrelevant to nearly all legitimate state objectives . . . ."); Fullilove v. Klutznick, 448 U.S. 448, 525 (1980) (Stewart J., dissenting) ("The color of a person's skin . . . bear[s] no relation to ability, disadvantage, moral culpability, or any other characteristics of constitutionally permissible interest to government."); *cf.* Edmonson v. Leesville Concrete Co., 500 U.S. 614, 643 (1991) (O'Connor, J., dissenting) ("Racism . . . is irrational, destructive and mean."); Mark Strasser, *The Invidiousness of Invidiousness: On the Supreme Court's Affirmative Action Jurisprudence*, 21 HASTINGS CONST. L.Q. 323, 327 (1994) (noting that "[t]he use of race as a criterion for who may vote or who may marry whom is obviously irrational").

341. INS v. National Ctr. for Immigrants' Rights, 502 U.S. 183, 194 (1991) (quoting Sure-Tan, Inc. v. NLRB, 467 U.S. 883, 893 (1984)).

342. And that may be enough for rational basis review. *See, e.g.*, McDonald v. Board of Election Comm'rs, 394 U.S. 802, 809 (1969) (stating that classification fails only if it is "totally unrelated" to the problem); Railway Express Agency v. New York, 336 U.S. 106, 109–10 (1949) (upholding a regulation in which the Court could not say that means had "no relation" to perceived evil).

343. *See* Flemming v. Nestor, 363 U.S. 603, 611–12 (1960) (upholding a classification in which there was a conceivable basis for it, and it was "constitutionally irrelevant whether this reasoning in fact underlay the legislative decision").

344. It would be rational to favor immigrants with high educational status because of the benefits to the United States or to seek unskilled workers for foreign policy reasons; similarly, it would be rational to seek young immigrants because of their greater likelihood of adjustment and ability to seek relevant education or to seek older people because of their maturity and probable higher level of education and training paid for by another country.

nationality and racial group, and probably every other group of humans sufficiently identifiable that they could be named in legislation.

But in recent rational basis cases, the Court has not resorted to the broadest possible reading of the many precedents. In *Romer v. Evans*,[345] rational basis review was sufficient to invalidate a classification motivated by "animus toward the class it affects."[346] The Court held irrational a Colorado constitutional amendment designed to exclude homosexuals from the protection of antidiscrimination statutes. The Court explained that, "[b]y requiring that the classification bear a rational relationship to an independent and legitimate legislative end, we ensure that classifications are not drawn for the purpose of disadvantaging the group burdened by the law."[347] When a classification is supported only by "animosity toward the class of persons affected," it will not be founded on a legitimate government interest.[348]

This understanding of rational basis review should serve to invalidate racial classifications in the immigration context. In *Cleburne Living Center*,[349] the Court said that classifications by race, alienage, or national origin "are so seldom relevant to the achievement of any legitimate state interest that laws grounded in such considerations are deemed to reflect prejudice and antipathy—a view that those in the burdened class are not as worthy or deserving as others."[350] In *Palmore v. Sidoti*,[351] the Court observed that "[c]lassifying persons according to their race is more likely to reflect racial prejudice than legitimate public concerns; the race, not the person, dictates the category."[352] This kind of inquiry is appropriate even within the premises of rational basis review. If, for example, Congress prohibited all Central Americans from immigrating, it might argue that it acted not out of any animosity toward that group, but because of the higher incidence of, say, malaria, in Central America than in other regions of the world. But, as in *Romer*, it would be obvious to a reviewing court why the classification was not more tightly crafted, for example, by excluding only aliens with malaria or even providing that all Central Americans had to prove their freedom from disease.[353]

345. 517 U.S. 620 (1996).
346. *Id.* at 632.
347. *Id.* at 633.
348. *Id.* at 634.
349. 473 U.S. 432 (1985).
350. *Id.* at 440.
351. 466 U.S. 429 (1984).
352. *Id.* at 432.
353. *See* Orleans Parish Sch. Bd. v. Bush, 242 F.2d 156, 163 (5th Cir. 1957) (Tuttle, J., joined by Rives & Brown, JJ.) (While "classification of students according to their proficiency or

There is also a practical ground for reading *Fiallo* to prohibit racial classifications. By providing for some judicial review, the Court implied that some imaginable classification would be unconstitutional. However, if racial classifications pass the test, there may be none that would fail. It would be surprising for the Court to have created a form of judicial review that upholds each and every action that it tests. One treatise says that the "apparent thrust of the Court's suggestion [in *Fiallo*] is that the 'limited judicial review' might be invoked, at the behest of an appropriate party, to challenge the most blatant type of discrimination, e.g., exclusions based solely on religion or race."[354]

The recent relaxation of the nondiscrimination provision of the INA, then, is partially unconstitutional. From 1952 until 1996, the INA stated that except as specifically provided by law, "no person shall receive any preference or priority or be discriminated against in the issuance of an immigrant visa because of the person's race, sex, nationality, place of birth or place of residence."[355] In 1996, Congress added the following provision: "Nothing in this paragraph shall be construed to limit the authority of the Secretary of State to determine the procedures for processing of immigrant visa applications or the locations where such applications will be processed."[356] This seemingly innocuous paragraph successfully terminated a case then pending in the Supreme Court, brought by Vietnamese asylum seekers who objected to a special rule applicable only to Vietnamese nationals.[357] Although the *granting* of the visas still apparently must be nondiscriminatory, procedures and application locations can be tremendously important.

The statute was intended to make sure that the State Department could draw lines on the basis of nationality in its visa application and

health traits" might be constitutional, "it is unthinkable that an arbitrary classification by race because of a more frequent identification of one race rather than another with certain undesirable qualities would be such reasonable classification.").

354. CHARLES GORDON & ELLEN GETTEL GORDON, IMMIGRATION AND NATIONALITY LAW 2-8 to -9 (1984).

355. Immigration and Nationality Act § 202(a)(1)(A), 8 U.S.C. § 1152(a)(1)(A).

356. *Id.* § 202(a)(1)(B), 8 U.S.C. § 1152(a)(1)(B) (added by the Illegal Immigration Reform and Immigrant Responsibility Act of 1996, Pub. L. No. 104-208, subtit. c, § 633, 110 Stat. 3009-701).

357. *See* Legal Assistance for Vietnamese Asylum Seekers v. Department of State, 45 F.3d 469, 472–73 (D.C. Cir.) (holding that the State Department regulation requiring Vietnamese asylum seekers to apply in Vietnam rather than Hong Kong violates the INA's prohibition on national origin discrimination), *on remand*, 909 F. Supp. 1, 2 (D.D.C. 1995) (dismissing case as moot), *rev'd*, 74 F.3d 1308 (D.C. Cir.) (reversing finding of mootness and remanding), *vacated and remanded for consideration of mootness*, 519 U.S. 1 (1996) (vacating decision for reconsideration in light of new statute), *on remand*, 104 F.3d 1349, 1353–54 (D.C. Cir. 1997) (holding the policy to be constitutional and consistent with statutory authorization).

processing procedures, an action that will sometimes be supported by a reasonable justification.[358] Instead of tailoring the statute to address the problem that gave rise to it, the statute permits the State Department to establish locations and procedures on the basis of other grounds, including race and religion. This breadth was apparently intentional,[359] but even under rational basis review, regulating visa application procedures on the basis of race or religion would be difficult to connect to any foreign policy rationale or other legitimate justification, and thus would be irrational.[360]

During the public agitation that led to the Chinese Exclusion Act, mobs exclaimed "the Chinese must go!" Over a century later, Professor Henkin turned the slogan to his own use, arguing passionately that "*Chinese Exclusion*—its very name is an embarrassment—must go."[361] Professor Henkin's desire, it appears, has been largely achieved. Even on its own terms, the plenary power doctrine would no longer authorize racial discrimination.

Chae Chan Ping and *Fong Yue Ting* incorrectly expressed the substantive limitations on federal immigration power because they authorized racial discrimination that would now be prohibited. They incorrectly expressed due process limitations on the federal immigration power because they exiled residents in ways now readily seen as unfair. Under these circumstances, as well as the racial history, the Supreme Court would be justified in asking itself why these cases remain the principal explanatory sources of the federal immigration power.

358. A secretary of state arguably could channel immigrants from particular countries to certain consulates for security reasons because of the expertise of particular consular stations or based on a good-faith desire to discourage illegal immigration to countries where embassies are located. Alternatively, as a means of improving relations with a foreign government, the United States might choose to grant preferential admissions to citizens of that nation or, on the other hand, restrict immigration if the foreign government wishes to keep its citizens at home. While favors and punishments awarded to nations may cheer or insult Americans who trace their ancestry there, if the United States is to have foreign relations at all, this is inevitable.

359. *See* H.R. REP. NO. 104-828, at 248 (1996) (stating that the amendment clarifies "that the Secretary of State has non-reviewable authority to establish procedures for the processing of immigrant visa applications and the locations where visas will be processed").

360. Outside of the realm of paranoid fantasy, the United States does not have diplomatic or political conflicts with races or genders; it is unlikely that the United States, or any other country for that matter, will find itself the target of a worldwide race war or other concerted action on the basis of race. *Cf.* Peter J. Spiro, *The States and Immigration in an Era of Demi-Sovereignties*, 35 VA. J. INT'L L. 121, 121–23 (1994) (explaining that immigration policy is not always connected to foreign policy).

361. Henkin, *supra* note 42, at 863.

CONCLUSION

What would happen if immigration laws were treated like other acts of Congress? It is beyond the scope of this Article to suggest precisely what judicial review after plenary power should look like; however, scholars such as Frank Wu[362] and Victor Romero[363] have persuasively explained why the world would not come to an end if the Court moved beyond plenary power. The national government now has the power to make reasonable distinctions between aliens and citizens.[364] Moreover, even with respect to citizens, when policies genuinely implicate foreign relations, courts take that into account.[365] Even Justices Brennan and Marshall acknowledged that foreign policy concerns might justify national origin classifications;[366] for example, courts have found it reasonable for the United States to restrict the immigration of enemy aliens or aliens from a country perceived as hostile.[367]

Even when the plenary power doctrine is not employed to create invidious classifications, it has a baleful influence. The old cases have tremendous rhetorical power. Moreover, the plenary power doctrine's relief of

---

362. *See* Wu, *supra* note 42, at 48–54.

363. *See* Romero, *supra* note 42.

364. *See, e.g.*, Earl M. Maltz, *Citizenship and the Constitution: A History and Critique of the Supreme Court's Alienage Jurisprudence*, 28 ARIZ. ST. L.J. 1135 (1996).

365. Although since *Bolling v. Sharpe*, the Court has stated its "approach to Fifth Amendment equal protection claims has always been precisely the same as to equal protection claims under the Fourteenth Amendment," Weinberger v. Wiesenfeld, 420 U.S. 636, 638 n.2 (1975), *Hampton v. Mow Sun Wong*, 426 U.S. 88 (1976), qualified this point, holding, in a case involving an alienage classification, that the national government could discriminate in a manner prohibited to the states if there were "overriding national interests which justify selective federal legislation." *Hampton*, 426 U.S. at 100.

366. Justice Marshall argued that national origin discrimination in admission of excludable aliens to parole could sometimes be legitimate. *See* Jean v. Nelson, 472 U.S. 846, 881 (1985) (Marshall, J., dissenting). A national group could be denied parole if that action would reduce the flow of undocumented aliens while refusing parole to other groups would not. *See id.* at 880. Denying parole might also be legitimate if it reduced the time it took to process asylum applications. *See id.* at 881.

367. *See, e.g.*, Malek-Marzban v. INS, 653 F.2d 113, 116 (4th Cir. 1981) (stating that "[t]he United States is not bound to treat the nationals of unfriendly powers with the same courtesy and consideration it extends to nationals of friendly powers" and upholding the 15-day time limit for voluntary departures of deportable Iranian nationals even though no similar limit applied to other aliens); Narenji v. Civiletti, 617 F.2d 745, 747 (D.C. Cir. 1979). The court in *Narenji v. Civiletti* upheld a registration requirement for nonimmigrant Iranian students; the attorney general claimed the measure was "'an element of the language of diplomacy by which international courtesies are granted or withdrawn in response to actions by foreign countries. The action . . . is therefore a fundamental element of the President's efforts to resolve the Iranian crisis and to maintain the safety of the American hostages in Tehran.'" *Narenji*, 617 F.2d at 747 (quoting the attorney general's affidavit).

Congress from ordinary forms of judicial review has encouraged a casual attitude toward crafting the INA. Freedom from judicial review may have inadvertently encouraged a certain inartfulness in a statute that has grown large and unwieldy over the decades. For example, when amending the nondiscrimination provision of the INA, Congress could have given the State Department power to implement policies and procedures for visa applications with respect to particular national origins, instead of exempting visa application procedures entirely. A more thoughtful statute would have addressed the problems the State Department actually faced without authorizing racial and religious discrimination.

In short, ample federal power to regulate immigration and foreign policy will not come to an end when the plenary power doctrine is abolished, and the quality of law making may even improve. But our law should be free of the last of the Fuller Court's great mistakes, which caused so much harm to Americans of every race.

THE TRANSFORMATION OF IMMIGRATION LAW

*Peter H. Schuck**

Immigration has long been a maverick, a wild card, in our public law. Probably no other area of American law has been so radically insulated and divergent from those fundamental norms of constitutional right, administrative procedure, and judicial role that animate the rest of our legal system. In a legal firmament transformed by revolutions in due process and equal protection doctrine and by a new conception of judicial role, immigration law remains the realm in which government authority is at the zenith, and individual entitlement is at the nadir.

The distinctiveness of immigration law reflects a number of factors. Since the ideal of nationhood first fired the human imagination, a country's power to decide unilaterally who may enter its domain, under what conditions, and with what legal consequences has been regarded as an essential precondition of its independence and sovereignty.[1] Aliens lack full membership in the moral and political communities that create and sustain our system of justice. They stand outside the gates beseeching the nation to permit them to enter. The relatively few who are fortunate enough to be admitted must remain nonmembers for a time; their moral claims upon us during that hiatus are provisional, contingent and seldom compelling. Indeed, by calling them "aliens," our law affirms that they remain strangers, objects of our vigilance, our suspicion, and perhaps even our hostility.[2]

The place of immigration law in our larger legal and political system is also a source of uniqueness. Immigration law often implicates the nation's

---

* Professor of Law, Yale Law School. B.A. 1962, Cornell University; J.D. 1965, Harvard University; LL.M. 1966, New York University; M.A. 1969, Harvard University. The author wishes to acknowledge the fine research assistance provided by Suhn Hong, Yale Law School class of 1985. He also benefited from comments on earlier drafts by Bruce Ackerman, T. Alexander Aleinikoff, Ted Harwood, Michael Heilman, David Martin, Eben Moglen, Rogers Smith and the participants in the Yale Law School faculty workshop.

1. The early immigration cases in the Supreme Court adopted this premise unequivocally. See, e.g., Fong Yue Ting v. United States, 149 U.S. 698, 711 (1893) ("The United States are a sovereign and independent nation, and are vested by the Constitution with the entire control of international relations, and with all the powers of government necessary to maintain that control and to make it effective."); Nishimura Ekiu v. United States, 142 U.S. 651, 659 (1892) ("It is an accepted maxim of international law, that every sovereign nation has the power, as inherent in sovereignty, and essential to self-preservation, to forbid the entrance of foreigners within its dominions, or to admit them only in such cases and upon such conditions as it may see fit to prescribe.") (citation omitted).

2. Cf. M. Walzer, Spheres of Justice 32 (1983) (pointing out that in Latin and some other ancient languages, strangers and enemies were designated by a single word).

basic foreign policy objectives, a circumstance that has sometimes provoked the Supreme Court, even in nonimmigration contexts, to be less scrupulous in safeguarding constitutional values and more deferential to the other branches of government.[3] The domestic politics of immigration, and hence the laws and policies governing aliens, also reveal unusual patterns. Even lawfully admitted aliens do not vote, of course, and millions of undocumented workers and their families live largely outside the protection of any law.[4] At the psychological and emotional levels, immigration generates complex public attitudes toward outsiders and newcomers. Especially during periods of economic distress, a recurrent nativism has tapped the dark wellsprings of racial and religious bigotry in America, releasing forces that have profoundly shaped the character of immigration and national politics.[5]

A central theme of this Article is the epiphenomenal nature of immigration law. Throughout our history, its changing character has reflected more fundamental social and ideological structures. From the birth of the Republic until the 1880's, American society was consumed by the tasks of populating a vast unsettled continent and exploiting its untapped wealth. These activities were driven by a decidedly open economic and social system and by an ideology that corresponded to a remarkable degree to the tenets of Lockean liberal theory.[6] The liberalism of America's first century conceived of persons as autonomous, self-defining individuals possessing equal moral worth and dignity and equally entitled to society's consideration and respect. This entitlement was in principle universally shared, a natural right deriving not from the particularities of one's time, place, or status, but from one's irreducible humanity. The good society, in this view, was one in which each individual enjoyed maximum liberty to pursue his or her own conception of the good by deciding whether, and on what terms, to enter into contractual relationships with other equally free individuals.[7] Liberal ideology was reflected in a policy of essentially open borders, one that strongly encouraged, indeed actively recruited, mass immigration to the United States.[8]

By the 1880's, the American frontier was closing and the United States turned from agrarian expansion to urban and industrial development. The social milieu in which liberalism had flourished receded, to be replaced by a

---

3. See, e.g., Dames & Moore v. Regan, 453 U.S. 654 (1981).

4. Although undocumented aliens may sue in the courts and enjoy certain substantive rights, see Plyler v. Doe, 457 U.S. 202 (1982), many of their rights are unenforceable by them as a practical matter. See generally, e.g., Developments in the Law—Immigration Policy and the Rights of Aliens, 96 Harv. L. Rev. 1286, 1437 (1983) [hereinafter cited as Developments].

5. The classic study is J. Higham, Strangers in the Land: Patterns of American Nativism, 1860–1925 (1955).

6. L. Hartz, The Liberal Tradition in America 3–14 (1955).

7. See id. In practice, of course, a considerable portion even of American society—most notably slaves and, to some degree, females—were denied this liberty. See infra notes 27–31 and accompanying text.

8. See infra notes 25–32 and accompanying text.

set of new conditions and attitudes far less congenial to it. Liberal values were challenged by an array of exclusionary impulses—racist and class-based opposition to Chinese laborers, nativist xenophobia, religious bigotry, and political reaction against radical movements drawing upon new immigrant groups.[9] These changes spawned a different ideology, which I shall call "restrictive nationalism," and a corresponding legal order, which I shall call "classical immigration law."

The new ideology and the legal structure that embodied it celebrated norms and countenanced practices that were decidedly, sometimes grotesquely, illiberal. Aliens' natural rights to pursue their self-interest through migration here were now regulated according to considerations of national interest, sovereignty and power. Consent remained the source of an alien's legal rights and duties, but it was no longer liberalism's consent of freely contracting individuals exercising their natural rights to define their own relationships with others. Restrictive nationalism drew upon the ideological conception of consent-based obligation, but reshaped it to respond to the dictates of an exclusionary sentiment. The new order would be based upon the national government's consent to allow the alien to enter and remain, which consent could be denied or withdrawn on the basis of arbitrary criteria and summary procedures that often transgressed liberal principles.[10]

Classical immigration law proved to be remarkably durable. Indeed, as we shall see, important elements of that system remain in place today, a century after restrictive nationalism began to supplant liberalism as the dominant ideology of immigration policy. Nevertheless, its inconsistency with contemporary public law values has become increasingly evident. The growth of the modern administrative state has magnified its isolation from the dominant developments in American law, throwing its distinctive features into even sharper relief. The currents that have transfigured constitutional jurisprudence, administrative law, civil rights, and judicial ideology since the New Deal, and especially since the 1960's,[11] have largely passed immigration law by, leaving it to navigate its own course.

To be sure, the insularity of immigration law has not gone unnoticed by legal scholars. Many have commented upon its persistence and almost all have vigorously condemned it. Aliens, they have urged, should be brought within the shelter of our dominant legal traditions. Immigration law should be made to embody the more universalistic norms, the more spacious conception of legal and moral community, to which liberalism and contemporary public law aspire.[12]

---

9. See generally, J. Higham, supra note 5.

10. See infra notes 93-108, 130-48 and accompanying text.

11. Brown v. Board of Educ., 347 U.S. 483 (1954), may be viewed as initiating the changes that flowered during the 1960's.

12. See, e.g., L. Tribe, American Constitutional Law 277-84 (1978); Lopez, Undocumented Mexican Migration: In Search of a Just Immigration Law and Policy, 28 UCLA L. Rev. 615 (1981); Rosberg, The Protection of Aliens from Discriminatory Treatment by the National

Today, this chorus of criticism is beginning to find expression in the discourse and doctrines of immigration law. New principles based upon fundamentally different values are beginning to undermine the classical regime and to etch the outlines of a new legal order. I shall call these new principles "communitarian," for their central idea is that the government owes legal duties to all individuals who manage to reach America's shores, even to strangers whom it has never undertaken, and has no wish, to protect. Communitarian immigration values, of course, are linked to those of the past. Although rejecting traditional liberalism's emphasis upon consent as the basis for legal obligations to strangers, for example, communitarianism echoes liberalism's emphasis upon universal rights based upon individuals' essential and equal humanity. What is also striking is their departure from the classical values and their affinity to newly emergent public law norms. These new norms derive the government's legal obligations to individuals from the nature of their social interactions and commitments rather than from the government's consent to be bound as expressed through traditional legal forms and procedures.

Although the infusion of communitarian values may profoundly alter immigration law, their ultimate triumph is by no means assured. Their absorption is still tentative and fragmentary, not yet fully confirmed by the Supreme Court. Moreover, they introduce many contradictions into the immigration system—contradictions among those values, the liberal assumptions that still pervade American law, and the vestiges of restrictive nationalism embedded in the classical immigration structure; contradictions within the communitarian ethos itself; and contradictions between communitarian values and contemporary social realities.

This Article explores the nature and implications of this still-embryonic transformation of immigration law. It does so at a time, perhaps only a fleeting moment, when its dimensions and future direction are widely recognized as urgent questions on the national policy agenda. The Article consists of four Parts. In Part I, elaborate the essential elements of classical immigration law. In Part II, I suggest that a fundamental transformation of the classical system is under way and seek to account for this change by analyzing two interacting fields of force. The first involves a set of changed objective conditions, a new social reality, within which immigration law has developed and is being applied. The second involves a fundamental ideological shift, an altered legal consciousness, concerning the sources and nature of the federal government's legal obligations toward individuals generally. This shift has generated doctrinal currents that has washed over other areas of American law for decades. These currents, however, have left immigration

Government, 1977 Sup. Ct. Rev. 275; Developments, supra note 4. Evidence that these norms have taken root in other areas of the law is provided, for example, by the expanded scope of constitutional equal protection, the enactment of civil rights statutes protecting minorities and women, and the general broadening of tort liability. From another perspective, of course, these developments may be—and have been—viewed as moves away from universal norms toward a regime of group rights. See generally G. Garson, Group Theories of Politics ch. 2 (1978).

law, isolated in the remote eddies and backwaters of the stream, largely untouched. Together, I argue, these forces are beginning to penetrate classical immigration law, reducing its insularity and moving it closer to the mainstream of public law from which it has been separated by its distinctive conceptions and context. In Part III, I explore the character of the communitarian immigration law that I believe is emerging in the cases. The final Part discusses some unresolved tensions that this new development is creating, tensions whose resolution poses a fundamental problem for contemporary liberalism and American society.

This essay is as much an exercise in the interpretation of legal change as it is one of causal explanation. Immigration is too intimately linked to the basic processes and structures of American life, its legal doctrines too epiphenomenal, to permit easy causal inferences. Only an integrated work of historical, economic, political, sociological, and legal scholarship could begin to unravel the complex interrelationships and account for the new patterns that we are beginning to observe. This Article has a far more modest ambition. It is to crystalize the still-tentative, often ambiguous changes that have begun to roil the surface of the immigration case reports, and to extract from them some larger meaning for our contemporary public law.

I. The Classical Conception of Immigration Law

Classical immigration law emerged in the late nineteenth and early twentieth centuries as a liberal American polity confronted for the first time the felt need to regulate immigration in significant ways. Not surprisingly, classical immigration law mirrored the highly individualistic values that underlay the prevailing social and legal order.[13] But it added to these values several new and somewhat inconsistent ideological elements. Three features of that order help to reveal the character of the classical view. First, the ethnic, cultural and class composition of the immigrant population changed dramatically between the 1870's and World War I. Unlike the "old" immigrants, who had come to the United States primarily from Northern and Western Europe, most of the "new" immigrants came from Southern and Eastern Europe and—until their entry was sharply restricted in the 1880's—from the Orient.[14] This change triggered the explosive passions of racial and religious prejudice, fears of revolutionary contagion, class conflict, and other deep-seated animosities against the newcomers who appeared to threaten the value system known as "Americanism." Powerful pressures to limit both the level of immigration

13. "Most of our public law—both substantive and procedural—grows quite directly out of private law and corresponding efforts to treat the government as a defendant in a private lawsuit." Sunstein, Judicial Relief and Public Tort Law (Book Review), 92 Yale L.J. 749, 758 (1983) (reviewing P. Schuck, Suing Government: Citizen Remedies for Official Wrongs (1983)). Sunstein goes on to assert that "that model is breaking down," id., a proposition with which I agree.

14. Thus, between the 1870's and the second decade of this century, Northern and Western Europe's share of the immigrant population declined from 73.6% to 17.4%. See M. Bennett, American Immigration Policies: A History 31 (1963).

and the rights of aliens consequently developed.[15] Classical immigration law was launched upon this tide of hostility to strangers.

Second, American society was developing a sense of autonomy and leadership on the international stage. The nation's emergence as an industrializing world power, its acquisition of an overseas empire during the 1890's, and a succession of strong, aggressive presidents from McKinley through Wilson, all contributed to the growth of a "new nationalism" in both the domestic and foreign policy spheres.[16]

Third, the larger legal system, of which classical immigration law was a part, was committed to a very broad definition of individual and governmental sovereignty. Individual's sovereignty was protected through an array of private property rights. A landowner, for example, enjoyed a plenary right to exclude trespassers, including those who had entered with the owner's permission but had violated the conditions under which the permission was granted. The owner owed no duty to trespassers, except to refrain from willfully injuring them or using unreasonable force in their expulsion.[17] The owner was ruler over his domain; his home, as the saying goes, was his castle. Governments, for their part, were protected by analogous public property rights, as well as by immunity doctrines that severely limited the claims that could be asserted against them.[18] In general, tort law imposed few affirmative legal duties to strangers on governments or private individuals; their only obligations were those that they had voluntarily undertaken.[19]

In such a context, control over which strangers might enter was viewed as a powerful expression of the nation's identity and autonomy—in a word, of its sovereignty. Sovereignty entailed the unlimited power of the nation, like that of the free individual, to decide whether, under what conditions, and with what effects it would consent to enter into a relationship with a stranger. The power implied by that sovereignty was enjoyed by the Nation exclusively and was based wholly upon its explicit undertaking; the government simply owed no other legal obligation to those who sought to enter or remain without its consent. That a stranger was desperate to enter, and had invested a great deal in the effort, was as immaterial as the reasons that prompted the government to refuse her admittance. As the Supreme Court stated in 1892, "It is an accepted maxim of international law, that every sovereign nation has the power, as inherent in sovereignty, and essential to self-preservation, to forbid the entrance of foreigners within its dominions, or to admit them only in such cases and upon such conditions as it may see fit to prescribe."[20] In this sense,

15. Id. at 31-34, 36-39.

16. See W. Lafeber, The New Empire (1963); A. Schlesinger, The Age of Roosevelt: The Crisis of the Old Order 17-26 (1957).

17. F. Harper & F. James, The Law of Torts 1435-40 (1956).

18. See P. Schuck, Suing Government: Citizen Remedies for Official Wrongs 36-39 (1983).

19. For an explanation of the limited concept of duty in tort law during that period, see F. Harper & F. James, supra note 17, at 1430-32; see also Ames, Law and Morals, 22 Harv. L. Rev. 97, 103-05 (1908).

20. Nishimura Ekiu v. United States, 142 U.S. 651, 659 (1892) (citation omitted).

the classical idea of sovereignty implied a relationship between government and an alien that resembled the relationship in late nineteenth century private law between a landowner and a trespasser. The essential purpose of immigration law, then, like that of property law, was to preserve and enhance this sovereignty.

The individualistic premises of a liberal legal order, of course, did not lead ineluctably to a restrictive immigration policy and a conception of absolute sovereignty that utterly denied aliens' claims to legal protection. Liberalism, after all, does not merely affirm the primacy of individual consent as the bedrock of political and juridical relationships. It also locates fundamental legal rights in individuals by reason of their universal humanity and without regard to contingencies of their status or condition. Just as the sweeping rights of landowners against trespassers in private law were subject to an exception—quite an elastic one, as it turned out—for "attractive nuisances,"[21] so public international law theorists had long advanced competing principles, also derived from liberal premises, emphasizing the state's obligations to protect strangers who were drawn to enter it.[22] Moreover, a liberal economy, celebrating the laissez-faire values of unimpeded resource flow and exchange, would pursue a policy of free immigration. Indeed, American liberalism maintained essentially that policy from the colonial period until the 1880's. This aspect of liberalism, of course, would suggest a far weaker, more qualified notion of sovereignty, one that would tolerate relatively open borders and at least some protections even for aliens who enjoyed no rights under positive domestic law. In the end, the ideological shift from traditional liberalism to restrictive nationalism reflected the influence of external pressures more than the internal logic of principle.

In what follows, we shall see that these two different ideological threads—the one denying that a society owes aliens any obligation to which it does not consent, the other affirming the existence of certain obligations to aliens owed simply by reason of their humanity—are woven throughout the fabric of immigration law. In the classical period, the first of these, reinforced by a growing sense of nationhood and legitimated by then-current notions of limited public and private obligation, clearly dominated the pattern. As we shall see in Parts II and III, however, the second thread, now buttressed by more expansive, communitarian principles of obligation drawn from other moral premises and embedded in other areas of the law, is beginning to challenge the primacy of the first, thereby altering that pattern.

This transformation is proceeding around seven central elements of classical immigration law: (1) the restrictive ideal of national community; (2) the principle of judicial deference; (3) the extraconstitutional status of exclusion; (4) the broad federal power to classify aliens; (5) the civil nature of the deportation sanction; (6) the detention power; and (7) the integration of

21. F. Harper & F. James, supra note 17, at 1447-59.

22. See 2 R. Phillimore, Commentaries Upon International Law 3-7 (3d ed. 1882); E. de Vattel, The Law of Nations 174-78 (1797).

adjudication and enforcement. Although the emerging changes in these norms and doctrines generally parallel one another, there are important differences. Some, for example, are merely incipient, while others are relatively far advanced. But immigration law's gradual assimilation of the communitarian ideal is shaping all of them to some degree.

A. *The Restrictive Ideal of the National Community*

The concept of nationhood expresses political, legal, and social dimensions of community. It entails the mutual acceptance of certain norms, participation in a common public life, commitment to an overarching structure of law, and sharing of the benefits and burdens of national institutions.[23] At this abstract level, the notion of national community is definitional, not operational; it tells us very little about who may or may not become a member, and about what practical difference membership makes in individuals' lives. In American law, at least, the struggle over these questions has revolved around the control and conditions of more specific legal statuses—those of natural-born citizen, naturalized citizen, citizen born abroad, permanent resident alien, nonimmigrant alien, undocumented alien, and the like.[24] In the positive sense, therefore, our conception of nationhood is revealed by the aggregate of these legal categories and the positive rights and obligations that flow from them. More normatively, however, national community is an ideal toward which we aspire, a way of expressing what it is that we share as a people.

Immigration law necessarily mediates between these two meanings. American society has always prided itself on the inclusive, assimilative conception of nationhood that it embraces and offers to those who would join us. These qualities of openness have been a dominant feature of our self-definition and national myths. Emma Lazarus's verse inscribed on the Statue of Liberty celebrates not only the tens of millions of foreign-born who have flocked to our shores, but also the society that awaited them behind the "golden door."[25]

There is much truth in this account, yet it overlooks much truth as well. Insofar as the availability of the legal status of citizenship is concerned, the ease of acquiring membership in the American community has been striking. As early as the colonial period, citizenship was automatically conferred by birth within the colony; for those born elsewhere, that status could still be readily obtained. Anxious to attract labor and enhance property values, the colonies actively and imaginatively promoted immigration by Europeans.

---

23. See generally M. Gordon, Assimilation in American Life chs. 4, 5 (1964).

24. See generally J. Kettner, The Development of American Citizenship, 1608–1870 (1978) (discussing citizenship issues).

25. At least 48 million immigrants reached the United States between 1821 and 1978. See Select Comm'n on Immigration and Refugee Policy, U.S. Immigration Policy and the National Interest 96 (1981) [hereinafter cited as SCIRP Staff Report].

Despite opposition in London, colonial naturalization was converted into a relatively inexpensive, expeditious administrative process based upon consent and available to almost all.[26] When the Republic was created, the right of naturalization was, "generally speaking," immediately extended to all whites who could satisfy a brief residency requirement and make a declaration of loyalty.[27] Citizenship has been conferred automatically by birth in the United States (jus soli)[28] and by birth abroad to American parents (jus sanguinis).[29] The full legal equality of birthright citizens and naturalized citizens was well established early in the nineteenth century,[30] and with the adoption of female suffrage in 1920, political equality between all adult citizens was affirmed, at least as a matter of law.[31] Aliens enjoy the protection of the Constitution once they have "entered" the United States; most of the rights that it recognizes are conferred upon "persons" or "people," rather than only upon citizens or lawful resident aliens.[32]

---

26. See J. Kettner, supra note 24, ch. 4. Although there was some local opposition to the naturalization of Jews, a 1740 English statute allowed Jews and other groups to obtain the rights of British subjects. Id. at 116-17.

27. Id. at 236. The preconditions for naturalization for white persons generally have remained easy to satisfy throughout our history, with only a few exceptions. See E. Hutchinson, Legislative History of American Immigration Policy, 1798-1965, at 12-16 (1981) (the Alien and Sedition Act interlude); J. Kettner, supra note 24, at 245-46 (requirement to renounce aristocratic titles); SCIRP Staff Report, supra note 25, at 88 (literacy test, first imposed during World War I). In 1790, the first federal naturalization act conferred citizenship only on a "free white person." J. Kettner, supra note 24, at 236. In 1870, Congress extended the naturalization laws to "aliens of African nativity and to persons of African descent." E. Hutchinson, supra, at 58.

28. This principle long antedated the fourteenth amendment of the Constitution, see J. Kettner, supra note 24, at 287-88, but was confirmed by both the amendment and the Civil Rights Act of 1866, ch. 31, 14 Stat. 27. See United States v. Wong Kim Ark, 169 U.S. 649 (1898). It does not embrace certain categories of individuals born here but not subject to United States jurisdiction: diplomats, alien enemies in hostile occupation, persons born on foreign ships in American waters, and Indians owing direct allegiance to tribes. Id. at 693; see also J. Kettner, supra note 24, at 287-333.

29. 8 U.S.C. § 1401(c) (1982). But see Rogers v. Bellei, 401 U.S. 815 (1971) (fourteenth amendment citizenship clause does not apply to individuals born abroad to American citizens). The statute at issue in *Bellei* imposed a requirement that a foreign-born person must reside in the United States for five years before reaching an age of twenty-eight years, in order to claim citizenship. This restriction has now been deleted by amendment. See Act of Oct. 10, 1978, Pub. L. No. 95-432, 92 Stat. 1046.

30. Osborn v. Bank of the United States, 22 U.S. (9 Wheat.) 738, 827 (1824) (dictum); J. Kettner, supra note 24, at 287 (sole exception being eligibility for Presidency).

31. But note the long history of deprivation of black voting rights and of continuing permissibility of property requirements, discussed in the legislative history of the Voting Rights Act of 1965, 42 U.S.C. §§ 1971, 1973-1973p (1976). See H. Rep. No. 439, 89th Cong., 1st Sess. 8-13 (1965), reprinted in 1965 U.S. Code Cong. & Ad. News 2437, 2439-44.

32. See, e.g., Plyler v. Doe, 457 U.S. 202, 210 (1982); Yick Wo v. Hopkins, 118 U.S. 356, 369 (1886). Even during the classical period, aliens did not necessarily relinquish those rights when, after leaving the jurisdiction of the United States, they attempted to reenter. See, e.g., Kwong Hai Chew v. Colding, 344 U.S. 590, 596 n.5 (1953) (quoting Bridges v. Wixon, 326 U.S. 135, 161 (1945)).

The inclusiveness of the legal and political communities, thus defined, has often been celebrated; under the conditions of American culture and the ideals of traditional liberalism, this inclusiveness may well have been inevitable. Alexander Bickel, for example, maintained that citizenship was "a simple idea for a simple government," arguing that it could not and should not function as an exclusionary concept in a liberal society.[33]

The story of the steady, progressive enlargement of the national community, however, is more complicated and contradictory than Bickel's account suggests. We have already seen that traditional liberal values fail to account for the emergence and character of the classical immigration order. Moreover, the concept of American citizenship, which Bickel thought unimportant in our constitutional vision,[34] actually seems to have mattered a great deal from the very beginning. Although Bickel contends that the original Constitution made "precious few references to the concept" and "made nothing depend on it explicitly, aside from a few offices,"[35] Justice Rehnquist is probably nearer to the mark in emphasizing that citizenship is a constitutionally significant status "in no less than 11 instances in a political document noted for its brevity."[36] It is, after all, a prerequisite for all high elected officeholders; indeed, the President must be "a natural born Citizen."[37] Citizenship is also required for access to certain categories of federal court jurisdiction,[38] and even for access to the voting rights protected by the post-Civil War amendments.[39] It seems doubtful that at a time when Loyalist disaffection was very much on the minds of American politicians,[40] the Founding Fathers thought citizenship an inconsequential status.

Historically, even the category of citizen has obscured important differences in legal status, the Constitution's unqualified language notwithstanding. That some 125,000 individuals of Japanese descent were American citizens did

---

33. A. Bickel, The Morality of Consent 54 (1975). Rogers Smith has noted that

> the very universality of liberalism's moral principles, along with its couching of them in terms of personal human liberties, makes it difficult for liberal theory to give nationality and nationalism any central place in its notions of personal identity and political purpose. The craving for national self-determination may be seen by liberals as a natural part of human psychology, but one's particular nationality is viewed as a contingent characteristic, of secondary moral importance.

R. Smith, The Dilemmas of American Citizenship 1-2 (unpublished manuscript on file with author) (citing I. Berlin, Nationalism, *in* Against the Current 338-41 (1980)) (footnotes omitted).

34. A. Bickel, supra note 33, at 36. Bickel also failed to consider the possibility that under contemporary conditions, perpetuating a liberal policy of easy access to citizenship and legal rights may require a less liberal, more restrictive immigration policy. See infra notes 465-76 and accompanying text.

35. A. Bickel, supra note 33, at 35.

36. Sugarman v. Dougall, 413 U.S. 634, 651-52 (1973) (Rehnquist, J., dissenting).

37. U.S. Const. art. II, § 1, cl. 5.

38. U.S. Const. art. III, § 2, cl. 1.

39. U.S. Const. amends. XIV, § 2, XV, § 1.

40. See J. Kettner, supra note 24, ch. 7.

not prevent them from being driven from their homes into internment camps during World War II solely on the basis of their ancestry.[41] The long struggle over political equality for females, a majority of the population, reflected the Supreme Court's conviction, in Rogers Smith's words, "that participation is extraneous to American citizenship."[42] One century after Congress and the Constitution had overruled the *Dred Scott* decision[43] and decreed the freed slaves to be American citizens,[44] it remained necessary for Congress to enact the Voting Rights Act of 1965[45] to secure the franchise for blacks and other minorities. The legal condition of the American Indian was, if anything, even more degraded and "alienated" than that of blacks.[46]

Perhaps most revealing of the exclusionary possibilities concealed within the broad portmanteau of American citizenship is the legal status of residents of the territories owned by the United States. In *Downes v. Bidwell (The Insular Cases)*,[47] which Bickel unaccountably fails to mention in his article on citizenship, the Supreme Court held that Congress need not confer citizenship on residents of the conquered territories;[48] if it wished, it could create a second-class American citizenship. Indeed, Congress did just that with respect to Puerto Rico, the Virgin Islands, Guam and the Northern Marianas.[49] And in *Balzac v. People of Porto Rico*,[50] the Court held that United States citizenship for a Puerto Rican conferred no rights under the Constitution except the right to move to the mainland. There, as a state resident, she could "enjoy

---

41. See Korematsu v. United States, 323 U.S. 214 (1944); see also Report of the Commission on Wartime Relocation and Internment of Civilians, Personal Justice Denied (1982).

42. R. Smith, "One United People": Discriminatory Citizenship Laws and the American Quest for Community, 1800–1937, at 22–23 (unpublished manuscript on file with author).

43. Scott v. Sandford, 60 U.S. (19 How.) 393 (1856).

44. U.S. Const. amend. XIV; Civil Rights Act of 1866, ch. 31, 14 Stat. 27; see also J. Kettner, supra note 24, at 300–33. Categorical racial exclusions from naturalized citizenship continued, however. In 1922, for example, a unanimous Supreme Court construed a 1906 statute limiting naturalization to "free white persons" and aliens of African nativity or descent, to exclude a Japanese alien. Ozawa v. United States, 260 U.S. 178 (1922).

45. 42 U.S.C. § 1973 (1976).

46. James Kettner observes:

> The peculiar "domestic dependent nation" status of the tribes thus ultimately served the purposes of those who wished to maintain control over Indians without fully incorporating them into the community of citizens. Because the tribes were "domestic" and "dependent," white laws could be extended over them. Yet such extension did not constitute the kind of protection that elicited allegiance and sustained citizenship as long as political and judicial authorities considered the tribes, in some sense, "nations" whose members were aliens. The logic of combining dependency and wardship with the idea of a separate allegiance and nationality was perhaps inconsistent; but it sufficed to exclude the Native Americans from the status and the privileges of American citizenship.

J. Kettner, supra note 24, at 299–300.

47. 182 U.S. 244 (1901).

48. Id. at 279–80.

49. See Cabranes, Citizenship and the American Empire, 127 U. Pa. L. Rev. 391, 486 n.460 (1978) (giving statutory citations).

50. 258 U.S. 298 (1922).

every right of any other citizen of the United States, civil, social and political."[51]

If classical immigration law permitted Congress to differentiate between American citizens for the purpose of qualifying for full membership in the national community,[52] it is not surprising that the right of aliens to participate equally in American life was severely limited. It is true that aliens were not wholly excluded from constitutional protection. As early as 1886, for example, the Supreme Court strongly affirmed that the Constitution safeguards *all* individuals within the United States, including aliens, from invidious discrimination by the states.[53] Less than twenty years later, the Court ruled that even an excludable alien was entitled to at least a minimal form of administrative due process.[54] But while these holdings demonstrated that the universalistic character of traditional liberalism retained some force during the classical period, they hardly defined an expansive national community. Although the equal protection principle did apply to aliens, for example, the status of alienage applies to a multitude of individuals with very different relationships to American society. That status has always remained a constitutionally permissible ground for classifying individuals for many invidious purposes. Thus, the equal protection principle did not prevent governments at all levels from barring lawfully admitted permanent resident aliens from public employment,[55] employment on public works,[56] ownership of real property,[57] and many, many other kinds of activities through which political and legal communities customarily express their essential character and unity.[58] And while due process was owed to the excludable alien deprived of liberty, the process that was due did not even include a right to a judicial hearing unless the individual claimed that she was in fact a United States citizen.[59]

This restrictive conception of national community was manifested most vividly in the succession of basic federal statutes that regulated immigration

---

51. Id. at 308. Absent such a move, of course, Puerto Ricans cannot vote. *Balzac* seems fully consistent with Congress's intention in enacting statutory citizenship for Puerto Ricans. See Cabranes, supra note 49, at 428–33.

52. Another category of American citizens is sometimes excluded not simply from enjoying the rights normally associated with that status but even from the right to reside in the United States. This category includes citizen children of deportable alien parents. See, e.g., INS v. Jong Ha Wang, 450 U.S. 139 (1981) (per curiam).

53. Yick Wo v. Hopkins, 118 U.S. 356, 368–69 (1886) (Chinese laborers entitled to protection against racial discrimination by municipal authorities in administration of ordinance.).

54. The Japanese Immigrant Case, 189 U.S. 86, 100–01 (1903) (construing a statute to require that such an alien at least be afforded a rudimentary administrative hearing before being deprived of liberty).

55. See, e.g., Exec. Order No. 11,935, 5 C.F.R. § 7.4 (1983) (upheld in Mow Sun Wong v. Campbell, 626 F.2d 739 (9th Cir. 1980), cert. denied, 450 U.S. 959 (1981)).

56. See, e.g., Heim v. McCall, 239 U.S. 175 (1915).

57. Oyama v. California, 332 U.S. 633 (1948).

58. See, e.g., Developments, supra note 4, at 1400–33.

59. See Ng Fung Ho v. White, 259 U.S. 276, 281–85 (1922); The Japanese Immigrant Case, 189 U.S. 86, 101 (1903).

during the classical period. Although opposition to aliens has been common throughout American history[60] and certain categories of undesirables, such as prostitutes and public charges, were excluded as early as 1645,[61] blanket federal exclusions began only during the 1880's with laws limiting and then barring the immigration of Chinese laborers. Blanket exclusions continued with the 1917 statute imposing—after many earlier presidential vetoes—a literacy requirement and barring political radicals and some Asians.[62] The 1921 and 1924 laws excluded aliens statutorily ineligible for citizenship, including most Asians. They also imposed quotas based upon the "national origins" of the existing United States population, computed according to a formula decidedly favorable to British immigrants and unfavorable to immigrants from Southern and Eastern Europe.[63] The 1952 Act left this exclusionary structure essentially intact, except to add provisions making all races eligible for naturalization but barring putative subversives.[64] The 1965 amendments, while repealing the national origins system, established for the first time a ceiling on immigration from the Western Hemisphere, designed primarily to limit the influx of Caribbean Basin laborers that had occurred during the earlier Bracero program.[65] In 1978, the hemispheric ceilings were replaced by worldwide and per country limits.[66]

A final feature of the classical conception of national community that is of particular interest here is that it was preeminently a political, not a judicial,

---

60. See generally J. Higham, supra note 5; Hofstadter, The Pseudo-Conservative Revolt, *in* The Radical Right 63-80 (D. Bell ed. 1963). The Founding Fathers were certainly not exempt. In 1753, for example, Benjamin Franklin warned against assimilating the Germans in Pennsylvania:

> Those who came hither are generally the most stupid of their own nation, and as ignorance is often attended with great credulity, when knavery would mislead it, and with suspicion when honesty would set it right; and, few of the English understand the German language, and so cannot address them either from the press or pulpit, it is almost impossible to remove any prejudices they may entertain. . . . Not being used to liberty, they know not how to make modest use of it.

J. Vialet, A Brief History of U.S. Immigration Policy 6 (Library of Congress 1980).

61. Laws to prohibit the admission of poor and indigent immigrants were passed by Massachusetts colony as early as 1645 and 1655. E. Hutchinson, supra note 27, at 390. The Immigration Act of 1875, ch. 141, 18 Stat. 477, prohibited the importation of prostitutes.

62. Immigration Act of 1917, ch. 29, § 3, 39 Stat. 874, 877.

63. Immigration Act of 1924, ch. 190, § 11, 43 Stat. 153, 159-60; Immigration Act of 1921, ch. 8, § 2, 42 Stat. 5, 5-6.

64. Immigration and Nationality Act of 1952, ch. 477, § 212(a)(28), 66 Stat. 163, 184-86 (codified as amended at 8 U.S.C. § 1182(a)(28)(1982)); id. § 311, 66 Stat. 239 (codified as amended at 8 U.S.C. § 1422 (1982)).

65. For a summary of the statutory history, see E. Hutchinson, supra note 27, at 366-79. There were, of course, some blanket admissions, especially after World War II. These included the Displaced Persons Act of 1948, ch. 647, 62 Stat. 1009, and a number of special refugee programs. For discussions of the latter, see Congressional Research Service, U.S. Immigration Law and Policy: 1952-1979, at 15-24, 46-48, 58-59, 76-78 (1979).

66. Immigration and Nationality Act of 1978, Pub. L. No. 95-412, § 1, 92 Stat. 907 (amending Immigration and Nationality Act of 1952, § 201(a)) (codified as amended at 8 U.S.C. § 1151 (1982).

artifact. Congress defined the qualifications for, and attributes of, each of the legal statutes that comprised it, and the Attorney General administered that definition. Until the 1960's, the federal government created few positive entitlements even for citizens. Moreover, the egalitarian potential of the due process and equal protection safeguards of the fifth and fourteenth amendments,[67] especially as applied to aliens, remained largely undiscovered. Except for a relatively few cases construing open-ended constitutional principles,[68] the courts' role in defining national membership was essentially limited to interpreting and enforcing the requirements for naturalization and other statutory statuses and occasionally construing the relatively determinate citizenship clause.[69]

Because it was Congress that essentially shaped the legal dimensions of the national community, its contours reflected the parochial, temporal political values so characteristic of the legislative process rather than the more cosmopolitan values of universality and pluralism. Citizenship in the liberal American polity, while almost always accessible to residents who desired it, was not available to all. And even citizen status did not always carry with it the full panoply of constitutional protections, either de facto or de jure. For many aliens, full membership was only a distant goal, not a palpable reality. Classical immigration law did not close the golden door, but it narrowed the passageway through it and limited the ways in which one might participate in the society after one had entered.

B. *Judicial Deference*

In the canon of classical immigration law, judges should be seen—if absolutely necessary—but not heard. For almost a century, the Court has abjured any significant judicial role in the area of immigration policy. Justice Field penned the classic statement of this position in *The Chinese Exclusion Case*:[70] "[If Congress] considers the presence of foreigners of a different race in this country, who will not assimilate with us, to be dangerous to its peace and security, . . . its determination is conclusive upon the judiciary."[71] Such self-abnegation may strike the contemporary reader, accustomed to lusty assertions of judicial machismo, as a bit faint-hearted, especially when applied to a measure that managed to combine blatant racism, retroactive lawmaking, and a flagrant denial of due process in a noisome statutory stew. Several years later, Justice Gray insisted that courts may not review an immigration inspector's decision concerning the facts underlying an alien's imprisonment and exclusion even in habeas corpus proceedings unless Congress expressly pro-

---

67. U.S. Const. amends. V, XIV.

68. E.g., Kwong Hai Chew v. Colding, 344 U.S. 590 (1953); Bridges v. Wixon, 326 U.S. 135 (1945).

69. U.S. Const. amend. XIV, § 2.

70. 130 U.S. 581 (1889).

71. Id. at 606.

vides otherwise.[72] This statement has a quaint, antique ring when read in an era in which even clear statutory indications of administrative finality have not managed to preclude judicial review.[73] Yet both of these pronouncements apparently remain good law in the Supreme Court.[74] Moreover, the Court has not hesitated to extend this "special judicial deference to congressional policy choices in the immigration context"[75] to administrative officials as well as to Congress.[76]

Judicial rhetoric about deference to the "political branches" and to administrative expertise, of course, is commonplace. Such talk is a venerable aspect of the courts' protective coloration, and should not necessarily be taken too seriously. The rhetoric of judicial deference in immigration cases prior to the 1980's, however, was striking in that courts almost invariably meant what

---

72. Nishimura Ekiu v. United States, 142 U.S. 651, 660 (1892).

73. See Johnson v. Robison, 415 U.S. 361, 373-74 (1974) (an administrative decision is judicially reviewable despite a statutory no-review provision where the statutory language and legislative history do not provide "the 'clear and convincing' evidence of congressional intent required by this Court before a statute will be contrued to restrict access to judicial review") (citing Abbott Laboratories v. Gardner, 387 U.S. 136, 141 (1967)).

74. Consider this utterance in a 1977 decision by the Court:

> At the outset, it is important to underscore the limited scope of judicial inquiry into immigration legislation. This Court has repeatedly emphasized that "over no conceivable subject is the legislative power of Congress more complete than it is over" the admission of aliens. Our cases "have long recognized the power to expel or exclude aliens as a fundamental sovereign attribute exercised by the Government's political departments largely immune from judicial control. . . . Our recent decisions have not departed from this long-established rule.

Fiallo v. Bell, 430 U.S. 787, 792 (1977) (citing *The Chinese Exclusion Case*, 130 U.S. 581 (1889)) (citations omitted). The opinion, after quoting Justice Frankfurter's observation in Galvan v. Press, 347 U.S. 522 (1954), that in support of Congress's plenary power over aliens, "there is not merely 'a page of history' . . . but a whole volume," went on to state that the Court would not "reconsider this line of cases." 430 U.S. at 793 n.4.

75. *Fiallo*, 430 U.S. at 793.

76. Consular decisions to deny visas, for example, are highly discretionary yet wholly immune from judicial review. See, e.g., Pena v. Kissinger, 409 F. Supp. 1182 (S.D.N.Y. 1976). The State Department, however, has created a system of administrative review. See 22 C.F.R. § 42.130 (1983); Note, Consular Discretion in the Immigrant Visa-Issuing Process, 16 San Diego L. Rev. 87, 101 (1978).

As recently as 1981, the Court upheld the refusal of the Attorney General to suspend the deportation of alien parents and their two children, both of whom happened to be American citizens. In doing so, it emphasized that the immigration authorities were entitled to construe the statutory standard for suspension narrowly if they wished, and criticized the court of appeals for encroaching on their discretion. INS v. Jong Ha Wang, 450 U.S. 139 (1981) (per curiam).

Even more recently, the Court refused to estop the INS from deporting an alien after having failed for 18 months to act upon an application for a visa, during which time his eligibility for the visa had evaporated. "Enforcing the immigration laws, and the conditions for residency in this country is becoming more difficult," the court noted, and "[a]ppropriate deference must be accorded" to the INS as "the agency primarily charged by Congress to implement the public policy underlying these laws." INS v. Miranda, 103 S. Ct. 281, 284 (1982).

they said. With a few exceptions,[77] the Supreme Court reflexively confirmed the deference principle with a decision on the merits in favor of the government rather than using that principle, as it did in many other administrative law contexts, merely as a disarming prelude to judicial self-assertion.[78]

It is worthwhile pausing to consider why this deference persisted for so long. One might suspect that judicial reluctance to challenge the power of the other branches reflected the weak political support for aliens. Yet even the Warren Court, whose case reports are filled with decisions checking governmental authorities on behalf of politically vulnerable groups,[79] was abjectly deferential in the context of immigration law and thus firmly within the classical tradition.[80] A lack of political influence, therefore, fails to explain why courts that eagerly picked up the cudgels to protect convicted criminals did not do so for aliens.

Other possible explanations for judicial deference seem equally dubious. The Immigration and Naturalization Service (INS), whose administrative competence and fairness have often been harshly criticized,[81] would seem an

---

77. In The Japanese Immigrant Case, 189 U.S. 86 (1903), for example, the Court construed the immigration statute to require a hearing before deportation. See also Kwong Hai Chew v. Colding, 344 U.S. 590 (1953); Bridges v. Wixon, 326 U.S. 135 (1945).

78. At least one commentator speaks of "a striking discontinuity between discourse and doctrine" in the exclusion area. Developments, supra note 4, at 1314. For reasons discussed infra notes 86–92 and accompanying text, I find the discontinuity during the classical period to have been slight, but hardly striking. Another commentator has pointed out that this deference has not taken the form, common in "political questions" cases, of declining to reach the merits. "The deference shown legislative and executive action is based on a finding that such action was within their constitutional powers, not on a determination that the subject matter was beyond judicial competence." Note, Constitutional Limits on the Power to Exclude Aliens, 82 Colum L. Rev. 957, 969 n.86 (1982).

79. See, e.g., In re Gault, 387 U.S. 1 (1967) (juveniles); Miranda v. Arizona, 384 U.S. 436 (1966) (criminal suspects); Brown v. Board of Educ., 347 U.S. 483 (1954) (blacks).

80. See, e.g., Boutilier v. INS, 387 U.S. 118 (1967) (upholding Congress's power to exclude homosexual aliens); Abel v. United States, 362 U.S. 217 (1960) (use of INS warrant to facilitate FBI search of suspected spy's hotel room does not violate fourth amendment if agencies cooperated in "good faith"). In Flemming v. Nestor, 363 U.S. 603 (1960), the Court, though sharply divided, upheld a statute that denied social security retirement benefits to an alien who had legally resided in the United States for 43 years, benefits to which he had contributed for two decades. The statute denied these benefits because he had been deported for being a Communist Party member during the 1930's—a time when mere membership was not illegal. Id. at 621 (Black, J., dissenting).

It seems highly doubtful that the Court would have upheld such a retroactive confiscation for Party membership had it been imposed upon a citizen rather than an alien. Although the Warren Court did not decide such a case itself, the lower federal courts and some state courts of that era repudiated somewhat analogous—and arguably less objectionable—efforts to deprive subversives of certain program benefits. See, e.g., Wellman v. Whittier, 259 F.2d 163 (D.C. Cir. 1958) (finding no express congressional intent to require Communist Party members to forfeit veterans benefits); Lawson v. Housing Auth., 270 Wis. 269, 70 N.W.2d 605, cert. denied, 350 U.S. 882 (1955) (public housing).

81. See, e.g., J. Crewdson, The Tarnished Door: The New Immigrants and the Transformation of America 113–41 (1983); Developments, supra note 4, at 1364–66.

odd repository for judicial trust. Nor has the congressional record in the immigration area been one to inspire much judicial confidence;[82] legislative threats to constitutional values have been common.[83] Another rationale for extraordinary judicial deference in classical immigration cases—the presence of foreign policy, national security, or other questions of an essentially nonadjudicatory nature—has explicitly been disavowed by the Court.[84] And even if one remains skeptical of this disclaimer—even if one believes that immigration cases often present questions of legislative fact and foreign policy with which courts are and should be profoundly uncomfortable—that rationale can hardly explain the striking pattern of judicial deference in cases decided as late as the 1970's in which daunting questions of that kind were either inconsequential or wholly absent.[85] Indeed, judicial deference cannot even be explained on the grounds that only the claims of aliens are involved: the plaintiffs in some of these decisions asserted claims—unsuccessfully, as it turned out—based primarily upon the rights of United States *citizens*, not of aliens.[86]

A wholly satisfactory answer to this puzzle still eludes us. I suspect, however, that the pronounced judicial passivity in classical immigration law has been subtly linked to the powerful conception of national sovereignty that lies at the heart of that tradition, and arises from the impact of that conception on the judge. In a constitutional system marked by an extraordinary degree of political, institutional and social fragmentation, manifestations of solidarity and nationhood can exercise a potent hold over the judicial, as well as the lay, imagination. The flag, the President, the American moon landings—these are compelling symbols of our national spirit and collective will, all the more awesome for their rarity in our atomized lives.[87] The idea of sovereignty, so elusive in our domestic constitutional structure, may come closest to being reified and recognizable when a unified national government deploys its laws against one who is plausibly seen as an outsider—as, quite literally, alien.

Other considerations reinforce the power of the sovereignty concept as an explanation for judicial deference. For over a century, the states have been consistently excluded from any significant role in the administration of immi-

---

82. Congress, for example, often passes private-bill legislation to provide immigration benefits for particular named aliens. See, e.g., P. Schuck, The Judiciary Committees 255-65 (1975).

83. See, e.g., Immigration and Nationality Act of 1952, § 212, 8 U.S.C. § 1182 (1982); El-Werfalli v. Smith, 547 F. Supp. 152 (S.D.N.Y. 1982) (exclusion of alien on reentry on basis of confidential information); cf. Immigration and Nationality Act of 1952, § 215, 8 U.S.C. § 1185 (1982).

84. See Fiallo v. Bell, 430 U.S. 787, 796 (1977).

85. See id.; Mathews v. Diaz, 426 U.S. 67 (1976); Kleindienst v. Mandel, 408 U.S. 753 (1972).

86. See Fiallo v. Bell, 430 U.S. 787 (1977) (plaintiffs included citizens making constitutional equal protection claim); Kleindienst v. Mandel, 408 U.S. 753 (1972) (constitutional claim based on citizen coplaintiffs' first amendment rights).

87. M. Edelman, The Symbolic Uses of Politics (1964).

gration policy.[88] Moreover, immigration policy has enjoyed a considerable consensus as among the President, Congress and the bureaucracy; one searches the immigration cases in vain for a titanic interbranch struggle like those that have occurred in other areas of public law.[89] One measure of this consensus is that Congress has chosen to confer exceedingly broad discretion over the most farreaching immigration decisions,[90] has delegated this discretion not merely to the executive branch but to a cabinet official who traditionally is a close political confidant of the President, and has done so not only for matters obviously requiring the weighing of many administrative considerations but also for those impinging upon fundamental civil liberties.[91]

In the face of broad, express congressional delegations of authority to the executive branch in the area of external relations, in which the President also enjoys substantial constitutional power in his own right, in view of the strong consensus between the political branches, and in light of the minimal moral claims upon us of those deemed to be outside the national community, judicial power is most problematic and the President's authority, in Justice Jackson's words, "is at its maximum." There, "[he may] be said (for what it may be worth) to personify the federal sovereignty."[92]

C. *The Extraconstitutional Status of Exclusion*

In the classical tradition, the exclusion process—the process for determining which aliens may enter the United States for the first time, or which aliens, having entered previously, may reenter after a significant absence or may remain after having been "paroled" into the country—is essentially freed from any significant constitutional constraints. Concerning excludable aliens, classical immigration law holds that "the decisions of . . . administrative officers, acting within powers expressly conferred by Congress, are due proc-

---

88. See, e.g., Chy Lung v. Freeman, 92 U.S. 275 (1875) (state may not tax or regulate immigrants). But see, e.g., Cabell v. Chavez-Salido, 454 U.S. 432 (1982) (state may require citizenship as a condition of public employment as probation officer); De Canas v. Bica, 424 U.S. 351 (1976) (state may restrict private employment of illegal aliens).

89. See, e.g., United States v. Nixon, 418 U.S. 683 (1974) (executive and judicial branches; issue of presidential privilege); Youngstown Sheet & Tube Co. v. Sawyer, 343 U.S. 579 (1952) (legislative and executive branches; issue of lawmaking power); Myers v. United States, 272 U.S. 52 (1926) (legislative and executive branches; issue of presidential power to remove executive officer).

90. See, for example, the adjustment of status authority under the Immigration and Nationality Act of 1952, § 245, 8 U.S.C. § 1255 (1982). For a discussion of the way in which the INS has exercised this authority, see Diver, The Optimal Precision of Administrative Rules, 93 Yale L.J. 65, 92–97 (1983); Sofaer, The Change-of-Status Adjudication: A Case Study of the Informal Agency Process, 1 J. of Legal Stud. 349 (1972). The INS has further expanded INS discretion by its practice of publishing only certain of its decisions as "precedents." See Diver, supra, at 95.

91. See, e.g., 8 U.S.C. § 1225(c) (1982). The predecessor of § 1225(c) was the statute upheld in United States ex rel. Knauff v. Shaughnessy, 338 U.S. 537 (1950).

92. Youngstown Sheet & Tube Co. v. Sawyer, 343 U.S. 579, 635–36 (1952) (Jackson, J., concurring).

ess of law."[93] Here, at least, the Constitution, not just politics, seems to stop at the water's edge.[94]

Congress and the Attorney General, by statute and regulation, long ago provided certain procedural rights to excludable aliens equivalent to those enjoyed by aliens who have "entered" the United States and who therefore can claim the Constitution's protection.[95] Still, the extraconstitutional status of exclusion proceedings remains far more than a formal or academic matter. Some important differences between exclusion and deportation procedures persist.[96] Moreover, what Congress and the Attorney General have given, they may also take away—unless, of course, that procedure is constitutionally required. Such a withdrawal of procedural rights is not merely a theoretical possibility. As unprecedented numbers of excludable aliens seek entry,[97] immigration officials desperately seek to manage their caseloads by economizing on procedures. The Simpson-Mazzoli immigration reform legislation recently

---

93. Nishimura Ekiu v. United States, 142 U.S. 651, 660 (1892). Or in the more pointed formulation: "Whatever the procedure authorized by Congress is, it is due process as far as an alien denied entry is concerned." United States ex rel. Knauff v. Shaughnessy, 338 U.S. 537, 544 (1950).

94. See L. Tribe, supra note 12, at 284. The Harvard Developments Note contends, however, that during the classical period the Court recognized some constitutional limitations on the exclusion power, its rhetoric to the contrary notwithstanding. Developments, supra note 4, at 1314-23. In fact, the cases that it cites do not refute the position taken here. Kwock Jan Fat v. White, 253 U.S. 454 (1920), Tang Tun v. Edsell, 223 U.S. 673 (1912), and Chin Yow v. United States, 208 U.S. 8 (1908), while exclusion cases, involved claims of United States citizenship, which have long been considered sui generis. See, e.g., Ng Fung Ho v. White, 259 U.S. 276, 284-85 (1922). Wong Wing v. United States, 163 U.S. 228 (1896), was a deportation case and involved not simply exclusion but expulsion plus *punishment*. Elting v. North German Lloyd, 287 U.S. 324 (1932), involved not an alien challenging her exclusion but a vessel challenging a fine that had been administratively imposed. Gegiow v. Uhl, 239 U.S. 3 (1915), holds only that judicial review of legal issues involving construction of the immigration statute is permissible. There is no suggestion in that opinion that such a holding is required by due process, although such a result is surely possible. See Crowell v. Benson, 285 U.S. 22, 49-50 (1932).

95. E.g., Landon v. Plasencia, 103 S. Ct. 321, 325-26 (1982).

96. In exclusion proceedings, for example, the burden of proof is on the alien, not the government, see 8 U.S.C. § 1361 (1982), and the availability of discretionary relief is far more limited. Compare id. § 1226 (appeal to the Attorney General from final exclusion decisions) and id. § 1183 (aliens excludable because likely to become public charges may be admitted at the discretion of the INS upon posting a bond) and id. § 1182(h) (aliens excludable for having committed crimes may secure waivers on account of hardship to a legal resident or citizen spouse or child in the United States) with, e.g., id. § 1254(a) (discretion of Attorney General to suspend deportation for hardship) and id. § 1253(h) (discretion of Attorney General to withhold deportation because of threat of persecution). The excludable alien, moreover, cannot designate the country to which she will be deported, as she can in deportation proceedings. Compare id. § 1253(a) (deportation) with id. § 1227(a) (exclusion). The appellate process exhibits yet another difference between deportation and exclusion: an alien in deportation proceedings can appeal to a federal court of appeals, while an alien in exclusion proceedings can seek review only through habeas corpus. Id. § 1105a(a), (b).

97. See Immigration and Naturalization Serv., U.S. Dep't of Justice, 1979 Statistical Yearbook 66-67; infra notes 212-27 and accompanying text.

derailed in Congress, for example, would authorize summary exclusion—without administrative hearing or court review—of aliens seeking to enter without adequate documentation.[98]

Finally, exclusion's extraconstitutional status has encouraged and legitimated some of the most deplorable governmental conduct toward both aliens and American citizens ever recorded in the annals of the Supreme Court. In *United States ex rel. Knauff v. Shaughnessy*,[99] to cite a notorious example, the Court allowed the wife of an American citizen who had served in the United States Army and been honorably discharged to be permanently excluded from the United States without any hearing. Her exclusion was based upon an administrative finding that admission would be "prejudicial to the interests of the United States"; this finding in turn was based upon confidential information disclosed neither to the wife, her husband, nor the Court. In an even more infamous case, *Shaughnessy v. United States ex rel. Mezei*,[100] the Court countenanced the exclusion and two-year incarceration[101] of a lawful United States resident of twenty-five years' standing on the basis of these same ex parte procedures.[102]

While these decisions are easy to denounce and their reasoning is not difficult to demolish,[103] they cannot simply be dismissed as McCarthy-era

98. See S. 529, 98th Cong., 1st Sess., § 121(b)(1), 129 Cong. Rec. S6970, S6973 (daily ed. May 18, 1983) (passed by the Senate on May 18, 1983). The version of Simpson-Mazzoli that has passed the Senate would, inter alia, establish sanctions for employers who knowingly employ undocumented aliens; provide for a system to verify work eligibility; create a summary exclusion process without appeal for those who enter without documents and do not claim asylum; limit judicial review of asylum claims and exclusion cases not subject to summary procedures; alter the standard of judicial review in deportation and other proceedings to "substantial evidence"; confer legal status upon illegal aliens who have continuously resided in the United States for certain periods of time; and limit the availability of adjustment-of-status relief for visa abusers. The bill appeared to have died in the House in the face of Democratic opposition. See Tolchin, Democrats Bar Action in House on Immigration, N.Y. Times, Oct. 2, 1983, at A1, col. 5. But Speaker of the House Thomas P. O'Neill, Jr. has recently stated that he is willing to support compromise legislation. See Tolchin, O'Neill Now Favors Revision of Immigration Laws, N.Y. Times, Nov. 30, 1983, at A32, col. 1.

99. 338 U.S. 537 (1950).

100. 345 U.S. 206 (1953).

101. Mr. Justice Jackson in dissent noted that "if the Government has its way he seems likely to be detained indefinitely, perhaps for life, for a cause known only to the Attorney General." Id. at 220. Mezei was finally paroled after some three years of detention following an inquiry by a special panel of distinguished attorneys that recommended release. See W. Gellhorn, Individual Freedom and Governmental Restraints 36–38 (1956).

102. Indeed, the Attorney General refused to permit even the trial court to examine the "confidential information" in camera. 345 U.S. at 228 n.9 (Jackson, J., dissenting).

103. See, e.g., Hart, The Power of Congress to Limit the Jurisdiction of Federal Courts: An Exercise in Dialectic, 66 Harv. L. Rev. 1362, 1387–96 (1953), reprinted in P. Bator, P. Mishkin, D. Shapiro & H. Wechsler, Hart & Wechsler's The Federal Courts and the Federal System 330, 348–56 (2d ed. 1973); Martin, Due Process and Membership in the National Community: Political Asylum and Beyond, 44 U. Pitt. L. Rev. 165, 173–80 (1983).

aberrations, anachronistic monuments to cold war hysteria.[104] The Supreme Court continues to cite them with apparent approval.[105] Indeed, their teaching remains enshrined in the current immigration statute. If, for example, a consular officer just out of Foreign Service training "has reason to believe" that an alien seeks to enter "incidentally to engage in activities which would be prejudicial to the public interest,"[106] she may deny the alien entry; her decision is final, subject only to administrative review.[107] Moreover, if an alien "may appear" to an immigration officer to be excludable on that or several other political grounds and the Attorney General agrees that the alien should be excluded on the basis of confidential information that he thinks should not be disclosed, that essentially ends the matter.[108]

D. *The Power to Classify Aliens*

The extraconstitutional status of exclusion procedures has been extended and strengthened by yoking it to another distinctive feature of the classical conception—the virtually unlimited power of the federal government to classify aliens on the basis of *substantive* considerations.[109] The equal protection principle enshrined in the fifth and fourteenth amendments prohibits the irrational or invidious exercise of legislative and executive power. Its requirement that legal classifications be related to a proper governmental objective[110]

---

104. Recent efforts to portray these decisions as anomalies include, for example, Rodriguez-Fernandez v. Wilkinson, 654 F.2d 1382, 1388 (10th Cir. 1981); Developments, supra note 4, at 1322-23. Compare supra note 74.

105. See, e.g., Landon v. Plasencia, 103 S. Ct. 321, 329-30 (1982); Fiallo v. Bell, 430 U.S. 787, 792 (1977).

106. 8 U.S.C. § 1182(a)(27) (1982).

107. See supra note 93.

108. 8 U.S.C. § 1225(c) (1982). The Court has never decided on the extent to which this decision is judicially reviewable. Recently, an excellent federal district court judge who possesses a detailed knowledge of immigration law, see Sofaer, supra note 90, ruled that a court may "inquire as to the Government's reasons" but may not "prob[e] into their wisdom or basis. If the Court finds that the Government acted on a facially legitimate and bona fide reason, its inquiry is complete." El-Werfalli v. Smith, 547 F. Supp. 152, 153 (S.D.N.Y. 1982).

109. Some classifications to which this discussion applies are not, strictly speaking, made on the basis of alienage; instead of distinguishing between citizens and aliens, classifications often distinguish between subcategories of aliens. See, e.g., Toll v. Moreno, 458 U.S. 1, 25 (1982) (Rehnquist, J., dissenting) (domiciled nonimmigrant aliens who hold G-4 visas); Mathews v. Diaz, 426 U.S. 67 (1976) (upholding classification of aliens on basis of residency status and length of residence); Narenji v. Civiletti, 617 F.2d 745 (D.C. Cir.) (upholding classification of aliens on basis of nationality), cert. denied, 446 U.S. 957 (1980).

110. See, e.g., Schweiker v. Wilson, 450 U.S. 221 (1981) (statutory classification of mentally ill patients by whether cared for in a public or private institution held a rational basis for denying certain welfare benefits); Parham v. Hughes, 441 U.S. 347 (1979) (statutory classification precluding father of an illegitimate child from suing for child's wrongful death held a rational means of proving paternity). How close the means-end relationship must be when the classification is on the basis of alienage is a question that need not be addressed here. For a discussion of the rational basis/strict scrutiny controversy surrounding this still unresolved question, see Note, A Dual Standard for State Discrimination Against Aliens, 92 Harv. L. Rev. 1516 (1979). The issue is

has, in other areas of public law, annihilated a formidable number of statutes, regulations and other official actions of every kind and description. Yet classical immigration law has essentially neutralized this principle.

In the immigration context Congress "regularly makes rules that would be unacceptable if applied to citizens."[111] It would be difficult indeed to make sense of the history of American immigration law were that observation not manifestly true. There is little question, for example, that the squalid prejudice that led Congress summarily to exclude Chinese laborers,[112] to preserve the ethnic distribution of an earlier era through national origin quotas,[113] and to preclude Asian-Americans for decades from becoming naturalized United States citizens,[114] only withstood equal protection scrutiny because they involved aliens and were instituted by the federal government.[115] Yet the Supreme Court did not hesitate to bless these and other such practices.[116]

In fact, the equal protection principle has been applied less rigorously to federal alienage classifications than to the states';[117] the Court has not re-

---

further complicated by uncertainty concerning whether it is to be addressed under equal protection principles or under a preemption analysis. See, e.g., Note, State Burdens on Resident Aliens: A New Preemption Analysis, 89 Yale L.J. 940 (1980) [hereinafter cited as Note, State Burden]; Note, The Equal Treatment of Aliens: Preemption or Equal Protection, 31 Stan. L. Rev. 1069 (1979).

111. Mathews v. Diaz, 426 U.S. 67, 80 (1976); see also Fiallo v. Bell, 430 U.S. 787, 792 (1977).

112. See The Chinese Exclusion Case, 130 U.S. 581 (1889).

113. The Immigration Act of 1924, ch. 190, § 11, 43 Stat. 153, 159–160, established a quota system based on national origins, replacing the interim system set up by the 1921 statute. See Immigration Act of 1921, ch. 8, § 2, 42 Stat. 5, 5–6. The national origins quota scheme continued in force, with modification, see Immigration and Nationality Act of 1952, ch. 477 §§ 201–207, 66 Stat. 163, 175–81 until 1965. See Act of Oct. 3, 1965, Pub. L. No. 89-236, § 21(e), 79 Stat. 911, 920 (repealing national origin quotas and substituting separate quotas for Eastern and Western Hemispheres). The 1952 act provided that total annual immigration should not exceed one sixth of one percent of the number of inhabitants in the United States in 1920, Immigration and Nationality Act of 1952, § 201(a), 66 Stat. at 175, and it set a special limit on aliens from the "Asia-Pacific Triangle," id. § 202(b), 66 Stat. at 177.

114. See E. Hutchinson, supra note 27, at 294–95.

115. See, e.g., Examining Bd. of Eng'rs v. Flores de Otero, 426 U.S. 572 (1976) (citizenship requirement for practice in civil engineering held unconstitutional); Miranda v. Nelson, 413 U.S. 902 (1973) (mem.) (state's dismissal of employees on account of alienage held unconstitutional); Yick Wo v. Hopkins, 118 U.S. 356 (1886) (city ordinance discriminating on basis of race held unconstitutional).

116. See, e.g., Kleindienst v. Mandel, 408 U.S. 753 (1972) (upholding denial of a visa to Marxist scholar from Belgium).

117. Compare Sugarman v. Dougall, 413 U.S. 634 (1973) (citizenship requirement for state civil service held unconstitutional), and Graham v. Richardson, 403 U.S. 365 (1971) (state denial of welfare benefits to resident aliens held unconstitutional), with Mow Sun Wong v. Campbell, 626 F.2d 739 (9th Cir. 1980) (presidential order excluding noncitizens from the federal civil service held constitutional), and Matthews v. Diaz, 426 U.S. 67 (1976) (alien ineligibility provision of Medicare program held constitutional). Classifications relating to American Indians comprise a separate category.

quired the former to demonstrate even a "rational basis," much less the "compelling" governmental interest required of the latter.[118] Although there are substantial arguments based upon constitutional text, structure, and intent in favor of a more general differentiation of equal protection standards as between federal and state classifications,[119] the Court has taken that dual standard approach for only a few classifications. The most notable of these is alienage.

As with judicial deference in the immigration area generally, it is not at all obvious why federal alienage classifications have been singled out for kid-glove treatment. One theory might hold that aliens are more "discrete and insular" in state politics and therefore more deserving of special judicial solicitude than they are in federal politics. This argument is plausible but not convincing. It is true that federal responsibilities for foreign affairs, national economic growth, and immigration generate certain policies that almost all aliens, despite their heterogeneity, might be expected to favor. Moreover, certain specialized organs of the federal government, including the INS and the relevant House and Senate judiciary subcommittees, are especially concerned about such matters and—at least theoretically—provide a forum in which aliens' interests can be advanced.[120] On the other hand, aliens and their ethnic compatriots who are citizens are concentrated in a handful of states, a fact of great political significance.[121] In at least some of those states, such as California, Texas, and New York, these ethnic groups, sometimes even including the disenfranchised aliens themselves, exert considerable influence upon local, state, congressional, and even presidential politics.[122] Although the situation might well be different in other states in which aliens' interests are less well-represented, it is far from clear that the political influence of aliens will generally be greater at the national level where their voting and organizational powers are more diffuse.

---

118. See, e.g., Fiallo v. Bell, 430 U.S. 787, 792 (1977); Mathews v. Diaz, 426 U.S. 67, 81-82 (1976); Rosberg, supra note 12.

119. See Note, A Madisonian Interpretation of the Equal Protection Doctrine, 91 Yale L.J. 1403, 1419-25 (1982).

One suspects that this doctrinal discontinuity, as well as confusion over whether alienage classifications under state law are "suspect" or not, see supra note 110, are important factors inclining the Court increasingly to emphasize preemption rather than equal protection principles in such cases. Such an analysis finds state law unconstitutional under the supremacy clause, U.S. Const. art. VI, cl. 2, on the theory that the issue has been preempted by federal law. See, e.g., Toll v. Moreno, 458 U.S. 1, 17 (1982).

120. For an account of those subcommittees during the 92d and 93d Congresses, see P. Schuck, supra note 82.

121. As of 1979, the four states with the largest alien population were California (26%), New York (15%), Texas (8%), and Florida (7%). Immigration and Naturalization Serv., U.S. Dep't of Justice, 1979 Statistical Yearbook 84 (1979).

122. See, e.g., Reinhold, Hispanic Leaders Open Voter Drive, N.Y. Times, Aug. 9, 1983, at A15, col. 1.; Hispanic Power Arrives at the Ballot Box, Bus. Wk., July 4, 1983, at 32; Gurwitt, Widespread Political Efforts Open New Era for Hispanics, 40 Cong. Q., 2707 (1982); see also Miller, The Political Impact of Foreign Labor: A Re-evaluation of the Western European Experience, 16 Int'l Migr. Rev. 27 (1982) (describing political influence of aliens who lack voting rights).

A second argument in favor of the essentially unlimited federal power to classify aliens might be grounded in the constitutional text itself. The difficulty, however, is that the only explicit reference in the Constitution to this question, apart from references to United States citizenship[123] and the slave trade,[124] is the provision authorizing Congress "[t]o establish an uniform Rule of Naturalization."[125] This hardly qualifies as a sweeping grant of power over immigration, much less as a ringing endorsement of invidious federal alienage classifications having nothing whatever to do with the naturalization process. Although the precise source of the exclusive congressional power to regulate immigration is far from clear,[126] it is nonetheless well-established.[127] But because this power is implicit rather than textual, it tells us little about the appropriate relationship between the equal protection principle and the federal government's power to classify.

To account for the privileged status that federal alienage classifications enjoy in classical immigration law, we must look, I suspect, neither to the alien's condition nor to the Constitution's text. Rather, the explanation can be found in the classical tradition's self-consciously political definition of national community and in its norm of extraordinary judicial deference to that choice. By the very nature of this definition, citizens and aliens are almost *never* "similarly situated," while the federal government's interests in emphasizing that difference—for example, giving preference to citizens in order to encourage aliens to naturalize and thus join the national community[128]—is almost *always* deemed compelling.[129]

E. *The "Civil" Nature of Deportation*

A fundamental tenet of classical immigration law holds that deportation is a civil, administrative proceeding, not a criminal prosecution. Deportation,

---

123. See supra notes 36-39 and accompanying text.

124. U.S. Const. art. I, § 9, cl. 1. For an historical analysis of the dispute over whether this provision was intended to affect the power of Congress to regulate free immigration and/or the domestic slave trade, see Berns, The Constitution and the Migration of Slaves, 78 Yale L.J. 198 (1968).

125. U.S. Const. art. I, § 8, cl. 4. See L. Henkin, Foreign Affairs and the Constitution (1972).

126. See, e.g., Note, State Burden, supra note 110, at 944-46.

127. See, e.g., Chy Lung v. Freeman, 92 U.S. 275, 280 (1875).

128. For an enumeration of several additional governmental interests in differentiating on the basis of alienage, see, e.g., Hampton v. Mow Sun Wong, 426 U.S. 88, 104 (1976). None seems terribly persuasive, much less compelling. Even the most plausible, the incentive-to-naturalize interest, could be achieved by a narrower classification that would not disqualify aliens who had formally declared their intentions to naturalize pursuant to 8 U.S.C. § 1445(f) (1982).

129. This deference to governmental interests is perhaps especially evident with respect to admission criteria. Given a strictly limited visa policy in the face of essentially unlimited demand for visas, the government is obliged to choose among putative immigrants, sometimes on rather arbitrary grounds. What rational principle, for example, could justify Congress establishing a different numerical ceiling for the Eastern and Western Hemispheres? See Pub. L. No. 89-236, §

Justice Holmes wrote without elaboration, is not "a punishment; it is simply a refusal by the Government to harbor persons whom it does not want."[130] This distinction possesses little logical power: at least concerning aliens who have established a foothold in American society, it is a legal fiction with nothing, other than considerations of cost and perhaps administrative convenience, to recommend it.[131] Nevertheless, it has proved to possess the staying power that Holmes knew to be far more important in law than logic.[132]

The legal consequences of classifying deportation as civil and not punitive are considerable. Certain constitutional rights, such as those protected by the sixth amendment,[133] are expressly limited to criminal proceedings. Others, such as the prohibition against ex post facto laws,[134] have been held to apply only to criminal or quasi-criminal "punishments."[135] In the immigration context, at least, the applicability of the exclusionary rule, the allocation of the burden of proof, and even the right to judicial review itself, may depend upon how the action is classified.[136]

With so many legal rights turning upon the assignment of the proceeding to one category or the other, we should expect that the hoary characterization

---

21(e), 79 Stat. 911, 920 (1965), amended by Act of Oct. 5, 1978, Pub. L. No. 95-412, 92 Stat. 907, 907 (replacing hemisphere quotas with a single, worldwide quota) (codified at 8 U.S.C. § 1151(a) (1982)).

130. Bugajewitz v. Adams, 228 U.S. 585, 591 (1913). Earlier, in Fong Yue Ting v. United States, 149 U.S. 698, 730 (1893), the Court had written:

> The order of deportation is not a punishment for crime. It is not a banishment, in the sense in which that word is often applied to the expulsion of a citizen from his country by way of punishment. It is but a method of enforcing the return to his own country of an alien who has not complied with the conditions upon the performance of which the government of the nation, acting within its constitutional authority and through the proper departments, has determined that his continuing to reside here shall depend.

131. The Court has sometimes insisted that considerations of administrative cost and convenience cannot diminish constitutional rights, e.g., Bounds v. Smith, 430 U.S. 817, 825 (1977) ("[T]he cost of protecting a constitutional right cannot justify its total denial."), but it is perfectly clear that in fact and in law they do, e.g., Mathews v. Eldridge, 424 U.S. 319 (1976) (administrative burdens a factor in determining whether Social Security benefit termination proceeding comports with due process).

132. The Court has never questioned this doctrine, reaffirming it at least as recently as 1977. See Ingraham v. Wright, 430 U.S. 651, 668 (1977).

133. U.S. Const. amend. VI.

134. U.S. Const. art. I, § 9, cl. 3.

135. See, e.g., Galvan v. Press, 347 U.S. 522, 530-32 (1954) (ex post facto clause does not apply to deportation because it is not punishment); Harisiades v. Shaughnessy, 342 U.S. 580, 594-95 (1952) (same); see also Woodby v. INS, 385 U.S. 276, 284-86 (1966) (no requirement of proof beyond a reasonable doubt); Dymytryshyn v. Esperdy, 285 F. Supp. 507, 510 (S.D.N.Y.) (prohibition against bills of attainder inapplicable to noncriminal proceeding), aff'd, 393 U.S. 77 (1968). See generally United States v. Ward, 448 U.S. 242, 251-54 (1980).

136. See infra note 146 and accompanying text (exclusionary rule); Developments, supra note 4, at 1385, 1386-89 (burden of proof). In the Japanese Immigrant Case, 189 U.S. 86, 100-02 (1903), the Court held that due process did not require judicial review of summary deportation. Had the Court instead conceived of the proceeding as criminal in nature, judicial review would have been constitutionally required. See U.S. Const. amend. VI.

of deportation as civil rather than punitive would be firmly rooted in the importance of its real world consequences to aliens. Deportation, however, has little in common with civil sanctions. Not only is it often imposed for conduct that also constitutes a crime, a feature that civil sanctions only occasionally exhibit, but it often results in incarceration.[137] Deportation, in fact, serves as an important adjunct and supplement to criminal law enforcement, and it reflects judgments, essentially indistinguishable from those that the criminal law routinely makes, concerning the moral worth of individual conduct.[138] For those deportees who cannot acquire American citizenship[139] and will therefore remain stateless, deportation may be catastrophic.[140] Even deportees who will not be rendered stateless may nevertheless be unable to return to their country of nationality due to perilous economic or political conditions prevailing there,[141] while deportation to a third country may mean having to live as a stranger in a strange land. Even those who can return to their country of nationality may be expelled from the only society to which they feel some attachment and be obliged to go instead to a place with which they have only the most tenuous link.[142] In addition to causing profound estrangement,[143] deportation, unlike conventional criminal punishment, imposes upon the alien a more or less permanent disability. Once deported, an alien may be prevented from ever returning legally; indeed, merely reapplying

137. See, e.g., 8 U.S.C. § 1252(a) (1982). In Kennedy v. Mendoza-Martinez, 372 U.S. 144, 168 (1963), the Court identified incarceration as a factor supporting the conclusion that the sanction is punitive in nature. See also United States v. Janis, 428 U.S. 433, 447 n.17 (1976). But see United States v. Ward, 448 U.S. 242, 249-51 (1980) (a penalty is not criminal where Congress does not so intend).

138. See Note, Deportation and Exclusion: A Continuing Dialogue Between Congress and the Courts, 71 Yale L.J. 760, 789 n.143 (1962).

139. For example, an alien deported because of her affiliation with the Communist Party may be precluded from acquiring American citizenship. See 8 U.S.C. §§ 1251(a)(b), 1424(c) (1982).

140. See, e.g., Kennedy v. Mendoza-Martinez, 372 U.S. 144, 160-61 (1963).

141. Under 8 U.S.C. § 1253(h) (1982), a deportable alien may not, with some exceptions, be returned to her country if the Attorney General determines that her life or freedom would be threatened there on account of race, religion, nationality, membership in a particular social group, or political opinion.

142. 8 U.S.C. § 1253(a)(1)-(7) (1982) enumerates the categories of countries to which a deportable alien must or may be deported, and gives the Attorney General much flexibility concerning the countries to which she can be sent. 8 U.S.C. § 1227(a) (1982) gives the Attorney General similar flexibility regarding the deportation of excluded aliens.

143. Justice Douglas, no friend to euphemism, likened deportation to the ancient practice of banishment:

> Banishment is punishment in the practical sense. It may deprive a man and his family of all that makes life worth while. Those who have their roots here have an important stake in this country. Their plans for themselves and their hopes for their children all depend on their right to stay. If they are uprooted and sent to lands no longer known to them, no longer hospitable, they become displaced, homeless people condemned to bitterness and despair.

Harisiades v. Shaughnessy, 342 U.S. 580, 600 (1952) (Douglas, J., dissenting).

for admission constitutes a felony unless she first obtains the Attorney General's express consent.[144]

In view of what is inevitably and personally at stake, then, it is undeniable that deportation punishes the alien and punishes her severely.[145] To maintain, as classical immigration law consistently has done, that deportation resembles a sanction like being ejected from a national park rather than that of being banished or sentenced to jail, suggests that something deeply symbolic, not dryly logical, has been at work in the shaping of the doctrine. In condoning the deportation of the alien without the safeguards that government must ordinarily afford before it can impose grave punishment,[146] the law affirms the contingent nature of her claims on the community.[147] The government's obligations to the alien are viewed as resting upon her formal status rather than upon her actual relationship to the society. Since under the classical order the alien's entry was conceived of as a privilege whose continued enjoyment was conditional upon her compliance with the formal terms that the government prescribed, deportation was simply the revocation of her license, a reversion to the status quo ante. No special procedural safeguards for this reversion were thought to be necessary.[148]

---

144. 8 U.S.C. § 1326 (1982). In practice, few of these cases are prosecuted unless the alien is thought to be a smuggler, has been deported several times before, or other aggravating factors are present, such as assault on a police officer. Letter from Edwin Harwood to Peter H. Schuck (July 28, 1983) (copy on file at the offices of the Columbia Law Review).

145. Even the Court has occasionally blushed while reaffirming the pretense that deportation is not punishment. See Galvan v. Press, 347 U.S. 522, 531 (1954) ("[S]ince the intrinsic consequences of deportation are so close to punishment for crime, it might fairly be said also that the *ex post facto* Clause, even though applicable only to punitive legislation, should be applied to deportation. But the slate is not clean.") (footnote omitted).

146. The statutes and regulations governing deportation proceedings afford many due process protections. The government bears the burden of proof. The alien has the right to counsel and the right to adduce evidence and to cross-examine witnesses. The alien also is entitled to notice, a hearing, a decision on the record, and judicial review. See 8 U.S.C. §§ 1105a, 1252 (1982); 8 C.F.R. §§ 242.1-.23 (1983). But the deportation laws fail to provide some protections that due process has been thought to require in other quasi-criminal contexts, such as trial before an independent judge and a jury of one's peers, see infra notes 167–89 and accompanying text, free legal assistance for indigents, see, e.g., Aguilera-Enriquez v. INS, 516 F.2d 565 (6th Cir. 1975), cert. denied, 423 U.S. 1050 (1976), the right to compel the production of relevant evidence, see supra notes 99–102, 108 and accompanying text, the exclusion of illegally obtained evidence, see, e.g., Abel v. United States, 362 U.S. 217 (1960), and the right to clear notice of the substantive standards to be applied, see, e.g., Boutilier v. INS, 387 U.S. 118 (1967).

147. The absence of such safeguards places the government in a position analogous to that of a traditional creditor or lessor who could enforce the voluntarily undertaken obligations of his debtor or lessee through self-help remedies, such as ex parte replevin, attachment and execution, or eviction. See, e.g., Brunswick Corp. v. J & P, Inc., 424 F.2d 100 (10th Cir. 1970) (replevin); Wheeler v. Adams Co., 322 F. Supp. 645 (D. Md. 1971) (same). In the classical tradition, the risk of error, unfairness, or indignity characteristic of such procedures, see Fuentes v. Shevin, 407 U.S. 67 (1972), is subordinated to the interest of the government in speedy enforcement of its claim. See generally Mashaw, Due Process and Its Discontents (unpublished manuscript on file with the author).

148. See infra notes 264–67 and accompanying text.

F. *The Power to Detain*

In classical immigration law, the detention of aliens[149]—often involving incarceration—has been an important tool of program administration and law enforcement: it ensures that aliens will appear for their hearings, commit no offenses, be inclined to cooperate with the immigration authorities, and, if in the country illegally, not compete for jobs with lawful aliens and citizens. The statute therefore expressly authorizes the Attorney General to detain aliens in a variety of circumstances.[150] But the detention authority is more than a programmatic resource, ancillary to the power to exclude and deport. Detention is also an awesome power in its own right.

Shaughnessy v. United States ex rel. Mezei[151] most vividly illustrates this point.[152] Mezei, a long-time resident of the United States who had journeyed abroad to visit his mother, was excluded on undisclosed security grounds when he returned. After some sixteen months in custody, the government announced that it had abandoned its futile efforts to find a country to which he could be deported. Thereafter, Mezei remained on Ellis Island in what Justice Black in dissent described as an "island prison" in which Mezei must "stay indefinitely, maybe for life."[153] The Supreme Court summarily rejected Mezei's challenge to his exclusion. Having never seen the government's evidence,[154] there was of course little for the Court to say. Mezei, it noted, was actually the beneficiary of "legislative grace"; as an excludable alien, the authorities could have kept him aboard the vessel that brought him here rather than extending "temporary harborage" at Ellis Island.[155] Mezei's complaint

---

149. Although the INS and the courts routinely employ the term "detention" to describe the practice of holding aliens pending initiation and completion of exclusion and deportation proceedings, the length of many detentions and the conditions of confinement suggest that the term "imprisonment" more accurately depicts reality. See Jean v. Nelson, 711 F.2d 1455, 1500 n.51, reh'g en banc granted, 714 F.2d 96 (11th Cir. 1983). The following discussion will use the less inflammatory term. Cf. Korematsu v. United States, 323 U.S. 214, 223 (1944) ("[W]e deem it unjustifiable to call [assembly and relocation centers] concentration camps with all the ugly connotations that term implies . . . .").

150. E.g., 8 U.S.C. §§ 1222-1227, 1252 (1982).

151. 345 U.S. 206 (1953).

152. The trio of cases involving the curfew, evacuation, and detention imposed upon American citizens of Japanese descent during World War II are not discussed here because they were exercises of the war power, not the power to regulate immigration. See Ex parte Mitsuye Endo, 323 U.S. 283 (1944); Korematsu v. United States, 323 U.S. 214 (1944); Hirabayashi v. United States, 320 U.S. 81 (1943).

153. *Mezei*, 345 U.S. at 217. Justice Jackson, also dissenting, did Justice Black one better:

> Government counsel ingeniously argued that Ellis Island is his "refuge" whence he is free to leave in any direction except west. That might mean freedom, if only he were an amphibian! Realistically, this man is incarcerated by a combination of forces which keep him as effectually as a prison, the dominant and proximate of these forces being the United States immigration authority. It overworks legal fiction to say that one is free in law when by the commonest of common sense he is bound.

Id. at 220.

154. Id. at 209.

155. Id. at 215.

was with the countries that would not accept him, not with the American government: "[A]n alien in [Mezei's] position," the Court concluded, "is no more ours than theirs."[156]

Mezei was unusual only in the potentially indefinite duration of the detention. Prior to 1954, INS policy was to detain almost all aliens at the port of entry pending a determination of admissibility.[157] Although the Supreme Court expressly upheld this policy, at least regarding Communists, who were denied any bail,[158] some lower courts resisted it.[159] In 1954, the Justice Department decided to change its detention policy, closed Ellis Island, and began to parole aliens into the country pending determination of their admissibility. Detention was reserved for situations in which the alien was likely to abscond or considerations of national security or public safety required continued confinement.[160] In the Court's words, detention thereafter became "the exception, not the rule"; although parole was a matter of administrative grace that in no way altered the alien's status as an outsider seeking entry, the new policy reflected "the humane qualities of an enlightened civilization."[161]

If that was true, humanity and enlightenment were more transient than the Court supposed. In 1981, when South Florida was besieged by aliens from the Caribbean Basin, the pre-1954 policy of routine detention with limited exceptions was reestablished.[162] The effect of this influx upon detention was dramatic and immediate; during 1982, more than one million person-days were spent in INS detention, almost double the figure for 1980.[163] The mass, routine incarceration of aliens, sometimes for long periods of time and often under overcrowded and unpleasant conditions, had once again become a fundament of our immigration policy.[164] It reaffirmed two classical concep-

156. Id. at 216.

157. During 1952, a total of almost 39,000 aliens were detained; in the New York district, the average length of detention was 39 days. Maslow, Recasting Our Deportation Law: Proposals for Reform, 56 Colum. L. Rev. 309, 360 (1956).

158. Carlson v. Landon, 342 U.S. 524 (1952).

159. See, e.g., United States ex rel. Daniman v. Shaughnessy, 117 F. Supp. 388 (S.D.N.Y. 1953); United States ex. rel. Klig v. Shaughnessy, 94 F. Supp. 157 (S.D.N.Y. 1950) (reasoning in substance disapproved in United States ex rel. Yaris v. Esperdy, 202 F.2d 109 (2d Cir. 1953)).

160. See Jean v. Nelson, 711 F.2d 1455, 1469, reh'g en banc granted, 714 F.2d 96 (11th Cir. 1983).

161. Leng May Ma v. Barber, 357 U.S. 185, 190 (1958). Even if parole were arbitrarily denied, however, the courts, as deferential to the immigration authorities as ever, see supra notes 74-92 and accompanying text, would not intervene to grant it. See, e.g., In re Cahill, 447 F.2d 1343 (2d Cir. 1971).

162. See Jean v. Nelson, 711 F.2d 1455, 1464-65, reh'g en banc granted, 714 F.2d 96 (11th Cir. 1983). The 1981 detention policy, invalidated on several grounds in *Jean* and reimposed on a different legal footing in 1982, has survived initial legal challenge. Ishtyaq v. Nelson, No. 82-2288 (E.D.N.Y. Oct. 4, 1983).

163. Telephone interview with Howard Brown, Office of Planning and Analysis, INS (May 10, 1983).

164. See, e.g., Jean v. Nelson, 711 F.2d 1455, 1500 n.51 (quoting from the Special Master's Report on the Haitians in detention), reh'g en banc granted, 714 F.2d 96 (11th Cir. 1983).

tions. First, it implied that aliens remain wholly outside the national community unless and until the formal requirements of membership are satisfied. Second, it implied a strong view of national sovereignty, one holding that the nation cannot be obliged against its will to enter into a continuing relationship with an intruder.[165] Under the classical model, such an individual has thrust herself upon the nation and may therefore be treated as the utter stranger that she is; she may be unceremoniously separated from our midst. In the *Mezei* Court's revealing phrase, she is "no more ours than theirs."[166]

G. *Non-Independent Adjudication*

Under our system of public administration, it is common for agencies to integrate investigative, prosecutorial, and adjudicatory functions within one bureaucracy. Regulatory commissions and boards are typically organized in that fashion, and due process is not thereby offended.[167] But of all the federal agencies that dispense "mass justice" to individuals,[168] the INS has been unusual, if not unique, in operating a system in which those who receive evidence and formally adjudicate individuals' rights are subject to the direct administrative supervision of officials empowered to interrogate, arrest, detain, and impose severe sanctions upon individuals.[169] Traditionally, the immi-

---

165. Indeed, in Leng May Ma v. Barber, 357 U.S. 185 (1958), in a remarkably niggardly construction of the statute, the Court refused even to protect the alien's right of *nonrefoulement*, one of the most basic rights conferred by international law. See Final Act and Convention of the United Nations Conference of Plenipotentiaries on the Status of Refugees and Stateless Persons, art. 33, U.N. Doc. A/CONF. 2/108, U.N. Sales No. 1951.IV.4 (1951), to which the United States acceded in 1968, see infra note 400.

Although *Mezei* and *Leng May Ma* were exclusion cases, there is no reason to believe that classical immigration law viewed protracted detention pending deportation any differently, so long as due process was observed prior to confinement. See United States ex rel. Hyndman v. Holton, 205 F.2d 228 (7th Cir. 1953); United States ex rel. Bryant v. Shaughnessy, 122 F. Supp. 326 (S.D.N.Y. 1954).

166. 345 U.S. at 216.

167. See, e.g., Withrow v. Larkin, 421 U.S. 35 (1975). Mixture of functions in particular cases, however, may be so unfair as to violate due process or equal protection principles. See, e.g., Gibson v. Berryhill, 411 U.S. 564 (1973); Cinderella Career and Finishing Schools, Inc. v. FTC, 425 F.2d 583 (D.C. Cir. 1970).

The administrative law judges who hear disability claims under the Social Security Act, 42 U.S.C. §§ 301-1397 (1976 & Supp. V 1981), arguably combine these functions to some degree in that they gather evidence and develop the facts as well as adjudicate the claims. Id. § 405(b), (g). This procedure was upheld, albeit without much reasoning, over due process and Administrative Procedure Act challenges. See Richardson v. Perales, 402 U.S. 389, 408-10 (1971).

168. The INS caseload is staggering. Each year, the INS adjudicates more than 1.5 million petitions for various benefits. Immigration and Naturalization Serv., U.S. Dep't of Justice, Asylum Adjudications: An Evolving Concept and Responsibility for the Immigration and Naturalization Service 24 n.* (internal memo 1982) (copy on file at the offices of the Columbia Law Review) [hereinafter cited as INS Asylum Study].

169. Approval by an independent magistrate prior to arrest and detention, constitutionally required of the police, is ordinarily not required of the INS. 8 U.S.C. §§ 1252(a), 1357(a)-(c) (1982); Abel v. United States, 362 U.S. 217 (1960). But see Blackie's House of Beef, Inc. v. Castillo, 467 F. Supp. 170 (D.D.C. 1978), aff'd, 659 F.2d 1211 (D.C. Cir. 1981), cert. denied, 455 U.S. 940 (1982).

gration judges[170] who must decide which aliens shall be excluded and deported were an integral part of the INS and, like the INS law enforcement agents, were subject to the direct and indirect influence of its political leadership.[171]

In the immigration setting, adjudications by nonindependent decisionmakers raise especially troublesome questions. In *Wong Yang Sung v. McGrath*,[172] the Court seemed to agree. The case came before the Court in 1950 under the then-applicable 1917 immigration statute, which required deportation hearings to be conducted by immigration inspectors whose duties were not limited to these adjudications; in addition, they investigated similar cases and might have performed some prosecutorial functions in the very cases that came before them for adjudication. The Court denounced this system, finding it "particularly" objectionable in deportation proceedings, which involve "a voteless class of litigants" unfamiliar with American law and custom, and often, the English language.[173] Noting that the Administrative Procedure Act (APA)[174] did not permit this practice,[175] the Court seemed to hold that a deportation hearing before an independent decisionmaker was constitutionally required.[176] But this concern for independence soon vanished—or at least was easily mollified. For when Congress responded to the Court's decision shortly thereafter by establishing a system of immigration judges who performed only adjudicatory duties but worked under the administrative aegis of the more enforcement-oriented INS district directors, the Court found no constitutional defect.[177] What had occurred to change the Court's view? Therein lies a confusing tale.

As early as 1948, Congress had considered legislation[178] to exempt the INS from the APA provisions that required trial-type procedures, including a substantially independent hearing officer, for certain kinds of adjudica-

170. In the vernacular of the statute, they are called "special inquiry officer[s]." 8 U.S.C. § 1101(b)(4) (1982).

171. For a discussion of some of the informal patterns of prosecutorial influence over the immigration judge corps, see Roberts, Proposed: A Specialized Statutory Immigration Court, 18 San Diego L. Rev. 1, 7-11 (1980).

172. 339 U.S. 33, modified, 339 U.S. 908 (1950).

173. Id. at 46.

174. 5 U.S.C. §§ 551-576 (1982).

175. 339 U.S. at 46.

176.

> A deportation hearing involves issues basic to human liberty and happiness and, in the present upheavals in lands to which aliens may be returned, perhaps to life itself. It might be difficult to justify as measuring up to constitutional standards of impartiality a hearing tribunal for deportation proceedings the like of which has been condemned by Congress as unfair even where less vital matters of property rights are at stake.

Id. at 50-51.

177. Marcello v. Bonds, 349 U.S. 302 (1955). The statute did allow a hearing officer a limited prosecutorial role. See infra note 183.

178. See H.R. 6652, 80th Cong., 2d Sess., 94 Cong. Rec. 6374 (1948). The bill, H.R. 6652, was returned to the House Judiciary Committee, 94 Cong. Rec. 6374 (1948), which reported the bill out without amendment within two weeks. H.R. Rep. No. 2140, 80th Cong., 2d Sess. (1948). No further action was taken.

tions.[179] Advocates of such exemption had argued, inter alia, that creating a corps of independent decisionmakers would be too costly and that immigration was too closely linked to political functions and foreign affairs to be subject to the APA's more scrupulous procedures.[180] After the Court in *Wong Yang Sung* suggested that the APA's separation-of-functions provisions were constitutionally required for deportation adjudications, Congress in 1950 adopted a rider to the Justice Department's supplemental appropriation that exempted exclusion and deportation proceedings from those APA requirements.[181] In 1952, Congress enacted a new and comprehensive immigration statute that included a separation-of-functions provision only slightly broader than that in the 1917 law, which had provided only that a hearing officer could not decide the same case that he had investigated,[182] but narrower than that in the APA. The 1952 provision, which remains in effect today, prohibits a hearing officer from conducting a proceeding in a case in which she has performed either investigative or prosecuting functions.[183] This procedure is to be "the sole and exclusive" deportation procedure.[184]

A sharply divided Court upheld these new provisions three years later in *Marcello v. Bonds*,[185] but its reasoning defies comprehension even today. An alien, Carlos Marcello, had challenged the 1952 Act's deportation procedures on the grounds that, inter alia, that they left the hearing officer subject to the supervision and control of INS investigators and prosecutors, thereby violat-

179. 5 U.S.C. §§ 554, 556–557 (1982). With certain enumerated exceptions, the hearing officer must be independent in at least three senses. First, she cannot "consult a person or party on a fact in issue, unless on notice and opportunity for all parties to participate." Second, she cannot "be responsible to or subject to the supervision or direction of an [agency] employee or agent engaged in the performance of investigative or prosecuting functions." Third, no agency employee "engaged in the performance of investigative or prosecuting functions . . . in a case may . . . in that or a factually related case, participate or advise in the decision . . . except as witness or counsel in public proceedings." Id. § 554(d).

180. In addition to these policy considerations, there were legal arguments as well centering around the contention that Congress had never intended for the APA to apply to immigration hearings and that certain exceptions from the separation-of-functions provisions were applicable. H.R. Rep. No. 2140, 80th Cong., 2d Sess. 1–6 (1948).

181. Supplemental Appropriation Act of 1951, ch. 1052, 64 Stat. 1044, 1048 (1950) (exempting deportation and exclusion proceedings from the Administrative Procedure Act, ch. 324, §§ 5, 7–8, 60 Stat. 237, 239 (1946), repealed by Immigration and Nationality Act of 1952, ch. 477, § 403(a)(47), 66 Stat. 163, 280.

182. Immigration Act of 1917, ch. 29 §§ 16–17, 39 Stat. 874, 885–87, repealed by Immigration and Nationality Act of 1952, ch. 477, § 403(a)(13), 66 Stat. 163, 279.

183. 8 U.S.C. § 1252(b) (1982). The statute, however, expressly permits the hearing officer, inter alia, to "present and receive evidence, interrogate, examine, and cross-examine the alien or witnesses." Id.

184. Id. This was not the last time that Congress would reject the APA model for deportation proceedings. In 1961, it overruled court decisions applying the APA's judicial review provisions to immigration cases. See Woodby v. INS, 385 U.S. 276, 282–83 (1966) (Clark, J., dissenting).

185. 349 U.S. 302 (1955).

ing the APA's separation-of-functions requirement and Marcello's due process rights.[186] The Court characterized the statutory issue as whether Congress in the 1952 immigration act had reversed its decision in the 1950 appropriations rider to exempt deportation cases from APA procedural requirements and had in effect reinstated *Wong Yang Sung* by making the APA's hearing provisions directly applicable to deportation proceedings. Concluding that Congress had not done so, it rejected Marcello's claim.[187]

The Court, however, failed even to acknowledge the real question: how could the 1950 appropriations rider possibly have overruled *Wong Yang Sung*? That decision, although ambiguous about precisely how much separation of functions due process required, was clear that the *Constitution itself* demanded a more impartial tribunal than Congress or the INS had yet afforded.[188] That being so, the Court's comparison of the APA provisions and 1952 Act's deportation procedures was quite beside the point. For purposes of constitutional analysis, the relevant comparison was between the 1952 Act's procedures and the earlier ones invalidated in *Wong Yang Sung*, for the issue was whether the new procedures met the Court's constitutional objections to the old ones.

The most generous interpretation of the Court's performance in *Marcello* would be that the Court, reasoning sub silentio, had made that comparison and had discerned sufficient differences between the two statutes to render *Wong Yang Sung* irrelevant. To have done so explicitly, of course, would have required the Court to say just how much independence the Constitution demanded of adjudicators in deportation cases. But the Court was not then prepared to confront that question, nor has it been since. Indeed, when the Court turned to Marcello's due process claim, which invoked *Wong Yang Sung*, the Court could manage only a one-sentence evasion of the issue:

> The contention is without substance when considered against the long-standing practice in deportation proceedings, judicially approved in numerous decisions in the federal courts, and against the special considerations applicable to deportation which the Congress may take into account in exercising its particularly broad discretion in immigration matters.[189]

Classical immigration law, the Court seemed to be saying, sanctioned a system of adjudication in which the hierarchy of legitimate legal succession is com-

---

186. U.S. Const. amend. V.

187. 349 U.S. at 306–07.

188. "When the Constitution requires a hearing, it requires a fair one, one before a tribunal which meets *at least* currently prevailing standards of impartiality." *Wong Yang Sung*, 339 U.S. at 50 (emphasis added). Because the deportation tribunal did not meet even the recently enacted APA's standards of impartiality, id. at 36, the *Wong Yang Sung* Court could invalidate the INS procedure without reaching the question of whether the Constitution mandated some level of impartiality, as reflected in a particular tribunal structure, beyond that then required by the APA. This was the question the Court evaded in *Marcello*.

189. 349 U.S. at 311.

pletely reversed. Here, where sovereignty confronts strangers, the Constitution can be subordinated to a congressional statute, indeed, to mere administrative practice. The Red Queen could not have said it better.

II. Pressures for Change

Each of these sturdy redoubts of classical immigration law—a restrictive notion of national community; extraordinary judicial deference; the extraconstitutional status of exclusion; the unlimited federal power to classify; the civil nature of deportation proceedings; broad detention power; and nonindependent adjudication—is under siege today. Capitulation seems only a matter of time for some of these doctrines, while for others the outcome remains highly uncertain. The forces of change, however, are insistently hammering at the gate, threatening the autonomy and insularity that have long sheltered classical immigration law from developments elsewhere in the legal culture.

The assaults upon the classical legal order in immigration come from two different kinds of pressure. The first includes what might be called *structural* changes. These are shifts in the basic economic, demographic, social and political institutions and relationships that inescapably shape immigration policy. The second consists of changes in *ideology* or legal consciousness. These are altered beliefs about the meaning of justice, the rightness or wrongness of certain actions, and the proper role of law and of particular legal institutions in society.

It is important to recognize that the legal and policy implications of these changes are to some degree indeterminate. Each structural change, for example, poses problems for immigration law that might generate a variety of possible responses. Thus, it is not at all inevitable, though I think it likely, that the recent massive influx of undocumented aliens would lead to expanded, rather than reduced, legal protections for them. By the same token, new values concerning the sources and nature of obligations to strangers in public and private law contain internally contradictory elements that may tend to drive immigration law and policy in quite different directions. We saw earlier that this contradiction can be traced to a tension within the interwoven strands of liberalism and restrictive nationalism upon which classical immigration law has drawn.[190]

In addition to those ambiguities, both structure and ideology bear a complex relationship to the system of immigration law and administration that is already in place. As the experience with the Caribbean immigration of the 1980's suggests, structure and ideology are autonomous to some degree, influenced by their own ebbs and flows and by conditions over which immigration law can exercise little effective control. Yet as that same experience shows, structure and ideology are also shaped by the existing system. Finally, the precise relationship between these two factors, legal ideology and social

---

190. See supra notes 21-22 and accompanying text.

structure, is uncertain, as is the causal contribution that each makes to observed changes in immigration policy.[191]

A. *Structural Changes*

Structural conditions, by their nature, do not appear as discrete, clearly bounded phenomena. Instead, they are multifaceted, interrelated, and causally complex. For analytical purposes, however, one can distinguish a number of structural changes that are beginning to impinge upon immigration law in important ways. I shall discuss five: the constraining of American foreign policy; changes in economic conditions; mass refugee and asylum claims; the influx of undocumented aliens; and shifts in political forces. Taken together, these structural changes call into serious question the continued plausibility and relevance of the assumptions concerning consent-derived obligation, strong sovereignty, and restrictive national community that linked and legitimated the elements of classical immigration law.

1. *The Constraining of American Foreign Policy.*—Classical immigration law developed during a period in which the United States emerged from relative isolation into a position of international power and, for a time, unchallenged supremacy. Only toward the end of this period, in the 1960's, did the nation begin to face any serious threats of military and diplomatic vulnerability. The last decade, however, has witnessed a significant erosion of the power of the United States to work its will in the world. During the post-Vietnam period, a succession of economic, political, and military crises has occurred in areas of the world of great strategic importance to the United States. Examples include the Arab oil embargo; wars in the Middle East; the Iranian revolution; confrontations with the Soviet Union over Afghanistan, Poland, Africa and Southeast Asia; continuing instability and repression by Caribbean basin and other Third World regimes; economic and political weakness among some of our allies; the emergence of Japan, Korea and other Asian countries as effective competitors in world markets; the near-collapse of the Mexican economy; and protracted world recession. These crises have left the United States more dependent than ever upon the goodwill of other nations for the attainment of its policy objectives. We probably enjoy less freedom of action today on the world stage than we have since we came to maturity as a world power at the end of the last century, when the ideology of restrictive nationalism and the structure of classical immigration law took root.[192]

This adjustment in the international balance of power inevitably affects our immigration policy. For example, the United States' thirst for Mexican oil and gas and its desire to sustain stable, politically moderate regimes in the volatile Caribbean basin have almost certainly discouraged the INS from

---

191. R. Merton, Social Theory and Social Structure 510–42 (1968).

192. See H. Kissinger, American Foreign Policy 55–58 (1969); C. Vance, Hard Choices: Critical Years in America's Foreign Policy (1983).

controlling the southern border as tightly as it might. Similarly, strong United States opposition to Soviet-sponsored regimes such as those in Afghanistan, Cuba, and Kampuchea, coupled with its firm support for authoritarian regimes in Central America and the Philippines, has made the administration of the Refugee Act of 1980 seem incoherent and arbitrary.[193] The growing reluctance of the United Kingdom, Australia, and other previously "open" countries to admit Third World immigrants[194] has intensified pressures on the United States to relax its own restrictions.

Dependency, of course, is an exceedingly volatile, complex, and poorly-understood psychological phenomenon at the level of individuals.[195] It is surely no less so at the level of nations. With respect to immigration law, a number of quite different, perhaps inconsistent, responses by the political branches and by the courts are possible. Congress might react defensively, hoping to restore our diminished sense of autonomy and well-being by restricting both immigration levels and alien rights, much as the United Kingdom has recently done.[196] Courts might see the matter differently. They might regard the interdependence between nations as a justification for a more permissive stance, softening the sharp boundary between subject and object, government and alien, us and them, that has characterized classical immigration law. What seems clear, however, is that feelings of greater dependency can modify our concepts of national sovereignty and community in subtle ways, altering immigration law in the process.

2. *Economic Changes.* — Classical immigration law emerged from a society in which rapid economic growth, with the exception of the Depression era, was more or less constant and taken for granted.[197] Immigration has consistently been viewed as a valuable device for influencing the labor market.[198] Prior to the 1920's, that generally meant relatively unlimited immigration—except for Orientals—to supply cheap manpower for an industrializing economy. In more recent years, it has meant limiting nonfamily legal immigration to workers capable of meeting specialized needs,[199] adopting special

193. See, e.g., Asylum Adjudication: Hearings Before the Senate Subcomm. on Immigration and Refugee Policy, 97th Cong., 1st Sess. 46-61 (1981) (statement of I. Kurzban); Scanlan, Regulating Refugee Flow: Legal Alternatives and Obligations under the Refugee Act of 1980, 56 Notre Dame Law. 618, 627-29 (1981).

194. Great Britain admitted approximately 7,000 refugees in 1980, but only about 3,000 in 1981. Australia admitted about 22,000 in 1981-1982 and about 17,000 in 1982-1983. Telephone interview with U.N. High Comm'n on Refugees (Oct. 26, 1983).

195. See, e.g., R. Burt, Taking Care of Strangers (1979).

196. See L. Fransman, British Nationality Law and the 1981 Act 45-49 (1982); supra note 194.

197. Bureau of Census, U.S. Dep't of Commerce, Historical Statistics of the United States: Colonial Times to 1970, at 224 (1975). See generally E. Denison, Accounting for the United States Economic Growth: 1929-1969, at 11 (1974).

198. See, e.g., Reimers, Recent Immigration Policy: An Analysis, *in* The Gateway: U.S. Immigration Issues and Policies 13 (B. Chiswick ed. 1982) [hereinafter cited as The Gateway].

199. For discussion of occupational preferences and labor certification, see Goldfarb, Occupational Preferences in the U.S. Immigration Law: An Economic Analysis, *in* The Gateway, supra note 198, at 412.

programs to supply agricultural laborers for farms in the Southwest,[200] and tolerating large numbers of undocumented workers who fill the lowest rungs of the occupational ladder.[201]

During the last decade, however, economic conditions have deteriorated. Low rates of growth, high unemployment and underemployment, and chronic inflation[202] have created a climate of economic uncertainty in which immigration policy is being pulled in several different directions. In part, this reflects the fact that immigration now creates a more intricate pattern of effects upon the domestic economy than ever before. The issue of undocumented aliens is an example. Labor economists are sharply divided concerning whether and to what extent undocumented workers displace domestic labor.[203] There seems little doubt, however, that at least in particular economic sectors, such as low-margin service establishments and certain agricultural industries, undocumented workers willing to work at or below the minimum wage rate preserve and create many jobs that would not survive in a more restricted, higher-wage labor market.[204]

For the vast majority of Americans who do not have to compete with undocumented aliens for jobs, housing, or services, this productive activity almost certainly constitutes a significant net benefit by creating anti-inflationary pressures and generating greater wealth throughout the society.[205] By helping American products to meet low-wage competition from abroad, such labor is an important national resource. At a time of fiscal stringency, American taxpayers are also probably advantaged on balance by the presence of undocumented aliens, who appear to pay a good deal more in direct state and local sales taxes, Social Security and withholding tax, and indirect taxes than they receive in governmental benefits.[206]

---

200. For discussion of the Bracero program, see The H-2 Program and Non-Immigrants: Hearing Before the Subcomm. on Immigration and Refugee Policy of the Senate Comm. on the Judiciary, 97th Cong., 1st Sess. (1981); Morgan & Gardner, Potential for a U.S. Guest-Worker Program in Agriculture: Lessons from the Braceros, *in* The Gateway, supra note 198, at 361.

201. See Fogel, Illegal Aliens: Economic Aspects and Public Policy Alternatives, 15 San Diego L. Rev. 63, 68-69 (1977); Wachter, The Labor Market and Illegal Immigration: The Outlook for the 1980's, 33 Indus. & Lab. Rel. Rev. 342, 350-54 (1980).

202. See generally E. Denison, Accounting for Slower Economic Growth: The United States in the 1970's (1979).

203. Compare, e.g., Piore, Impact of Immigration on the Labor Force, 98 Monthly Lab. Rev. 41, 43 (May 1975) (arguing that undocumented aliens are hired to fill gaps in the secondary labor market), with, e.g., M. Wachter, supra note 203, at 343, 350-54 (arguing that undocumented workers depress the wages of the low-skilled American work force).

204. See Piore, supra note 203, at 43.

205. See Simon, The Overall Effect of Immigration on Natives' Incomes, *in* The Gateway, supra note 198, at 314.

206. The classic study is D. North & M. Houston, The Characteristics and Role of Illegal Aliens in the U.S. Labor Market: An Exploratory Study 142-49 (1976). Compare Reinhold, Taxes Aliens Pay to Texas Found to Top Benefit, N.Y. Times, Nov. 15, 1983, at A17, col. 1 ("A new study in Texas has found that [illegal immigrants] contribute far more to the state's coffers in taxes than they take out in health, welfare and other publicly financed services."), with The Jobs

Other interactions between a changing economy and immigration are also important. For example, immigration is already altering both the short-term and long-term demographic profiles of the United States in ways that may have enormous implications for the future shape of the economy.[207] Aliens admitted to the United States under labor certifications also increase the society's wealth and efficiency by meeting labor market shortages and providing the means for flexible adjustments to new economic conditions. In certain occupational areas, such as the staffing of municipal hospital systems, they have filled crucial gaps, permitting some breathing space for longer-term development of specialized domestic manpower.[208]

These examples, of course, do not begin to tell the whole economic story. For the many domestic workers who do not enjoy any easy insulation from the competitive effects of a large pool of undocumented aliens, the presence of relatively docile, desperate aliens willing and even anxious to accept very low wages and often illegal working conditions may be nothing less than a calamity. The easy availability of such workers has probably retarded technological innovation in some industries, encouraged employers to violate labor standards, and accelerated the decline of labor unions. Certain public services in particular localities have been severely burdened by undocumented aliens. Other economic effects are even more ambiguous. Demographic changes, and their economic consequences, are notoriously difficult to predict.[209] Even the system of labor certification for alien workers has had mixed consequences. It is said to have been abused by employers and putative employees alike, to suffer from administrative and other inefficiencies, and to have produced low-quality services in some important instances.[210]

If the precise relationship between immigration and economic changes is opaque, however, the importance of that relationship is transparently clear. Classical immigration law, as we have seen, responded in part to the rhythms of the labor market.[211] In this respect, the 1980's are likely to be no different.

Illegal Aliens Take From Americans, N.Y. Times, Dec. 3, 1983, at 22, col. 4 (letter to the editor disputing Texas study and arguing on the basis of recent INS research that, assuming only 21% of illegal aliens displace Americans from jobs, they cost over $13 billion more per year than they pay in taxes).

207. Immigration to the United States: Hearings Before the Select Comm. on Population, 95th Cong., 2d Sess. 183–90 (1978).

208. See Goldfarb, supra note 199, *in* The Gateway, supra note 198, at 412, 432–40 (discussion of the relationship between doctor shortage and immigration policy). But see Kaye, Danilov & McDonald, Alien Physicians and Their Admission into the United States, 16 San Diego L. Rev. 61, 62–64 (1978) (congressional discussion of the foreign medical graduate).

209. See U.S. Comptroller General, Report to the Senate Committee on the Budget: Impact of Illegal Aliens on Public Assistance Programs: Too Little Is Known, app. 2, at 17 (1977).

210. See, e.g., INS Efficiency Legislation: Hearing on H.R. 2043 Before the Subcomm. on Immigration, Refugees, and International Law of the House Comm. on the Judiciary, 97th Cong., 1st Sess. 174–183 (1981); Goldfarb, supra note 199, *in* The Gateway, supra note 198, at 412, 417–21; Kaye, Danilov & McDonald, supra note 208, at 66–69.

211. See supra notes 8–9 and accompanying text.

The new economic order that the United States now confronts, especially the intensified competitive pressures that are spurring it at a time of lagging productivity growth, constitutes a structural change of which the political branches and the courts will surely take account in deciding the kind of society that immigration law is to fashion and defend.

3. *Refugee and Asylum Claims.* — Unlike the world in which classical immigration law first evolved, today's world is teeming with refugees—individuals displaced from their homes and countries by convulsive events such as civil war, persecution, natural disaster, and unimaginable destitution.[212] There have always been refugees, of course, but the brutality of modern warfare, the ferocity of political struggle, and the disruptiveness of social and economic changes have dramatically altered the scale of displacement and devastation. In 1982, there were some ten million refugees, and the number may well be increasing.[213] Hundreds of thousands are in overseas refugee camps awaiting resettlement in the United States; during fiscal 1984, the United States is expected to accept no more than 72,000.[214] What is unprecedented, however, is not the number of refugees abroad who seek resettlement here—the upheaval of World War II displaced at least eight million persons[215]—but the number of refugees who manage to reach American shores to seek asylum, and the kinds of societies from which they have come.

Prior to 1972, relatively few asylum claims—under 500 on average each year—were filed in the United States; these were often prominent defectors from Soviet-bloc countries. These claims were handled informally and according to ad hoc procedures; neither the immigration statute nor INS regulations addressed the question of asylum.[216] INS regulations did not establish procedures and standards for resolving asylum claims until January 1975,[217] and only the enactment of the Refugee Act of 1980[218] created a legislative basis for an asylum status and a systematic basis for determining it.

During the middle and late 1970's, the number and national origins of asylum claimants began to change. In 1977, the INS received 3702 asylum

---

212. The legal definition of refugees, of course, is much narrower. See 8 U.S.C. § 1101(a)(42) (1982).

213. U.S. Comm. for Refugees, World Refugee Survey 1982, at 40-41 (1982). The number of refugees in Africa, for example, increased from 750,000 in 1967 to 6.5 million in 1981. Id.

214. The Presidential determination for fiscal 1983 was 90,000, see Presidential Determination No. 83-2 (Oct. 11, 1982), reprinted in 2 Fed. Imm. L. Rep. (CCH) ¶ 14,010 (Aug. 1983), but only about 60,000 are expected to be accepted, see N.Y. Times, Sept. 27, 1983, at A3, col. 2. For fiscal 1984, the administration has proposed a refugee limit of 72,000. Presidential Determination No. 83-11 (Oct. 7, 1983).

215. This figure includes only those who were under the guardianship of the allied armies: about seven million were eventually repatriated but approximately one million remained in the allied zones. See S. Rep. No. 950, 80th Cong. 2d Sess., reprinted in 1948 U.S. Code Cong. & Ad. News 2028, 2035.

216. INS Asylum Study, supra note 168, at 3-5.

217. 39 Fed. Reg. 41,832 (1974).

218. 8 U.S.C. §§ 1157-1159 (1982).

applications and two years later, the number was 5801, a significant increase, to be sure, but still manageable. Indeed, the Refugee Act that President Carter signed on March 17, 1980 had little to say about asylum; to the drafters of the Act, it was essentially an afterthought.[219] Within weeks of the signing, however, Fidel Castro expelled large numbers of Cubans, the Mariel boatlift brought them to the United States, and the asylum question occupied center stage. Between March 1980 and July 1981, over 53,000 asylum claims were filed, as Mariel Cubans and undocumented Haitians streamed into southern Florida.[220] Since then, they have been joined by asylum claimants from other Caribbean Basin trouble spots, the Middle East, and elsewhere. The INS has estimated the number of pending claims at the end of fiscal 1983 at about 166,500; new claims were being filed at the rate of more than 25,000 a year.[221] At that time, the INS was completing action on asylum claims at a rate of approximately 30,000 cases a year; one measure of the unprecedented administrative response by the INS to the migrations of the 1980's is that this number was almost *fifty* times the total completions during 1980. Despite these unprecedented efforts, then, it will require decades to eliminate the backlog. Indeed, of the more than 120,000 asylum claims filed between 1979 and February 1983, the INS had decided fewer than 25,000 by the later date.[222]

These numbers may somewhat exaggerate both the seriousness and the uncontrollable nature of the backlog, especially if one assumes that the status of the Mariel Cubans eventually will be regularized through legislation, and recognizes that United States policy actually encouraged the filing of many of these claims.[223] Nevertheless, there is no question that the flood of asylum claimants presents extremely difficult problems for immigration law and policy. Most come not from the Soviet bloc but from nations with which the United States has, or until recently had, close political ties. These nations, such as Haiti and El Salvador, would bitterly resent our granting asylum to their citizens. Many asylum claimants come from nearby Caribbean nations and allege patterns of persecution that are presumably common in those countries. Migrating from largely rural, preindustrial societies that are cultur-

---

219. INS Asylum Study, supra note 168, at 6; Martin, The Refugee Act of 1980: Its Past and Future, 1982 Mich. Y.B. of Int'l Legal Stud. 91, 109.

220. INS Asylum Study, supra note 168, at 18.

221. Telephone interview with Margaret Sullivan, Director of Statistics, INS (Nov. 29, 1983).

222. Telephone interview with Howard Brown, Office of Planning and Analysis, INS (May 10, 1983); Telephone interview with Margaret Sullivan, Director of Statistics, INS (Nov. 29, 1983). "Completions" include grants, denials and withdrawals. The reasons for the backlog are many, including the need to await State Department reports on each individual claim.

223. See Aleinikoff, Aliens, Due Process and 'Community Ties': A Response to Martin, 44 U. Pitt. L. Rev. 237, 253–56 (1983). That assumption becomes more dubious with the increasingly uncertain fate of the Simpson-Mazzoli bill, S. 529, 98th Cong. 1st Sess., 129 Cong. Rec. S6970 (daily ed. May 18, 1983), the recently proposed immigration bill discussed supra note 98 and accompanying text.

ally quite distinct from ours, many face serious obstacles to assimilation here. Most probably cannot qualify for asylum status under the existing law.[224] These features have created considerable concern by the INS that favorable action on these claims would open the floodgates even wider.

Regardless of their prospects for success, the incentives for aliens to claim asylum are powerful, often overwhelming. Merely by filing the claim, an alien apprehended by the INS automatically wins a delay in deportation until all avenues of administrative and judicial review have been exhausted. For one who wishes to remain as long as possible, this delay can be extremely valuable in itself. Beyond that, the claimant, if not detained, may be able to obtain work, melt into the surrounding population, and avoid detection. She may be able to establish ties that ultimately prevent the INS from ever deporting her. In any event, she asserts constitutionally protected life and liberty interests so great that extensive procedural rights are probably triggered by her claim. In any event, she asserts constitutionally protected life and liberty interests so great that extensive procedural rights are probably triggered by her claim. The stakes in delay being so high and the filing of a claim being essentially costless to the alien, she has little to lose and much to gain by doing so. Indeed, even if the INS denies her claim and is in a position to expel her, it may nevertheless decide to defer deportation indefinitely, as it has done in the past with respect to Poles, Ethiopians, and many other groups.[225]

In sum, the ease of asserting asylum claims has made illegal immigration far more difficult to deter. It now imposes serious strains on the INS's administrative system and enforcement strategies quite apart from the substantial resources required simply to adjudicate the claims themselves. Finally, mass asylum claims have encouraged the INS to adopt an explicit policy of mass incarceration of undocumented aliens, reversing a long-standing, judicially ratified preference for parole,[226] and raising urgent legal and policy issues.[227]

The large, steadily increasing number of asylum claims, then, presents classical immigration law with unprecedented dilemmas of a legal, moral, and administrative nature. This development exerts massive pressure upon the notions of sovereignty, consent-based obligation, and restrictive national community that have supported the structure of classical immigration law. It does so, moreover, in the context of uniquely poignant, compelling human claims to justice.

4. *Undocumented Aliens.* — Perhaps the central fact about immigration with which social policymakers must deal today is the presence of an immense

224. See, e.g., In re Sibrun, 19 I. & N. Dec. No. 2932 (BIA Jan. 20, 1983) ("Generalized oppression by a government of virtually its entire populace does not come within those specified grounds."); see also Immigration Reform: Hearings Before the Subcomm. on Immigration, Refugees, and International Law of the House Comm. on the Judiciary, 97th Cong., 1st Sess. 622-701 (1981) (hearings on asylum) [hereinafter cited as 1981 House Hearings].

225. See infra notes 321-39 and accompanying text. 1981 House Hearings, supra note 224, at 622-701; INS Asylum Study, supra note 168, at 66.

226. See supra notes 156-61 and accompanying text.

227. See infra notes 382-406 and accompanying text for further discussion.

population of aliens in the United States in apparent violation of the immigration laws. Conservative extrapolation from a recent Census Bureau study suggests a current figure of at least 3.5 million.[228] The vast majority of these entered illegally; the remainder have violated the legal conditions of their original entry by staying beyond the time limit prescribed in their visas.[229] The number of undocumented aliens who reside here appears to be steadily growing. The standard estimate puts the annual increase at 500,000.[230]

The complex effects of undocumented aliens upon the domestic economy were discussed earlier. But the massive influx of undocumented aliens challenges classical immigration law on grounds far more fundamental than those of economics. When a routine external event such as the decline in value of the Mexican peso can generate a torrent of illegal border crossings that the INS is largely powerless to prevent,[231] confidence in the efficacy of our law, in our capacity to control events, is threatened. Indeed, as the INS has shifted its personnel to the Mexican border and the South, illegal entries across the far longer and essentially unpatrolled Canadian border, hitherto only a very minor problem, have increased significantly.[232] Significantly, this deluge is occurring at a time when INS border apprehensions are actually at record levels.[233] In view of these developments, it has become something of a political

228. The study suggests a figure of at least two but no more than six million as of 1980. R. Warren & J. Passel, Estimates of Illegal Aliens from Mexico Counted in the 1980 United States Census (unpublished manuscript on file with author). If 500,000 more enter and remain each year, see infra text at note 230, this would imply an increase in the United States population of some 1.5 million undocumented aliens in the three years since the 1980 census, increasing the Warren-Passell total to at least 3.5 million, leaving aside deportations and voluntary departures.

Calculating the number of undocumented aliens has become something of a parlor game, without any criteria for selecting a winner. It need hardly be added that counting undocumented aliens is a methodological nightmare. See the discussion and summary of studies in SCIRP Staff Report, supra note 25, at 36–37. For another review and critique of the studies, see Keely, Illegal Migration, Sci. Am., March 1982, at 41. One reasonably firm datum in this otherwise soft area is that INS has apprehended close to a million deportable aliens each year since 1977. Id. at 44. The total has hovered there ever since. The most recent available total, 970,246, was for the year 1982. Letter and Attachments from John E. Nahan, Director, Office of Plans and Analysis, INS, to Peter H. Schuck (October 25, 1983) (copy on file at the offices of the Columbia Law Review).

229. See, e.g., Vialet, Cong. Research Serv., Illegal/Undocumented Aliens 1 (Jan. 4, 1983) (93% of those apprehended by INS entered illegally, 7% violated conditions of entry).

230. After providing this estimate, Keely notes that "it has no empirical basis." Keely, supra note 228, at 41.

231. See, e.g., King, Mexican Money Crisis Impels New Surge of Aliens to Texas, N.Y. Times, Mar. 19, 1983, at A1, col. 2. ("The Border Patrol is inundated, and the 2,000-mile border has turned into something more like a series of commuter stations than an international boundary.").

232. Winerip, Smuggling of Aliens by Canadian Route to U.S. Is Increasing, N.Y. Times, May 1, 1983, at A1, col. 3.

233. During the early part of 1983, Border Patrol apprehensions averaged 30% above the previous year. Departments of Commerce, Justice, and State, The Judiciary, and Related Agencies Appropriations for 1984: Hearings Before a Subcomm. of the House Comm. on Appropriations, Department of Justice, 98th Cong., 1st Sess. 690 (1983) (testimony of Alan C. Nelson, Comm'r, INS); see also N.Y. Times, Apr. 9, 1983, at A12, col. 4 (apprehensions during March 1983 at the southern border the highest for any single month).

cliché, indispensable for congressional testimony and press conferences alike, to affirm that our ability to regulate entry and exit is perhaps the basic attribute of national sovereignty, and that the easy, continual and largely unimpeded entry of undocumented aliens demonstrates that "we have lost control of our borders."[234]

The central social fact about undocumented aliens is this: once they enter illegally, and especially if they decide to remain more or less permanently,[235] they quickly begin to establish significant relationships with the individuals and institutions in the locales to which they gravitate. Some bring or are joined by their spouses, others marry here. If they bring children, those children usually attend local schools and acquire American customs; if their children are born here, the children are automatically United States citizens.[236] Virtually all find employment, pay taxes, and have access to certain public benefits.[237] Many establish ties with churches and other voluntary associations.[238] And a substantial proportion will continue to reside in the United States for the rest of their lives, the immigration law notwithstanding.

Suppose that an anthropologist or sociologist, intimately familiar with American culture but innocent of any knowledge concerning aliens' legal status, were to observe life in the Haitian section of Miami or the barrios of El Paso. She would undoubtedly notice that some residents tend to live furtively

---

234. On July 30, 1981, for example, the Attorney General addressed a joint congressional subcommittee, testifying that "[w]e have lost control of our borders. We have pursued unrealistic policies. We have failed to enforce our laws effectively." Administration's Proposals on Immigration and Refugee Policy: Joint Hearing Before the Subcomm. on Immigration, Refugees, and International Law of the House Comm. on the Judiciary, and the Subcomm. on Immigration and Refugee Policy of the Senate Comm. on the Judiciary, 97th Cong., 1st Sess. 6 (1981) (testimony of Att'y Gen. William French Smith).

235. It has been estimated that those who reside temporarily in the United States stay for six to 48 months. SCIRP Staff Report, supra note 25, at 544.

236. See United States v. Wong Kim Ark, 169 U.S. 649 (1898) (fourteenth amendment grounds). British law, in contrast, states that a person born in Britain will be a citizen if one parent at birth is a citizen or is settled in Britain, or can become a citizen if either parent does so while the child is a minor. British Nationality Act, 1981, ch. 61, §§ 1, 3. Previous British law conferred citizenship at birth unconditionally. See British Nationality Act, 1948, 11 & 12 Geo. 6, ch. 56, § 4; see also Sandifer, A Comparative Study of Laws Relating to Nationality at Birth and to Loss of Nationality, 29 Am. J. Int'l L. 248, 253-59 (1935) (discussing the widespread application of the *jus sanguinis* principle in civil law countries).

237. Piore, supra note 203; see also Immigration to the United States: Hearings before the Select Comm. on Population, 95th Cong., 2d Sess. 183-90 (1978) (statement of Burdette Wright). For an interesting example of an important public benefit newly extended to undocumented aliens, see Cummings, California Law Gives Aliens Lower College Costs, N.Y. Times, Dec. 6, 1983, at A22, col. 2 (resident illegal aliens may attend University of California system at reduced tuition rates, considerably lower than what out-of-state American citizens must pay).

238. For this reason, the proposed Simpson-Mazzoli legislation would rely heavily upon voluntary groups to implement the legalization program. See S. 529, 98th Cong., 1st Sess. §§ 301(c)(1), 301(f), 129 Cong. Rec. S6970, S6982-83 (daily ed. May 18, 1983). For a discussion of the role of churches in harboring and supporting illegal aliens, see Austin, More Churches Join in Offering Sanctuary for Latin Refugees, N.Y. Times, Sept. 21, 1983, at A18, col. 1.

and seem especially vulnerable to exploitation. But she would surely also conclude that these same individuals are neither strangers nor outcasts but are vital elements in their neighborhoods, sinking deep roots into the local society and participating in many aspects of community life.

The challenge that this development poses to the continued coherence and integrity of classical immigration law can scarcely be exaggerated. Just as the hard facts of hopelessly porous borders demand a new understanding of sovereignty, the reality of undocumented aliens' humanity and their steady integration into communal life here undermines some of the traditional conceptions that that law reflected. New "social contracts" between these aliens and American society are being negotiated each day, and these cannot easily be nullified with invocations of sovereignty, as classically understood. These understandings often represent commitments valued not only by the aliens who seek to acquire legal status but also by the substantial number of Americans who apparently benefit from their contribution to our society.[239] It is also true, however, that most undocumented aliens have come here uninvited,[240] that they remain here in knowing violation of our laws, and that any moral claims they may have must be compared to those of the numerous would-be immigrants who wait patiently, often at great personal cost, for legal admission. Conventional ideas about morality and community simply cannot adequately capture the complex, ambiguous character of the relationship between undocumented aliens and American society.

5. *Political Shifts.* — The character and magnitude of the structural changes that I have been discussing are certain to engender far-reaching alterations in demography, public attitudes and power relationships. These alterations have themselves become part of the context and background against which the adequacy of classical immigration law must be appraised. Three seem particularly significant: the rise of Hispanic and Asian Americans as formidable political forces; the skewed geographical distribution of aliens; and the unusually broad public consensus on the need for legal change.

In a democracy, demography is political destiny. In the American context, that destiny is one in which the traditionally dominant ethnic groups of European ancestry will gradually yield political ground to those of Hispanic and Asian descent.[241] In 1980, the white, non-Hispanic percentage of the

239. For an analysis of recent polling data bearing on this question, see Harwood, Alienation: American Attitudes toward Immigration, Pub. Op., June/July 1983, at 49.

240. But see Lopez, supra note 12, at 639-72 (historically, United States government and American employers actively encouraged Mexicans to immigrate here).

241. Obviously, Americans do not always vote along ethnic lines. Nevertheless, ethnic identification continues to be a powerful force in American politics. See Hispanic Power Arrives at the Ballot Box, Bus. Wk., July 4, 1983, at 32; Into the Mainstream: Widespread Political Efforts Open New Era for Hispanics, 40 Cong. Q. 2707 (1982). But the implication of the growing Hispanic political power for immigration policy is ambiguous. See Graham, Illegal Immigration and the New Restrictionism, 12 Center Mag., May-June 1979, at 54-64. Also, polls taken among blacks and Hispanics, both citizen and noncitizen, show they favor restriction of legal immigration and harsher policies toward illegal immigrants. 1 Fed. Imm. L. Rep. (CCH) Update No. 29, at 3-4 (Aug. 8, 1983).

population was about eighty percent; as a result of anticipated immigration patterns[242] and differential fertility rates,[243] that share is expected to decline to seventy-seven percent in 1990 and seventy-five percent in 2000.[244] More than half of the population growth during that period will come from minority groups, and Hispanic groups alone will account for twenty-five percent of the growth.[245] These projections, of course, are highly fallible and sensitive to assumptions about difficult-to-predict variables, especially fertility and migration rates. Nevertheless, their orders of magnitude suggest that extremely important political changes impend, changes that are certain to affect future struggles over the direction of immigration law. Indeed, these changes are already evident, for example, in the growing number of Hispanic elected officials,[246] and in the considerable influence exerted by Hispanic groups over the shape and likely fate of current legislative proposals for immigration reform.[247]

Political influence, however, is a function not simply of aggregate votes but also of the geographical location of those voting, not simply a matter of the benefits and costs of a policy change but also of the distribution of those benefits and costs.[248] And the voters most directly concerned with immigration policy, as well as the benefits and costs of that policy, are concentrated in a small number of politically important states, especially California, Texas, Florida, New York, New Jersey and Illinois. Indeed, one estimate is that of the 5.5 million eligible Hispanic voters, ninety-three percent reside in nine states that account for three-fourths of the electoral votes needed to elect a Presidential candidate.[249] The great majority of undocumented aliens appar-

242. INS statistics break down the number of immigrants admitted in 1980 by region of birth as follows: Asia (236,097) 44%; Americas (except Canada) (190,880) 35%; total population 530,639. 1980 Statistical Yearbook of the INS 4.

243. In 1980, the fertility rate for white, non-Hispanic women was 1.76 children; for Hispanics, it was 2.6. A Non-Anglo America by 2080?, 14 Nat'l J. 2006 (1982).

244. Study Sees Shift in U.S. Population, N.Y. Times, July 19, 1983, at A10, col. 1 (early ed.).

245. Id. The study, done by the Center for Continuing Study of the California Economy, indicates that by the year 2080—a time span, to be sure, well beyond the concerns of statesmen, not to say mere politicians—the white, non-Hispanic share is expected to decline to under 50%, based on current trends. The study is based upon very conservative estimates of illegal immigration.

246. See, e.g., Schmidt, Denver Election Widens Circle of Hispanic Leaders, N.Y. Times, June 23, 1983, at A16, col. 1. For an analysis of Hispanic voting power, see Kirschten, The Hispanic Vote—Parties Can't Gamble that the Sleeping Giant Won't Awaken, 15 Nat'l J. 2410 (1983).

247. See, e.g., Pear, Immigration and Politics: Hispanic Bloc's Fears Key to Failure of Bill, N.Y. Times, Oct. 6, 1983, at A1, col. 1; Hispanic Leaders Oppose Sanctions, 1 Fed. Imm. L. Rep. (CCH) Update No. 1, at 3 (Feb. 28, 1983); Coleman, Hispanics Hit Immigration Revision, Wash. Post, Feb. 26, 1983, at A9, col. 1.

248. See generally Schuck, Book Review, 90 Yale L.J. 702, 717-20 (1981) (reviewing The Politics of Regulation (J. Wilson ed. 1980)).

249. Smothers, Two Parties Woo Votes of Hispanic Americans, N.Y. Times, Sept. 17, 1983, at A7, col. 1.

ently go to those states as well,[250] and despite efforts to distribute them more evenly, most refugees have also migrated there.[251]

This geographic concentration, of course, is hardly surprising. Immigrants, like most other people, tend to go where there are jobs and people of similar background who can help them to find work, establish social ties, and begin the difficult transition into the American culture. For present purposes, however, the more far-reaching implications are that immigration policy will acquire an increasing political urgency, and that aspects of immigration law that impose significant costs upon those regions are not likely to be tolerated for long.

Perhaps the most important political change is related to the above factors but is not fully explained by them. There now exists a shared conviction that the traditional immigration system, especially in light of the structural changes described earlier, has become fundamentally incoherent; its assumptions and values seem to diverge from the evolving moral basis of the larger legal system of which it has become an increasingly insular, incongruous and discordant part. Public opinion polls indicate a growing anxiety over the inability of the government to control the influx of undocumented workers.[252] The first comprehensive legislative overhaul of immigration law in more than thirty years, the Simpson-Mazzoli bill,[253] demonstrated broad political support before its recent derailment.[254] This legislation, to be sure, would leave much of the body of classical immigration law intact; in some respects, such as the provisions for summary exclusion,[255] it would further entrench some of the classical assumptions. Nevertheless, it would fundamentally alter the way in which the law addresses the major contemporary challenges to the classical immigration system, especially the problems of undocumented aliens and asylum.[256] Another piece of evidence that a perceived incongruity exists between traditional assumptions and the changing social order lies in the recent judicial efforts to dismantle the structure of classical immigration law. This development, which is the subject of Part III of this Article, not only reflects but is helping to transform the politics of immigration.

250. American Enterprise Institute for Public Policy Research, Illegal Aliens: Problems and Policies 12 (1978).

251. An estimated 60% of refugees who have come from Southeast Asia since 1975 are concentrated in forty counties. Placement Policy Task Force, Office of Refugee Resettlement, Dep't of Health and Human Servs., Concept Paper on Refugee Placement Policy (1981). The concentration of Cuban, Haitian and Salvadoran refugees in south Florida and New York, of course, is especially striking.

252. See Harwood, supra note 239.

253. S. 529, 98th Cong., 1st Sess., 129 Cong. Rec. S6970 (daily ed. May 18, 1983).

254. The Senate passed the Simpson-Mazzoli bill on May 18, 1983, by a 76-18 vote. See 129 Cong. Rec. S6969-70 (daily ed. May 18, 1983). In the House, the bill has not yet reached the floor for debate. See supra note 98 and accompanying text.

255. S. 529, 98th Cong., 2d Sess. § 121(a), 129 Cong. Rec. S6970, S6973 (daily ed. May 18, 1983).

256. Id. tit. I, pt. C; id. tit. III.

B. *Ideological Changes*

We saw in Part I that classical immigration law arose within a legal order in which nationalistic, exclusionary values were superimposed upon the earlier individualistic and universalistic ideology of traditional liberalism. The resulting ideology of restrictive nationalism emphasized three related aspects of the government-alien relationship. The government's legal obligation to aliens rested almost entirely upon the terms and conditions upon which its consent to their entry had been granted and to which they had at least implicitly agreed. This consent principle legitimated, and in turn was legitimated by, a strong conception of sovereignty, in which the alien's relationship with American society was defined in legal, formal terms prescribed by the government. Finally, membership in the national community, while nominally universal for those within the United States, was in fact and in law graduated and restrictive for aliens. In recent years, however, profound changes have been wrought in the individualistic underpinnings of these basic conceptions, especially the idea of consent-based legal obligation. In the process, the moral and legal foundations of classical immigration law have eroded.

1. *Individualism and the Right-Privilege Distinction.* — Duncan Kennedy's analysis of the individualistic legal order affords some insight into the evolving legal meanings of consent, sovereignty and national community in classical immigration law.[257] Although his work concerns private-law contract rules, it isolates certain fundamental values that have also shaped the ideology of immigration law. To Kennedy, conflicts in the law over three recurring issues reveal the nature of individualistic values. The first concerns the extent to which one party should be obliged against its will to defer to another's interest. Individualism assumes that people are independent and self-sufficient. It seeks to limit reciprocal obligations of sharing and sacrifice, the scope of legal duties, and the force of equitable claims.[258] The second concerns the extent to which the parties' bargaining power should determine legal outcomes. An individualistic system recognizes only limited constraints upon the balance of forces in the state of nature, and accepts the outcomes that even highly unequal bargaining power generates.[259] The third issue concerns the extent to which a court is justified in overriding a party's expressed intention in the interest of some broader notion of justice. Individualism denies any such justification.[260]

The fundamental elements of classical immigration law analyzed in Part I substantially reflected an individualistic ideology emphasizing the ideas of consent, sovereignty and restrictive national community. Not being bound by any transcendent conception of community, for example, the government was

257. Kennedy, Form and Substance in Private Law Adjudication, 89 Harv. L. Rev. 1685 (1976).

258. Id. at 1733-35.

259. Id. at 1735-36.

260. Id. at 1736-37.

free to define the American commonwealth as it saw fit, acknowledging only those claims or interests to which it was prepared, out of considerations of national interest, to consent. As *Downes v. Bidwell (The Insular Cases)*[261] and the qualified legal rights of women and racial minorities during this period suggest,[262] this design was often restrictive in practice even when it was expansive in principle. A concept of national sovereignty that applied the individualistic values of autonomy, self-determination, and consent in the immigration context, where nationalistic goals challenged traditional liberal ideals and altered their meanings,[263] would also find congenial the other substantive elements of the classical tradition—the extraconstitutional status of exclusion, plenary federal power to classify aliens, "civil" deportation, broad detention power, and nonindependent adjudication. Rhetorically, it would seek some principle to harmonize these doctrines and render them coherent.

Restrictive nationalism, the ideology underlying classical immigration law, found such a principle in the "right-privilege" distinction. This idea was sententiously formulated by Justice Holmes, who rejected a policeman's first amendment challenge to his discharge by observing that "[t]he petitioner may have a constitutional right to talk politics, but has no constitutional right to be a policeman. . . . The servant cannot complain, as he takes the employment on the terms which are offered him."[264] Like many other areas of public law, immigration doctrine found the power of this principle to be irresistible. If government employment was not a "right" but only a "privilege," defeasible at the instance of the employer without constitutional impediment, how much more contingent was the alien's ability to enter the United States and enjoy equal benefits there, especially under a regime of sharply restricted immigration?[265] By conceiving of entry as a privilege conditioned upon the alien's consent to the decidedly limited rights and inferior status that classical immigration law offered her, restrictive nationalism sought to reconcile the illiberal elements of classical immigration law with the individualistic values that it drew from traditional liberalism. Indeed, when the "demise" of the right-privilege distinction in the law generally was announced during the 1960's[266]—quite prematurely, as it turned out[267]—immigration law took little notice. In

261. 182 U.S. 244 (1901), discussed supra text accompanying note 47.

262. See supra notes 29-34 and accompanying text.

263. See supra notes 9-11 and accompanying text.

264. McAuliffe v. Mayor of New Bedford, 155 Mass. 216, 220, 29 N.E. 517, 517-18 (1892).

265. That initial entry by an alien is quintessentially a privilege, explicitly characterized as such, has been a cardinal principle from the Court's earliest immigration decisions, see, e.g., The Chinese Exclusion Case, 130 U.S. 581 (1889), down to the Court's last Term, see Landon v. Plasencia, 103 S. Ct. 321 (1982).

266. See Van Alstyne, The Demise of the Right-Privilege Distinction in Constitutional Law, 81 Harv. L. Rev. 1439 (1968), and cases discussed therein.

267. To oversimplify a bit, the distinction continues to do service, albeit often sub rosa, in the Court's determinations concerning whether a protected property or liberty interest exists and

that isolated realm, and especially with respect to doctrines relating to exclusion, the distinction remained a seductive principle through which the dominant ideas of consent, sovereignty, and national community could be vindicated.

Restrictive nationalism's use of individualistic values to validate whatever outcomes existing power differentials happened to produce—its refusal to disturb the conditions imposed by the government upon aliens—left little independent role for the courts. Rather than legitimating judicial infusion of equitable, extraconsensual, or communitarian values into immigration law, individualism—or at least that philosophical thread of individualism that emphasized consent as the source of legal obligation—seemed to demand the kind of abject judicial deference to the sovereign's immigration authorities that, as Part I suggested, became a hallmark of the classical tradition.

2. *Toward a Communitarian Legal Order.*—During the last decade, however, the ideological stability and coherence of classical immigration law have been threatened simultaneously from two different directions. First, our legal system has dramatically expanded the circumstances under which the law imposes legally enforceable duties upon both government and private individuals or groups. Second, there has been a related movement, especially but not exclusively in public law, from a preoccupation with the rights of individuals to an increasing emphasis upon the vindication of group rights.[268]

In the classical view, as we have seen, government owed essentially no duty to aliens that it had not already imposed upon itself through positive law. This idea of autonomy married a central theme in liberal ideology—the primacy of consent as the source of legal obligation—to a growing nationalistic concern with America's sovereignty in its burgeoning international relations. But liberalism has always contained another, somewhat contradictory theme—the universality of certain natural rights.[269] Such rights are anchored not in the accidental contingency of one's birthplace or previous nationality, but in the common humanity, rationality and dignity of all individuals.

The legal order, animated by the emergence of new, "communitarian" public law norms,[270] has gradually begun to generalize from this natural rights dimension of liberalism. These norms are expanding and transfiguring the sources of, and justifications for, legal obligation to individuals whom public and private law traditionally conceived of as "strangers." These norms,

---

if so, what process is due before it can be infringed. See, e.g., Smolla, The Reemergence of the Right-Privilege Distinction in Constitutional Law: The Price of Protesting Too Much, 35 Stan. L. Rev. 69 (1982), and cases discussed therein.

268. The historical character of this phenomenon is traced in Yeazell, Group Litigation and Social Context: Toward a History of the Class Action, 77 Colum. L. Rev. 866 (1977).

269. See Fishkin, The Boundaries of Justice, 27 J. Conflict Resolution 355, 358-61 (1983) ("To the extent one gives independent weight . . . to a special class of persons within the [nation's] boundary, one is departing from the liberal paradigm for theories of justice."); supra note 33 and accompanying text.

270. See supra text following note 12.

however, are grounded in a fundamentally nonliberal intuition—the perception that individuals, societies and nations are bound to each other by pervasive interdependencies. These interdependencies are thought to imply certain moral and legal consequences that neither liberalism nor restrictive nationalism would accept. They imply that socially accepted values should augment consent as a basis for imputing legal duties; that the conception of national sovereignty should be weakened in order to define the relationship between the United States and aliens in terms of morally significant, informal social interactions; and that membership in our national community should depend not upon formalistic criteria but upon the functional social linkages actually forged between aliens and the American people. Thus, communitarian norms grow out of the universal human rights aspect of traditional liberalism. They diverge, however, both from liberalism's individualistic ethos and from the restrictive nationalistic values that underlie classical immigration law.[271]

a. *The Scope of Obligation in Private Law.* — Well before immigration law began to feel the influence of communitarian values, these values had already limited the sway and reshaped the contours of the consent principle in private law. The most notable effect occurred in contract law, where the bargain principle, which imposed legal obligation only for consensual, intentional exchanges, was augmented by more encompassing theories of obligation that have nothing to do with consent, such as detrimental reliance and unjust enrichment. This movement from contract to tort [272] premises liability not only upon the expressed will of the parties but also upon widely shared norms concerning what is fair. It recognizes that individuals are neither independent nor self-sufficient, and that spillover effects pervade not only the marketplace but the moral domains of our collective life. Our concern for the welfare of others clearly transcends the realm of contract, extending to non-market values such as procedural fairness and humane treatment as well. Courts increasingly eschew the individualistic liberal values of autonomy and self-determination in favor of communitarian values emphasizing paternalistic protection and social justice. For example, they now commonly refuse to enforce contracts of adhesion,[273] and often relieve promisors of the consequences of mistakes.[274] Ian MacNeil has sought to capture the essence of these

271. Kennedy's "altruist" ethos, which he opposes to individualism, includes values similar to those that I have termed "communitarian." Altruism, however, implies a particular kind of motivation that may or may not underlie communitarian values. Moreover, Kennedy applies his theoretical framework only to the private law of contracts. Kennedy, supra note 257, at 1685.

272. G. Gilmore, The Death of Contract 65, 87-90 (1974). Historically, one might regard this as the next stage of an earlier movement, that from status to contract. See H. Maine, Ancient Law 163-65 (1864).

273. See Rakoff, Contracts of Adhesion: An Essay in Reconstruction, 90 Harv. L. Rev. 1174, 1190-97 (1983); Lublin, Legal Challenges Force Firms to Revamp Ways They Dismiss Workers, Wall St. J., Sept. 13, 1983, at 1, col. 6.

274. Compare G. Gilmore, supra note 272, at 42-44 (discussing the traditional view), with Kronman, Mistake, Disclosure, Information and the Law of Contracts, *in* The Economics of Contract Law 114-121 (1979) (discussing the emerging view).

developments, writing of a change from "discrete contract" to "relational contract" conceptions, in which social structures, relationships and values play a crucial role in defining the scope and content of legal obligation.[275] More generally, an increasing preoccupation with substantive, rather than formal, criteria of justice is leading courts to emphasize flexible, tort-like standards such as "reasonableness" and "fairness," standards more capable of assimilating communitarian norms.[276]

In tort law proper, an analogous evolution has occurred. During most of the classical period, tort liability was relatively confined, reflecting a highly focused, restrictive definition of the individual injury-producing event and employing narrow criteria of relevance. Increasingly, however, tort obligation grows out of the larger social context in which such interactions occur. The entire matrix of social relationships—the expectations and nonmarket values that they generate, such as distributive justice, economic efficiency, social equality, and human dignity—have become relevant to the scope and content of legal duties. Landlords, by reason of their presumably superior knowledge, access, and power, must compensate losses due to crimes perpetrated on their property.[277] Property owners, for much the same reasons, are obliged to protect trespassers against certain injuries sustained on their premises.[278] Similarly, psychiatrists have a duty to prevent assaults by their patients,[279] physicians must obtain informed consent,[280] and novel fiduciary obligations have been extended to new groups such as insurers.[281]

b. *The Scope of Obligation in Public Law.*—Thus, the domain within which private autonomy, action and property may be exploited without regard to the interests of strangers has been progressively abridged by judicially imposed duties. It does not require a great leap of imagination to suppose that the same moral ideas and legal consciousness that generated such a pronounced restriction of individual sovereignty in private law have begun to influence the evolving conception of governmental sovereignty in the public law of immigration as well. Such an influence has been reinforced by parallel developments in public law itself, where a counterpart to the movement from contract to tort has fundamentally altered administrative and constitutional law conceptions of obligation during recent decades.

Richard Stewart has chronicled an evolution in administrative law away from what he calls the "traditional" model. In the traditional model, private

---

275. I. MacNeil, The New Social Contract: An Inquiry into Modern Contractual Relations (1980).

276. See Kennedy, supra note 257, at 1751.

277. See, e.g., Trentacost v. Brussel, 82 N.J. 214, 412 A.2d 436 (1980).

278. Restatement (Second) of Torts § 335 (1965); Harper & James, supra note 17, at 1443-44.

279. See, e.g., Tarasoff v. Board of Regents, 17 Cal. 3d 425, 551 P.2d 334 (1976).

280. See, e.g., Cobbs v. Grant, 8 Cal. 3d 229, 502 P.2d 1 (1972).

281. See, e.g., Employee Retirement Income Security Act of 1974, §§ 401-404, 29 U.S.C. §§ 1101-1104 (1976 & Supp. V 1981) (fiduciary duties of insurers).

property rights are taken as a given. The state may intrude upon those rights only to the extent that the legislature has expressly commanded it, and administrative decision procedures and judicial review must be tailored to facilitate the enforcement of that limitation.[282] The palpable inadequacy of this model, Stewart argues, lies in its failure to come to grips with the inescapably problematic nature of administrative discretion in the contemporary welfare state.[283] This failure has recently given rise to what Stewart calls an "interest representation" model of administrative law. In this model, the state, before deciding whether and how to act, is obliged to establish and enforce administrative and judicial procedures that are designed to generate a plausible approximation of the "public interest"—that is, procedures that facilitate participation by all interests that may be affected by those decisions.[284]

This "interest representation" model has assumed many doctrinal forms, such as judicial and statutory rules expanding opportunities for standing, intervention, and fee-shifting, and more searching judicial review that holds administrative officials to higher standards of rationality and procedural justice.[285] For purposes of the present discussion, perhaps the most far-reaching consequence of this reconceptualization of administrative law is the emergence of group litigation as a crucial strategy of legal change. This development has enlarged not only the number and types of claimants but the kinds of claims that may be pursued against government in the courts.[286] The vast expansion of group rights and remedies against governmental entities and officials has generated new forms of injunctive relief,[287] and secured new conceptions of property rights.[288]

Even more than its counterpart in private law, this evolution of adminstrative law doctrine toward the vindication of group rights has increasingly emphasized relational, social expectancy, group, and other communitarian values, in addition to individual rights based upon traditional notions of private property.[289] Group litigation has not only made it economically possible to challenge governmental policies more effectively and to press new kinds of values on courts; it has also raised the stakes in judicial decisions, making it more difficult for courts to regard cases as isolated disputes on the private-law model. Such cases have helped to reshape judges' conceptions of their role, of the sources of governmental misconduct, of the functions of litigation, and of

282. Stewart, The Reformation of American Administrative Law, 88 Harv. L. Rev. 1669, 1671-76 (1975).

283. Id. at 1676-88.

284. Id. at 1711-802.

285. Chayes, The Role of the Judge in Public Law Litigation, 89 Harv. L. Rev. 1281, 1304-13 (1976); Stewart, supra note 282, at 1723-56.

286. See, e.g., Motor Vehicle Mfrs. Ass'n v. State Farm Mut. Auto. Ins. Co., 103 S. Ct. 2856 (1983).

287. P. Schuck, supra note 18, ch. 7; Chayes, supra note 285, at 1298-301.

288. See generally Reich, The New Property, 73 Yale L.J. 733 (1964).

289. See generally Chayes, supra note 285; Stewart, supra note 282.

the remedial obligations of government.[290] As we shall see in Part III, these transformations wrought by the rise of group litigation and group rights have been especially significant in a number of the important class action immigration cases of the early 1980's.[291]

The reformation of administrative law that Stewart describes has been paralleled by a radical change in constitutional law, particularly in the Supreme Court's understanding of what constitutional due process and equal protection requires. These developments generally have received enormous scholarly attention[292] and need not be repeated here. Two aspects, however, are particularly relevant to immigration law. First, constitutional law has accorded greater prominence to precisely those values—procedural fairness and equal treatment—that the consent, sovereignty and restrictive national community conceptions of classical immigration law tended to depreciate. Second, it extended protection to many groups, including criminal defendants, convicted prisoners, and students, that are not obviously more deserving or needful of special judicial protection than the aliens to whom classical immigration law denied significant protection.

c. *The Implications for Immigration Law.* — Each of these ideological changes, then—in private law, administrative law, and constitutional law—has contributed to the erosion of the moral and legal foundations of classical immigration law. Each has responded to deeper changes in social structures and patterns of legitimation. Each reflects the waning of the individualistic emphasis upon consent as the principal basis of legal obligation, the rise of communitarian values, and the rapid growth of the administrative state. Just as the imposition of private obligation has come to be justified by fluid, relational, tort-like norms, so the closed, formalistic, property-oriented values of administrative and constitutional law have yielded to those emphasizing open processes and the protection and advancement of diffuse nonmarket values.[293] These changes do not simply constitute significant evolutions in legal doctrine. They exemplify a fundamentally changed understanding of moral, political, and social reality, an understanding from which immigration law can no longer be isolated.

290. See P. Schuck, supra note 18, at 1-29; Chayes, supra note 285, at 1284; Fiss, The Supreme Court, 1978 Term—Foreword: The Forms of Justice, 93 Harv. L. Rev. 1, 24-28 (1979).

291. See, e.g., infra notes 294-339, 391 and accompanying text. Group litigation on behalf of aliens was made possible by, and itself facilitated, the creation of "public interest" law centers specializing in aliens' rights, such as the Haitian Refugee Center in Florida.

292. See, e.g., J. Ely, Democracy and Distrust 14-22 (1980).

293. MacNeil calls these "supracontract norms," MacNeil, supra note 275, at 70, while Kennedy, with much the same phenomena in mind, refers to "altruistic" values, Kennedy, supra note 257, at 1717-22. For discussions of nonmarket values in administrative law, see Stewart, Regulation in a Liberal State: The Role of Non-Commodity Values, 92 Yale L.J. (July 1983 forthcoming); Schuck, Regulation, Non-Market Values and the Administrative State, 92 Yale L.J. (July 1983 forthcoming).

III. The Communitarian Conception of Immigration Law

In Part I, I distilled the fundamental, distinctive ideological basis and doctrinal elements of classical immigration law that were developed by the Supreme Court as early as the 1880's and confirmed by it well into the 1970's. In Part II, I suggested some structural and ideological changes that are beginning to cause different principles, already well-established in other areas of law, to penetrate immigration law. It remains to examine the lines along which the basic elements of immigration law are being transformed today, a task undertaken in this Part, and to consider in Part IV some of the implications of these transformations for the future character and consequences of immigration law.

A. *Broadening the National Community:* Plyler v. Doe

The Supreme Court's recent decision in *Plyler v. Doe*,[294] compelling a state to provide the children of undocumented aliens with a free public education, may ultimately come to have the same epochal significance for that group that *Brown v. Board of Education*[295] has had for black Americans. In one sense, of course, *Plyler* may be read narrowly as nothing more than an equal protection case in which the state failed to adduce a "substantial purpose" for treating the children of undocumented aliens differently than it treated their documented alien or citizen peers.[296] In this view, the question of how the national community ought to be defined was not even at issue: *Plyler*, after all, invalidated a state law, not a federal one, and only the latter can prescribe the contours of that community. But in another, broader sense, the decision may mark a fundamental break with classical immigration law's concept of national community and of the scope of congressional power to decide who is entitled to the benefits of membership.

Plyler effects this change on three different levels. First, it inducts into the national community a new group of uncertain size and composition. These uncertainties do not simply reflect our ignorance about which aliens are here illegally and how numerous they are, for even the defining characteristic of the new group remains opaque. Previous expansions of the national community, most notably the fourteenth amendment's extension of citizenship to the freed slaves[297] and the nineteenth amendment's extension of suffrage to women,[298]

294. 457 U.S. 202 (1982).

295. 347 U.S. 483 (1954).

296. *Plyler*, however, appears to depart from conventional equal protection analysis. The Court acknowledged that, strictly speaking, the case involved neither a "suspect class" nor a "fundamental right." 457 U.S. at 223. Under the Court's precedents, therefore, it should have simply asked whether the state statute had a "rational basis." See, e.g., San Antonio Indep. School Dist. v. Rodriguez, 411 U.S. 1 (1973); Railway Express Agency v. New York, 336 U.S. 106 (1949). Instead, the Court insisted that Texas could sustain its statute only by showing that "it furthers some substantial goal of the State." *Plyler*, 457 U.S. at 223. The doctrinal anomaly thus presented lends some credence to a broader interpretation of *Plyler*.

297. U.S. Const. amend XIV.

298. Id. amend XIX.

admitted well-defined, undifferentiated categories of individuals; within those categories, membership was universal. But neither *Plyler*'s holding about public benefits nor its more general spirit necessarily embraces undocumented aliens as an undifferentiated group. To be sure, the Court reaffirmed a principle that had been clearly established for a century[299]—that all aliens "within" the United States, whatever their legal status, are entitled to some constitutional protection. But it was the "innocent" children, not their "guilty" parents, to whom Texas denied equal protection; although it is children, not parents, who attend public school, the unconstitutionality of the state law involved more than the withholding of educational benefits from aliens in general. Had Texas adopted an analogous system of free public *adult* education or job training, for example, the Court apparently would not have upheld an equal protection claim by the *parents*.[300] The enduring importance of education benefits for *children*, moreover, accounted both for the level of scrutiny selected and the extent of the right created.[301]

These narrowing distinctions—between undocumented alien parents and their children, and between education and other services—raise a number of significant problems with the Court's reasoning, and suggest that those distinctions may be difficult to sustain in future cases. For example, the Court failed to explain why denying educational benefits to an innocent child differs from the denial of other governmental benefits to her undocumented parent, upon whose income and well-being the child's welfare ultimately depends. If Texas may not harm innocent children by depriving them of access to public education, why may it harm these same innocents—who may actually be United States *citizens*—even more grievously by denying their parents access to welfare benefits or public or private employment, which provides their essential economic support?[302] Does *Plyler* mean that a state may not exclude from public housing a family in which an adult has committed a crime? Can the Court's "innocent children" rationale be harmonized with its willingness to allow the routine deportation of children who are not only innocent of their parents' wrongdoing but are United States citizens?[303] Are not these depriva-

299. See The Japanese Immigrant Case, 189 U.S. 86 (1903), discussed supra note 77.

300. The plaintiff children in these cases are special members of this class of undocumented aliens. Persuasive arguments support the view that a state may withhold its beneficence from those whose very presence within the United States results from their own unlawful conduct. These arguments do not apply with the same force to classifications imposing disabilities on the minor children of such illegal entrants. Those who elect to enter our territory by stealth and in violation of our law should be prepared to bear the consequences. But their children are not comparably situated. "[Parents] 'have the ability to conform their conduct to societal norms,' and presumably the ability to remove themselves from the State's jurisdiction; but [their] children . . . 'can affect neither their parents' conduct nor their own status.'" *Plyler*, 457 U.S. at 220 (quoting Trimble v. Gordon, 430 U.S. 762, 770 (1977)).

301. Id. at 221-23.

302. See De Canas v. Bica, 424 U.S. 351 (1976).

303. See supra note 52.

tions as directly a result of parental status or wrongdoing as the deprivation that the Court refused to countenance in *Plyler*? Because the Court addressed none of these questions, the potential scope of *Plyler* remains an open question.

Second, *Plyler* not only enlarged the national community to uncertain dimensions and on the basis of uncertain principles, but did so in the face of a congressional policy to exclude undocumented aliens from the country *altogether*. Some have doubted that such a policy can even be said to exist, given the historical ambivalence of Southern States toward the entry of low-cost agricultural labor.[304] In *Plyler*, the Court went so far as to deny that the Texas statute "corresponds to any identifiable congressional policy," or "operate[s] harmoniously within the federal program."[305]

This proposition is demonstrably false, at least under current conditions. Unauthorized entry, after all, is a federal crime.[306] In recent years, considerable law enforcement effort has been devoted to preventing it.[307] Congress has excluded undocumented aliens from numerous federal benefit programs.[308] Indeed, the Court only six years earlier in *De Canas v. Bica*[309] had deemed Congress's policy against illegal aliens to be strong enough to sustain a state statute barring their employment; it was enough for the Court that the statute was consistent with federal law.[310] The *Plyler* majority's effort to distinguish *De Canas* is utterly unpersuasive.[311] Its other arguments—that because some undocumented aliens might ultimately escape deportation and because denial of benefits did not stem the influx of illegal aliens, Texas's approach was thereby rendered irrational[312]—stand on no firmer ground. If substantial effectiveness were the criterion, most regulatory statutes would probably be

304. See, e.g., Lopez, supra note 12, at 645-46 (noting historical pattern of recruiting of Mexican aliens in Southwestern States; Developments, supra note 4, at 1453 (discerning an "unspoken policy of inviting undocumented aliens into the country to contribute to the economy").

305. *Plyler*, 457 U.S. at 225-26.

306. 8 U.S.C. § 1325 (1982).

307. In fiscal 1983, the Border Patrol apprehended one million aliens attempting to cross the border illegally. Wash. Post, Oct. 22, 1983, at A3, col. 1. On October 21, 1983, the INS Commissioner announced that next year he will seek supplemental appropriations for a 50% increase in Border Patrol personnel. Id.

308. See General Accounting Office, Report to the Congress by the Comptroller General of the United States: Illegal Aliens: Estimating their Impact on the United States 22 (1980); General Accounting Office, Report to the Senate Committee on the Budget by the Comptroller General of the United States: Impact of Illegal Aliens on Public Assistance Programs: Too Little is Known (1977).

309. 424 U.S. 351 (1976).

310. Id. at 356-63.

311. *Plyler*, 457 U.S. at 225-26. Indeed, in *De Canas*, unlike in *Plyler*, there was a plausible ground for arguing that the state law was in fact inconsistent with the immigration statute, yet the Court sustained the former as "consistent" with the latter. 424 U.S. at 363-65.

312. *Plyler*, 457 U.S. at 226.

doomed.[313] Certainly, there are few areas in which the public policy options are as problematic as in the control of illegal immigration. What is disturbing about *Plyler* is not that it makes the policymaker's task more difficult—any judicial invalidation of a statute does that. *Plyler*'s peculiar vice is that it uses the inherent difficulty of a problem as a justification for making it even more intractable.[314]

Third, *Plyler* could perhaps be understood and even defended as a case of federal preemption that merely reaffirms the exclusive, plenary power of Congress to regulate immigration. This view, however, is undercut by the Court's previous decision in *De Canas*. That case announced a broad tolerance for state legislation that discourages immigration in areas traditionally of local concern, such as labor markets and public education, in ways that are generally consistent with federal policy. Unless this aspect of *De Canas* has been overruled sub silentio, *Plyler* must be seen as the germination of a new and quite different principle. This principle seems to be that a state may not seek to discourage illegal entry by means of disincentives that may harm the children of those who, because the disincentives are ineffective, decide to enter anyway. This principle, of course, would have dictated a different result in *De Canas*, since it would have invalidated the statute in that case.[315] Even more important, the principle would seem to require that result even if *Congress*, rather than a state, enacted the statute.[316]

---

313. The employer sanction provision that is the centerpiece of the pending immigration reform law probably cannot meet such a test either. General Accounting Office, Report to the Senate Subcommittee on Immigration and Refugee Policy, Committee on the Judiciary by the Comptroller General of the United States: Administrative Changes Needed to Reduce Employment of Illegal Aliens 22 (1981). For comparison with foreign countries, see General Accounting Office, Report to the Senate Subcommittee on Immigration and Refugee Policy, Committee on the Judiciary by the Comptroller General of the United States: Information on the Enforcement of Laws Regarding Employment of Aliens in Selected Countries (1982).

314. See also Ruiz v. Blum, 549 F. Supp. 871 (S.D.N.Y. 1982) (invalidating restrictions on participation by illegal aliens in publicly subsidized day care).

315. By denying parents employment, the statute in *De Canas* surely harmed their innocent children. It is possible, of course, to argue that the relationship between the statute and the harm to children was significantly less direct in *De Canas* than in *Plyler*. But both harms are eminently foreseeable, intended as disincentives to illegal immigration, and therefore direct. On the other hand, a distinction might be drawn between the immediate economic harm caused by the statute in *De Canas*, see supra note 302, and the real but more, attenuated economic harm that would result from a statute that denied, for example, educational benefits to undocumented alien parents, see supra note 300.

316. To be sure, the Court analyzes *Plyler* as a fourteenth amendment case, continues to cite approvingly Mathews v. Diaz, 426 U.S. 67 (1976), and emphasizes the plenary power of Congress in this field, see *Plyler*, 457 U.S. at 210-25. Nevertheless, much of the *Plyler* Court's equal protection analysis would seem to cast doubt upon a congressional enactment that barred undocumented children from public schools receiving federal assistance. This is particularly true of the arguments about innocent children, id. at 219-20, the uncertainty of deportation for any particular illegal alien—what the Court calls "an inchoate federal permission to remain," id. at 226—the policy's ineffectiveness, id. at 228, the de minimis effect on educational quality, id. at 229, and the long-term effects of deprivation, id. at 230.

If that is the meaning of *Plyler*, it constitutes a conceptual watershed in immigration law, the most powerful rejection to date of classical immigration law's notion of plenary national sovereignty over our borders. If those to whom we have refused entry are entitled by their mere presence—together with the presence of their children—to claim not only constitutional procedural protections,[317] but also the significant substantive entitlements that legislatures grant to lawful residents,[318] then immigration law's ideal of national community has also been transformed. In the classical view, the political branches of government defined the boundaries of that community and the consequences of exclusion. That definition, responsive to the political values, local interests and policy concerns that animate Congress, has tended to exclude those whose entry was believed to threaten these interests. The most important meaning of *Plyler* may be that the courts are beginning to assert a coordinate, if not supervisory, role in defining the dimensions and meaning of national community in the immigration context. Courts are expositors of a constitutional tradition that increasingly emphasizes not the parochial and the situational, but the universal, transcendant values of equality and fairness imminent in the due process and equal protection principles. In that capacity, they have also asserted a larger role in the creation and distribution of opportunities and status in the administrative state. In *Plyler*, the Supreme Court moved boldly on both fronts. In doing so, the Court seems to have begun to redefine the community to include all those whose destinies have somehow, even in violation of our law, become linked with ours.

B. *Judicial Self-Assertion*

On this reading of *Plyler*, the days of extraordinary judicial deference to Congress in the immigration field may be over. Still, by setting *De Canas* to one side, it is possible to view *Plyler* not as a case about the contours of community or the equal protection principle but simply as a straightforward federal preemption case dressed up in equal protection garb, the kind of state regulation of immigration that courts have long rejected. But even if one rejects the view that such a reading of *Plyler* may be too narrow, the judicial boldness apparent in the increasing number of recent decisions confronting *federal* exercises of power over immigration cannot be dismissed so lightly.

To be sure, the emergent judicial assertiveness in the face of federal regulation of immigration essentially remains a lower court phenomenon,

---

317. Earlier decisions, beginning with The Japanese Immigrant Case, 189 U.S. 86 (1903), had conferred certain procedural rights upon excludable or deportable aliens, but not substantive rights.

318. The Texas statute in *Plyler* extended education benefits to lawfully admitted aliens and citizens, including children born here to illegal aliens, 457 U.S. at 245 n.4 (Burger, C.J., dissenting), and aliens who were in the process of obtaining documentation, id. at 205 n.2 (Opinion of the Court).

Plyler aside, as does most of the transformation of immigration law to date.[319] Moreover, some of these cases have arisen in a deportation context, in which due process rights of aliens are well-established. Significantly, however, many innovative rulings involve exclusion proceedings,[320] where the powers of Congress and the Attorney General are at their greatest.

One very recent district court decision, *Hotel & Restaurant Employees Union, Local 25 v. Smith*,[321] suggests the extraordinary flavor of these cases. The unconventional procedural posture in which the court decided the case is itself a sign of deeper changes. *Employees Union* did not arise in an exclusion or deportation proceeding, nor did it involve a detained plaintiff. Indeed, the plaintiff was not an alien at all but a union local, and it challenged not a particular adjudication but a general administrative practice known as "extended voluntary departure" (EVD). Under this practice, the Attorney General may accede to a State Department request that no action be taken to deport nationals of a particular country due to dangerous, unsettled conditions there, such as civil war, invasion or repression.[322] EVD is generally used in situations in which returning the alien to her country of origin might be hazardous but in which asylum probably cannot legally be granted.[323] EVD,

319. The Supreme Court's epochal decision in INS v. Chadha, 103 S. Ct. 2764 (1983), cannot be counted an exception. *Chadha* invalidated the legislative veto provision contained in 8 U.S.C. § 1254(c)(2) (1982), part of the immigration statute's suspension-of-deportation section, holding it severable from the rest of the act. But the decision was clearly preoccupied with broad constitutional principles rather than with anything peculiar to immigration law. The Court's criticisms of the legislative veto are unquestionably stronger with respect to the narrow, nonpolicy, adjudicative decision presented in the deportation setting than they are in most other statutory contexts, where broad national policy is at issue. See, e.g., Elliott, INS v. Chadha: The Administrative Constitution, the Constitution, and the Legislative Veto, 1983 Sup. Ct. Rev. (forthcoming). Indeed, were it not for the conventional characterization of deportation as a "civil" proceeding, see supra notes 130-36 and accompanying text, the legislative veto as applied in *Chadha* might well constitute a bill of attainder. Nevertheless, the Court resisted the suggestion, see 103 S. Ct. at 2788-89 (Powell, J., concurring); id. at 2792-93 (White, J., dissenting) (both objecting to the breadth of the holding), that it limit its holding to the particular statute before it. If all legislative veto provisions violate the Constitution, as the Court seems to believe, no plausible basis exists for excepting the immigration statute.

For what it is worth, the Court reaffirmed "the plenary authority of Congress over aliens." It may or may not be significant that the Court then immediately added that Congress has plenary authority *whenever* it has legislative jurisdiction. Id. at 2779.

320. See e.g., Fernandez-Roque v. Smith, 567 F. Supp. 1115 (N.D. Ga. 1983).

321. 563 F. Supp. 157 (D.D.C. 1983).

322. 8 C.F.R. § 244.2 (1983). Plaintiff also challenged another alleged practice—the routine recommendation of denial of El Salvadoran asylum claims without regard to their individual merits. *Employees Union*, 563 F. Supp. at 161; Plaintiff's Memorandum of Points and Authorities in Opposition to Motion to Dismiss at 16-17. The court failed to discuss this claim.

323. Asylum may be unavailable for any of a number of reasons. The threatening conditions, for example, may not constitute a kind of "persecution" enumerated by the Refugee Act of 1980, § 201, 8 U.S.C. § 1101(a)(42) (1982). They may be too generalized or random to satisfy the Act's requirement that asylum be granted only for persecution directed at the particular individual who claims that status. 8 U.S.C. § 1158. The claimant may have been "firmly resettled" elsewhere and therefore ineligible for asylum hereunder. § 1101(a)(42). See, e.g., Rosenberg v. Yee Chien Woo, 402 U.S. 49 (1971); In re Portales, 18 I. & N. Dec. No. 2905 (BIA 1982).

then, provides temporary refuge for an indefinite period to all nationals of that country who are here at a particular time. Since 1960, it has been granted to nationals of fifteen countries; today, it is enjoyed by Poles, Afghans, Ethiopians, and Ugandans, among others.[324]

In *Employees Union*, the plaintiff challenged the Attorney General's decision denying EVD status to Salvadorans while granting it to other groups.[325] It alleged violations of section 4 of the Administrative Procedure Act,[326] and of the constitutional guarantees of equal protection and due process.[327] The Attorney General moved to dismiss the complaint, contending that a decision to grant or deny EVD was a highly delicate, discretionary judgment based upon a variety of difficult and confidential foreign policy considerations. To be compelled to articulate those considerations publicly, he maintained, would hamper the President's conduct of foreign policy and jeopardize the interests of Americans and other innocent people outside the United States. EVD decisions, he insisted, are made in close consultation with Congress; for the courts to second-guess such decisions would constitute an unwarranted intrusion into the delicate realm of high-stakes foreign policy in which it was essential that the United States speak with one voice.[328]

In the context of EVD decisions, these foreign policy and national sovereignty arguments have great force and plausibility, certainly far more than the similar arguments advanced by the government and readily embraced by the Supreme Court in some deportation cases decided during the 1970's.[329] The purpose of EVD, after all, is to create a flexible alternative to conventional legal categories—deportation, exclusion, asylum, voluntary departure. It is designed to be an additional administrative tool capable of responding to rapidly changing, politically charged events. EVD, far from depriving an alien of her opportunities to pursue the traditional avenues of relief from exclusion or deportation, actually creates a new remedy for situations in which the others may be unavailable or inadequate. To surround this device with constitutional constraints not applicable to excludable aliens generally could reduce and perhaps even eliminate its usefulness to the government and to aliens. If granting it to one group meant that the Attorney General was legally obliged to grant it to another, he might grant it more sparingly in the first instance. In that event, excludable aliens faced with dangerous conditions at home might well be made worse off. As is sometimes the case, more law might in the end mean less justice.[330]

---

324. INS Asylum Study, supra note 168, at 66.

325. Plaintiff's Memorandum of Points and Authorities in Opposition to Motion to Dismiss at 23, *Employees Union*.

326. 5 U.S.C. § 553 (1982) establishes certain procedures for agency rulemaking.

327. 563 F. Supp. at 161–63; Plaintiff's Memorandum of Points and Authorities in Opposition to Motion to Dismiss at 16–24.

328. 563 F. Supp. at 160; Defendants' Memorandum of Points and Authorities in Support of Motion to Dismiss at 7–9.

329. See Fiallo v. Bell, 430 U.S. 787 (1977); Kleindienst v. Mandel, 408 U.S. 753 (1972).

330. See, e.g., Schuck, Suing Our Servants: The Court, Congress, and the Liability of Public Officials for Damages, 1980 Sup. Ct. Rev. 281, 368.

The district judge, however, would have none of this. He found no difficulty in discerning a judicially manageable standard for evaluating grants or denials of EVD, despite the sensitive nature of the political and foreign policy issues presented, and despite considerable doubt that such a standard exists or can even be formulated. Indeed, the standard that the judge relied upon—"'widespread fighting, destruction and breakdown of public services and order'"[331]—has been disclaimed by the government as a workable basis for EVD decisions;[332] moreover, it appears to require data not easily obtainable[333] and calls for a comparative assessment of international conditions that would be most difficult for a court to make.[334]

Undaunted by such obstacles, however, the judge went on to suggest that "prudential considerations" did not preclude judicial review. He speculated that "since there has been criticism of American treatment of Salvadoran refugees," reversing the EVD denial might enhance American credibility abroad.[335] This non sequitur is singularly unpersuasive. There is probably no United States foreign policy that does not ignite some criticism; certainly, those domestic and foreign interests that support our current EVD policy would protest its reversal. The judge also failed to explain why he refused to credit the contention that the United States must "speak with one voice" on this matter.[336] Instead, brushing aside all other objections, the court concluded that plaintiffs had stated a legal claim and could proceed to trial.[337]

Perhaps *Employees Union* is an isolated, unrepresentative case—a single district judge doing no more than interpreting plaintiffs' claims generously, as

331. *Employees Union*, 563 F. Supp. at 160. This phrase was actually extracted from a letter written by a State Department official in response to Senator Kennedy's letter to Secretary of State Haig criticizing the Administration's EVD policy:

> While fighting in some areas has been severe, El Salvador has not suffered the same level of *widespread fighting, destruction and breakdown of public services and order* as did for example, Nicaragua, Lebanon or Uganda at the time when voluntary departure was recommended by the Department and granted by INS for nationals of those countries.

Letter from A. Drischler, Acting Ass't Secretary for Congressional Relations, to Senator Edward Kennedy (Apr. 17, 1981) (emphasis added) (explaining why EVD was not granted to El Salvadorans), reprinted in 128 Cong. Rec. S831 (daily ed. Feb. 11, 1982).

332. Government's Motion to Dismiss at 7-9, *Employees Union*.

333. First, independent objective indices for the standard would be necessary, such as casualties, extent of destruction, change in the quality of public services and order, and political conditions. Then, available data from other countries would have to be gathered and compared to these indices.

334. Even if the necessary indices and data were made available to a court, neither the alien plaintiff nor the court would be likely to possess the resources to independently assess them. Moreover, in evaluating plaintiff's equal protection claim, the court must decide what other country or countries are "similarly situated" to the country to whose nationals the EVD, if granted, would apply. Yet it is not clear how the court would make such a judgment.

335. *Employees Union*, 563 F. Supp. at 161.

336. Id. Although the judge stated that the government had failed to allege that the case "specifically presents a situation in which it is essential that the United States speak with one voice," id., he earlier noted that the government had contended that prudential considerations "such as a need for the government to speak with a single voice," in combination with other factors, rendered the case nonjusticiable, id. at 160.

337. Id. at 161-62.

he must, on a motion to dismiss. Yet until the early 1980's, it was almost inconceivable that a court would haul the government into court primarily at the behest of excludable or deportable aliens and require it to defend a policy such as that concerning EVD, a policy that is plainly and appropriately discretionary in nature, intimately intertwined with the most delicate and volatile aspects of our foreign policy, and reflective of political negotiation between Congress and the President.[338] In embryonic form, *Employees Union* bears the distinctive marks of much contemporary public law judging—deep skepticism if not indignation about the propriety of political compromises reached by the other branches, boundless faith in the courts' ability to analyze, comprehend, and predict the real-world consequences of complex policy alternatives, and serene confidence about the capacity of constitutional litigation and procedural reforms to harmonize the relevant public values.[339] What is surprising is not that this judicial style has become commonplace in other areas of the law; the truly remarkable development is that it is beginning to appear for the first time in immigration cases.

C. *The Constitutionalizing of Exclusion*

It is only since 1980 that the courts have started to bring the exclusion process into the mainstream of constitutional law. Litigation challenging the mass detention of excludable aliens in the wake of the recent Cuban, Haitian and Salvadoran influxes has generated most of this change. As we shall see below, courts have begun to create extensive procedural and even substantive rights to avoid or terminate detention.[340] Glimmerings of this new approach, however, are also discernible in some nondetention cases.

The Supreme Court's own contribution to this development has been quite modest. In its most recent exclusion decision, *Landon v. Plasencia*,[341] the Court reaffirmed earlier holdings that a lawful permanent resident who returns to the United States after a very brief trip outside the country is entitled to a due process hearing when the government seeks to exclude her on the theory that she has made a new "entry."[342] Similarly, the Court also

---

338. A nonbinding resolution calling upon the Reagan Administration to grant Salvadorans extended voluntary departure was recently adopted by Congress as part of the State Department appropriations bill for 1984–1985. See H.R. 2915, § 1012, 129 Cong. Rec. H10,236, H10,249 (daily ed. Nov. 17, 1983). Bills to bind the Administration in that regard have been introduced. S.2131, H.R. 4447, 98th Cong., 1st Sess. (1983).

339. See P. Schuck, supra note 18, ch. 7.

340. See infra notes 393–404 and accompanying text.

341. 103 S. Ct. 321 (1982).

342. Id. at 327–29; see also Rosenberg v. Fleuti, 374 U.S. 449 (1963); Kwong Hai Chew v. Colding, 344 U.S. 590 (1953). One commentary seems to regard *Landon* as a significant and welcome departure from earlier doctrine, particularly the Court's decisions in Shaughnessy v. United States ex rel. Mezei, 345 U.S. 206 (1953), discussed supra notes 100–02, 151–61 and accompanying text; and United States ex rel. Knauff v. Shaughnessy, 338 U.S. 537 (1950), discussed supra text accompanying note 99. See Developments, supra note 4, at 1319–20, 1323.

reaffirmed the classical doctrine on the legal status of aliens seeking initial entry, the far more common and important situation.[343]

The lower courts, however, have begun to stir. Of the recent nondetention cases, two are of particular interest. In *In re Phelisna*,[344] the court held that the burden of proof on the question of whether the alien, when apprehended, had already made an initial "entry"—and therefore is entitled to the due process that deportation proceedings afford—rests upon the government, not the alien.[345] Because earlier immigration decisions made "entry" turn in part upon the alien's subjective state of mind,[346] a matter obviously difficult for the government to prove, such a shift in the burden of proof could enable aliens to obtain deportation hearings—and thus eligibility for immigration benefits, including discretionary relief from deportation under a variety of statutory provisions—in many cases now governed by exclusion procedures.

Traditionally, as we have seen, the substantive grounds for exclusion have been even more immune from constitutional review than the procedural ones.[347] Even here, however, that immunity has begun to erode, at least in some lower courts. In *Lesbian/Gay Freedom Day Committee, Inc. v. INS*,[348] the court, reasoning in the shadow cast by an earlier Supreme Court case[349] that had rejected a constitutional challenge to the long-standing policy

---

This view, however, neglects the clear language of *Rosenberg*, 374 U.S. at 460, the fact that the absence in *Mezei* was lengthy and therefore distinguishable, and that *Knauff* did not involve a returning permanent resident at all. It is of interest that the Court in *Landon* cites both *Knauff* and *Mezei*. 103 S. Ct. at 329-30. Indeed, in one respect, *Landon* may limit the procedural protections available to the returning permanent resident, for the Court there holds for the first time that the question of her "entry"—that is, of excludability—can be decided in an exclusion rather than deportation proceeding. 103 S. Ct. at 325-29. The Court strongly suggests that the process which is due in that proceeding may be less than that which would apply in the deportation setting. Id. at 330. For a very recent lower court decision extending the *Fleuti* approach to a nonimmigrant seeking to remain in the United States after an INS-approved departure and return, see Joshi v. INS, No. 83-1614 (4th Cir. Oct. 27, 1983) (deportation rather than exclusion proceedings appropriate for addressing merits of claim).

343. *Plasencia*, 103 S. Ct. at 327-30. In so doing, the Court cited with apparent approval United States ex rel. Knauff v. Shaughnessy, 338 U.S. 537 (1950). 103 S. Ct. at 329-30. The Court also distinguished, on the basis of the length of absence, Shaughnessy v. United States ex rel. Mezei, 345 U.S. 206 (1953). 103 S. Ct. at 330. Finally, the Court once again embraced the rights-privilege distinction: "an alien seeking initial admission . . . requests a privilege and has no constitutional rights regarding his application, for the power to admit or exclude aliens is a sovereign prerogative." Id. at 329.

344. 551 F. Supp. 960 (E.D.N.Y. 1982), appeal dismissed sub nom. Phelisna v. Sava, No. 83-2034 (2d Cir. Apr. 29, 1983).

345. Id. at 963.

346. See id. at 962-63 (discussing Thack v. Zurbrick, 51 F.2d 634 (6th Cir. 1931), and citing United States ex rel. Giacone v. Corsi, 64 F.2d 18 (2d Cir. 1933)).

347. See supra notes 109-29 and accompanying text.

348. 541 F. Supp. 569 (N.D. Cal. 1982), aff'd sub nom. Hill v. INS, 714 F.2d 1470 (9th Cir. 1983). For a contrary holding see In re Longstaff, 716 F.2d 1439 (5th Cir. 1983) (since homosexual could have been admitted only through error, he had not been "legally" granted residency).

349. Boutilier v. INS, 387 U.S. 118 (1967). The district court in *Lesbian/Gay Freedom Day*, 541 F. Supp. at 584-85, distinguished *Boutilier* on the ground that Congress had not intended to

of excluding homosexuals, proceeded to invalidate that policy on statutory construction grounds. It went on to hold that even if the statute were interpreted to bar alien homosexuals, it would violate the first amendment rights of homosexuals in the United States to communicate and associate with such aliens,[350] a theory criticized, although not clearly rejected, a decade earlier by the Supreme Court.[351]

These decisions may be harbingers of a view of exclusion that differs radically from that represented by the relevant Supreme Court precedents.[352] The classical notion drastically limited the circumstances under which the government would be subjected to a procedural or substantive duty that it did not wish to acknowledge. Its explicit willingness to enter into a relationship with the alien, and its express consent to particular limitations upon its power to expel, were the sole measures of that duty. Today, however, obligation is increasingly imputed to the government on the basis of more abstract principles gleaned from the congeries of domestic law norms, including constitutional due process and equal protection, the Refugee Act of 1980,[353] and judicially elaborated communitarian values. Under these norms, even strangers can, by a kind of proxy relationship to citizens, claim some legal protections.

The Supreme Court in 1972 discerned and warned against the implications of this new view: "[C]ourts in each case would be required to weigh the strength of the audience's interest against that of the Government in [excluding] the particular alien applicant, according to some as yet undetermined standard."[354] While this caution may be amply warranted, it is also somewhat disingenuous. For that, of course, is precisely the kind of interest-balancing process that judges, with the encouragement of the Court and most legal scholars, routinely perform today in areas of the law as disparate as abortion regulation[355] and first amendment cases.[356] Once the excludable alien's rights

---

freeze the exclusion of homosexuals into the statutory category—aliens afflicted with sexual deviation, or a mental defect—but wanted the exclusion to be subject to changing medical opinion.

350. 541 F. Supp. at 585–88.

351. Kleindienst v. Mandel, 408 U.S. 753, 768–70 (1972). The district court distinguished *Mandel* on the ground that here, unlike in *Mandel*, the government had advanced no "facially legitimate and bona fide reason" for the exclusion. *Lesbian/Gay Freedom Day*, 541 F. Supp. at 585–86 (quoting *Mandel*, 408 U.S. at 770).

352. Kleindienst v. Mandel, 408 U.S. 753 (1972); Shaughnessy v. United States ex rel. Mezei, 345 U.S. 206 (1953); United States ex rel. Knauff v. Shaughnessy, 338 U.S. 537 (1950); Nishimura Ekiu v. United States, 142 U.S. 651 (1892).

353. Pub. L. No. 96-212, § 201(a), 94 Stat. 102 (1980).

354. Kleindienst v. Mandel, 408 U.S. 753, 769 (1972).

355. See, e.g., City of Akron v. Akron Center for Reproductive Health, Inc., 103 S. Ct. 2481, 2495–2516 (1983); Christensen v. Wisconsin Medical Bd., 551 F. Supp. 565, 568–70 (W.D. Wis. 1982).

356. See, e.g., Bolger v. Youngs Drug Prods. Corp., 103 S. Ct. 2875, 2881–85 (1983); United States v. Dickens, 695 F.2d 765, 772–74 (3d Cir. 1982), cert. denied, 103 S. Ct. 1792 (1983).

are defined not by her compliance with government-prescribed forms and conditions but in terms of her relationship with and benefit to American citizens,[357] the genie is out of the bottle. Under that approach, the sovereign's consent no longer constitutes the limiting principle from which the government's duty can be derived, and the courts' role can no longer be confined to enforcing the government's willingness or refusal to acknowledge the existence of rights in the alien.

D. *Limiting the Federal Power to Classify Aliens*

The classical view of Congress's power to classify aliens—one of essentially unlimited scope—would seem to be securely entrenched. The Supreme Court's most recent decisions affirming this view, *Mathews v. Diaz*[358] and *Fiallo v. Bell*,[359] are less than a decade old. And the one threat to this principle—a novel constitutional theory announced by a bare majority of the Court in *Hampton v. Mow Sun Wong*[360] to condemn a century-old bar to federal employment of aliens—has not prospered.[361]

Still, the seeds of judical intervention have been planted, and under the new conditions described in Part II, they may yet take root. *Diaz* seemed to concede that even federal alienage classifications were reviewable, albeit under a "narrow standard of review,"[362] and *Fiallo* affirmed that it must be based upon a "facially legitimate and bona fide reason."[363] Great and imposing structures of judicial power have been built upon far less prepossessing foundations than these.[364] It seems likely that, just as the Court earlier imposed

357. The Court in Kleindienst v. Mandel, 408 U.S. 753, 769-70 (1972), seemed to approve of this approach, however reluctantly. Thus, while upholding the government's position, the Court invited the kinds of challenges presented by *Lesbian/Gay Freedom Day* by recognizing the government's obligation to justify its exclusion and by analyzing those justifications, however superficially, for their facial legitimacy and bona fides.

358. 426 U.S. 67 (1976).

359. 430 U.S. 787 (1977).

360. 426 U.S. 88 (1976).

361. For criticism of the Court's approach in this case, see Schuck, Organization Theory and the Teaching of Administrative Law, 33 J. of Legal Educ. 13, 17 (1983). Even the result in *Hampton* was ephemeral; it was quickly reversed. See Exec. Order No. 11,935, 5 C.F.R. § 13 (1983), reprinted in 5 U.S.C. § 3301 app. at 384 (1976) (upheld in Mow Sun Wong v. Campbell, 626 F.2d 739 (9th Cir. 1980), cert. denied, 450 U.S. 959 (1981)). The Court's decision in *Diaz*, announced on the same day as *Hampton*, thus seems to be of hardier stock. Indeed, even state laws barring aliens from state employment seem to enjoy considerable constitutional legitimacy today, at least where the classification can be depicted, however implausibly, as manifesting the state's conception of its "political community." See Cabell v. Chavez-Salido, 454 U.S. 432, 447 (1982).

362. *Diaz*, 426 U.S. at 82.

363. *Fiallo*, 430 U.S. at 794-95 (quoting Kleindienst v. Mandel, 408 U.S. 753, 769-70 (1972), discussed supra notes 351-54, 357).

364. The classic example, of course, is the doctrine of heightened scrutiny of classifications affecting "discrete and insular minorities," rooted in a mere footnote. United States v. Carolene Prods. Co., 304 U.S. 144, 152 n.4 (1938). See generally J. Ely, supra note 292; Lusky, Footnote Redux: A *Carolene Products* Reminiscence, 82 Colum. L. Rev. 1093 (1982); Powell, *Carolene Products* Revisited, 82 Colum. L. Rev. 1087 (1982).

constitutional limits on Congress's broad power to regulate the acquisition and loss of American citizenship,[365] the Court will not continue indefinitely to tolerate, as it did in *Diaz*, unprincipled discrimination against aliens who sociologically, economically, morally—indeed, in all respects except legally—have become vital members of the community. I have suggested that *Plyler v. Doe*,[366] although involving a state classification, may evince an increasing judicial hostility to federal alienage classifications as well, at least where the power to define "political community" is not plausibly at issue.[367] Less ambiguously, the decision in *Hotel & Restaurant Employees Union, Local 25 v. Smith*,[368] and the detention cases discussed below suggest a growing judicial solicitude for those whom Congress may unquestionably exclude but who may now draw upon other sources of legal obligation once they make it inside our gates.

E. *Deportation and Due Process*

The principle that deportation is a civil, administrative proceeding, with its corollary that certain procedural safeguards need not be provided, is a fixture of classical immigration law. It shows no signs of being jettisoned. This result was unexceptionable when individualistic premises emphasizing consent and formalism animated the legal system, when deportation plausibly could be viewed as the revocation of a conditional "privilege" or license, and when such remedies did not trigger due process requirements.[369] In today's altered ideological and doctrinal setting, however, deportation appears in a rather different light. In a system that increasingly recognizes communitarian values, courts can be expected to take notice of the fact that deportation wrenches and uproots the alien from her most profound human attachments, attachments valued by the larger community from which she is torn as well as by the alien herself. They are unlikely to ignore the fact that even where she has failed to fulfill the conditions under which she gained entry, her presence here has spawned new social relationships, generating expectations and obligations that may transcend in importance those that she violated.

The lower courts have begun to respond to this new normative environment. In 1983, for example, the Ninth Circuit, overruling the Board of Immigration Appeals, applied the exclusionary rule to bar introduction of

365. See, e.g., Rogers v. Bellei, 401 U.S. 815, 830 (1971); Afroyim v. Rusk, 387 U.S. 253, 257 (1967).

366. 457 U.S. 202 (1982), discussed supra notes 294-318 and accompanying text.

367. That criterion has been viewed by the Court as crucial to rationalize the series of cases, beginning with Graham v. Richardson, 403 U.S. 365 (1971), that involve *state* alienage classifications. The most recent of these cases is Cabell v. Chavez-Salido, 454 U.S. 432, 447 (1982).

368. 563 F. Supp. 157 (1983), discussed supra notes 321-39 and accompanying text.

369. See supra notes 145-48, 264-67 and accompanying text. Deportation under the individualistic ideology can also be analogized to a creditor's summary enforcement of contract rights. See supra note 147.

illegally obtained evidence into deportation proceedings.[370] The same court has also applied the fourth amendment to limit the INS's "area control" method of questioning deportable aliens.[371] Another court has suggested that due process requires free, appointed counsel for indigents at deportation hearings "[whenever] an unrepresented indigent alien would require counsel to present his position adequately to an immigration judge."[372] Courts have also found various due process violations in the mass, summary processing of deportable aliens with potential asylum claims.[373]

Despite some congressional resistance,[374] this judicial assimilation into the deportation process of rights drawn from criminal procedure has been spasmodic and fragmentary[375] but seems likely to continue. Doctrinally, this evolution may take any of several forms: expanded notions of what "fundamental fairness" requires; classification of deportation as a "quasi-criminal" proceeding;[376] or application to deportation decisions of equal protection standards[377] or general administrative law principles governing the decision-making process.[378] Normatively, this evolution reflects the courts' growing conviction, manifested in the reformation of administrative law that Stewart describes and in the increasing influence of communitarian values generally, that our society has as important a stake as aliens do in a fair, accurate and dignified public law process for deciding who shall remain among us.

Empirically, however, it remains unclear whether judicializing deportation procedures further would actually yield such a process. The question is

---

370. Lopez-Mendoza v. INS, 705 F.2d 1059 (9th Cir. 1983) (en banc), petition for cert. filed, 52 U.S.L.W. 3294 (U.S. Sept. 22, 1983) (No. 83-491).

This decision is especially significant in light of the Supreme Court's growing opposition to the exclusionary rule. See, e.g., Rawlings v. Kentucky, 448 U.S. 98 (1980); United States v. Payner, 447 U.S. 727 (1980); see also Illinois Migrant Council v. Pilliod, 531 F. Supp. 1011 (N.D. Ill. 1982).

371. International Ladies' Garment Workers' Union v. Sureck, 681 F.2d 624 (9th Cir. 1982), cert. granted sub nom. INS v. Delgado, 103 S. Ct. 1872 (1983).

372. Aguilera-Enriquez v. INS, 516 F.2d 565, 569 n.3 (6th Cir. 1975), cert. denied, 423 U.S. 1050 (1976). One commentator has concluded that a right to appoint counsel has been created in effect for asylum claimants facing deportation. See infra note 402.

373. See infra notes 391-95 and accompanying text.

374. The Simpson-Mazzoli immigration bill, S. 529, 98th Cong., 1st Sess., 129 Cong. Rec. S6970 (daily ed. May 18, 1983), for example, would alter the government's standard of proof in deportation cases from "clear, unequivocal and convincing evidence," Woodby v. INS, 385 U.S. 276, 285 (1966), to "preponderance of the evidence," S. Rep. No. 62, 98th Cong., 1st Sess. 37 (1983).

375. See Developments, supra note 4, at 1384-95.

376. United States v. Ward, 448 U.S. 242, 251 (1980), created such a quasi-criminal category in a nonimmigration context. See Wheeler, The Constitutional Case for Reforming Punitive Damages Procedures, 69 Va. L. Rev. 269, 322-51 (1983); see also Woodby v. INS, 385 U.S. 276, 285 (1966) (a deportation proceeding requires a higher degree of evidence despite its noncriminal nature).

377. See, e.g., Francis v. INS, 532 F.2d 268 (2d Cir. 1976); Noel v. Chapman, 508 F.2d 1023 (2d Cir.), cert. denied, 423 U.S. 824 (1975).

378. See, e.g., Developments, supra note 4, at 1395-98.

complex for several reasons. The sheer number of deportable aliens apprehended each year suggests that more formal procedures might require an enormous reallocation of resources from other activities to adjudication; total apprehensions in 1982 were almost one million.[379] If even a modest fraction of these aliens had to be processed through judicially mandated formal procedures, the INS would almost certainly be obliged to rely upon informal deportation practices to an even greater extent than it does now. Yet these practices could be rendered ineffective if illegal aliens could, by invoking more formal deportation procedures, impose substantial costs and delay upon the enforcement process.[380] In addition, a further formalization of deportation might affect decisions on aliens' applications for discretionary relief from deportation in unpredictable, perhaps undesirable ways.

Finally, it is far from clear that more formal procedures would actually enhance the accuracy or fairness of deportation decisions. The central reality of immigration administration is the overwhelming caseload on the fifty-five immigration judges; in fiscal 1983, they were expected to decide an average of 5.35 deportation and exclusion cases *per day*, yet this assembly-line justice would not even keep the backlog from growing.[381] It is not difficult to imagine either the quality of the procedural forms that can be provided to aliens under such conditions or the effect upon the process of requiring new ones. In short, until we know much more about the actual relationships between procedural forms, substantive outcomes and process values in the peculiar setting of the deportation sanction, the law's treatment of that sanction as a merely "civil" one may be a necessary compromise with the uncertainties surrounding the dynamics of immigration administration.

F. *Detention and Due Process*

No single development has animated and shaped the current transformation of immigration law more powerfully than the massive influx and subsequent detention of aliens from Cuba, Haiti, El Salvador and other Caribbean Basin countries since 1980. The prolonged incarceration of thousands of aliens, most of them innocent victims of severe economic deprivation, indiscriminate armed conflict, or intense political persecution, has seared the judicial conscience as few events since the civil rights struggles of the 1950's and 1960's have done.[382]

379. See Letter from John E. Nahan, Director, Plans and Analysis, INS, to Peter H. Schuck (Oct. 25, 1983) (copy on file at the offices of the Columbia Law Review).

380. See infra notes 437-39 and accompanying text.

381. See Petition for Writ of Certiorari at 17 n.10, INS v. Lopez-Mendoza, filed, 52 U.S.L.W. 3268 (U.S. Sept. 22, 1983) (No. 83-491). For a discussion of the effect of procedures on decision quality in the context of the social security disability program, see J. Mashaw, Bureaucratic Justice 106-23 (1983). Professor Mashaw has also analyzed this question in the context of the Aid to Families with Dependent Children program during the early 1970's. See Mashaw, The Management Side of Due Process, 59 Cornell L. Rev. 772, 776-91, 811-15 (1974) (due process hearings of little value to AFDC claimants); see also infra notes 457-58 and accompanying text.

382. See, e.g., Louis v. Nelson, 544 F. Supp. 973, 975-76 (S.D. Fla. 1982), aff'd sub nom. Jean v. Nelson, 711 F.2d 1455, reh'g en banc granted, 714 F.2d 96 (11th Cir. 1983).

Several aspects of this mass detention have especially troubled the courts. First, the conditions of confinement are often harsh and oppressive. These include chronic overcrowding, lack of recreation, violent disturbances, legal uncertainties, and the volatile mixture of different age groups and of as many as forty different nationalities in a given facility.[383] These individuals are not being held for criminal prosecution yet are detained without bond, a situation that clearly raises the most serious questions of denial of due process.[384] In addition, many of those so confined have at least colorable claims to various kinds of legal protections under domestic and international law notwithstanding their undocumented status. These include claims to asylum under the Refugee Act of 1980,[385] withholding of deportation if return would threaten the alien's life or freedom due to persecution,[386] extended voluntary departure,[387] so-called "deferred action,"[388] and relief through private legislation.[389] It also appears that certain actions by both the Carter and Reagan Administrations have actually encouraged aliens to assert these claims.[390] Finally, the INS has approached its undeniably difficult administrative task with an exceedingly heavy and arbitrary hand. The list of INS's adjudicated, post-1980 violations of the constitutional and statutory rights of aliens and of its own regulations is long, varied and profoundly dispiriting.[391]

Under these circumstances, the courts are understandably less willing to credit the conventional administrative law presumption that the agency, if left

---

383. See supra note 164.

384. Although the eighth amendment's prohibition of "cruel and unusual punishment," U.S. Const. amend. VIII, does not apply prior to conviction, fifth amendment due process requirements do. U.S. Const. amend. V. See City of Revere v. Massachusetts Gen. Hosp., 103 S. Ct. 2979 (1983).

385. 8 U.S.C. § 1158 (1982). Indeed, the provisions of the Refugee Act have recently been held to require INS to provide an asylum hearing even to stowaways, who are denied an exclusion hearing under 8 U.S.C. § 1182(a)(18) (1982) of the immigration statute. See Yin Sing Chun v. Sava, 708 F.2d 869 (2d Cir. 1983).

386. 8 U.S.C § 1253(h) (1982).

387. See supra notes 322-24 and accompanying text.

388. See Nicholas v. INS, 590 F.2d 802 (9th Cir. 1979) (INS district director has affirmative duty to recommend deferred action status where she feels it is appropriate; court will review her decision for abuse of discretion). But cf. Pasquini v. Morris, 700 F.2d 658 (11th Cir. 1983) (deferred action guidelines confer no substantive right and court will only review for abuse of discretion).

389. See P. Schuck, supra note 82, at 255-65.

390. See INS Asylum Study, supra note 168, at 20-26; Aleinikoff, supra note 223, at 254-56.

391. These transgressions include violations of due process and equal protection under the fifth amendment, U.S. Const. amend. V, and of 8 U.S.C. §§ 1252(b), 1253(h) (1982). See Haitian Refugee Center v. Smith, 676 F.2d 1023 (5th Cir. 1982) (rights of alien seeking political asylum held sufficient to invoke due process guarantees); Fernandez-Roque v. Smith, 567 F. Supp. 1115 (N.D. Ga. 1983) (INS violated due process clause by not providing detained excludable aliens with procedurally adequate hearings); Louis v. Nelson, 544 F. Supp. 973 (S.D. Fla. 1982) (INS detention program violated equal protection clause), aff'd sub. nom. Jean v. Nelson, 711 F.2d 1455, reh'g en banc granted, 714 F.2d 96 (11th Cir. 1983); Orantes-Hernandez v. Smith, 541 F. Supp. 351 (C.D. Cal. 1982) (INS violated due process clause by not advising aliens of their right to apply for political asylum); Nunez v. Boldin, 537 F. Supp. 578 (S.D. Tex.) (same), appeal dismissed, 692 F.2d 755 (5th Cir. 1982).

to its own devices and to only sporadic congressional oversight, will comply with the law. As the lower court precedents granting injunctive relief against the INS accumulate, judges are naturally less hesitant to intervene. In the past, courts exercising their habeas corpus jurisdiction occasionally intervened to limit detention of individual deportable or excludable aliens.[392] Recently, however, judges have demonstrated a willingness to grant broad, class relief to large numbers of aliens, enjoining INS's efforts to use detention as a way of deterring what the courts themselves acknowledge to be a tidal wave of undocumented entrants. Thus, within the last two years alone, the courts have invalidated some of INS's key detention initiatives. These include a general policy decision to detain all undocumented aliens who do not establish a prima facie case for admission;[393] its program of mass deportation of detained, undocumented Haitians coupled with mass processing and denial of their asylum claims;[394] its mass detention, deportation, and coerced voluntary departure of Salvadorans, coupled with failure to notify them of their right to apply for asylum;[395] and its prolonged detention of excludable Cubans pending deportation.[396]

In recent detention cases, the courts have transmuted classical immigration law's conception of the nature of the government-alien relationship into a rather different one in which rights against the government accrue to aliens without the government's consent and without the formal conditions for immigration having been observed. This conception lies at the core of the recent district court decision in *Fernandez-Roque v. Smith.*[397] In *Fernandez-Roque*, the individuals being detained were not simply aliens subject to future exclusion proceedings. Some apparently had already received statutory exclusion hearings and were subject to final orders of exclusion; the others had been administratively determined to have had criminal records in Cuba. Many of the plaintiffs had been detained since their initial arrival here.[398] Plainly, if any aliens could be considered total strangers to American society the plaintiffs in *Fernandez-Roque* could. Despite this fact, the court extended to these excludable aliens an array of constitutionally derived procedural rights that were actually far more extensive than those enjoyed by deportable aliens—indeed more extensive than those enjoyed by American citizens who chal-

---

392. See Developments, supra note 4, at 1366-68.

393. See Jean v. Nelson, 711 F.2d 1455, 1472, reh'g en banc granted, 714 F.2d 96 (11th Cir. 1983).

394. See Haitian Refugee Center v. Smith, 676 F.2d 1023 (5th Cir. 1982).

395. See Orantes-Hernandez v. Smith, 541 F. Supp. 351 (C.D. Cal. 1982); Nunez v. Boldin, 537 F. Supp 578 (S.D. Tex.), appeal dismissed, 692 F.2d 755 (5th Cir. 1982).

396. See Rodriguez-Fernandez v. Wilkinson, 654 F.2d 1382 (10th Cir. 1981); Fernandez-Roque v. Smith, 567 F. Supp. 1115 (N.D. Ga. 1983). But see Palma v. Verdeyen, 676 F.2d 100 (4th Cir. 1982) (upholding an indefinite detention of an alien after an unsuccessful attempt to deport him).

397. 567 F. Supp. 1115 (N.D. Ga. 1983).

398. Id. at 1120, 1140.

lenged revocation of parole after a criminal conviction. It also created "a presumption of releasability" with the government bearing the burden of proof on the ultimate issue.[399] As a practical matter, this ostensibly procedural decision amounts to granting a *substantive* right to remain in this country indefinitely to many aliens who are clearly excludable under our laws.

This new conception of the government-alien relationship is also revealed in the emerging law of asylum. Far from viewing the alien as classical immigration law did, as an utter stranger with no legal claims upon the government other than those the government has agreed to recognize, the emerging law of asylum now enables any alien to acquire rights against the government to which the latter has not expressly consented, rights about which the Refugee Act of 1980 is silent.[400] These new rights—for example, timely notice of the right to apply for asylum[401]—reflect what David Martin has called "the procedural exuberance of the lower courts."[402] But some of them—for example, the rights to stay deportation proceedings pending final resolution of an asylum claim[403] and to be free from unnecessary or prolonged detention in the interim[404]—may have far-reaching substantive effects. These rights may as a practical matter amount to the right to work and live in the United States indefinitely. To enjoy them, all the alien need do is reach our shores. If she can also qualify as a refugee, she may then claim further statuses to which the government *has* expressly consented—asylum and the probability of adjusting to permanent legal residence.[405] Even if she cannot qualify as a refugee, she may be able to obtain withholding of deportation under section 1253(h).[406]

---

399. Id. at 1139.

400. It is arguable that Congress intended to create these new rights when it acceded to the 1967 Protocol Relating to the Status of Refugees, Nov. 1, 1968, 19 U.S.T. 6224, T.I.A.S. No. 6577, 606 U.N.T.S. 267. But neither the Protocol nor the Handbook on Procedures and Criteria for Determining Refugee Status (1979), which is recognized as an authoritative explication of the Protocol, mentions these rights. Even article 26 of the 1951 Convention relating to the Status of Refugees, 189 U.N.T.S. 137, which the United States in the 1967 Protocol has undertaken to apply, provides for freedom of movement only for "refugees," who are "lawfully within the country," not for those whose status has not yet been determined. See Bertrand v. Sava, 684 F.2d 204, 219 (2d Cir. 1982) ("the Protocol affords the petitioners no rights beyond those they have under our domestic law").

401. Orantes-Hernandez v. Smith, 541 F. Supp. 351 (C.D. Cal. 1982).

402. Martin, supra note 103, at 171. Martin also suggests that the lower courts have in effect, if not in law, created a right to free appointed counsel by requiring expeditious hearings while also enjoining deportation hearings in which the alien lacks counsel. Id. at 219–20.

403. 8 C.F.R. § 208.3(b) (1983) ("asylum requests shall also be considered as requests for withholding exclusion or deportation").

404. Fernandez-Roque v. Smith, 567 F. Supp. 1115 (N.D. Ga. 1983).

405. 8 U.S.C. § 1158 (1982).

406. Id. § 1253(h)(1) (precluding deportation "if the Attorney General determines that such alien's life or freedom would be threatened in such country on account of race, religion, nationality, membership in a particular social group, or political opinion"). The showing that an alien must make to qualify for withholding of deportation under this provision is somewhat different, and less demanding, than that for asylum. Cf. 8 U.S.C. § 1101(a)(42)(A) (asylum requires showing of "persecution" or a "well-founded fear of persecution").

In limiting the permissible detention of aliens who have not yet "entered" the country,[407] then, the courts have found new nonconsensual sources of legal rights in an excludable alien's constitutionally protected interest in liberty, a potential asylum applicant's statutory interest in protection from persecution, and more general communitarian values. To define such rights and obligations, a court does not simply content itself with assessing the government's compliance with legal forms or its expressed intention, but appraises the alien's social context—her relationships, fears, expectations, and alternatives. *Mutatis mutandis*, a similar kind of inquiry directed at similar kinds of questions underlies modern theories of liability in private law and other areas of public law. Substantive justice, not the forms of consent and sovereignty, has become the principal touchstone of legal analysis.

G. *The Separation of Adjudication and Law Enforcement*

We saw earlier that the structural integration of adjudication and law enforcement has been a distinctive element of classical immigration law.[408] The government, by integrating these functions, has sought to enhance policy control and administrative efficiency. Like civil deportation and extraconstitutional exclusion, this decision structure is simply seen as part of the price that the alien must pay—literally, the price of admission—in order to enjoy the privilege that the government offers. If the alien is dissatisfied, she is free to try to do better by going elsewhere.[409] But when the individualistic, consent-based legal order gives way to one that seeks to vindicate universal human rights, communitarian values, and imputed social duties, the legitimacy of the original structure is seriously undermined. A society—or at least a federal judiciary—committed to these norms expects the government to exercise its awesome power over aliens fairly, in a way that preserves their dignity, protects their interest, and minimizes the risk of arbitrariness and error. This is true even if meeting this expectation makes life more difficult or costly for the government.

It is not surprising, then, that like the other features of classical immigration law, the integration of adjudication and enforcement is now under attack. Objections to this practice are neither theoretical nor purely historical. For example, in *Haitian Refugee Center v. Smith*,[410] a 1982 decision, the Court of Appeals affirmed the trial court's finding that the immigration judges had cooperated with the INS prosecutors in administering a program

407. Under Leng May Ma v. Barber, 357 U.S. 185 (1958), aliens paroled into the country are not deemed to have "entered." See supra notes 160-61 and accompanying text.

408. See Marcello v. Bonds, 349 U.S. 302 (1955), discussed supra notes 185-89 and accompanying text.

409. The distinctions between the rights of asylum claimants and other aliens becomes understandable in these terms as well. One with a valid claim to asylum, almost by definition, has no other place to go.

410. 676 F.2d 1023 (5th Cir. 1982).

of mass, summary injustice.[411] Other abuses in earlier times have often been alleged.[412]

Limited reforms have already been instituted. In 1983, an administrative reorganization formally moved the immigration judges outside of the administrative control of the INS and placed them, along with the Board of Immigration Appeals (BIA), into a new Executive Office for Immigration Review.[413] This new unit is not independent of the Justice Department, for it remains under the supervision of the Deputy Attorney General. The administrative adjudication of immigration cases, therefore, continues to be conducted under the direct supervision of the Attorney General, the Nation's chief prosecutorial official. Moreover, training and career lines increase the possibility of bias; most immigration judges come from many years of service in the ranks of the INS. The pending Simpson-Mazzoli immigration bill would essentially confirm this reorganization, giving it statutory status.[414] But more thoroughgoing changes, such as an independent administrative law judge system, have successfully been resisted.[415] In this respect, even more than in others that we have discussed, the transformation of immigration law remains embryonic, tentative, and incomplete.

IV. The Future of Immigration Law

Although still partial and fragmentary, the signs of incipient change are abundant and unmistakable. The courts' almost complete deference to Congress and the immigration authorities, long a keystone of the classical structure, is beginning to give way to a new understanding and rhetoric of judicial role, as unusual in immigration cases as it is now commonplace in other domains of public law.[416] Perhaps emboldened by the declining legitimacy of the existing system and the compelling human dilemmas presented in recent cases, lower courts are testing and sometimes transcending the confines of the classical canons. The Supreme Court, it must be emphasized, has not yet spoken on the post-1980 cases. Nevertheless, the Court has seemed to provide the lower courts with ample leeway within which to innovate.[417] The adminis-

411. Haitian Refugee Center v. Civiletti, 503 F. Supp. 442, 519-23 (S.D. Fla. 1980), aff'd sub nom Haitian Refugee Center v. Smith, 676 F.2d 1023 (5th Cir. 1982).

412. See Roberts, supra note 171, at 7-11; Wasserman, Some Defects in the Administration of Our Immigration Laws, 21 Law & Contemp. Probs. 376, 380-81 (1956); Developments, supra note 4, at 1364-66.

413. 48 Fed. Reg. 8038-40 (1983).

414. See S. 529, 98th Cong., 1st Sess., § 122, 129 Cong. Rec. S6973-74 (daily ed. May 18, 1983). According to the committee report, the change is designed to "upgrade the existing structure and provide increased independence, authority, and stature." S. Rep. No. 62, 98th Cong., 1st Sess. at 35-36 (1983).

415. The exception from APA adjudication procedures would continue. S. Rep. No. 62, 98th Cong., 1st Sess. 35-36 (1983).

416. See supra notes 294-407 and accompanying text. For examples and a discussion of the causes and consequences of this development, see P. Schuck, supra note 18, at 35-53.

417. See, e.g., Kleindienst v. Mandel, 408 U.S. 753, 770 (1972). For a discussion of the "law of leeways," see K. Llewellyn, The Bramble Bush 156-57 (1930).

trative apparatus of the immigration system, virtually overwhelmed by the numbers and novelty of the problems it confronts, is in such serious disarray that the courts appear increasingly unwilling to countenance INS's business as usual.[418] Congress, which only manages to reassess immigration policy in a fundamental way every thirty years or so, has just abandoned a half-decade's effort—perhaps the last for years to come—to enact the most far-reaching changes in immigration law since the McCarran-Walters Act of 1952.[419] In an important sense, the lower courts are on their own. If *Plyler v. Doe* is any guide, the Supreme Court is not inclined to rein them in.

To say that the individualistic structure of the classical immigration system is gradually being supplanted by one increasingly based upon communitarian norms is not to say that the particular forms that change is taking are inevitable. The relatively open immigration policy and easy access to citizenship that largely prevailed prior to the 1880's[420] represented one model of the government-alien relationship; the restrictive nationalism embodied in the 1917 and 1924 statutes[421] represented quite another. The Bracero and H-2 programs, which allowed foreign workers to enter the country on a limited basis,[422] and the employer sanction legalization, and summary exclusion provisions of the recent Simpson-Mazzoli immigration bill[423] represented still others. The United Kingdom, although sharing a common legal tradition, has recently taken a very different, still more restrictive path.[424] These choices and others were and remain open to us so long as we affirm the somewhat inconsistent ideals of consent-based obligation, universal human rights, and strong national sovereignty.

Why, then, has immigration law begun to move in the particular directions that Part III described? Although a fully satisfactory causal explanation lies well outside the scope of this essay, I have tried to suggest the beginnings of an answer. I have noted that the foundations of classical immigration law were laid at a time when the assumptions of traditional liberalism, qualified and reshaped by the very different assumptions of restrictive nationalism,

---

418. Government Accounting Office, Prospects Dim for Effectively Enforcing Immigration Laws i–v (1981); see also supra notes 391–96 and accompanying text.

419. Immigration and Nationality Act of 1952, ch. 477, 66 Stat. 163 (codified as amended in 8 U.S.C. §§ 1101–1503 (1982)).

420. See supra notes 6–8, 26–31 and accompanying text.

421. Immigration Act of 1924, ch. 139, § 11, 43 Stat. 153, 159 (repealed 1952); Immigration Act of 1917, ch. 301, § 3, 39 Stat. 874, 877. See supra notes 62-63 and accompanying text.

422. See S. 529, 98th Cong., 1st Sess. § 211(b), 129 Cong. Rec. S6970 (daily ed. May 18, 1983) (H-2 program); Morgan & Gardner, Potential for a U.S. Guest-Worker Program in Agriculture: Lessons from the Braceros, *in* The Gateway, supra note 198, at 361.

423. S. 529, 98th Cong., 1st Sess. §§ 101(a)(1), 121(b)(1), 129 Cong. Rec. S6970, S6973 (daily ed. May 18, 1983).

424. See British Nationality Act, 1981, ch. 61. The 1981 Act, for example, restricts the extension of citizenship to persons born in the United Kingdom to those born to a British citizen or resident. Id. § 1. Prior law extended citizenship based on birth without restriction. See British Nationality Act, 1948, ¶ 908.

dominated legal thought about the meaning of national sovereignty and about the sources and scope of the government's obligation to individuals in general and to aliens in particular. Important economic and social changes, however, have rendered both ideologies increasingly incoherent and irrelevant as applied to immigration problems. Consequently, they gradually have been supplanted by theories of obligation derived not from the forms of consent by a sovereign government, but from a new communitarian ethos that grounds obligation in social relationships and notions of substantive justice.

Courts adjudicating under these changed structural and ideological conditions and drawing upon principles and moral intuitions that have long dominated private, administrative, and constitutional law, are now beginning to fuse these ideas into a new and radically different system of immigration law. In doing so, judges are both responding to and accelerating the penetration of communitarian values into *all* areas of contemporary law and life, as collectively imposed duties and bureaucratic institutions replace the individualistic values and decentralized structures of the market.[425] Immigration law, with its political and institutional insularities and eccentricities, has simply assimilated these pervasive changes in the legal culture more slowly and equivocally than have other areas of law.

If these structural and ideological conditions help to explain the course that immigration law has followed and to identify the coordinates of its present location, they suggest little about its ultimate destination. Legal history, like all history, reveals few continuities or uninterrupted patterns.[426] In the social sciences, projections into the future on the basis of predictive models and extrapolations from current trends are notoriously unreliable.[427] And law, it need hardly be added, is far from being a science.

In fact, the transformation of immigration law has significantly increased its indeterminacy by aggravating certain tensions and contradictions long embedded within the classical structure, and by creating new ones peculiar to a communitarian legal system. Three facets of this transformation reveal its problematic nature most starkly, and also seem likely to affect the future development of immigration law in particularly important ways. These relate to the legitimacy of immigration law, the role of the courts in developing the law, and the conception of community that the law embodies.

425. "[I]n a world of hierarchy, bureaucracy, and constituency," MacNeil notes, "[t]he law . . . is constitutional and administrative law and the law of political relations, more than the law of contract to which generations of American lawyers have been accustomed." I. MacNeil, supra note 275, at 84.

426. Even the expansion of fault-based tort liability, as powerful a trend as one finds in our law, has not been a steady one. See, e.g., M. Horwitz, The Transformation of American Law, 1780-1860, at 67-108 (1977) (describing nineteenth century tort law as limiting liability in order to provide a subsidy to economic interests). But see Schwartz, Tort Law and the Economy in Nineteenth-Century America: A Reinterpretation, 90 Yale L.J. 1717, 1722-34 (1981) (nineteenth century tort law was solicitous toward victims).

427. See generally, e.g., W. Ascher, Forecasting: An Appraisal for Policy-Makers and Planners (1979).

A. *Legitimacy*

Law's ability to influence human conduct derives ultimately from its moral legitimacy. This legitimacy in turn is significantly affected by the incentive structure—the array of costs and benefits—that the law in fact creates.[428] Immigration law deals with private activity that is quintessentially deliberate and calculating, activity therefore in which incentives are highly and directly salient to motivation.[429] Classical immigration law, as we have seen, took an uncompromising but detached view of the relationship between government and aliens. If admitted, an alien was bound by the terms of her privilege or license, as prescribed by the visa or parole conditions. When the INS sought to enforce those conditions against her in deportation or exclusion proceedings, the INS almost always prevailed. If equities in favor of the alien had developed subsequent to her admission, of course, she might seek relief by way of a private bill or, after 1952, by administrative suspension of deportation. But the legislative route was unpromising,[430] administrative relief was highly discretionary, and the courts characteristically deferred to that discretionary judgment.[431]

Today, the incentive structure of immigration law has shifted—to the marked advantage of these aliens who are undocumented or out of status but also wish to remain. Evasion, delay, surreptitious reentry, and administrative overload in the immigration system are the most precious resources of such aliens, and a more communitarian law is increasingly making those resources abundantly available to them. As the gap widens between the standards of living and working in the United States and in the countries that are the major sources of undocumented aliens, the rewards for entering and remaining are correspondingly greater. In a classic demonstration of how procedure and substance are intimately related, the courts have expanded aliens' procedural rights in the various ways discussed in Part III, thereby increasing the probability that aliens will in fact obtain the ultimate substantive benefits—avoidance of deportation and receipt of the right to work here. They can utilize the numerous avenues of discretionary administrative relief that exist, invoke administrative and judicial review of each administrative decision, and move to stay and reopen proceedings at each stage of each administrative and judicial process—while steadily accumulating new relationships and equities

428. See generally P. Schuck, supra note 18, at 257.

429. In this respect, it resembles contract law and differs from most of tort law.

430. See INS v. Chadha, 103 S. Ct. 2764, 2774-75 (1983) (Congress established legislative veto over deportation suspensions to alleviate "intolerable" burden of private bills); id. at 2805-06 (White, J., dissenting) (same).

431. See, e.g., Jay v. Boyd, 351 U.S. 345 (1956) (suspension of deportation is not a matter of right, but a matter of grace); Comment, Congressional Review of Suspension of Deportation and the Doctrine of Separation of Powers, 19 San Diego L. Rev. 177 (1981); Comment, A Suspension of Deportation: A Revitalized Relief for the Alien, 18 San Diego L. Rev. 65 (1980); Comment, Suspension of Deportation: Illusory Relief, 14 San Diego L. Rev. 229 (1976).

that will enable them to delay deportation, obtain work authorization, and perhaps permanently legalize their status.[432] Concerning those who are ultimately expelled from this country, the vast majority choose to leave "voluntarily" and can subsequently—often only hours later—return unnoticed, only to begin the process once again.[433] "Beating the system" has become a game, it seems, that almost any resourceful alien equipped with easily obtained fraudulent documents[434] or a competent lawyer can successfully play.

The judicial innovations that have begun to transform immigration law cannot properly be evaluated without considering these intractable realities. It is not simply that these reforms have increased the cost of administering the immigration system; greater cost, after all, almost inevitably attends new procedural rights and cannot alone be a decisive factor. The more important concern is the loss of credibility and deterrence that a legal apparatus suffers when it cannot effectively execute its own rules and intentions, and when its impotence is both widespread and apparent to those who are expected to comply. In the case of immigration law, this includes aliens, employers, smugglers, and others who know of or participate in violations of that law.

Strengthened incentives to "beat the system" would be of relatively little concern if that system possessed the capacity to defend itself against this threat. Increasingly, however, it does not. Immigration law is being altered in ways that make it far more time-consuming, costly, and difficult for the INS to apprehend, interrogate, search, arrest, detain, deport, assign status to, and allocate benefits among aliens. At the same time, the administrative resources of the INS, never adequate to its responsibilities, are actually being reduced in real—and in some respects, even in nominal—terms.[435] One crude but revealing index of one important dimension of the INS's enforcement program is that the number of Border Patrol officers is little more than twice the number

432. Perhaps the most celebrated practitioner of this strategic art is Carlos Marcello, who at this writing remains in the United States despite an order of deportation entered affirmed by the Supreme Court in 1955, and based upon a 1938 narcotics conviction. For the latest chapter in this never-ending epic, see marcello v. INS, 694 F.2d 1033 (5th Cir.), cert. denied, 103 S. Ct. 3112 (1983), and for a synopsis of events up to 1981, see United States ex rel. Marcello v. District Director, 634 F.2d 964, 973-79 (5th Cir. 1981).

433. See Harwood, Can Immigration Laws Be Enforced?, 72 Pub. Interest 107, 109 (1983).

434. See, e.g., King, Flood of Bogus I.D. Papers Imperils Plan to Curb Aliens, N.Y. Times, Sept. 9, 1983, at A1, col. 1. The prevalence and importance of document fraud surely would be increased by enactment of the Simpson-Mazzoli immigration bill, which would create programs of employer sanctions and amnesty for illegal aliens based upon personal identification documents. See infra note 452.

435. INS expenditures increased from $358 million during 1980, the year during which the Mariel boatlift occurred, to $495 million in fiscal 1983. This constituted an increase of only 17% in real terms, during a three-year period in which extraordinary and massive new demands were being made on the agency. Perhaps more relevant to the effectiveness of the agency's program, authorized personnel in fact *declined* during that period (from 10,943 to 10,501), as did actual staffing (from 9982 to 9857). Within that aggregate category, enforcement staff levels remained essentially unchanged (from 4764 to 4499). Letters from INS to Peter H. Schuck (May 12, 1983 and Oct. 25, 1983) (copies on file at the offices of the Columbia Law Review).

of guards patrolling the United States Capitol.[436] There are few powerful political constituencies seeking stronger border enforcement.

In such a situation, an effective enforcement program simply is not feasible. When fewer staff must perform far more complex and demanding tasks, when more formal, individualized, humane, and accurate decisions must be made with respect to more numerous and strategically sophisticated aliens, the gap between the demands of the law and the realities of administration may pass beyond the inevitable and tolerable, attaining truly critical dimensions.

Edwin Harwood has suggested another measure of the magnitude of this moral and administrative predicament. In 1961, over twelve percent of all illegal aliens who were ordered by the INS to leave were formally deported rather than allowed "voluntary departure"; by 1982, fewer than two percent were deported.[437] As the number of undocumented aliens entering the United States has grown, INS border enforcement has come to depend almost entirely upon its ability to persuade the vast majority of the deportable aliens whom it apprehends that they should admit to illegal status and depart voluntarily. If a substantial fraction of them were to begin to stand mute, claim to be citizens, or invoke their right to a due process deportation hearing, the INS enforcement machinery would grind to a virtual halt, at least without substantial infusions of additional budget and manpower resources.[438] As Harwood puts it, "[e]nforcement works to the extent it does only because aliens are either insufficiently informed of the law's vulnerability in this regard, or because they are basically honest and unable to persist in a deception."[439] In short, the viability of immigration enforcement today depends largely upon the continuation of two conditions that may well prove transitory—aliens' ignorant waiver of their legal rights and their candor in the face of powerful incentives to dissemble.

The total collapse of a credible enforcement posture could result from legal developments that enhance the strategic position of undocumented aliens, developments the administrative implications of which are only now beginning to be perceived. First, whatever their compensating benefits, new procedural protections, such as the application of the exclusionary rule to immigration enforcement proceedings and a right, in effect, to have free appointed counsel,[440] must inevitably strengthen the bargaining position of the

436. See N.Y. Times, Sept. 14, 1983, at A26, col. 1 (2800 Border Patrol officers); Telephone interview with the Office of the Sergeant at Arms of the Congress (Dec. 22, 1983 (1222 Capitol building security guards). The INS is seeking a significant increase in Border Patrol resources. See supra note 307.

437. See Harwood, supra note 433, at 109–10.

438. Id. at 111, 116.

439. Id. at 116.

440. See Lopez-Mendoza v. INS, 705 F.2d 1059, 1089–92 (9th Cir. 1983) (en banc) (Alarcon, J., dissenting), petition for cert. filed, 52 U.S.L.W. 3294 (U.S. Sept. 22, 1983) (No. 83-491). To be sure, the exclusionary rule has often been applied in immigration cases. Id. at 1064–65. But its use under current conditions could raise novel administrative problems. With respect to appointed counsel, see supra note 402.

numerous illegal aliens who wish to resist or delay deportation. Second, the dramatic increase in asylum applications growing out of the chaotic conditions in Central America in part reflects legal and strategic advice rendered to aliens by lawyers and organizations that, quite appropriately, seek to delay or defeat deportation of their clients. Many international developments outside of anyone's control could unleash new waves of undocumented entries and asylum petitions. Third, expanded procedural rights, such as those extended to excludable aliens in *Fernandez-Roque v. Smith*[441] also enhance their bargaining power by making the expulsion process more costly, protracted, and uncertain. There is some evidence suggesting that such a situation can exacerbate these problems by attracting additional aliens who see strategic opportunites in the system's marasmus.[442] Finally, several of the pending Simpson-Mazzoli legislative proposals, particularly the amnesty for illegal aliens, would not only create additional opportunities to delay and prevent legitimate expulsions of aliens already here; in addition, they probably would attract new illegal entrants who hope, through use of fraudulent documents, to qualify for amnesty and legal permanent residence status.[443] While that legislation may not pass Congress, any reform package that is enacted in the future will almost certainly have to contain an amnesty provision that will create these risks. Nor is it clear that a workable amnesty program that avoids them can be devised.

At some point, administrative incapacity prefigures a critical loss of legitimacy. In the case of immigration, it is entirely possible that this point has already been reached. In this sense, the lamentation that "we have lost control of our borders" is more than an expression of profound frustration, more than a rhetorical *cri de coeur*. The tidal wave of undocumented aliens who cannot be effectively deterred from crossing into the United States has swept away the credibility of INS enforcement. At the same time, only three years after enactment of the Refugee Act, a backlog of more than 166,000 pending asylum applications[444] is encouraging, if not requiring, policies of mass detention. These policies have led to judicially mandated mass releases into the community and pressures for crude categorical relief.[445] These simultaneous, parallel explosions in the number of putative refugees and in the number of illegal aliens have, as Alexander Aleinikoff puts it, imposed humanitarian and

---

441. 567 F. Supp. 1115 (N.D. Ga. 1983). See discussion supra notes 397–99 and accompanying text.

442. See Martin, supra note 402, at 180–81 (data from West Germany).

443. See, e.g., Harwood, Immigrant Bill Alien to Reality, N.Y. Times, Sept. 5, 1983, at 19, col. 4; King, supra note 434.

444. See supra note 222 and accompanying text.

445. See, e.g., Jean v. Nelson, 711 F.2d 1455, 1462–63, reh'g en banc granted, 714 F.2d 96 (11th Cir. 1983); Fernandez-Roque, 567 F. Supp. 1115 (N.D. Ga. 1983). A substantial fraction of these asylum cases would be disposed of by the Simpson-Mazzoli bill, which would extend categorical permanent resident status to Cubans and Haitians who arrived before January 1, 1980. S. 529, 98th Cong., 1st Sess. § 301, 129 Cong. Rec. S6970, S6982–83 (daily ed. May 18, 1983).

process demands regarding respect to the former that, if applied to the latter, will effectively immobilize enforcement.[446]

These contradictions do not simply confirm that the law's promises have outstripped its capacity to deliver. Far more important, they are sapping immigration law of its moral force in the eyes of aliens, employers, officials and the general public. Law without legitimacy is little more than naked force, the power, as it has been said, that comes out of the barrel of a gun. And because immigration law reflects some of our most deeply held values concerning community, self-definition, national autonomy, and social justice, any diminution of its legitimacy entails a profound, perhaps irretrievable, loss.

Unlike some problems that government faces, the problem of restoring legitimacy to immigration law is partly one of money. Additional resources and staff could accelerate exclusion, deportation, and asylum decisions, reducing the need for prolonged detention of aliens and reducing the period during which they can acquire the ties that a more communitarian immigration law is most reluctant to sever. More funds could also improve the INS's border control and investigation activities, although serious difficulties would remain under even the most optimistic foreseeable budgetary conditions.[447] Coupled with other reforms, such as effective employer sanctions and an amnesty program,[448] it might at least contain the influx of undocumented aliens. Until Congress appropriates funds and authorizes personnel levels that are adequate to the scale of the challenge, however, immigration law will continue to teeter on the brink of moral and programmatic bankruptcy.

But even dramatically increased funds could not wholly solve immigration law's legitimacy problem, at least in the short run. The more spacious conception of community revealed in cases such as *Plyler v. Doe*[449] is also working to undermine the legitimacy of immigration policies directed at the apprehension and exclusion of undocumented aliens. Two very recent examples suggest the dimensions of this conflict. The first involves the insistence by the police chief of Santa Ana, California, probably repeated less candidly in many other communities near the Mexican border, that federal immigration officials who seek to apprehend illegals are not welcome in that city. Accord-

446. Letter from T. Alexander Aleinikoff to Peter H. Schuck (Aug. 25, 1983) (copy on file at the offices of the Columbia Law Review). Aleinikoff adds: "When we add to this the perceived political necessity of labelling refugees from friendly countries as 'illegal aliens' (Salvadorans) and illegal aliens from enemies as 'refugees' (Cubans), no wonder things are as confused as they are." Id.

447. See, e.g., supra notes 432–46 and accompanying text.

448. The Simpson-Mazzoli bill would establish both programs. Immigration Reform and Control Act of 1983, S. 529, 98th Cong., 1st Sess. §§ 101(a)(1), 301–02, 129 Cong. Rec. S6970, S6982–83 (daily ed. May 18, 1983). Their effectiveness, however, remains in serious doubt. For example, both programs could be subverted by massive document fraud. See Comment, Illegal Immigration: Employer Sanctions and Related Proposals, 19 San Diego L. Rev. 149, 164 (1981); King, supra note 434.

449. 457 U.S. 202 (1982).

ing to another Santa Ana official, "we never invited the undocumented alien population to settle in our city but now that they have, we are going to work with them. You can't afford to have 25 percent of the population hostile towards the Police Department."[450]

The second example involves the willingness of a steadily increasing number of churches and individuals, now estimated at up to 30,000, to defy the immigration authorities by openly offering sanctuary to undocumented aliens, especially those from Central America and Caribbean countries who are unlikely to be granted asylum status in the near future.[451] By providing this sanctuary, these individuals almost certainly violate criminal provisions that impose possible sanctions of fine and imprisonment.[452] Such attitudes betray a growing sense of moral commitment and obligation to strangers. But while these attitudes and the acts of personal witness that they inspire may be morally admirable, they bode ill for the effective enforcement of immigration law.

Indeed, communitarian values are themselves antithetical to vigorous immigration enforcement, especially when reinforced by traditional liberalism's human rights dimension, which Americans continue to cherish. By regarding all who arrive here not as strangers but as members, along with Americans, of a universal moral community in which abstract principles of exclusion yield to the more palpable claims of actual and potential human linkage, communitarian values deprive administrative deportation efforts of the moral legitimacy that the classical order managed to sustain for so long. If that is so, the collapse of immigration enforcement may actually generate a profound public reaction against aliens and the communitarian values that increasingly protect them, reviving the nativist impulses that have always been an important, albeit often deplorable, element of our national character.

B. *Judicial Role*

The transformation of immigration law also casts into bold relief a set of contradictions surrounding the parallel, intimately related transformation of the function of courts in the American legal system. Few areas of public law are so susceptible to administrative abuse and lawlessness as immigration law. The INS is among the most insular and chronically understaffed of federal agencies; it is vulnerable to manipulation and neglect by Congress and to

---

450. N.Y. Times, Sept. 11, 1983, at A29, col. 1.

451. See Austin, More Churches Join in Offering Sanctuary for Latin Refugees, N.Y. Times, Sept. 21, 1983, at A18, col. 1.

452. An example is the so-called "Texas" proviso which provides in pertinent part "that for the purposes of this section, employment (including the usual and normal practices incident to employment) shall not be deemed to constitute harboring." 8 U.S.C. § 1324(a)(4) (1982). This proviso is not abolished by the Senate version of the Simpson-Mazzoli bill, S. 529, 98th Cong., 1st Sess. § 101, 129 Cong. Rec. S6970 (daily ed. May 18, 1983). The effect of the proviso, however, would certainly be affected by the sanctions against employers in the bill. See S. Rep. No. 98-62, 98th Cong., 1st Sess. 7-11 (1983).

political reprisals by powerful employer interests opposed to vigorous law enforcement. Aliens, the nominal clients of the system, are politically and economically weak, unfamiliar with legal forms, procedures, and the language, and often reluctant or unable to assert their rights. Unlike welfare recipients, school children, and mental hospital patients, for example, aliens often exist in a kind of *social* vacuum, outside any structures of institutional, programmatic, administrative, or professional support.

INS decisions, moreover, have low visibility; they occur in isolated adjudicatory contexts in which their larger policy consequences, if any, are fragmented and thus difficult to discern or monitor. Decision errors are likely to be common, and the distribution of costs and benefits of such errors makes their detection unlikely. Thus positive errors—in which the agency erroneously admits or grants a benefit to one who is not entitled to it—do not affect the agency's budget and will not be challenged unless Congress does so.[453] Negative errors—in which the INS erroneously excludes, deports or denies a benefit to an alien who is entitled to it—only affect the agency's budget if it is challenged, and then only to the extent of the resources necessary to contest the appeal. Negative errors, of course, are highly salient to aliens, perhaps even life threatening, but they often are not in a position to challenge such errors in administrative or judicial review proceedings or in Congress.

Federal judges must be presumed to be aware of these realities, and it would not be surprising if many of them concluded that such a decisional context cries out for judicial intervention. Courts of general jurisdiction might be expected to hold the INS to the same constitutional and administrative standards to which other federal agencies are held—indeed, to higher standards in view of the life and liberty interests so often at stake in immigration cases, especially those involving detention, asylum, and EVD.[454] In conjunction with the changes discussed in Part II, these factors surely help to explain and perhaps even to justify the emergent judicial assertiveness that is both a hallmark and a necessary condition of immigration law's incipient transformation.

Perhaps paradoxically, however, an active, intrusive judicial role in immigration cases raises serious dilemmas, the resolution of which will shape the future of immigration law. Some of the manifest difficulties of devising a new constitutional order in an area of law that has long defied one are revealed in *Plyler v. Doe*,[455] in which the Court felt obliged to turn conventional legal

---

453. For a rare example of such a challenge, see INS v. Chadha, 103 S. Ct. 2764 (1983).

454. As David Martin puts it, the private interests at stake are "off the charts," in terms of the balancing test of Mathews v. Eldridge, 424 U.S. 319, 334-35 (1976). Martin, supra note 103, at 190. This is also true in § 243(h) cases, which involve withholding of deportation because of the threat of persecution, 8 U.S.C. § 1253(h) (1982). See, e.g., Stevic v. Sava, 678 F.2d 401 (2d Cir. 1982) (deportation); Fernandez-Roque v. Smith, 567 F. Supp. 1115 (N.D. Ga. 1983) (detention); Hotel & Restaurant Employees Union, Local 25 v. Smtih, 563 F. Supp. 157 (D.D.C. 1983) (extended voluntary departure).

455. 457 U.S. 202 (1982).

categories and precedents inside out in order to reach a morally appealing result. Others are revealed in *Fernandez-Roque*, in which the court created an entire procedural system out of whole cloth, one that places some excludable aliens in an even better legal and strategic position than deportable aliens, many of whom have already become more or less integrated into our society. Still other difficulties are apparent in *Hotel & Restaurant Employees Union, Local 25 v. Smith*,[456] in which the court proceeded as if the justiciability of EVD for particular nationality groups were no different than the justiciability of an individual detention or adjudication case, as if the intricacies of contemporary foreign policy in Central America were simply an irrelevant smokescreen thrown up by the government in order to confuse the court.

In addition to the courts' problematic efforts to fashion a new order in an area that historically has been resistant to judicial involvement, obstacles to effective intervention inhere in immigration law's own administrative structure. The broad substantive discretion that the statute confers upon the Attorney General in almost all areas of immigration policy suggests the distinct possibility that judicially mandated procedural changes would not significantly affect outcomes. Instead, they may simply make the agency's journey to the same substantive result more costly and cumbersome. By adversely affecting the agency's bargaining position vis-à-vis illegal aliens, such changes may succeed in incapacitating the informal enforcement process as well. Yet if courts attempt to avoid these difficulties by ordering reform of a more structural, systemic nature, they will encounter not only the same kinds of pitfalls that have plagued judicial efforts to constitutionalize other areas of governmental activity[457]—lack of manageable standards, inadequate policymaking tools, limited inducements, implementation problems, questions of institutional legitimacy, and the like—but also some others that are more or less peculiar to the immigration field.[458]

If this analysis is correct, the conflicts over judicial role engendered by the transformation of immigration law can probably be best resolved in the broad middle ground between the abject judicial deference of the classical tradition and the sweeping structural relief sometimes granted in other policy areas, precursors of which are beginning to appear in immigration cases as well.[459] Courts should actively scrutinize INS adjudications, demanding deci-

---

456. 563 F. Supp. 157, 160–61 (D.D.C. 1983).

457. See, e.g., P. Schuck, supra note 18, chs. 6 & 7; Altman, Implementing a Civil Rights Injunction: A Case Study of *NAACP v. Brennan*, 78 Colum. L. Rev. 739 (1978); Diver, The Judge as Political Powerbroker: Superintending Structural Change in Public Institutions, 65 Va. L. Rev. 43 (1979); Frug, The Judicial Power of the Purse, 126 U. Pa. L. Rev. 715 (1978). Moreover, unlike the typical structural injunction case, in which a state agency confronts the combined power of the federal government, a federal court ordering immigration reforms would oppose the United States Department of Justice, and perhaps the Congress as well. In that posture, successful implementation of the decree would be especially problematic.

458. See supra notes 429, 432–56 and accompanying text.

459. See supra notes 70–92, 377–79 and accompanying text.

sionmaking procedures of high accuracy, fairness and dignity. Although courts can certainly do this in the context of reviewing those individual adjudications that are appealed, the peculiarities of the immigration decision process—especially the weakness of politics and litigation as controls over administrative abuse, and the high costs to aliens, and citizens, of negative errors—argue strongly that courts should insist that administrators install more systematic quality-assurance techniques at the INS and BIA levels over decisions,[460] rather than demanding ever more formalized, extensive hearing procedures whose effects, apart from increased cost and delays, are uncertain and quite possibly perverse.[461]

Courts should also be more discriminating in their approach to reviewing immigration decisions. In particular, they should distinguish between congressional and executive branch decisions that do not involve delicate foreign policy judgments, such as many deportation decisions,[462] and those that do, such as policies on EVD. They should feel much freer to apply conventional constitutional and administrative law principles to the former than to the latter. Where life and liberty interests collide with powerful foreign policy, law enforcement, or other governmental interests, courts should seek relatively flexible solutions, such as "clear statement" requirements or remands to the agency with instructions to develop new approaches, rather than finding refuge in rigid constitutional rulings.[463] Similarly, courts in appropriate cases should require the INS to issue standards that narrow and guide its vast discretion, thereby putting aliens and low-level officials on notice concerning what the law requires and enabling administrative superiors, Congress and the courts to hold the INS to those standards.[464]

In urging the courts to take these steps, however, we should not imagine that any set of immigration adjudications are likely to constitute an adequate solution to the contradictions and problems that this transformation is engendering. The crucial determinants of INS policy—resources, political support, international developments, the inherent vulnerability of illegal aliens—lie

---

460. See generally Chassman & Rolston, Social Security Disability Hearings: A Case Study in Quality Assurance and Due Process, 65 Cornell L. Rev. 801 (1980).

461. See supra notes 379–81, 440–46 and accompanying text.

462. See, e.g., Fiallo v. Bell, 430 U.S. 787 (1977) (upholding the definition of "child" under 8 U.S.C. § 1101(b)(1) (1982) as excluding illegitimate children who seek preference by virtue of the child's relationship with the father); Kleindienst v. Mandel, 408 U.S. 753 (1972) (upholding the refusal of the Attorney General to waive ineligibility for entry into the United States of a foreign journalist who was barred under 8 U.S.C. §§ 1182(a)(28)(D), (G)(v) & (d)(3)(A) (1982) as an advocate or publisher of the "doctrines of communism").

463. For a general exploration of remedial flexibility, see P. Schuck, supra note 18, ch. 8. For an application of the clear statement approach in a quasi-immigration case, see Kent v. Dulles, 357 U.S. 116 (1958) (holding that the Secretary of State may not refuse to issue a passport because of the applicant's political views, absent an explicit congressional provision).

464. See generally P. Schuck, supra note 18, at 143–46. According to Professor Diver's analysis, many categories of INS decisions are susceptible to this form of control. Diver, supra note 90, at 92–97, 106–09.

well outside judicial control. Because judges can affect these matters only marginally, if at all, they cannot really ease the profoundly difficult moral and policy choices that make immigration policy so unmanageable today. Those choices involve tradeoffs that only Congress can effectively address. Unfortunately, as the recent derailment of the Simpson-Mazzoli legislation vividly suggests, Congress remains politically stalemated on precisely these questions.

Regardless of whether or not Congress is prepared to act, however, a communitarian immigration law will demand a great deal of courts. They must maintain their traditional concern for the protection of individual rights and procedural fairness values. Indeed, if the INS cannot be made to respect these values, judicial vigilance is bound to become more intense. At the same time, however, courts must recognize the dangers to political control and effective administration of immigration policy posed by judicial pursuit of communitarian ideals that, however appealing in the abstract, are still only weakly rooted in our collective attitudes toward contemporary immigration.

C. *Community*

Community, as Carl Friedrich observed, is "the central concept of politics."[465] It follows, then, that immigration law is a fulcrum of our political system. By seeking to define, mold and protect the American community, it undertakes to answer the first questions that any society must put to itself: What are we? What do we wish to become? How shall we reach that goal? And most fundamentally, which individuals constitute the "we" who shall decide these questions?[466] A liberal polity asks these questions with a special intensity and self-consciousness. To most Americans, they cannot be neatly resolved by invoking divine guidance, settled traditions, or some imminent, organic principle. Instead, these questions can only be addressed through the never-ending processes of rational argument, irrational commitments and political choice.

Liberalism has never satisfactorily answered these questions and probably never will. It regards any fixed or exclusive definition of community with profound suspicion. Indeed, in a truly liberal polity, it would be difficult to justify a restrictive immigration law or perhaps any immigration law at all. National barriers to movement would be anomalous. Criteria of inclusion and exclusion based upon accidents of birth, criteria that label some individuals as insiders and others as outsiders, would be odious. Wealth, security and freedom would not be allocated on such grounds, especially in a world in which the initial distribution of those goods is so unequal.[467] Instead, individuals

---

465. Friedrich, The Concept of Community in the History of Political and Legal Philosophy, *in* Community (Nomos II) 20 (1959).

466. See B. Ackerman, Social Justice in the Liberal State 89–95 (1980).

467. See C. Beitz, Political Theory and International Relations 161–76 (1979); Fishkin, supra note 269, at 358–61.

would remain free to come and go, to form attachments, and to make choices according to their own aspirations, consistent with the equal right of others to do likewise. No self-defining, self-limiting group could deny to nonmembers the individual freedom of action that liberalism distinctively celebrates. In essence, this was the conception of community—one embracing all who wished to come and remain here—that prevailed until the 1880's. Indeed, just as the period of open borders was coming to a close, the Supreme Court breathed additional constitutional life into that communal ideal by extending the succor of the equal protection principle, and thus the protection of the constitutional community, to all "persons" in the United States, whether citizen or alien.[468]

In the second American century, immigration law, shaped by a burgeoning sense of American nationhood and international stature, narrowed this ideal in the interests of defining a *national* community. As the flood of "new immigrants" prior to World War I attests, this definition was not always or inevitably a static one. Nevertheless, it eventually placed severe limits upon the potential character and membership of that community, constricting it from an essentially universalistic one to a legal artifact self-consciously constricted by Congress. The felt necessity to restrict the community compelled liberalism for the first time to try to develop criteria of exclusion. But the idea of a restrictive community based not upon universal human rights but upon contingent national interests proved to be a very hard one for liberalism to integrate. Its meaning shifted depending upon who was defining it, for what purposes and in what context.

If immigration law were to promote some ideal of community, then, it must first identify what the grounds of community formation should be. Those grounds, however, are as varied as the patterns and motives of human association. Earlier idealizations of community—the intimacy of the Greek *polis*[469] or the spiritual unity of a universal religion,[470] for example—were obviously irrelevant to conditions in America. Liberal theory provided no guidance either. For all its universality, liberalism essentially viewed society as a contrivance animated solely by individuals' self-interest, by their need for protection against strangers and against each other. It denied the natural sociability and shared values that the ancients had taken to be the fundamental basis of political life.

Indeed, it is no exaggeration to say that traditional liberalism articulated no real theory of community at all. Its vision of the good society was one of privatistic acquisitiveness, not social solidarity and common purpose.[471] Unable to stake out any middle ground between the utopian community of all

468. Yick Wo v. Hopkins, 118 U.S. 356, 369 (1886).

469. Aristotle, Politics, Bk. I, ii 8-12, at 111-13 (W. Bollard trans. 1877).

470. See discussion in Friedrich, supra note 465, at 7-12.

471. T. Hobbes, Leviathan 56-58 (Oakeshott ed. 1962); J. Locke, The Second Treatise of Government 62-64 (J. Gouch ed. 1946).

mankind and the ahistorical community of individuals freely contracting in the state of nature, liberalism could not furnish immigration law with a coherent definition of the relevant community. In that ideological vacuum, the decidedly illiberal, nationalistic features of classical immigration law—the national origins quota and the fortress mentality that underlay summary, extraconstitutional exclusion procedures, for example—were able to flourish.

As America's third century begins, the struggle over the dominant conception of community—and thus over the contemporary meaning, relevance and relative significance of liberal, nationalistic and collectivist values—continues. But this struggle is now being waged in a new legal and political context. Today, large numbers of migrants can easily, inexpensively, and surreptitiously enter United States territory, where they can readily form social and economic attachments that the government cannot easily sever. Here, they find a society in which private property increasingly consists of advantages—including public benefits and legal alien status—that only the government can confer, a society in which the stakes in defining community, in inclusion or exclusion, are therefore higher than ever. They also find a legal order that is in great flux, one whose premises are shifting from individualistic values, which emphasize consent as the source of duty, to communitarian values, which ground duty in expanding social relationships, interdependencies and expectations. From the point of view of the undocumented alien, who under present conditions often cannot be apprehended much less expelled until she has had an opportunity to form these attachments, these developments add up to good news indeed.

These brute facts present liberalism with a poignant predicament. Committed to the rule of law but confronted by individuals who, sociologically speaking, have found community in America only after flouting that law, liberalism cannot legitimate their presence. Committed to the moral primacy of consent, liberalism cannot embrace those who enter by stealth. Committed to universal human rights, liberalism cannot secure those rights in the real world without rooting itself in political institutions that are actually capable of instantiating its values. Today and for the foreseeable future, those political institutions are the institutions of the *nation*.

As Part I revealed, moreover, even a liberal nation has powerful propensities toward illiberal, exclusionary practices. The very idea of nationhood implies a coherence of shared tradition, experiences and values—a national community. This community is inevitably parochial and inward-turning; just as it affirms a core of common commitments, it also sets the nation apart from, if not above, the rest of humanity.[472] Moreover, the nation has ordained an activist government that seeks to assure at least a minimal level of economic security and well-being for individuals. It cannot possibly provide this assurance to mankind in general; instead, it must restrict its primary concerns to those to whom it has undertaken a special political responsibility. Even this

472. J. Rousseau, The Government of Poland 5-9 (W. Kendall trans. 1972).

more limited task is impossible if masses of destitute people, many ill-equipped to live and work in a postindustrial society, may acquire legally enforceable claims against it merely by reaching its shores. Finally, the nation inevitably seeks to mobilize its people's passions and energies against real or imagined threats, as in wartime or domestic crisis. To accomplish this, it must make powerful emotional appeals to national unity and sacrifice. Whether the goal is the realization of universal liberal values—"making the world safe for democracy," for example—or more parochial ones—such as "manifest destiny" or "the great society"—the individual wills that liberalism glorifies must be merged into an almost mystical embodiment of national character and patriotic purpose.

For liberal values to triumph, then, liberalism is obliged to accommodate a competing, illiberal conception of community that threatens those values even as it promises to actualize them—albeit for only a limited portion of humanity. By investing the distinction between insiders and outsiders with moral and political significance, nationalism—the ideal of national community—rejects liberalism's own communal visions. The universal brotherhood of man enjoying natural rights and the society of strangers linked by little more than contract are both too impractical and alienating, and represent too impoverished a view of what our society and political natures require, to fully realize our humanity. Both ignore our basic need for what Michael Walzer has called "communities of character, historically stable, ongoing associations of men and women with some special commitment to one another and some special sense of their common life."[473] For the liberal welfare state to enlist the active public support necessary if it is to do its affirmative, individuality-enhancing work, some such community is essential. Our psychological natures also seem to link mutual commitments to some degree of human propinquity. Thus, as Walzer observes, the very notion of a meaningful community, liberal or otherwise, probably implies some exclusion for at least some purposes. The relevant questions are how much exclusion the sense of community requires, and upon what basis it should proceed.

Liberalism rules out certain answers to these questions—exclusion based upon race, for example. In practice and probably in principle, however, it also leaves open many possibilities. It remains to be seen whether the communitarian values that are increasingly shaping our legal order can supply immigration law with appropriate criteria of inclusion and exclusion. Certainly, these new values reflect important sociological facts about how social linkage and expectations evolve, facts that no just legal order, no ideal of legal community, can ignore.

Yet a communitarian conception of immigration law provides some grounds for concern. Once society's duty to aliens is no longer moored to the classical norms of mutual consent and compliance with publicly sanctioned legal procedures but is derived from vague, even circular, notions of social

473. M. Walzer, supra note 2, at 62.

expectations and relationships, the legal order is cast adrift upon a sea whose ungovernable tides may carry it to realms unknown, unimagined, and fraught with dangers.[474] The problem is not simply one of a limited American capacity to assimilate newcomers. Our history testifies eloquently to the greatness and durability of that capacity. It is not inexhaustible, of course, and as noted earlier, the conditions of a postindustrial society may make the effort to assimilate masses of migrants from premodern societies especially problematic today. Still, when one recalls that more than fourteen percent of the United States population in 1910 was foreign-born compared to four and a half percent in 1970 and that eight other industrialized countries now have a higher percentage of foreign-born than the United States does,[475] the prospect of even several million new immigrants joining a nation of 226.5 million people,[476] especially one that is rapidly aging and has an historically low fertility rate, seems considerably less daunting.

The sheer *number* of immigrants, then, is unlikely in itself to create unmanageable difficulties, even assuming a range of admissions far more generous than current policy permits. Two other considerations, however, are more troubling. First, there is a risk that a communitarian immigration law will encourage further social and cultural fragmentation, intergroup hostility, distributional inequities, and intensified political conflict, and that these conditions will at some point degrade the quality of American democracy. The discord between English-speaking and French-speaking Canadians, not to say the civil strife that afflicts so many other polyglot societies, looms as a haunting reminder of this danger.

Second, the tension between liberalism's universal aspirations and our need as a society to achieve the degree of solidarity that effective activist government requires must be resolved at *some* level of exclusion. Although it remains unclear what this level should be, it is imperative that America make this choice self-consciously rather than having it preempted by the self-interested actions of others. When courts wielding some conception of communi-

---

474. Mutatis mutandis, it begins to approach Roberto Unger's "countermision" of contract law, in which

> the initial, tentative definition of any entitlement must now be completed by a second stage. Here the boundaries are drawn and redrawn in context according to judgments of both the expectations generated by the specific situation of interdependence and the impact that a particular exercise of a right might have upon other parties to the relationship or upon the relationship itself.

Unger, The Critical Legal Studies Movement, 96 Harv. L. Rev. 561, 639-40 (1983). Indeed, to some commentators, even clear positive law is no limitation upon the vast transformational power of communitarian values. See, e.g., Lopez, supra note 12, at 696 ("No matter how strongly our formal laws deny it, our conduct creates the obligation [to undocumented Mexican workers]."). For several thoughtful efforts to refine the concept of community as applied to contemporary immigration law, see id.; Aleinikoff, supra note 223; Martin, supra note 103.

475. See SCIRP Staff report, supra note 25, at 296.

476. U.S. Bureau of the Census, Statistical Abstract of the United States: 1982-1983 (1982).

tarian values decide to empower strangers unilaterally to create and enforce substantial claims against American society, that process of self-definition is seriously impeded. If the American community's power to define its common purposes and obligations is no greater than the power of strangers to cross our borders undetected and to acquire interests here, our capacity to pursue liberal values—to decide as individuals and as a society what we wish to be—may be critically impaired.

Conclusion

Social conditions and a changed legal consciousness have begun to undermine the foundations of classical immigration law. In its growing decrepitude, we can glimpse both the remnants of the individualistic legal order that once gave it life and legitimacy, and the outlines of the communitarian one that promises to transform it. The courts are busily razing the old structure and designing the new one, largely along the lines laid down by the contemporary administrative and constitutional orders. Immigration is gradually rejoining the mainstream of our public law.

Immigration Law After a Century of Plenary Power: Phantom Constitutional Norms and Statutory Interpretation

Hiroshi Motomura†

CONTENTS

---

† Associate Professor of Law, University of Colorado School of Law. I would like to thank Alex Aleinikoff, Richard Collins, Richard Delgado, Philip Frickey, Glenn George, Stephen Legomsky, Mark Loewenstein, David Martin, James Nafziger, Robert Nagel, Gene Nichol, Peter Schuck, Steven Smith, Arthur Travers, and Charles Wilkinson for making many helpful comments and suggestions on ideas and earlier drafts. Elizabeth Griffin and Katherine Rice Mackey provided not only excellent research assistance, but also countless ideas and insights that contributed greatly to this Article's organization and substance.

> [W]e are in the never-never land of the Immigration and Nationality Act, where plain words do not always mean what they say.[1]

INTRODUCTION

The 1980's witnessed a resurgence of interest in immigration law, especially in the role played by constitutional principles. Much of this interest is attributable to events that tested the limits of immigration law in unprecedented ways. The influx of Cubans in the Mariel boatlift and the large number of Haitians fleeing the Duvalier regime were the most prominent, but civil strife and economic conditions throughout the world further swelled the numbers of would-be immigrants. The result was an increased and often chaotic demand for a commodity in very short supply—lawful admission to the United States. And when the federal government tried to enforce the immigration laws in this setting, long and complicated litigation was the predictable result. These events forced courts to search for new answers to the fundamental questions that immigration law poses—who are "we" as a community, and what do we demand of those who want to join?

The most prominent case arising out of these developments is *Jean v. Nelson*,[2] a 1985 decision of the United States Supreme Court that captures much of what is significant about the immigration law cases of the past decade. *Jean* involved the influx of Haitians into south Florida during the early 1980's, most of whom sought political asylum. The United States government adopted

1. Yuen Sang Low v. Attorney Gen., 479 F.2d 820, 821 (9th Cir. 1973).
2. 472 U.S. 846 (1985).

a general practice of detaining these Haitians pending a decision on their asylum claims, and did not grant their requests to be released temporarily on "parole."[3] The plaintiffs in *Jean* alleged that this detention policy denied equal protection because it discriminated on the basis of national origin and race against these Haitians, almost all of whom were Black. The case eventually reached the Supreme Court.

As I use the term, "immigration law" refers to the body of law governing the admission and expulsion of aliens.[4] It should be distinguished from the more general law of aliens' rights and obligations, which includes, for example, their tax status, military obligations, and eligibility for government benefits and certain types of employment.[5] *Jean* could have been the occasion for the Court to confront squarely the "plenary power doctrine." This doctrine has dominated immigration law since the Court adopted it almost one hundred years ago in rejecting constitutional objections to Congress' first immigration statutes. The plenary power doctrine's contours have changed over the years, but in general the doctrine declares that Congress and the executive branch have broad and often exclusive authority over immigration decisions. Accordingly, courts should only rarely, if ever, and in limited fashion, entertain constitutional challenges to decisions about which aliens should be admitted or expelled.

By the time *Jean* reached the Supreme Court, the doctrine had long been under heavy fire from many quarters. Critics expressed deep concern over the continuing isolation of an entire body of law from the mainstream of American public law—isolation not only from the process of constitutional judicial review, but also from the constitutional norms and principles developed through that process over the years.[6] Even though the Court had endorsed some version of the plenary power doctrine in cases decided in the 1970's,[7] a number of observers had predicted the gradual demise of the doctrine and a corresponding reintegration of our usual expectations regarding judicial review into immigration law.[8]

The *Jean* decision, however, was anticlimactic because the majority did not reach the constitutional issue. Instead, it found that any discrimination against the Haitians based on race or national origin would be unlawful as contrary to the statute and the government's regulations. Many knowledgeable observers

3. *See* Immigration and Nationality Act of 1990, § 212(d)(5), 8 U.S.C. § 1182(d)(5) (1988) [hereinafter INA]; 47 Fed. Reg. 30,044 (1982) (as amended at 47 Fed. Reg. 46,493 (1982)) (amending 8 C.F.R. §§ 212.5, 235.3).

4. The term "alien" is standard usage, but I use it advisedly because of the word's distancing effect and somewhat pejorative connotation. I also admit to some hypersensitivity on this point as a former "alien" myself.

5. This definition of "immigration law" is common to the literature. *See, e.g.*, Legomsky, *Immigration Law and the Principle of Plenary Congressional Power*, 1984 SUP. CT. REV. 255, 256.

6. Schuck, *The Transformation of Immigration Law*, 84 COLUM. L. REV. 1 (1984) (immigration law isolated from other legal areas).

7. *See* Fiallo v. Bell, 430 U.S. 787 (1977); Kleindienst v. Mandel, 408 U.S. 753 (1972).

8. *See, e.g.*, Legomsky, *supra* note 5; Schuck, *supra* note 6.

saw *Jean* as a potential landmark decision that never was. Understandably, those observers had focused on the plenary power doctrine's *substance*—its denial to aliens of an opportunity to vindicate constitutional claims in immigration cases. I believe, however, that this focus on the most obvious consequence of plenary power has obscured its subtle, more fundamental effects on the *process* of growth and change in this field of law.

The broader subject suggested by *Jean*—but one never before explored in any depth—is the relationship between constitutional and subconstitutional immigration law. As I use the term here, "constitutional immigration law" means the application of constitutional norms and principles to test the validity of immigration rules in subconstitutional form, including statutes, regulations, and administrative guidelines. In contrast, "subconstitutional immigration law" means the interpretation and application of those statutes, regulations, administrative guidelines, and the like. Several commentators have noted that aliens tend to receive more favorable treatment when judges render subconstitutional decisions than when they render constitutional ones, but these observations have been made almost in passing, with little explanation or detail.[9] This Article will attempt to provide the missing analysis.

I begin with a point that was central to Justice Marshall's dissent in *Jean*: nothing in the parole statute or regulations expressly prohibited the government from considering race or national origin in its parole decisions.[10] He is correct about the language of the statute and regulations.[11] Also correct is his more general point that parole is one of many immigration matters committed by statute to the broad discretion of the Attorney General.[12] What, then, prohibits the government from considering race or national origin when regulating the conduct of aliens? If this prohibition has a constitutional source, then was *Jean* a constitutional decision even though the Court purported to rule exclusively on statutory grounds? And if *Jean* was really a constitutional decision that implicitly upheld an equal protection claim, to what extent did this decision limit plenary power?

My ultimate purpose is to examine immigration law for what it tells us about a general problem of legal process—the relationship, usually symbiotic, between statutory interpretation and constitutional law. The usual pattern in our law, most observers would agree, is that constitutional norms manifest themselves in two principal ways. First, they govern expressly constitutional deci-

---

9. *See* S. LEGOMSKY, IMMIGRATION AND THE JUDICIARY: LAW AND POLITICS IN BRITAIN AND AMERICA 233-35, 239-41, 252-53, 259, 300 (1987); D. MARTIN, MAJOR ISSUES IN IMMIGRATION LAW 19, 25-27 (1987); Anker, Jean v. Nelson*: Neutral Principles in the Supreme Court Without the Constitution*, 8 IMMIGR. J. 9 (1985). I am indebted to these commentators, whose observations on the relationship between constitutional and subconstitutional immigration cases first raised in my mind the questions that this Article tries to answer.

10. Jean v. Nelson, 472 U.S. 846, 858-68 (1985) (Marshall, J., dissenting).

11. The governing regulation at that time was 8 C.F.R. § 212.5 (1982).

12. *See* 472 U.S. at 862-64.

sions; this is the intuitive definition of "constitutional." Second, in a less intuitive but equally correct use of the term, "constitutional" norms provide the background context that informs our interpretation of statutes and other sub-constitutional texts. This second use of constitutional norms explains and reflects the time-honored canon that courts ought to interpret statutes so as to avoid constitutional doubts.

Immigration law, as it has developed over the past one hundred years under the domination of the plenary power doctrine, represents an aberrational form of the typical relationship between statutory interpretation and constitutional law. The aberrant quality is attributable to the prolonged nature of the contradiction between these two sets of "constitutional" norms in immigration law. The constitutional norms that courts use when they directly decide constitutional issues in immigration cases are not the same constitutional norms that inform interpretation of immigration statutes. To serve the latter function, many courts have relied on what I call "phantom constitutional norms,"[13] which are not indigenous to immigration law but come from mainstream public law instead. The result has been to undermine the plenary power doctrine through statutory interpretation.

In my view, any fair assessment of phantom norm decisionmaking should reflect deep ambivalence. On the one hand, it has been an understandable and perhaps even noble response to the shortcomings of the plenary power doctrine, and for that reason I do not intend the term "phantom" pejoratively. More generally, the use of phantom norms during a transitional phase may be a healthy form of constitutional change. On the other hand, statutory interpretation confuses and contorts the law when the interpreting court relies for an extended period on constitutional norms that are doctrinally "improper" in the sense that they do not control in cases which explicitly involve interpreting the Constitution. I suggest that the only way out of the dilemma posed by the prolonged reign of phantom norm decisionmaking in immigration law is to bring the transitional phase to an end—in short, to undertake a direct and candid reassessment of plenary power as constitutional doctrine.

In addition to the light that I hope to shed on the relationship between statutory interpretation and constitutional law generally, I also will try to answer a more specific operational question, probably the most important one in immigration law today—what is left of the plenary power doctrine? A widely accepted view, with which I generally agree, is that the doctrine is in some state of decline. Yet the apparent conflicts and contradictions among the cases, and sometimes their open ambivalence, have frustrated the efforts of courts and commentators to be more precise about the doctrine's current vitality. I will suggest that the gradual demise of the plenary power doctrine is best understood as a function of the tension in immigration cases between constitutional doctrine

13. I am grateful to Alex Aleinikoff for suggesting the term "phantom norm."

and statutory interpretation. The fact that the subconstitutional cases play such a central role in understanding plenary power means, in turn, that the doctrine's current vitality is the most important question not only in constitutional immigration law, but in immigration law as a whole. In short, this Article endeavors to provide a coherent theoretical and practical framework for all of immigration law, both constitutional and subconstitutional.

The discussion proceeds in four stages. Part I explains the advent and persistence of the plenary power doctrine as the dominant principle of constitutional and subconstitutional immigration law. Part II discusses the emergence of phantom norm decisions in the 1950's and 1960's. Part III analyzes more recent phantom norm decisions—most prominently *Jean v. Nelson*, but also similar cases in the lower courts. Part IV examines more closely the problems that arise when courts choose phantom constitutional norms to guide their interpretation of statutes.

I. PLENARY POWER AS CONSTITUTIONAL LAW

A. *Classical Immigration Law*

The story of the plenary power doctrine's role in constitutional immigration law begins with the Supreme Court's 1889 decision in the *Chinese Exclusion Case*.[14] Chae Chan Ping, a Chinese laborer, came to the United States in 1875, in a period when the Burlingame Treaty of 1868 seemed to guarantee unrestricted immigration from China. The United States had negotiated the treaty when it was interested in obtaining cheap Chinese labor and improving trade with China.[15] But soon afterwards, racist and nativist anti-Chinese sentiment became widespread in California, and gradually gained influence on the national political scene.

In 1880, the two countries negotiated a supplemental treaty that allowed the United States to "regulate, limit or suspend" immigration of Chinese laborers.[16] Those already here in November 1880 could continue "to go and come of their own free will and accord."[17] In 1882, Congress suspended immigration of Chinese laborers for ten years. Those here who wished to leave and return could obtain certificates to show that they had come before November 1880,

14. Chae Chan Ping v. United States, 130 U.S. 581 (1889)

15. Justice Field's opinion gives a contemporary account of Chinese immigration to the United States. *Id.* at 590-98; *see also* Fong Yue Ting v. United States, 149 U.S. 698, 715-19 (1893). For more historical background, see T. ALEINIKOFF & D. MARTIN, IMMIGRATION: PROCESS AND POLICY 1-5 (1985); R. TAKAKI, STRANGERS FROM A DIFFERENT SHORE 79-131 (1989).

16. The treaty authorized the United States to "regulate, limit or suspend" immigration of Chinese laborers whenever their entry or residence in the United States "affects or threatens to affect the interests of that country, or to endanger the good order of [the United States] or of any locality within the territory thereof." 130 U.S. at 596 (citing treaty). The United States could not "absolutely prohibit" immigration, and its power to suspend was limited to Chinese laborers. *Id.*

17. *Id.*

and in 1884 these certificates became necessary for reentry.[18] Chae Chan Ping obtained a certificate in 1887 and returned to his native China. The next year, while he was overseas, Congress barred the return of Chinese laborers, even those with certificates. Held on a ship in San Francisco Bay, Chae Chan Ping asserted two challenges to the 1888 statute in a petition for habeas corpus. His first argument was that the statute violated the 1880 treaty provision that Chinese laborers already here could leave and return. Justice Field acknowledged that the statute and treaty conflicted, but found that they were on an equal footing, and thus that the statute, being later in time, prevailed.

The second issue in the *Chinese Exclusion Case* was whether, as a constitutional matter, the 1888 statute was "beyond the competency of Congress to pass it."[19] Here the Court had its first chance to consider directly the federal government's power to exclude aliens,[20] which the Constitution does not address explicitly.[21] In setting out what was to become the conceptual framework for classical immigration law, the Court's concern was the federal government's power under the Constitution;[22] it did not consider whether the rights of an individual alien might somehow limit Congress' immigration power. We must bear in mind that this was an earlier era of constitutional law, when equal protection was well on its way to "separate but equal," and judicial recognition of the substantive and procedural rights of individuals was still far beyond the constitutional horizon.

Justice Field's opinion established that the federal government has the power to regulate immigration, and it further suggested that the political branches could exercise this power without being subject to judicial review.[23] He wrote that Congress' power to regulate immigration was based on national

---

18. *Id.* at 598-99.

19. *Id.* at 603.

20. *Cf.* Head Money Cases, 112 U.S. 580 (1884) (upholding federal tax on vessel owners of 50¢ for each alien brought into United States); The Passenger Cases, 48 U.S. (7 How.) 283 (1849) (invalidating state taxes on immigrants).

21. *Cf.* U.S. CONST. art. I, § 8, cl. 4 (power to "establish an uniform Rule of Naturalization").

22. *Chae Chan Ping*, 130 U.S. at 604 ("While under our Constitution and form of government the great mass of local matters is controlled by local authorities, the United States, in their relation to foreign countries and their subjects or citizens are one nation, invested with powers which belong to independent nations, the exercise of which can be invoked for the maintenance of its absolute independence and security throughout its entire territory.").

23. Some of the most often quoted language from the many cases that express this view is from Oceanic Steam Navigation Co. v. Stranahan, 214 U.S. 320 (1909), which held constitutional a levy on carriers who had brought inadmissible aliens to the United States: "[O]ver no conceivable subject is the legislative power of Congress more complete than it is over [the admission of aliens]." *Id.* at 339, *quoted with approval*, Fiallo v. Bell, 430 U.S. 787, 792 (1977); Kleindienst v. Mandel, 408 U.S. 753, 766 (1972); *see also* Lloyd Sabaudo Societa v. Elting, 287 U.S. 329, 334-35 (1932); Mahler v. Eby, 264 U.S. 32, 40 (1924); Lapina v. Williams, 232 U.S. 78, 88 (1914); Bugajewitz v. Adams, 228 U.S. 585, 591 (1913); Keller v. United States, 213 U.S. 138, 143-44 (1909); United States *ex rel.* Turner v. Williams, 194 U.S. 279, 289-92 (1904); Fok Yung Yo v. United States, 185 U.S. 296, 302 (1902); Li Sing v. United States, 180 U.S. 486, 495 (1901); United States v. Wong Kim Ark, 169 U.S. 649, 699-701 (1898); *cf.* United States v. Ju Toy, 198 U.S. 253, 263 (1905) (applicant for admission has no right to de novo judicial review of his claim of citizenship); Lee Lung v. Patterson, 186 U.S. 168, 175-77 (1902) (upholding refusal to admit wife and daughter of returning resident of Chinese descent).

security, sovereignty over its own territory, and self-preservation.[24] "[If Congress] considers the presence of foreigners of a different race in this country, who will not assimilate with us, to be dangerous to its peace and security . . . its determination is conclusive upon the judiciary."[25] Justice Field also characterized the Chinese laborers' right to return as a license, revocable by the United States government.[26] Suggesting that immigration cases might be nonjusticiable as political questions, Field wrote that any remedy for the aggrieved alien must be sought on the alien's behalf by the Chinese government from the political branches of the United States government.[27]

Two years later, in *Nishimura Ekiu v. United States*,[28] the Court expanded the notion of plenary power in two significant ways. First, it rejected a challenge based on a claim of individual constitutional right. An administrative officer had found that a Japanese immigrant was likely to become a public charge, which in turn made her excludable under the statute. She argued that due process required a judicial proceeding. As in the *Chinese Exclusion Case*, the Court relied on maxims of international law and sovereignty[29] to conclude that the immigration power belonged to the federal government and to the political branches, and thus was exempt from judicial review.[30] *Nishimura* also broadened plenary power in a second way. While the *Chinese Exclusion Case* had upheld Congress' substantive admission and exclusion rules, *Nishimura* applied plenary power to Congress' procedures for enforcing immigration laws, including procedures that entrusted final decisionmaking power to agency officials.[31]

---

24. 130 U.S. at 604 ("'Any restriction upon [the power to exclude aliens], deriving validity from an external source, would imply a diminution of its sovereignty to the extent of the restriction, and an investment of that sovereignty to the same extent in that power which could impose such restriction.'" (quoting *The Schooner Exchange v. M'Fadden*, 11 U.S. (7 Cranch) 116, 136 (1812))); *see also id.* at 608 ("The power [to exclude paupers, criminals and persons afflicted with incurable diseases] is constantly exercised; its existence is involved in the right of self-preservation.").

25. 130 U.S. at 606.

26. *Id.* at 609.

27. *Id.* at 606 ("If the government of the country of which the foreigners excluded are subjects is dissatisfied with this action it can make complaint to the executive head of our government, or resort to any other measure which, in its judgment, its interests or dignity may demand; and there lies its only remedy."); *see also id.* at 609 ("If there be any just ground of complaint on the part of China, it must be made to the political department of our government, which is alone competent to act upon the subject.").

28. 142 U.S. 651 (1892).

29. *Id.* at 659 ("It is an accepted maxim of international law, that every sovereign nation has the power, as inherent in sovereignty, and essential to self-preservation, to forbid the entrance of foreigners within its dominions, or to admit them only in such cases and upon such conditions as it may see fit to prescribe.").

30. *Id.* (Power over admission and exclusion "belongs to the political department.").

31. *Id.* at 660 ("[T]he final determination of those facts [establishing an alien's right to land] may be entrusted by Congress to executive officers; and in such a case, as in all others, in which a statute gives a discretionary power to an officer, to be exercised by him upon his own opinion of certain facts, he is made the sole and exclusive judge of the existence of those facts, and no other tribunal, unless expressly authorized by law to do so, is at liberty to reexamine or controvert the sufficiency of the evidence on which he acted"); *see also* Lem Moon Sing v. United States, 158 U.S. 538, 541-47 (1895) (following *Nishimura Ekiu*).

In *Fong Yue Ting v. United States*,[32] decided in 1893, the Court further extended plenary power to the deportation of resident aliens already in the United States. An 1892 statute had extended the ban on Chinese immigration for ten years. Those already resident when the ban took effect were allowed to stay, but only if they could prove pre-1892 residency through a white witness.[33] Fong and two other Chinese laborers claimed pre-1892 residency, but could not produce white witnesses. Justice Gray's majority opinion rejected Fong's constitutional challenge, based on procedural due process, to the white witness rule. Gray declined to distinguish between the power to deport and the power to exclude,[34] dismissing the idea that deportation should trigger the more substantial constitutional safeguards associated with "punishment."[35] Relying on the *Chinese Exclusion Case* and *Nishimura Ekiu*, he concluded that the political branches could regulate immigration, immune from judicial review unless provided for by Congress.[36]

The three dissenters in *Fong* advanced arguments that would prove influential in later attempts to limit the plenary power doctrine. Justice Brewer argued that deportation should prompt greater concern than exclusion because constitutional protections apply within United States territory even if they do not apply outside it.[37] All three dissenters suggested that resident aliens have ties to the United States that justify greater legal protection,[38] and that Fong's constitutional challenge merited judicial attention because the "credible white witness" rule violated procedural due process.[39] Justices Brewer and Fuller took the position that deportation was "punishment" to which constitutional safeguards

32. 149 U.S. 698 (1893).

33. Under the statute, residency was to be proved by a certificate, which under the regulations required the "affidavit of at least one credible witness." *See id.* at 701 n.1 (citing statute and regulations). The statute allowed for deportation of those without certificates unless they showed good cause for not obtaining one and proved pre-1892 residency through a "white" witness. *Id.* at 700 n.1.

34. *Id.* at 713 ("The power to exclude aliens and the power to expel them rest upon one foundation, are derived from one source, are supported by the same reasons, and are in truth but parts of one and the same power.").

35. *Id.* at 730 ("The order of deportation is not a punishment for crime. It is not a banishment, in the sense in which that word is often applied to the expulsion of a citizen from his country by way of punishment. It is but a method of enforcing the return to his own country of an alien who has not complied with the conditions upon the performance of which the government of the nation, acting within its constitutional authority and through the proper departments, has determined that his continuing to reside here shall depend.") (distinguishing Yick Wo v. Hopkins, 118 U.S. 356 (1886), as concerning "the power of a State over aliens continuing to reside within its jurisdiction, not of the power of the United States to put an end to their residence in the country").

36. 149 at 715. *See also id.* at 706 ("conclusive upon the [J]udiciary"); *id.* at 711 ("The United States are a sovereign and independent nation, and are vested by the Constitution with the entire control of international relations, and with all the powers of government necessary to maintain that control and to make it effective.").

37. *Id.* at 738 (Brewer, J., dissenting).

38. *Id.* at 734-37 (Brewer, J., dissenting); *id.* at 746 (Field, J., dissenting); *id.* at 762 (Fuller, J., dissenting).

39. *Id.* at 741-42 (Brewer, J., dissenting); *id.* at 754 (Field, J., dissenting); *id.* at 762-63 (Fuller, J., dissenting). Justice Gray's majority opinion characterized the "credible white witness" rule as a mere rule of evidence. *Id.* at 729-30.

should apply.[40] Interestingly, Fong and the dissenters all seemed to assume that Congress' decision to single out Chinese aliens was not open to challenge.[41]

The first sign that the *Fong* dissents would be influential came in a 1903 decision, *Yamataya v. Fisher*,[42] also known as the *Japanese Immigrant Case*. The government arrested and tried to deport Yamataya four days after she had landed in the United States. Although the Court did not question the validity of the plenary power doctrine in general,[43] in refusing to allow deportation it relied on two of the grounds suggested by the *Fong* dissenters. According to the Court, aliens inside the United States can invoke more constitutional safeguards than aliens seeking admission, and courts reviewing deportation orders should examine procedural due process questions more closely than they should examine substantive immigration rules.[44]

These four cases established a classical immigration law[45] with two significant dimensions—the alien's location and the nature of the constitutional challenge to an adverse immigration decision. The *Chinese Exclusion Case* and *Nishimura Ekiu* suggested that no constitutional objection by an alien outside the United States would be successful. All of the *Fong Yue Ting* opinions assumed that an alien in the United States who challenged substantive deportation rules would likewise be unsuccessful. Finally, *Yamataya* suggested that an alien in the United States who objected to deportation procedures might have some success.

B. *Plenary Power in the Early Modern Era*

As Peter Schuck has pointed out, the plenary power cases of classical immigration law reflect an emphasis on territoriality that was consistent with the sovereignty and consent-based legal culture of the nineteenth century.[46] It is noteworthy, if not striking, that the doctrine, a product of the same era as

---

40. *Id.* at 740 (Brewer, J., dissenting); *id.* at 763 (Fuller, J., dissenting). *See generally* Schuck, *supra* note 6, at 66-68.

41. The Court upheld the constitutionality of substantive discrimination against specified classes of aliens in Lees v. United States, 150 U.S. 476, 480 (1893). *See also* United States *ex rel.* Turner v. Williams, 194 U.S. 279, 294 (1904) (deportation of anarchist withstands First Amendment challenge).

42. 189 U.S. 86 (1903).

43. *Id.* at 97-98.

44. *Id.* at 101 ("[I]t is not competent for . . . any executive officer . . . arbitrarily to cause an alien, who has entered the country, and has become subject in all respects to its jurisdiction, and a part of its population, although alleged to be illegally here, to be taken into custody and deported without giving him all opportunity to be heard upon the questions involving his right to be and remain in the United States.").

45. I borrow this term from Schuck, *supra* note 6.

46. *See id.* at 7-8. On the historical and theoretical bases of the plenary power doctrine, see S. LEGOMSKY, *supra* note 9, at 177-222; Legomsky, *supra* note 5, at 260-78; Note, *Constitutional Limits on the Power to Exclude Aliens*, 82 COLUM. L. REV. 957 (1982). *Cf.* Wani, *Truth, Strangers, and Fiction: The Illegitimate Uses of Legal Fiction in Immigration Law*, 11 CARDOZO L. REV. 51, 59-83 (1989) (criticizing "fiction of sovereignty").

Plessy v. Ferguson,[47] has faded so little with the passage of time. A number of decisions in the first half of this century might be read as ameliorating the doctrine's harshest aspects, especially by building on *Yamataya*'s apparent requirement of some procedural due process in deportation.[48] Yet the Supreme Court reinvigorated the doctrine in a series of cases decided in the McCarthy Era, three of which are particularly important.

The first, *United States* ex rel. *Knauff v. Shaughnessy*,[49] involved Ellen Knauff, who fled her native Germany after the Nazi seizure of power and served in Great Britain's Royal Air Force during World War II. She worked with the Allied occupation forces in Germany after the war, and in 1948 she married a United States citizen. Later that year she sought to enter the United States to be naturalized. The government, without a hearing, ordered her permanently excluded on the ground that her admission would be prejudicial to the interests of the United States. Not knowing what had prompted this order, she challenged it by seeking a writ of habeas corpus; the lower courts rejected her petition. Over vigorous dissent, the Supreme Court affirmed the exclusion order in language reminiscent of the nineteenth-century cases. The Court held that the power to exclude aliens, even the wife of a citizen, is fundamental to sovereignty. Unless Congress provides otherwise, that power is beyond judicial review, especially during national emergency, and it includes procedure as well

---

47. 163 U.S. 537 (1896).

48. Deportation cases to this effect include United States *ex rel.* Vajtauer v. Commissioner of Immigration, 273 U.S. 103, 106 (1927) ("Deportation without a fair hearing or on charges unsupported by any evidence is a denial of due process."); United States *ex rel.* Tisi v. Tod, 264 U.S. 131, 133 (1924) (similar); Lewis v. Frick, 233 U.S. 291, 300 (1914) (similar); Zakonaite v. Wolf, 226 U.S. 272, 275 (1912) (deportation based on agency factfinding "after a fair though summary hearing, may constitutionally be made conclusive"); Low Wah Suey v. Backus, 225 U.S. 460, 468 (1912) (similar); *cf.* Ng Fung Ho v. White, 259 U.S. 276, 281-85 (1922) (due process requires judicial proceeding before person claiming to be citizen, and presenting evidence to that effect, can be deported).

Similar exclusion cases all involved claims of United States citizenship. For example, Kwock Jan Fat v. White, 253 U.S. 454 (1920), held that an administrative decision to exclude a returning resident of Chinese descent, who claimed United States citizenship, was final and conclusive upon United States courts unless such a finding were

> "manifestly unfair," were "such as to prevent a fair investigation," or show "manifest abuse" of the discretion committed to the executive officers by the statute, Low Wah Suey, [225 U.S. 460, 468 (1912)], or that "their authority was not fairly exercised, that is, consistently with the fundamental principles of justice embraced within the conception of due process of law." Tang Tun v. Edsell, [223 U.S. 673, 681-81 (1912)]. The decision must be after a hearing in good faith, however summary, Chin Yow v. United States, 208 U.S. 8, 12 (1908), and it must find adequate support in the evidence.

Id. at 457-58; *see also Tang Tun*, 223 U.S. at 675-82 (constitutional to refuse admission after "careful and fair" administrative hearing to returning resident of Chinese descent who claimed United States citizenship); Chin Yow v. United States, 208 U.S. 8, 12 (1908) ("The decision of the Department is final, but that is on the presupposition that the decision was after a hearing in good faith, however summary in form."); *cf.* Lloyd Sabaudo Societa v. Elting, 287 U.S. 329, 335-36 (1932) (fine imposed on steamship company for transporting aliens found inadmissible is valid if procedure "satisfies elementary standards of fairness and reasonableness"). *See generally* Hart, *The Power of Congress to Limit the Jurisdiction of the Federal Courts: An Exercise in Dialectic*, 66 HARV. L. REV. 1362, 1390-91 (1953) (noting growth of a body of law qualifying the plenary power doctrine); *Developments in the Law—Immigration and Nationality*, 66 HARV. L. REV. 643, 674-76 (1953) (similar).

49. 338 U.S. 537 (1950).

as substance.[50] The Court declared with a quiet, chilling brutality: "Whatever the procedure authorized by Congress is, it is due process as far as an alien denied entry is concerned."[51]

The second case, *Shaughnessy v. United States* ex rel. *Mezei*,[52] involved a permanent resident who had lived in the United States for twenty-five years. In 1948, leaving his wife at home, Mezei traveled to Eastern Europe, apparently to visit his dying mother. While there, Mezei had trouble obtaining the necessary exit papers to embark on his journey back home. When he finally arrived on this country's shores, in possession of an immigration visa issued by the American Consul in Budapest, the government excluded him without a hearing, on the "basis of information of a confidential nature, the disclosure of which would be prejudicial to the public interest."[53] No other country would take Mezei, so the government confined him indefinitely on Ellis Island. He remained in custody for twenty-one months before a district court first granted a writ of habeas corpus ordering his release.

One might have expected the law, in the form of either the immigration statute or constitutional immigration law, to treat a long-time permanent resident less harshly than it would treat a first-time entrant. But with respect to the statute, any such hopes were unfounded. The Immigration and Nationality Act (INA) and its predecessor statutes, like the constitutional immigration law of the classical period, have been structured to ask first if an alien is located inside or outside the country. The alien's location then determines which of two basic sets of rules apply—admission rules, which are tied to quotas and supplemented by exclusion rules, or deportation rules, which expel aliens who should not have been admitted or who have committed bad acts after admission.[54]

---

50. *Id.* at 542-43 ("Whatever the rule may be concerning deportation of persons who have gained entry into the United States, it is not within the province of any court, unless expressly authorized by law, to review the determination of the political branch of the Government to exclude a given alien.").

51. *Id.* at 544. Eventually, the government revealed the adverse information at a full hearing, and the immigration judge still found her excludable. The Board of Immigration Appeals (BIA) then reversed and admitted her. For more background, see E. KNAUFF, THE ELLEN KNAUFF STORY (1952).

52. 345 U.S. 206 (1953).

53. *Id.* at 208. The government relied on the Act of June 21, 1941, ch. 210, § 1, 55 Stat. 252 (currently codifed as amended at 8 U.S.C. § 1225(c), INA § 235(c), and 8 C.F.R. § 175.57).

54. Starting in 1875, immigration statutes began to exclude some aliens on the basis of national origin. Other statutes applied to all countries but excluded certain types of undesirables. *See* Act of Feb. 20, 1907, ch. 1134, § 2, 34 Stat. 898, 898-99; Act of Mar. 3, 1903, ch. 1012, § 2, 32 Stat. 1213, 1214; Act of Mar. 3, 1891, ch. 551, § 1, 26 Stat. 1084, 1084; Act of Aug. 3, 1882, ch. 376, § 2, 22 Stat. 214, 214 (lunatics, idiots, convicts, and those liable to become a public charge); Act of Mar. 3, 1875, ch. 141, § 5, 18 Stat. 477, 477 (excluding prostitutes and convicts). In 1917, Congress enacted legislation that codified the existing categories of excludable aliens, excluded illiterates for the first time, and banned almost all Asian immigration. Act of Feb. 5, 1917, ch. 29, § 3, 39 Stat. 874, 875-78.

Alongside these exclusion statutes, Congress enacted the predecessors of today's deportation grounds. At first, these laws were limited to removing aliens who had entered in violation of the admission and exclusion statutes. Beginning in 1907, however, Congress provided for the deportation of aliens for acts committed after arrival. Act of Feb. 20, 1907, ch. 1134, § 3, 34 Stat. 898, 899-900. *Cf.* Act of June 25, 1798, ch. 58, § 1, 1 Stat. 570, 571 (President may order any alien whom he judged "dangerous to the peace and safety of the United States" to leave country without hearing); Act of July 6, 1798, ch. 66, § 1, 1 Stat. 577, 577 (President during war may apprehend, restrain, secure and remove all enemy aliens without

The key statutory question has always been whether an alien has "entered" the United States, which can occur through the government's decision to admit her or through a surreptitious border crossing.[55] If an alien has not entered but seeks entry, the admission and exclusion rules apply; after entry, deportation rules apply.[56] To be sure, "entry" could be defined to include only an alien's first entry, thus categorizing returning permanent residents as deportable rather than excludable. But the Court had unmistakably rejected this view long before *Mezei* by establishing the "reentry doctrine," which subjects returning permanent residents to exclusion grounds and procedures each time they return to this country from a trip, however brief.[57] Congress largely incorporated the reentry

---

hearing). *See generally* Kleindienst v. Mandel, 408 U.S. 753, 761 (1972); Harisiades v. Shaughnessy, 342 U.S. 580, 588 n.15 (1952); T. ALEINIKOFF & D. MARTIN, *supra* note 15, at 348-49.

Then, in 1924, the National Origins Act, our first comprehensive immigration statute, established a quota system that tied immigration from each foreign land to its share of the existing United States population. Act of May 26, 1924, ch. 190, § 11, 43 Stat. 153, 159-60. In addition, the National Origins Act limited European immigration to 150,000 annually. *Id.* § 11. Without expressly mentioning Japan, it also barred Japanese immigration altogether by prohibiting the admission of any alien "ineligible to citizenship." *Id.* § 13(c), a phrase clearly intended to apply to Japanese. *See generally* R. TAKAKI, *supra* note 15, at 208-10. In 1921, Congress had adopted the first quotas in a provisional measure that capped European immigration and limited the number of immigrants of each nationality to three percent of their foreign-born countrymen resident in the United States as of the 1910 census. Act of May 19, 1921, ch. 8, § 2, 42 Stat. 5, 5. The McCarren-Walter Act of 1952, Pub. L. No. 82-414, 66 Stat. 163, consolidated previous immigration laws, and per-country limits replaced the national origins quotas in 1965. Immigration and Nationality Act Amendments of 1965, Pub. L. No. 89-236, § 2, 79 Stat. 911, 911-12. Through all of these changes, the basic immigration statute has continued to rely on admission rules tied to quotas and supplemented by exclusion rules, and deportation rules for aliens who should not have been admitted or who have committed bad acts after admission.

55. "Admission" is a term of art and does not include the granting of permission to enter the country on parole. *See* Leng May Ma v. Barber, 357 U.S. 185, 188 (1958). This article will not discuss the close questions that can arise when a surreptitious border crossing is claimed to constitute an "entry." *See generally* T. ALEINIKOFF & D. MARTIN, *supra* note 15, at 315-23; *id.* at 25-26 (Supp. 1987).

56. Generally, admission criteria are set forth in INA §§ 201-203, 8 U.S.C. §§ 1151-1153 (1988), exclusion grounds in INA § 212, 8 U.S.C. § 1182 (1988), and deportation grounds in INA § 241, 8 U.S.C. § 1251 (1988).

57. The most prominent of these cases is United States *ex rel.* Volpe v. Smith, 289 U.S. 422 (1933). Volpe entered the United States from Italy in 1906, when he was 16 years old. In 1925, he pleaded guilty to counterfeiting and was sentenced to prison. In 1928, he visited Cuba briefly, and two years later the government sought to deport him for having committed a crime of "moral turpitude" before entry. The deportation statute applied to "any alien who was convicted, or who admits the commission, prior to entry, of a felony or other crime or misdemeanor involving moral turpitude." Act of Feb. 5, 1917, ch. 29, § 19, 39 Stat. 874, 889. Volpe was deportable only if his return from Cuba in 1928 had been an "entry," as the government urged. Otherwise, if he had last entered in 1906, the statute would not apply because the crime was not before entry. Another deportation ground, for a single crime within five years after entry, also would not apply because 19 years had passed. 289 U.S. at 426. A unanimous Court agreed with the government's reading of "entry" and upheld the deportation order. *Id.* at 425. *See also* United States *ex rel.* Stapf v. Corsi, 287 U.S. 129, 132 (1932); United States *ex rel.* Claussen v. Day, 279 U.S. 398, 401 (1929); Lewis v. Frick, 233 U.S. 291, 297 (1914); Lapina v. Williams, 232 U.S. 78, 91 (1914); *cf.* Wani, *supra* note 46, at 89-96 ("entry" fiction is primary determinant of procedural due process despite its "acknowledged falsity and outrageousness"). *See also* Maslow, *Recasting Our Deportation Law: Proposals for Reform*, 56 COLUM. L. REV. 309, 327-29 (1956) (criticizing reentry doctrine for its "harsh consequences").

On the facts of *Volpe*, this decision did not directly distinguish between exclusion and deportation; Volpe had entered the United States in 1928, making it too late to exclude him, so the government had to deport him if it was to act against him. The underlying logic of the decision, however, was that Volpe's 22 years in the United States made no difference. If, for example, the government had stopped Volpe at the border upon his return from Cuba in 1928, he would have been "entering" at that time and thus subject

doctrine into the 1952 Immigration Act, when it defined "entry" to mean "*any* coming of an alien into the United States."[58]

Because the reentry doctrine dashed Mezei's hopes for success on statutory grounds, the Court had to decide if the statute could constitutionally apply to a returning permanent resident the same exclusion rules that the statute would apply to a first-time entrant, including denial of a hearing. Relying on the plenary power doctrine and national security considerations, but once again over vigorous dissent, the Court said that the statute could deny Mezei both substantive rights and procedural safeguards available to deportable aliens.[59] The Court characterized Mezei's constitutional position, like his position under the statute, as that of "an alien on the threshold of initial entry."[60] Mezei also sought to distinguish his case from *Knauff* by arguing that he, unlike Knauff, sought not just admission, but also release from indefinite detention. The Court rejected this argument as well, characterizing his incarceration as "temporary harborage, an act of legislative grace."[61] Eventually, Mezei was paroled into the United States under a special clemency measure, but only after spending four years in detention.[62]

The third case, the 1952 decision in *Harisiades v. Shaughnessy*,[63] involved substantive immigration categories, in contrast to the procedural claims in *Knauff* and *Mezei*. The Court upheld the deportations of Harisiades and two other resident aliens for having once belonged to the United States Communist Party, even though they were no longer members when that ground for deportation became law.[64] While *Knauff* and *Mezei* had shown that an alien's location outside the United States would place him in a very precarious constitutional position, and *Yamataya* had established that location inside the United States

---

to exclusion rather than deportation grounds and procedures.

58. Pub. L. No. 82-414, Act of June 27, 1952, ch. 477, § 101, 66 Stat. 163, 167 (emphasis added). The 1952 legislation put an end to arguments, *see, e.g.*, T. ALEINIKOFF & D. MARTIN, *supra* note 15, at 326-27, that in reading "entry" to include a permanent resident's return, *Volpe* conflicted with the overall structure of the statutory exclusion and deportation grounds. The present definition of "entry" is set forth in INA § 101(a)(13), 8 U.S.C. § 1101(a)(13) (1988).

59. *Mezei*, 345 U.S. at 210 ("Courts have long recognized the power to expel or exclude aliens as a fundamental sovereign attribute exercised by the Government's political departments largely immune from judicial control.").

60. *Id.* at 212 (distinguishing Yamataya v. Fisher, 189 U.S. 86 (1903), discussed in Part I.A). *See also id.* at 213 ("For purposes of the immigration laws, moreover, the legal incidents of an alien's entry remain unaltered whether he has been here once before or not. He is an entering alien just the same, and may be excluded if unqualified for admission under existing immigration laws.").

61. *Id.* at 215.

62. *See* Trop v. Dulles, 356 U.S. 86, 102 n.36 (1957).

63. 342 U.S. 580 (1952).

64. For years, Congress tried to deport aliens who were Communists or otherwise regarded as subversives. In 1939, the Supreme Court construed the deportation statute then in effect to allow the deportation of aliens who were members of organizations that advocated the overthrow of the government by force and violence, but not to allow deportation once membership had ceased. Kessler v. Strecker, 307 U.S. 22 (1939). In response, the Communist Party struck aliens from its membership rolls to spare them the risk of deportation. Congress in turn enacted a new statute that made aliens deportable if they had been members of the proscribed organizations at any time. *See* Alien Registration Act of 1940, Pub. L. No. 670, ch. 439, § 23, 54 Stat. 670, 673 (1940). The deportations in *Harisiades* were based on this second statute.

would enhance his ability to raise procedural constitutional challenges, *Harisiades* demonstrated that location inside the country would not necessarily facilitate substantive claims. The Court rejected arguments rooted in the Fifth Amendment due process clause that aliens, once admitted to permanent residence, have a "vested right" to remain, or, alternatively, that any deportation grounds for permanent residents must be "reasonable."[65] The Court's language suggested little if any judicial role: "[A]ny policy toward aliens is vitally and intricately interwoven with contemporaneous policies in regard to the conduct of foreign relations, the war power, and the maintenance of a republican form of government."[66] According to the Court, "[s]uch matters are so exclusively entrusted to the political branches of government as to be largely immune from judicial inquiry or interference."[67] The Court also rejected the argument that the First Amendment barred deportation, in reasoning that seemed to assume that the First Amendment did not protect the Communist Party membership of anyone, citizen or alien.[68] Finally, the Court rejected the claim that Harisiades' deportation rested on an unconstitutional *ex post facto* law, reasoning that statute was not retroactive, and that even if it were, the constitutional prohibition would not apply because deportation is a "civil" proceeding.[69]

65. *Harisiades*, 342 U.S. at 586-87 ("Under our law, the alien in several respects stands on an equal footing with citizens, but in others has never been conceded legal parity with the citizen. Most importantly, to protract this ambiguous status within the country is not his right but is a matter of permission and tolerance. The Government's power to terminate its hospitality has been asserted and sustained by this Court since the question first arose.").

66. *Id.* at 588-89.

67. *Id.* at 589 (footnote omitted). Two years later, in Galvan v. Press, 347 U.S. 522, 529-32 (1954), the Court relied on *Harisiades* to uphold the constitutionality of the Internal Security Act of 1950, Pub. L. No. 81-831, ch. 1024, tit. I, § 22, 64 Stat. 987, 1006 (1950) (substantially codified in INA § 241(a)(6)(C), 8 U.S.C. § 1251(a)(6)(C) (1988)), which dispensed with the need under the Alien Registration Act of 1940 to prove in each deportation that the Communist Party did "advocate the violent overthrow of the Government." Justice Frankfurter wrote "that the formulation of these [admission and expulsion] policies is entrusted exclusively to Congress has become about as firmly embedded in the legislative and judicial tissues of our body politic as any aspect of our government." 347 U.S. at 531. *See also* Jay v. Boyd, 351 U.S. 345, 357 n.21 (1956) (constitutionality of statute permitting Attorney General to deny suspension of deportation to former Communist Party member on basis of undisclosed confidential information "gives us no difficulty"); Carlson v. Landon, 342 U.S. 524, 534 (1952) ("So long . . . as aliens fail to obtain and maintain citizenship by naturalization, they remain subject to the plenary power of Congress to expel them under the sovereign right to determine what noncitizens shall be permitted to remain within our borders.") (footnote omitted).

68. 342 U.S. at 591-93. The Court's reliance on Dennis v. United States, 341 U.S. 494 (1951), suggests that it made this assumption. Because First Amendment law has changed significantly since *Harisiades*, later cases might pose the question whether an alien can be deported for activity that would be constitutionally protected for United States citizens. *See generally* T. ALEINIKOFF & D. MARTIN, *supra* note 15, at 365-66. The court in American-Arab Anti-Discrimination Comm. v. Meese, 714 F. Supp. 1060, 1077-82 (C.D. Cal. 1989), reasoned that an alien's First Amendment challenge to a deportation ground must be analyzed just as if a citizen had raised it, and struck down certain ideological deportation grounds as unconstitutional. For a more detailed discussion of *American-Arab Anti-Discrimination Committee* see *infra* text accompanying notes 352-57.

69. 342 U.S. at 593-95 (citing Mahler v. Eby, 264 U.S. 32, 39 (1924), and Bugajewitz v. Adams, 228 U.S. 585, 591 (1913)). *See also* Flemming v. Nestor, 363 U.S. 603, 612-21 (1960); Lehmann v. United States *ex rel.* Carson, 353 U.S. 685, 689-90 (1957); Marcello v. Bonds, 349 U.S. 302, 314 (1955); Galvan v. Press, 347 U.S. 522, 530-31 (1954).

Taken together, *Knauff*, *Mezei*, and *Harisiades* confirmed the modern importance of the two basic lines of inquiry in the early plenary power decisions: the alien's location and the type of constitutional challenge. Specifically, aliens "outside" the United States would continue to find it very difficult to raise any constitutional challenge to immigration decisions. Those "inside" the United States could have some success with procedural claims but would be likely to have none with substantive claims. At the same time, spirited dissents in all three cases expressed serious concern that aliens would be denied access to judicial review in such harsh and unremitting terms. In *Mezei*, the dissenters were particularly troubled by the majority's upholding as constitutional the statute's treatment of a returning permanent resident, effectively limiting Mezei's substantive and procedural rights to those of exclusion rather than deportation. In both *Knauff* and *Mezei*, the apparent disregard for even the most rudimentary procedural due process troubled the dissenters, just as it had been a primary focus of the *Fong Yue Ting* dissents, and of the Court's opinion in *Yamataya*.[70] And in *Harisiades*, Justice Douglas, relying heavily on Justice Brewer's dissent in *Fong Yue Ting*,[71] argued that resident aliens cannot be deported unless they are afforded constitutional protections, especially due process and equal protection.[72] Gradually, the concerns expressed in the dissents in *Mezei*, *Knauff*, and *Harisiades* found sympathetic ears. The result has been an expansion in the number and range of claims that courts, including the Supreme Court, would hear in immigration cases.

II. Phantom Norms, Statutory Interpretation, and Constitutional Change

A. *Constitutional Norms and Statutory Interpretation*

The principal decisions that have contributed to this expansion of judicial review in immigration cases have not been decisions of constitutional immigration law. Instead, they reached results favorable to aliens by interpreting statutes, regulations, or other forms of subconstitutional immigration law. Of course, there is nothing remarkable about a court decision that relies on a subconstitutional rather than constitutional ground. A time-honored canon of

---

70. *See Mezei*, 345 U.S. at 224 (Jackson, J., dissenting) ("Procedural due process is more elemental and less flexible than substantive due process. It yields less to the times, varies less with conditions, and defers much less to legislative judgment."). *See also Knauff*, 338 U.S. at 552 (Jackson, J., dissenting) (objecting to "a finding of serious misconduct against the wife of an American citizen without notice of charges, evidence of guilt and a chance to meet it"). In contrast, Jackson had written for the *Harisiades* majority in rejecting a substantive constitutional challenge. Sharp criticism of *Knauff* and *Mezei* was also expressed in Hart, *supra* note 48, at 1391-96 (*Knauff*'s statement of complete acquiescence in procedure prescribed by Congress is "a patently preposterous proposition").

71. For a discussion of Fong Yue Ting v. United States, 149 U.S. 698 (1893), see *supra* text accompanying notes 32-41.

72. 342 U.S. at 598 (Douglas, J., dissenting).

statutory interpretation, often invoked by citing Justice Brandeis' 1936 concurrence in *Ashwander v. Tennessee Valley Authority*,[73] says that judges should interpret statutes to avoid constitutional doubt.[74] While this canon is not uncontroversial,[75] and its predictive value is open to question,[76] reliance on the canon, whether stated or unstated, seems to be a fact of everyday judicial life.

The canon largely reflects the fact that constitutional norms usually manifest themselves in our law both directly and indirectly. First, constitutional norms may directly govern decisions that are expressly constitutional; this is the intuitive definition of "constitutional." In a less intuitive but equally correct use of the term "constitutional," these norms operate indirectly, by serving as the unstated background context that informs our interpretation of statutes and other subconstitutional texts. In other words, contemporary constitutional law is a significant element of the legal culture that judges inevitably, if often subconsciously, absorb and rely upon when acting in their judicial capacity, including those instances in which they engage in statutory interpretation.[77]

---

73. 297 U.S. 288, 347 (1936) (Brandeis, J., concurring).

74. *See* Jean v. Nelson, 472 U.S. 846, 854 (1985); *see also* Public Citizen v. United States Dep't of Justice, 109 S. Ct. 2558, 2572-73 (1989); Edward J. DeBartolo Corp. v. Florida Gulf Coast Bldg. & Constr. Trades Council, 485 U.S. 568, 575-77 (1988); United States v. Clark, 445 U.S. 23, 27 (1980); NLRB v. Catholic Bishop, 440 U.S. 490, 499-501 (1979); International Ass'n of Machinists v. Street, 367 U.S. 740, 749 (1961). *See generally* W. ESKRIDGE & P. FRICKEY, LEGISLATION: STATUTES AND THE CREATION OF PUBLIC POLICY 676-89 (1988); Eskridge, *Public Values in Statutory Interpretation*, 137 U. PA. L. REV. 1007, 1020-22 (1989). A closely related canon is the "rule of lenity" in criminal cases. *See* United States v. Campos-Serrano, 404 U.S. 293, 298-99 (1971); Eskridge, *supra*, at 1029-30.

75. Judge Posner, for example, objects that this practice allows courts to bend the meaning of statutes, thereby creating an inappropriate judge-made "penumbra" around the Constitution. *See* R. POSNER, THE FEDERAL COURTS 285 (1985). *See also* United States v. Monsanto, 109 S. Ct. 2657, 2664 (1989) ("such 'interpretive canon[s are] not a license for the judiciary to rewrite language enacted by the legislature'") (quoting United States v. Albertini, 472 U.S. 675, 680 (1985)); Public Citizen v. United States Dep't of Justice, 109 S. Ct. 2558, 2580 (1989) (Kennedy, J., concurring) ("The fact that a particular application of the clear terms of a statute might be unconstitutional does not provide us with a justification for ignoring the plain meaning of the statute."); NLRB v. Catholic Bishop, 440 U.S. 490, 510-11 (1979) (Brennan, J., dissenting) (to confine judiciary to its proper role, canon for construing statutes wherein constitutional questions may lurk must be limited to constructions that are fairly possible); H. FRIENDLY, BENCHMARKS 210-12 (1967); Posner, *Statutory Interpretation—In the Classroom & in the Courtroom*, 50 U. CHI. L. REV. 800, 815-16 (1983).

76. *See, e.g.*, Webster v. Reproductive Health Servs., 109 S. Ct. 3040, 3064-65 (1989) (Scalia, J., concurring) ("a sound general principle, but one often departed from when good reason exists. . . . I have not identified with certainty the first instance of our deciding a case on broader constitutional grounds than absolutely necessary, but it is assuredly no later than *Marbury v. Madison* . . . ."). *See also* Eskridge, *supra* note 74, at 1073 ("for every case like *Catholic Bishop*, which interprets statutes to avoid constitutional doubts, there are other cases where a statute is construed boldly, to face substantial constitutional troubles").

77. *See, e.g.*, Eskridge & Frickey, *Statutory Interpretation as Practical Reasoning*, 42 STAN. L. REV. 321, 352 (1990) ("Because we always approach texts from the perspective of our own historically situated horizon, we tend to project our 'preunderstandings' onto the text, viewed as a whole. That is essential to interpretation, because the preunderstandings are conditioned by tradition and, hence, help us link our horizon with that of the text."); Smith, *Law Without Mind*, 88 MICH. L. REV. 104, 106-07 (1989) ("The pragmatic judge would not necessarily spurn traditional legal authorities, like statutes or constitutional provisions, in part because such authorities may prove to be useful tools, and in part because they may in fact have given rise to practices and expectations which the pragmatic calculus must take into account."); Sunstein, *Interpreting Statutes in the Regulatory State*, 103 HARV. L. REV. 405, 411 (1989) (statutory interpretation relies on background principles drawn from context and, more generally, from legal culture); *id.* at 459 ("Interpretive principles are often a product of constitutional norms."); *id.* at 468 ("the Constitution

Bob Jones University v. United States[78] illustrates how constitutional norms inform statutory interpretation. A nearly unanimous Court held that a nonprofit private school with racially discriminatory admissions standards did not qualify for a federal income tax exemption for institutions "organized and operated exclusively for religious, charitable, . . . or educational purposes."[79] Reaching beyond the statutory text itself, the Court reasoned that the exemption was based on public services of value to society. It found that this purpose required denying the exemption to any organizations whose activities were "contrary to a fundamental public policy."[80] Citing *Brown v. Board of Education*,[81] the Court noted that "racial discrimination in education violates a most fundamental national public policy, as well as rights of individuals."[82] Even though the Congress that passed the original tax exemption statute probably did not contemplate future developments in the law against racial segregation, the Court's interpretation of the statute was intended to mirror standards of nondiscrimination embedded in the contemporary legal culture.[83]

If *Bob Jones* illustrates how constitutional norms act as background norms when judges interpret statutes, *Chevron U.S.A., Inc. v. Natural Resources Defense Council, Inc.*,[84] a 1984 Supreme Court decision, illustrates how constitutional norms can perform an analogous function for statutory interpretation by administrative agencies. *Chevron* endorsed substantial deference to agency interpretation of statutes by holding that, absent discernible congressional intent, judges may decide only whether the agency's view is reasonable, not whether it is "appropriate."[85] And when judges defer to agencies, they defer to inter-

---

provides the backdrop against which statutes are written and interpreted, and it furnishes the basic assumptions of interpretation"); White, *Law as Language: Reading Law and Reading Literature*, 60 TEX. L. REV. 415, 435 (1982) (interpreted text "must be fit together with other texts and with the rest of the culture to make what we call a field or body of law").

78. 461 U.S. 574 (1983).

79. I.R.C. § 501(c)(3) (1988).

80. 461 U.S. at 592.

81. 347 U.S. 483 (1954).

82. 461 U.S. at 593.

83. *See generally* Eskridge, *supra* note 74, at 1034-36 (public policy against racial segregation provided background context for interpreting statute in *Bob Jones*); Eskridge & Frickey, *supra* note 77, at 343-44 (similar). The same norm against racial discrimination also guided the Court's decision in Patterson v. McLean Credit Union, 109 S. Ct. 2363 (1989), in which the Court declined to overrule its interpretation of 42 U.S.C. § 1981 in Runyon v. McCrary, 427 U.S. 160 (1976). *See* 109 S. Ct. at 2371 (referring to "our society's deep commitment to the eradication of discrimination based on a person's race or the color of his or her skin"); *id.* at 2380 (Brennan, J., concurring in part and dissenting in part) (similar); *see also* Jones v. Alfred H. Mayer Co., 392 U.S. 409 (1968); *cf.* Pennhurst State School & Hosp. v. Halderman, 451 U.S. 1, 17 (1981) (background norm based on notions of federalism: "[I]f Congress intends to impose a condition on the grant of federal moneys, it must do so unambiguously. By insisting that Congress speak with a clear voice, we enable the States to exercise their choice knowingly, cognizant of the consequences of their participation." (citation omitted)).

84. 467 U.S. 837 (1984).

85. *Id.* at 842-43 ("If the intent of Congress is clear, that is the end of the matter; for the court, as well as the agency, must give effect to the unambiguously expressed intent of Congress. If, however, the court determines Congress has not directly addressed the precise question at issue, the court does not simply impose its own construction on the statute . . . as would be necessary in the absence of an administrative interpretation. Rather, if the statute is silent or ambiguous with respect to the specific issue, the question

pretations that are very likely to reflect, even if at an unspoken level, the contemporary legal culture, which inevitably includes its constitutional law.[86]

An outwardly different understanding of the canon—but one that is just the flip side of the view that constitutional norms serve as the background for subconstitutional interpretation—is to characterize its application as the "underenforcement" of constitutional norms for prudential reasons. Institutional constraints, especially the judiciary's sensitivity to its limited factfinding capability and attenuated electoral responsibility, make courts reluctant to issue a constitutional command to the political branches of government.[87] Even if such a command clearly would reflect an established constitutional norm, courts can sometimes vindicate that norm less intrusively, and thus perhaps more justifiably, through the indirect route of statutory interpretation.[88] Many such prudential considerations defy generalization, while others can be collected in bodies of doctrine—the political question[89] and standing doctrines come to mind—that tend to limit judicial review.

Indeed, the issue presented in *Bob Jones* could easily have arisen in litigation in which the identity of the parties or the particular facts would prompt the Court to deny constitutional relief, perhaps based on standing or state action requirements.[90] That these constitutional norms are real in spite of underenforcement in some cases is apparent in the fact that these prudential concerns operate only interstitially. The Court in *Bob Jones* clearly took seriously the constitutional doubt that would have arisen if the Internal Revenue Code were

for the court is whether the agency's answer is based on a permissible construction of the statute." (footnotes omitted)); *cf.* INS v. Cardoza-Fonseca, 480 U.S. 421, 446-48 (1987) (courts should not defer to agency interpretations on pure questions of law but only on mixed questions of law and fact).

86. *Cf.* Aleinikoff, *Updating Statutory Interpretation*, 87 MICH. L. REV. 20, 42-46 (1988) (*Chevron*'s rule of substantial deference to agency interpretation provides major route for updating statutes because agency interpretations are likely to reflect current policy considerations more than they reflect original intent of legislature).

87. *See* Sunstein, *supra* note 77, at 468 (courts reluctant to act with "vigor appropriate to governmental bodies with a better democratic and policymaking pedigree"); *see also* Sager, *Fair Measure: The Legal Status of Underenforced Constitutional Norms*, 91 HARV. L. REV. 1212, 1214 (1978) ("important difference between a true constitutional conception and the judicially formulated construct is that the judicial construct may be truncated for reasons which are based not upon analysis of the constitutional concept but upon various concerns of the Court about its institutional role").

88. This view addresses Judge Posner's objection to the judge-made "penumbra" around the Constitution. *See supra* note 75.

89. *See* Sager, *supra* note 87, at 1224 (political question doctrine rests on distinction "between a determination that there exist decisive reasons for the judiciary to decline to apply a norm of the Constitution to a given set of facts and a determination that the norm in question does not reach that set of facts").

90. The Court noted that its reading of the statute eliminated any need to address arguments by many of the amici curiae that it would be unconstitutional for the government to read the Internal Revenue Code as the private schools had urged. 461 U.S. at 599 n.24. On the standing questions raised by those arguments, the Court found one year later, in Allen v. Wright, 468 U.S. 737 (1984), that parents of Black children who were attending public schools undergoing desegregation in seven states lacked standing to sue the federal government to force it to deny tax exemptions to racially segregated private schools that were created or expanded in their communities during that period.

interpreted to permit the tax exemption.[91] Acceptance of the background constitutional norm of racial equality as part of mainstream public law is abundantly clear from other cases, among them *Brown*, in which the Court had to first invent and then apply such a constitutional norm explicitly for lack of any subconstitutional escape hatch.[92]

B. *The Emergence of Phantom Norm Decisionmaking in Immigration Law*

There is nothing wrong with borrowing constitutional norms to interpret statutes, but in choosing those norms courts must exercise care. I will try in the rest of Part II and in Part III to review an array of immigration cases decided in the past fifty years to show that immigration law has come to represent an aberrational form of the typical relationship between constitutional and subconstitutional law exemplified by *Bob Jones*. In immigration law, the "constitutional" norms that actually inform statutory interpretation—which are norms borrowed from public norms generally—conflict with the expressly articulated constitutional norm—unreviewable plenary power. The former are "phantom" rather than "real" constitutional norms in the sense that they do not serve the first function of "constitutional" norms—namely, direct application to constitutional issues raised in immigration cases. In fact, the plenary power cases provide, as a matter of explicit constitutional theory, that the immigration context is different, and that therefore we cannot directly apply mainstream constitutional norms in immigration cases. But "phantom constitutional norms" are "constitutional" in the sense that they, having been at least seriously entertained as a constitutional argument and in many cases actually adopted as an expressly constitutional decision in other areas of law, then carry over to immigration cases, where they are substantial enough to serve the limited function of informing interpretation of immigration statutes and other subconstitutional texts. Or, to use an image from the physical sciences, they have enough gravitational force to exercise a pull on these other sources of law.[93] In this context, phantom norms produce results that are much more sympathetic to aliens than the results that would follow from the interpretation of statutes in

---

91. *See* 461 U.S. at 599 n.24 (citing two then-recent leading cases invoking this interpretive canon to avoid constitutional doubts: United States v. Clark, 445 U.S. 23, 27 (1980), and NLRB v. Catholic Bishop, 440 U.S. 490, 504 (1979)).

92. That this norm is "real" is also clear from cases holding that Congress may enforce these norms, even against private activity, under section 5 of the Fourteenth Amendment or under Congress' power to provide for the "general Welfare," U.S. CONST. art. I, § 8, cl. 1. *See, e.g.*, Metro Broadcasting, Inc. v. FCC, 110 S. Ct. 2997 (1990); Fullilove v. Klutznick, 448 U.S. 448, 472 (1980). Similar reasoning underlies the interpretation given Title VII, 42 U.S.C. § 2000e (1988), in United Steelworkers v. Weber, 443 U.S. 193, 200-08 (1979). For a discussion of the use of background norms to interpret Title VII in *Weber* and Johnson v. Transportation Agency, 480 U.S. 616, 627-42 (1987), which used similar reasoning to uphold affirmative action programs for women, see Eskridge, *supra* note 74, at 1033-34.

93. *See* Eskridge, *supra* note 74, at 1018-19.

light of the expressly applicable constitutional immigration law based on plenary power.

The emergence of these more sympathetic results is easier to understand if we look beyond the field of "immigration law," defined as the law of admission and expulsion of aliens. Once we expand our inquiry to include the more general law of aliens' rights and obligations, the force of the plenary power doctrine diminishes considerably. In contrast to the harshness of classical immigration law, a long line of Supreme Court decisions has afforded a measure of protection to aliens that much more closely resembles the substantive and procedural rights of individuals in mainstream public law. By use of the term "mainstream," I do not mean to suggest that there is a set of transcendent constitutional norms "out there" that do not vary according to the type of case or the nature of the power that Congress is asserting. The Court clearly adopts constitutional norms with the highest degreee of sensitivity to substantive context. My point is a more limited one—namely, that in all of these other contexts involving aliens the Court has adopted approaches that, despite variations among them, consistently conflict with the begrudging attitude toward aliens reflected in the plenary power doctrine.[94]

A seminal case in this regard is *Yick Wo v. Hopkins*, decided by the Court in 1886, which held that the Constitution protects all individuals inside the United States, including aliens, from invidious discrimination at state hands.[95] The classification in question was not based on alienage per se, but rather was an ostensibly race-neutral regulation that was enforced so as to make it impossible for aliens of Chinese descent to operate laundries in San Francisco. The basic point of *Yick Wo*—and one that stands in sharp contrast to the possible implications of the *Chinese Exclusion Case*—is that aliens, at least those lawfully present in the United States, are within the constitutional fold.[96] Another seminal case is *Wong Wing v. United States*,[97] decided by the Court in 1896, which involved a statute providing that any Chinese national whom executive officials found to be in the United States illegally "shall be imprisoned at hard labor."[98] The Court, only three years after the *Chinese Exclusion Case*, relied on *Yick Wo* in holding that an alien is entitled to the full range of protections afforded by the Fifth and Sixth Amendments, and specifically that

94. I also do not intend to amalgamate artificially all other constitutional law into a unitary whole. Other areas of public law, among them federal Indian law and foreign affairs, to name only two, similarly stand outside the mainstream. But it is probably more accurate to say that those other areas are also isolated than it is to suggest that there is no such thing as "mainstream" constitutional law.

95. 118 U.S. 356, 369 (1886). The Supreme Court cited *Yick Wo* with approval in later cases. For example, in Johnson v. Eisentrager, 339 U.S. 763, 770 (1950), the Court said that "[t]he alien, to whom the United States has been traditionally hospitable, has been accorded a generous and ascending scale of rights as he increases his identity with our society."

96. *See* 118 U.S. at 369 ("The Fourteenth Amendment to the Constitution is not confined to the protection of citizens.").

97. 163 U.S. 228 (1896).

98. Act of May 5, 1892, ch. 60, § 6, 27 Stat. 25.

"a judicial trial to establish the guilt of the accused" would be required before such "infamous punishment" could be imposed.[99] Similarly, in the 1931 case, *Russian Volunteer Fleet v. United States*, the Court relied on *Yick Wo* and *Wong Wing*, among other cases, in applying the takings clause of the Fifth Amendment to benefit non-enemy aliens present within United States territory.[100]

By the 1950's, aliens' rights decisions beyond the scope of immigration law already conflicted with assumptions implicit in the plenary power doctrine.[101] As Alex Aleinikoff has pointed out, the *Yick Wo* approach differs from classical immigration law's model of aliens' rights, which was a "membership model of citizenship and alienage" that responded to the perceived need to draw a line between those who belong to the "national community" and those who do not.[102] The legacy of *Yick Wo* represents a different model based on "a notion of fundamental human rights that protects individuals regardless of their status."[103] This model's emergence, as Peter Schuck has noted, is attributable to fundamental cultural changes fostering the growth of the values that he calls "communitarian."[104] More specifically, the *Yick Wo* model draws sustenance from developments elsewhere in the constitutional law of individual rights and liberties, including the landmark 1954 decision in *Brown v. Board of Education*.[105]

These decisions, both in the law of aliens' rights and in the law of individual rights generally, have never fully coalesced into a coherent and comprehensive body of doctrine to guide courts' admission and expulsion decisions, and thus have never offered a fully textured alternative to the plenary power doctrine. Nor have they ever suggested that aliens be guaranteed the full range and force of constitutional protections enjoyed by citizens. But *Yick Wo* and its progeny took these constitutional claims seriously, in contrast to the cavalier treatment of constitutional claims in immigration law. Moreover, lawyers and judges, whether justified or not, have in fact relied on the values and the approach manifest in *Yick Wo*, and this reliance has influenced the course of

---

99. 163 U.S. at 237.

100. 282 U.S. 481, 489 (1931).

101. *Yick Wo*'s more recent progeny are discussed *infra* text accompanying notes 194-217.

102. *See* Aleinikoff, *Citizens, Aliens, Membership and the Constitution*, 7 CONST. COMMENTARY 9, 10 (1990) [*hereinafter* Aleinikoff, *Membership*] ("Immigration policy, conceived of as membership rules, is thought to lie at the core of national self-determination and self-definition."); *see also* Aleinikoff, *Federal Regulation of Aliens and the Constitution*, 83 AM. J. INT'L L. 862, 864 (1989) [*hereinafter* Aleinikoff, *Federal Regulation*]; Bosniak, *Exclusion and Membership: The Dual Identity of the Undocumented Worker Under United States Law*, 1988 WIS. L. REV. 955, 967-77.

103. Aleinikoff, *Membership*, *supra* note 102, at 19; *see also* Martin, *Due Process and Membership in the National Community: Political Asylum and Beyond*, 44 U. PITT. L. REV. 165, 180 (1983) ("[A]nyone inspired by the *Yick Wo* tradition—a certain pride in the broad reach of our Constitution's protections—cannot rest content with *Knauff-Mezei* nor with the pretense that excludable aliens really are not here.").

104. *See* Schuck, *supra* note 6, at 34-53.

105. 347 U.S. 483 (1954).

immigration law. For example, *Wong Wing*'s application of constitutional protections to any "punishment" of an alien explains the persistence of the argument, first heard no later than *Fong Yue Ting*[106] and since then often reattempted in vain,[107] that deportation is punishment.[108] In short, *Yick Wo* and its progeny, as well as other developments in the law of individual rights, have provided the normative foundation for results at odds with strict application of the plenary power doctrine.

The centrality of phantom norm decisionmaking in immigration law gradually emerged through several Supreme Court decisions from roughly the same period as *Mezei*, *Knauff*, and *Harisiades*. Their unifying characteristic is their propensity to use phantom norm constitutional reasoning to reach subconstitutional outcomes favorable to aliens. They differ from each other only in the boldness of their phantom norm decisionmaking.

The earliest of these cases is *Bridges v. Wixon*,[109] a 1945 decision that arose from the government's efforts to deport the union activist Harry Bridges. The pertinent deportation ground covered past membership or affiliation with "any organization, association, society, or group, that believes in, advises, advocates, or teaches . . . the overthrow by force of violence of the Government of the United States."[110] The Court, reading the statute, held that Bridges was not subject to deportation.[111] This was unremarkable; earlier decisions had relied on subconstitutional grounds to reach results favorable to aliens.[112] One way to understand the case, however, is to focus on the potentially constitutional character of the reasoning in Justice Douglas' majority opinion. That opinion

---

106. As discussed *supra* text accompanying note 35, Fong Yue Ting v. United States, 149 U.S. 698 (1893), distinguished *Yick Wo* in reaching this conclusion. *Fong Yue Ting*, 149 U.S. at 725.

107. *See* Carlson v. Landon, 342 U.S. 524, 537 (1952) ("Deportation is not a criminal proceeding and has never been held to be punishment."); Mahler v. Eby, 264 U.S. 32, 39 (1924); United States *ex rel.* Bilokumsky v. Tod, 263 U.S. 149, 154 (1923); Bugajewitz v. Adams, 228 U.S. 585, 591 (1913) (Holmes, J.) (deportation is not "a punishment; it is simply a refusal by the Government to harbor persons whom it does not want").

108. Aliens have also attempted to characterize deportation as punishment in order to challenge certain deportation grounds as unconstitutional ex post facto laws. *See* Harisiades v. Shaughnessy, 342 U.S. 580, 593-95 (1952), discussed *supra* text accompanying notes 63-69, and other sources cited *supra* note 69.

109. 326 U.S. 135 (1945).

110. Alien Registration Act of 1940, ch. 439, § 23, 54 Stat. 670, 671.

111. The Court first interpreted "affiliation" narrowly, saying that the term required more than the mere sympathy or cooperation that the government had shown in *Bridges*. 326 U.S. at 143-44 ("Whether intermittent or repeated, the act or acts tending to prove 'affiliation' must be of that quality which indicates an adherence to or a furtherance of the purposes or objectives of the proscribed organization as distinguished from mere cooperation with it in lawful activities. The act or acts must evidence a working alliance to bring the program to fruition."). It then held that the finding of "membership" that triggered deportation had rested on unsworn statements that were inadmissible under the government's own regulations. *Id.* at 149-56.

112. *See, e.g.*, Gegiow v. Uhl, 239 U.S. 3, 9-10 (1915) (Commissioner of Immigration exceeded his statutory authority in basing "public charge" determination on local labor market only); Chin Yow v. United States, 208 U.S. 8, 12 (1908) (returning resident of Chinese descent not given adequate opportunity to prove United States citizenship: "If one alleging himself to be a citizen is not allowed a chance to establish his right in the mode provided by those statutes, although that mode is intended to be exclusive, the statutes cannot be taken to require him to be turned back without more.").

focused on deportation's harsh consequences,[113] an argument which lawyers and judges had previously used in asserting the constitutional claim that deportation should be treated like criminal punishment.[114] Justice Murphy's concurrence made clear this theme's constitutional dimensions by going further and relying on the same set of harsh consequences to declare that "[t]he unconstitutionality of the statute in issue and the invalidity of the proceeding brought pursuant thereto are obvious."[115] Although the Court had consistently rejected that view as a constitutional matter,[116] it had sufficient gravitational force to affect the interpretation of the statute in *Bridges*.

No single decision established the method of using statutory interpretation to offset the disadvantaged position of aliens in constitutional immigration law, but several cases taken together trace the development of a statutory-interpretation pattern. The next key case in this progression is *Fong Haw Tan v. Phelan*.[117] In *Fong Haw Tan*, a 1948 decision, the Court was confronted with the issue of the proper construction of a statutory deportation ground. The statute in question provided for the deportation of any alien "sentenced [to imprisonment] more than once" for committing crimes involving moral turpitude.[118] Writing for a unanimous Court, Justice Douglas again focused on deportation's harsh consequences.[119] Even though those consequences had never led a court to sustain a constitutional challenge to a substantive deportation ground, Justice Douglas declared, as a general interpretive rule, that courts must read ambiguous deportation statutes or regulations in the light most favorable to the alien.[120]

113. *Bridges*, 326 U.S. at 147 ("[I]t must be remembered that although deportation technically is not criminal punishment, it may nevertheless visit as great a hardship as the deprivation of the right to pursue a vocation or a calling" (citing, *inter alia*, Ng Fung Ho v. White, 259 U.S. 276, 284 (1922))); *see also id.* at 154 ("Though deportation is not technically a criminal proceeding, it visits a great hardship on the individual and deprives him of the right to stay and live and work in this land of freedom. That deportation is a penalty—at times a most serious one—cannot be doubted.").

114. *See supra* text accompanying notes 106-08.

115. *Bridges*, 326 U.S. at 160 (Murphy, J., concurring).

116. *See supra* text accompanying note 107.

117. 333 U.S. 6, 7 (1948).

118. Immigration Act of 1917, ch. 29, § 19(a), 39 Stat. 874, 889 (current version codified at 8 U.S.C. § 155(a) (1988)).

119. 333 U.S. at 10 ("We resolve the doubts in favor of that construction because deportation is a drastic measure and at times the equivalent of banishment or exile. . . . It is the forfeiture for misconduct of a residence in this country. Such a forfeiture is a penalty. To construe this statutory provision less generously to the alien might find support in logic. But since the stakes are considerable for the individual, we will not assume that Congress meant to trench on his freedom beyond that which is required by the narrowest of several possible meanings of the words used." (citation omitted)).

120. More recent applications of this rule include INS v. Cardoza-Fonseca, 480 U.S. 421, 449 (1987) (referring to "longstanding principle of construing any lingering ambiguities in deportation statutes in favor of the alien"); INS v. Errico, 385 U.S. 214, 225 (1966); Costello v. INS, 376 U.S. 120, 128 (1964); Bonetti v. Rogers, 356 U.S. 691, 699 (1958); Barber v. Gonzales, 347 U.S. 637, 642-43 (1954). Similar canons of construction, discussion of which is beyond the scope of this article, ameliorate the inequalities associated with treaties between the federal government and Indian tribes. *See generally* F. COHEN, HANDBOOK OF FEDERAL INDIAN LAW 221-25 (R. Strickland ed. 1982); D. GETCHES & C. WILKINSON, FEDERAL INDIAN LAW: CASES AND MATERIALS 214-17 (2d ed. 1986); C. WILKINSON, AMERICAN INDIANS, TIME, AND THE LAW 46-52 (1987); Wilkinson & Volkman, *Judicial Review of Indian Treaty Abrogation: "As Long As*

In 1950, the Court decided *Wong Yang Sung v. McGrath.*[121] At that time, the statute and regulations let the same corps of immigration inspectors investigate some deportation cases and preside over deportation proceedings in others. They could preside over cases in which they had performed prosecutorial tasks, though not over cases that they had investigated.[122] The combination of adjudicative and prosecutorial roles conflicted with the Administrative Procedure Act's separation-of-functions provisions,[123] but the APA applied by its terms only to "adjudication required by statute."[124] Was a deportation hearing "required by statute"? The Court said that it was, even though the statute contained no express requirement, reasoning that construing it otherwise "might again bring it into constitutional jeopardy."[125]

That holding[126] belies *Wong Yang Sung*'s complexity. Deportation with no hearing at all was almost certainly unconstitutional. This constitutional norm, having already been established in *Yamataya v. Fisher*, was not a phantom.[127] Another constitutional norm, however, also played a role in reaching the subconstitutional conclusion that the APA applied.[128] This second norm, foreign to immigration law, would have required as a constitutional due process matter, and contrary to the immigration statute without the APA, that the deportation proceedings be conducted by an independent hearing examiner who had not acted as a prosecutor. That this constitutional norm was a phantom finds confirmation five years later in *Marcello v. Bonds*,[129] which rejected a due process challenge to similar separation-of-functions provisions in the 1952 immigration statute.[130]

Water Flows, or Grass Grows Upon the Earth"—How Long a Time Is That?, 63 CALIF. L. REV. 601 (1975); Note, *The Canons of Indian Treaty and Statutory Construction: A Proposal for Codification*, 17 U. MICH. J. LEGAL RES. 681 (1984). For the view that these sympathetic canons may not always have an ameliorative effect, see Frickey, *Scholarship, Pedagogy, and Federal Indian Law*, 87 MICH. L. REV. 1199, 1208-09 (1988).

121. 339 U.S. 33 (1950).

122. Immigration Act of 1917, ch. 29, §§ 16-17, 39 Stat. 874, 885-87, *repealed by* Immigration and Nationality Act of 1952, ch. 477, § 403(a)(13), 66 Stat. 163, 279. The then-governing regulations are summarized in *Wong Yang Sung*, 339 U.S. at 45-46.

123. Act of 1946, ch. 324, §§ 5(c), 11, 60 Stat. 237, 240, 244 (current version at 5 U.S.C. §§ 1004(c), 1010).

124. Act of 1946, ch. 324, § 5, 60 Stat. 237, 240 (current version codified at 5 U.S.C. § 1004).

125. 339 U.S. at 50.

126. *See id.* at 49 ("the difficulty with any argument premised on the proposition that the deportation statute does not require a hearing is that, without such hearing, there would be no constitutional authority for deportation"); *id.* at 49-50 ("It was under compulsion of the Constitution that this Court long ago held that an antecedent deportation statute must provide a hearing at least for aliens who had not entered clandestinely and who had been here some time even if illegally.") (citing Yamataya v. Fisher, 189 U.S. 86 (1903)).

127. 189 U.S. 86 (1903), discussed *supra* text accompanying notes 42-44.

128. *See id.* at 48-51.

129. 349 U.S. 302 (1955).

130. *Id.* at 311 ("The contention is without substance when considered against the long-standing practice in deportation proceedings, judicially approved in numerous decisions in the federal courts, and against the special considerations applicable to deportation which the Congress may take into account in exercising its particularly broad discretion in immigration matters."). *Cf.* Withrow v. Larkin, 421 U.S. 35, 46-55 (1975) (rejecting due process challenge against board of medical examiners); Richardson v. Perales,

In the end, *Wong Yang Sung*'s interpretation of the APA reflects two constitutional norms, one "real"—deportation without a hearing violates procedural due process—and one "phantom"—a deportation hearing before an official who mixes functions violates procedural due process. That case represents a further step in the development of phantom norm decisionmaking because its language is bolder than that found in *Bridges* or in *Fong Haw Tan*. In those two earlier cases, the Court's decision had compensated only in very general fashion for the relatively disadvantageous position of aliens in immigration law. Viewing only those two decisions, one could argue that they do not reflect phantom norms at all. They might simply be statutory interpretation cases that favor aliens, which we could explain without necessarily suggesting that they are the subconstitutional reflection of constitutional norms that the Court would not apply directly. But in *Wong Yang Sung*, the Court's constitutional reasoning, based in part on a phantom norm, was more specific to the question presented. This makes it harder to avoid characterizing the decision as a direct challenge to the result that the Court would have reached had it applied the "real" constitutional norm, which permitted combination of adjudicative, investigative, and prosecutorial functions.

Phantom norm decisionmaking reached maturity in *Kwong Hai Chew v. Colding*,[131] a 1953 case that the Supreme Court decided five weeks before *Mezei*, and which, like *Mezei*, concerned a returning permanent resident. After a five-month voyage as a seaman aboard a United States merchant ship, Chew sought to reenter the country. The government denied him entry, claiming it was "prejudicial to the public interest," and relying on the same enactments used to exclude Mezei.[132] The regulations further provided that such aliens could be excluded without a hearing where, as in Mezei's case, disclosure of the specific grounds for exclusion also would be "prejudicial to the public inter-

---

402 U.S. 389, 408-10 (1971) (rejecting due process challenge based on lack of separation of functions in social security disability hearings). *Marcello* arose after an appropriations rider had legislatively overruled *Wong Yang Sung* by expressly stating that the APA did not cover "exclusion or expulsion" of aliens. Act of Sep. 27, 1950, ch. 1052, § 1, 64 Stat. 1044, 1048. *Marcello* decided that the 1952 Immigration and Nationality Act continued this limitation on APA coverage; enforcement officials could also preside over hearings, but not do both in the same case. The Court then had to decide if the statute, so construed, violated constitutional due process, and the Court held without extended discussion that it did not. *Cf. Marcello*, 349 U.S. at 315-19 (Black, J., dissenting) (would read statute differently, so would not reach due process issue); *Wong Yang Sung*, 339 U.S. at 53 (Reed, J., dissenting) ("In this case no one questions the constitutionality of the hearing Wong received . . . ."). The constitutional issues in *Marcello* and *Wong Yang Sung* were similar but not identical to each other. The 1917 Act had separated adjudicative from investigative functions but not from prosecutorial functions in the same case. Under the 1952 law, hearing officers with either investigative or prosecutorial functions could not preside over the same case, though they remained under the supervision of officials with investigative and prosecuting functions. INA § 242(b), 8 U.S.C. § 1252(b) (1988). To see *Wong Yang Sung* as a statutory decision and *Marcello* as a constitutional one helps answer the question posed by Peter Schuck, "[H]ow could the 1950 appropriations rider possibly have overruled *Wong Yang Sung*?" Schuck, *supra* note 6, at 33 (*Wong Yang Sung* was a constitutional decision). *See also* *Wong Yang Sung*, 339 U.S. at 30-34, 72-73. For background from a first-hand perspective, see Rawitz, *From* Wong Yang Sung *to Black Robes*, 65 INTERPRETER RELEASES 453 (1989).

131. 344 U.S. 590 (1953).

132. *Id.* at 591-92; 8 C.F.R. § 175.53 (1953).

est."[133] Subconstitutionally, Chew argued that the regulation did not apply to him. Alternatively, he made the constitutional claim that excluding him without a hearing would deny his procedural due process rights as a returning permanent resident.

The Court's holding that the summary exclusion regulation did not apply to Chew was subconstitutional.[134] This was a highly questionable reading of the regulation's text that Louis Henkin rightly called one of the Court's "feats of creative interpretation" in immigration law.[135] The Court was careful to disclaim reaching the constitutional issue,[136] but as *Mezei* soon confirmed, *Chew*'s underlying reasoning was unmistakably constitutional.[137] The opinion's language, especially its references to *Yick Wo*, *Wong Yang Sung*, and Justice Murphy's concurring opinion in *Bridges*,[138] suggests that the Court favored a constitutional norm of procedural due process for returning permanent residents like Chew, even if the statute and regulations, by applying the reentry doctrine to a temporary departure, treated them no better than first-time entrants.[139] In 1953, however, any such constitutional norm was a phantom because, as the Court would soon confirm in *Mezei*, aliens seeking admission could not challenge immigration law on explicitly constitutional grounds. At the same time, the phantom norm had enough gravitational force to exercise a pull on the Court's interpretation of the regulation. The *Chew* opinion thus subconstitutionally reached the result that the expressly constitutional holding in *Mezei* seemed to preclude.

---

133. *Chew*, 344 U.S. at 591-92; 8 C.F.R. § 175.57(b) (1953).

134. 344 U.S. at 600 ("Section 175.57(b)'s authorization of the denial of hearings raises no constitutional conflict if limited to 'excludable' aliens who are not within the protection of the Fifth Amendment.").

135. Henkin, *The Constitution and the United States Sovereignty: A Century of* Chinese Exclusion *and Its Progeny*, 100 HARV. L. REV. 853, 861 n.40 (1987).

136. *Chew*, 344 U.S. at 602 ("We do not reach the issue as to what would be the constitutional status of 8 CFR § 175.57(b) if it were interpreted as denying to petitioner all opportunity for a hearing."). *See also Mezei*, 345 U.S. at 214 (in *Chew*, "to escape constitutional conflict we held the administrative regulations authorizing exclusion without hearing in certain security cases inapplicable to aliens so protected by the Fifth Amendment") (citing *Chew*, 344 U.S. at 600).

137. *See Mezei*, 345 U.S. at 213 ("a lawful resident alien may not captiously be deprived of his constitutional rights to procedural due process") (citing *Chew*, 344 U.S. at 601). The Court further confirmed the constitutional character of the reasoning of *Chew* in two important cases that Part II.C discusses. *See* Landon v. Plasencia, 459 U.S. 21, 33 (1982) ("the rationale [in *Chew*] was one of constitutional law"); Rosenberg v. Fleuti, 374 U.S. 449, 460 (1963) ("the returning resident alien is entitled as a matter of due process to a hearing on the charges underlying any attempt to exclude him") (citing *Chew*, 344 U.S. 590).

138. 344 U.S. at 596-98 nn.5-7.

139. "For purposes of his constitutional right to due process, we assimilate petitioner's status to that of an alien continuously residing and physically present in the United States." *Id.* at 596. *Mezei* distinguished *Chew* by remarking that Chew had "with full security clearance and documentation pursued his vocation for four months aboard an American ship," while Mezei had "apparently without authorization or reentry papers, simply left the United States and remained behind the Iron Curtain for 19 months." *Mezei*, 345 U.S. at 214. Scholarship from that era raised similar concerns about the treatment of returning permanent residents. *See, e.g.*, Hart, *supra* note 48, at 1392-93; Hesse, *The Constitutional Status of the Lawfully Admitted Permanent Resident Alien: The Inherent Limits of the Power to Expel*, 69 YALE L.J. 262 (1959); Hesse, *The Constitutional Status of the Lawfully Admitted Permanent Resident Alien: The Pre-1917 Cases*, 68 YALE L.J. 1578 (1959).

Another example of the influence of phantom norms is the Court's 1966 decision in *Woodby v. Immigration and Naturalization Service (INS).*[140] There, a majority held that the government must show deportability by "clear, unequivocal, and convincing evidence," rather than by a "preponderance of the evidence" as the government had urged.[141] While the Court's opinion claimed to turn solely on a statute, the Court, as in *Chew*, relied heavily on constitutional considerations of the kind that the Court would recognize in the procedural due process revolution that began a few years afterwards with the Court's 1970 decision in *Goldberg v. Kelly*.[142] Indeed, on at least one later occasion, the Court characterized *Woodby* as a constitutional decision based on notions of fundamental fairness.[143] In particular, the *Woodby* Court noted the harsh consequences of deportation and that many permanent residents had strong ties to this country.[144] Guided by a strong constitutional due process norm, the Court interpreted the immigration statute to require more procedural due process than was available in an ordinary civil proceeding. As in *Chew*, that norm was a phantom.[145] Earlier cases starting with *Yamataya*[146] had suggested some procedural due process minimum, but in fact the strictest real norm had imposed only the most modest requirements that have never begun to approach the constitutional protections of criminal procedure.[147]

140. 385 U.S. 276 (1966). David Martin has suggested that both *Fong Haw Tan*, D. MARTIN, *supra* note 9, at 19, and *Woodby*, *id.* at 25-27, were attempts to compensate for the relative unavailability of judicial review in classical immigration law.

141. 385 U.S. at 281-84. The government relied mainly on references in the statute to "reasonable, substantial, and probative evidence." INA §§ 106(a)(4), 242(b)(4), 8 U.S.C. §§ 1105a(a)(4), 1252(b)(4) (1988). The majority interpreted this language to set the standard for judicial review, not for the deportation proceedings in the first instance. Justice Clark, joined in dissent by Justice Harlan, would have decided for the government based on the language and legislative history of INA § 242(b). 385 U.S. at 287-91 (Clark, J., dissenting). *Cf.* Kungys v. United States, 485 U.S. 759, 772 (1988) (denaturalization proceedings); Schneiderman v. United States, 320 U.S. 118, 125 (1940) (same).

142. 397 U.S. 254 (1970). *See also* Mathews v. Eldridge, 424 U.S. 319 (1976); Perry v. Sinderman, 408 U.S. 593 (1972); Board of Regents v. Roth, 408 U.S. 564 (1972).

143. *See* Santosky v. Kramer, 455 U.S. 745, 756, 759 (1982). *But cf.* Vance v. Terrazas, 444 U.S. 252, 266 (1980) (*Woodby* did not purport to be constitutional holding). The Immigration Marriage Fraud Amendments of 1986, Pub. L. No. 99-639, 100 Stat. 3537, may force courts to decide what standard of proof the Constitution requires. INA § 216(b)(2), 8 U.S.C. § 1186a(b)(2) (1988), gives the INS the burden to prove by a mere preponderance of the evidence that an alien in deportation is not entitled to the conditional resident status created by IMFA.

144. 385 U.S. at 285 ("To be sure, a deportation proceeding is not a criminal prosecution. . . . But it does not syllogistically follow that a person may be banished from this country upon no higher degree of proof than applies in a negligence case. This Court has not closed its eyes to the drastic deprivations that may follow when a resident of this country is compelled by our Government to forsake all the bonds formed here and go to a foreign land where he often has no contemporary identification.").

145. *See* Marcello v. Bonds, 349 U.S. 302 (1955), discussed *supra* text accompanying notes 129-30. *See generally Developments in the Law—Immigration and Nationality*, 66 HARV. L. REV. 643, 680-82, 690-95 (1953).

146. *Yamataya*, 189 U.S. 86 (1903), discussed *supra* text accompanying notes 42-44.

147. Leading cases rejecting the argument that deportation amounts to criminal punishment are cited *supra* note 107.

A pattern emerges from *Bridges, Fong Haw Tan, Wong Yang Sung, Chew,* and *Woodby*.[148] In *Bridges, Fong Haw Tan,* and *Wong Yang Sung,* the Court allowed phantom constitutional norms to guide statutory interpretation by reading statutes in favor of aliens. In *Chew* and *Woodby,* phantom norm decisionmaking produced results that directly conflicted with those that the Court would have reached by applying the "real" constitutional norms.[149] Further evidence of the sway of phantom norms is found in the fact that dissenting opinions also relied on such norms. In *Knauff,* for example, Justices Frankfurter and Jackson did not criticize the majority for its constitutional holding. Rather, they urged the Court to interpret the ambiguous statute against summary exclusion, an interpretation that would have been in accordance with a phantom constitutional norm of procedural due process in exclusion.[150]

These phantom norm decisions represent an aberrational application of the canon that judges should interpret statutes to avoid constitutional doubts. The usual reason for applying the canon is to avoid defining what the Constitution means with virtually irrevocable finality, and yet to provide a just result. This rationale is valid only when the directly and indirectly applicable constitutional norms are roughly consistent with each other. Where direct and indirect norms harmonize with each other, "'underenforced' constitutional norms"[151] and

148. Some counterexamples from this period deserve mention. I believe that they do not undermine my basic thesis, since I do not claim that the Court seized every opportunity to interpret subconstitutional texts by relying on a phantom norm, only that enough decisions did so to allow us to identify it as a type of judicial decisionmaking in immigration cases. The Court rejected both subconstitutional and constitutional challenges in at least two other cases from the same general period. In Galvan v. Press, 347 U.S. 522 (1954), Justice Frankfurter first rejected the argument that the deportation ground for Communist Party membership, Internal Security Act of 1950, ch. 1024, § 22, 64 Stat. 987, 1006 (substantially codified at INA § 241(a)(6)(C), 8 U.S.C. § 1251(a)(6)(C) (1988)), should be read to exclude any member not "fully cognizant of the Party's advocacy of violence." 347 U.S. at 528. He then rejected the constitutional attack on the same statute. *Id.* at 529-32. *See also* Jay v. Boyd, 351 U.S. 345, 357 (1956) ("[W]e must adopt the plain meaning of [the suspension of the deportation] statute, however severe the consequences.") (citing Galvan v. Press, 347 U.S. 522, 528 (1954)). A third counterexample is *Knauff,* where the Court could have taken a statutory approach but did not. *See* Hart, *supra* note 48, at 1395 ("the War Brides Act was highly ambiguous on the point in issue of whether exclusion without a hearing was authorized"). *Mezei* and *Harisiades* are not counterexamples because subconstitutional solutions would have involved much more strained readings of the applicable statutes or regulations.

149. A similar case in a related area of law is Kent v. Dulles, 357 U.S. 116 (1958), which arose when several United States citizens sued to force the Department of State to issue them passports so that they could travel abroad. While Justice Douglas' reasoning for the majority was based on the constitutional right to travel, the decision was clearly subconstitutional—the Department's regulations, which elicited information on passport applicants' political affiliations, exceeded the scope of its statutory authority. "Since we start with an exercise by an American citizen of an activity included in constitutional protection, we will not readily infer that Congress gave the Secretary of State unbridled discretion to grant or withhold it." *Id.* at 129. Justice Clark, writing for the four dissenters, disagreed with the majority's reading of congressional intent and would have found that the Department of State acted within its statutory authority. *Id.* at 130-43 (Clark, J., dissenting). *Kent* is a phantom norm case like *Chew* and *Woodby* in that no tradition of constitutional judicial review of executive decisions regarding passports existed.

150. *See* United States *ex rel.* Knauff v. Shaughnessy, 338 U.S. 537, 549-50 (1950) (Frankfurter, J., dissenting); *id.* at 551-52 (Jackson, J., dissenting).

151. *See* Sager, *supra* note 87, at 1213-20; *see also* Sunstein, *supra* note 77, at 468 ("courts underenforce many constitutional norms").

"constitutional common law" sensibly and justifiably arise.[152] But neither this reasoning nor this terminology applies where the constitutional norms underlying subconstitutional decisions are phantom norms. This duality of constitutional norms present in immigration law makes *Chew* and *Woodby* quite different from *Bob Jones* and other examples of constitutionally influenced statutory interpretation in mainstream public law.

I recognize that describing norms as "real" is not a simple matter. The "racial equality" norm, for example, is derived from a series of concrete decisions, none of which need be explained that abstractly. If the norm is "real," how do we account for a case like *Washington v. Davis*,[153] which established the principle that a finding of unconstitutional race discrimination requires proof of discriminatory purpose in addition to proof of discriminatory effect? In my view, that case represents underenforcement of the norm, though not the particular type of underenforcement represented by avoidance of constitutional issues through statutory interpretation. The *Washington v. Davis* Court had prudential reasons for its decision; it was responding to perceived limits on the judiciary's institutional capacity to measure unlawful discrimination. While that underenforcement may have been unjustified, and while it may represent a practical retrenchment of race remedies doctrine,[154] *Washington v. Davis* and similar cases affirm the basic proscription against racial discrimination. "Norms" are by definition standards of some generality, much broader than the rule laid down in *Washington v. Davis*. In identifying constitutional norms that courts adopt as background for statutory interpretation, we must accept the degree of generality with which litigants and judges tend to invoke them. As with the state action, standing, and political question doctrines, the interstitial cases in which the norm is not fully enforced do not take away from the existence of the norm against racial discrimination if it can be established from other sources. Indeed, the norm has become so engrained that even a court hostile to civil rights could not plausibly call into question the fundamental principle against racial discrimination that many questioned in all seriousness before *Brown*.

Unlike the standing or political question doctrines, however, the plenary power doctrine smothers the entire field of immigration law so completely that it is difficult to find the benchmarks of "full enforcement" of real norms.[155] In 1953 no immigration decisions had established a constitutional norm prohib-

152. *See* Monaghan, *The Supreme Court, 1974 Term—Foreword: Constitutional Common Law*, 89 HARV. L. REV. 1 (1975).

153. 426 U.S. 229 (1976).

154. *See, e.g.*, Delgado, *On Taking Back Our Civil Rights Promises: When Equality Doesn't Compute*, 1989 WIS. L. REV. 579, 583-84.

155. Of the examples offered by William Eskridge of groups—"discrete and insular minorities"—favored by statutory interpretation, only aliens in immigration law cases stand out as unprotected by analogous judicial concern at the constitutional level. *See* Eskridge, *supra* note 74, at 1032-34 (1989); *see also* Frickey, *Congressional Intent, Practical Reasoning, and The Dynamic Nature of Federal Indian Law*, __ CALIF. L. REV. ___ (forthcoming 1990); *Sunstein, supra* note 77, at 472-73, 483-86.

iting exclusion of a returning permanent resident without a hearing. If there had been such a case, then *Chew*, like *Bob Jones*, might be an example of interpreting a statute to avoid constitutional doubt and thus of underenforcing a real constitutional norm to avoid unnecessary friction between the branches of government. Or more generally, if plenary power were repudiated, subconstitutional immigration cases would look a lot more like subconstitutional race cases. In *Chew* and *Woodby*, the majority's sense of what the constitutional law should be in the future guided the Court's interpretation of the statute. The Court could have disposed of the case consistent with the reigning constitutional norm, or it could have replaced the plenary power doctrine with a constitutional norm favorable to aliens. However, the Court was unwilling to adopt either course, and instead adopted the middle ground of interpreting the statute in accordance with a phantom constitutional norm.[156]

C. *Constitutional Change: From Phantom Norms to Real Norms*

In the past twenty-five years, one key aspect of the movement away from classical immigration law has been an increasing acceptance of the view that an alien's constitutional status when challenging immigration statutes should depend on more than just her geographic location. A returning permanent resident, even if outside the United States, typically has ties or a "stake" here that merit more exacting constitutional scrutiny than that available to first-time entrants.[157] As *Chew* suggests, the first stage in this development was the emergence of a phantom constitutional norm that recognized a resident alien's stake in this country and served as a guide to further statutory interpretation. Later cases suggest that through phantom norm decisionmaking, subconstitu-

156. When a decision relies on a real constitutional norm to guide statutory interpretation, it need not be as careful to avoid a constitutional pronouncement as a decision, such as *Chew*, that relies on a phantom norm. A good illustration is Yamataya v. Fisher, 189 U.S. 86 (1903), discussed *supra* text accompanying notes 42-44. *Yamataya* is usually read as a constitutional decision. *See, e.g.*, Shaughnessy v. United States *ex rel.* Mezei, 345 U.S. 206, 212 (1953) (*Yamataya* established that "aliens who have once passed through our gates, even illegally, may be expelled only after proceedings conforming to traditional standards of fairness encompassed in due process of law"); T. ALEINIKOFF & D. MARTIN, *supra* note 15, at 408 ("*The Japanese Immigrant Case* [*Yamataya*] stands for the proposition that deportation procedures must conform to the dictates of the due process clause of the Constitution."). The Court did speak constitutionally. *See* 189 U.S. at 100 ("[T]his court has never held, nor must we now be understood as holding, that administrative officers, when executing the provisions of a statute involving the liberty of persons, may disregard the fundamental principles that inhere in 'due process of law.'"). But the holding was also cast in subconstitutional terms. *See id.* at 101 ("The words here used do not require an interpretation that would invest executive or administrative officers with the absolute, arbitrary power implied in the contention of the appellant." The Court was not particularly careful to avoid deciding constitutional issues, in large part because it was willing to say that the constitutional norm was real. Two other examples of the Supreme Court's uncritical willingness to pronounce on constitutional issues may be found in the part of *Wong Yang Sung* that relies on a real constitutional norm, and in United States v. Witkovich, 353 U.S. 194, 201-02 (1957) (construing statute to limit Attorney General's authority to interrogate aliens in custody pending deportation because broader reading would raise constitutional doubts).

157. *See* T. ALEINIKOFF & D. MARTIN, *supra* note 15, at 453-63.

tional immigration law gradually became the engine that would drive constitutional change.

The Court's 1963 decision in *Rosenberg v. Fleuti* is one milestone in the transition from phantom to real of a norm recognizing a returning resident alien's stake.[158] Fleuti, a permanent resident since 1952, took a short trip of "about a couple hours" to Mexico in 1956.[159] In 1959, the Immigration and Naturalization Service ("INS") tried to deport him as an an alien "excludable by the law existing at the time of . . . entry."[160] The INS sought to declare Fleuti excludable on the ground that, by virtue of his homosexuality, Fleuti was "afflicted with psychopathic personality."[161] The Ninth Circuit set aside the deportation, holding that, as applied to homosexuality, the term "psychopathic personality" was unconstitutionally vague.[162] The Supreme Court held for Fleuti on the altogether different ground that he had not "entered" the United States in 1956. This conclusion was fatal to the government's case because "psychopathic personality" had become an exclusion ground between Fleuti's initial entry in 1952 and his possible "reentry" in 1956. If he last entered in 1952, he had not been excludable at entry and thus was not deportable. Modifying the reentry doctrine significantly, the Court defined "entry" so as to avoid application where an alien returned to the United States from a temporary absence that was not "meaningfully interruptive" of permanent residence.[163] The *Fleuti* Court reasoned that the 1952 Act, which had amended the statutorily imposed definition of "entry" expressly to exclude a return from an unexpected or involuntary absence,[164] must also have intended to exempt "innocent, casual, and brief" trips like Fleuti's.[165]

158. 374 U.S. 449 (1963).

159. *Id.* at 450.

160. INA § 241(a)(1), 8 U.S.C. § 1251(a)(1) (1988).

161. INA § 212(a)(4), 8 U.S.C. § 1182(a)(4) (1988). The INS had argued earlier that Fleuti had been excludable in 1956 because he had been "convicted of a crime involving moral turpitude," INA § 212(a)(9), 8 U.S.C. § 1182(a)(9) (1988), but the crime turned out to be a petty offense insufficient for exclusion. *See* Fleuti v. Rosenberg, 302 F.2d 652, 656 n.10 (9th Cir. 1962), *aff'd on other grounds*, 374 U.S. 449 (1963); *see also* 374 U.S. at 450.

162. Fleuti v. Rosenberg, 302 F.2d 652, 655-58 (9th Cir. 1962) ("psychopathic personality" unconstitutionally vague absent case law that might have clarified statute), *aff'd on other grounds*, 374 U.S. 449 (1963).

163. 374 U.S. at 462.

164. Act of June 27, 1952, Pub. L. No. 82-414, § 101, 66 Stat. 163, 167. The amendment codified the holdings of two cases that had avoided especially harsh applications of the reentry doctrine. *See* Delgadillo v. Carmichael, 332 U.S. 388 (1947) (alien on sleeping car from Buffalo to Detroit, unaware that train would travel through Canada, did not "enter" upon arrival in Detroit); DiPasquale v. Karnuth, 158 F.2d 878 (2d Cir. 1947) (alien rescued and taken to Cuba for one week from ship that had been torpedoed and sunk did not "enter" upon his return to the United States).

165. 374 U.S. at 457-58 ("[T]he major congressional concern in codifying the definition of 'entry' was with 'the status of an alien who has previously entered the United States and resided therein . . . .' This concern was in the direction of ameliorating the harsh results visited upon resident aliens by the rule of *United States ex rel. Volpe v. Smith* . . . . Congress, in approving the judicial undermining of *Volpe*, and the relief brought about by the *Di Pasquale* and *Delgadillo* decisions, could not have meant to limit the meaning of the exceptions it created in § 101(a)(13) to the facts of those two cases." (citation omitted)).

In large part, the majority's *Fleuti* opinion, like *Chew*, evinces sympathy for the view that immigration law inadequately recognized a permanent resident's stake in remaining in the United States. This observation explains several aspects of the opinion: the focus on the harshness of the leading reentry doctrine case, *United States* ex rel. *Volpe v. Smith*,[166] and on the hardships that Fleuti would suffer if deported;[167] citation to *Chew* for the observation that "a resident alien who leaves this country is to be regarded as retaining certain basic rights";[168] and the call for "more civilized application of our immigration laws . . . ."[169] But as in *Chew* and *Woodby*, the constitutional norm that guided *Fleuti*'s reading of "entry" was a phantom, which in turn explains why the Court decided the case by interpreting "entry" in the statute, rather than by striking down the statute as unconstitutional. No decision of the Court had openly recognized procedural due process rights for an alien seeking entry, not even for a returning permanent resident. Indeed, *Mezei* had suggested and perhaps even held to the contrary. Still, the phantom norm of procedural due process for returning permanent residents was close enough to the constitutional horizon that it is unsurprising that the Court adopted it as a guide to statutory interpretation. Using this norm, the Court construed "entry" to recognize stake, even if prevailing doctrine did not permit a direct constitutional holding.

The Court's recognition of Fleuti's stake through the definition of "entry," but not through other subconstitutional elements of the government's case against him, suggests that phantom norm decisions must rely on constitutional norms that are "real" enough to exercise gravitational pull, even if they have not yet appeared above the constitutional horizon. If the Court had found that Fleuti had "entered" in 1956, the case's center of gravity would have shifted to the statutory exclusion ground of "psychopathic personality" and the constitutional attack on that ground as void-for-vagueness. The term "psychopathic personality" and the term "entry" were similar in one respect—both required ignoring clear legislative intent before Fleuti could win a favorable subconstitutional outcome. Congress almost certainly intended the term "psychopathic personality" to apply to homosexuals.[170] And, as Justice Clark noted in dis-

166. 374 U.S. at 453 (noting that Volpe had been deported "after 24 years of residence in this country"). United States *ex rel.* Volpe v. Smith, 289 U.S. 422 (1933), is discussed *supra* note 57.

167. 374 U.S. at 458 (Fleuti's interests are "high and momentous"); *id.* at 459 ("'[d]eportation can be the equivalent of banishment or exile'") (citing Barber v. Gonzales, 347 U.S. 637, 642-43 (1954); Fong Haw Tan v. Phelan, 333 U.S. 6, 10 (1948); Bridges v. Wixon, 326 U.S. 135, 147 (1945); Ng Fung Ho v. White, 259 U.S. 276, 284 (1922)).

168. 374 U.S. at 460.

169. *Id.* at 462.

170. The statute was enacted pursuant to a Senate Judiciary Committee report and recommendation that "homosexuals and other sex perverts" be excluded. S. REP. NO. 1515, 81st Cong., 2d Sess. 345 (1950). The Senate Report accompanying the bill noted: "The Public Health Service has advised that the provision for the exclusion of aliens afflicted with psychopathic personality or a mental defect which appears in the instant bill is sufficiently broad to provide for the exclusion of homosexuals and sex perverts. This change of nomenclature is not to be construed in any way as modifying the intent to exclude all aliens who are sexual deviates." S. REP. NO. 1137, 82d Cong., 1st Sess. 9 (1950). *See also* H.R. REP. NO. 1365, 82d Cong.,

sent, Congress probably also intended the term "entry" to cover Fleuti's return from Mexico, having allowed only limited exceptions to the statutory definition of "entry." Indeed, Congress had even rejected amendments that would have covered Fleuti.[171] The Court's focus on "entry" reveals that a constitutional norm that would declare the term "psychopathic personality" void-for-vagueness was even more of a phantom, far off in the constitutional distance. In fact, seven years later the Court squarely rejected a vagueness challenge to the same statute.[172] A judicial interpretation of the term "psychopathic personality" that did not include homosexuality would have required reliance on an even less palpable constitutional norm than did the interpretation of the word "entry."[173]

Limits on what the Court could read into "entry" inevitably rendered *Fleuti* an incomplete solution. It did nothing for some returning permanent residents—those with a "meaningfully interruptive" absence—even if they could claim a "stake" of constitutional dimensions.[174] Cases were bound to arise in which the statutory solution was inadequate, adding to the pressure for open recognition of the underlying constitutional reasoning. In a 1982 case, *Landon v. Plasencia*,[175] the Court took the logical next step: it addressed the constitutional issue directly and made "real" the phantom constitutional norm that had previously been used only as a guide to statutory interpretation.

Maria Antonieta Plasencia, a citizen of El Salvador, had been a permanent resident of the United States since 1970. She lived in Los Angeles with her husband, who was a United States citizen, and their children. In 1975, the INS detained her upon her return from a brief trip she and her husband had taken

---

2d Sess. (1952), *reprinted in* 1952 U.S. CODE CONG. & ADMIN. NEWS 1652, 1701 (quoting and adopting Public Health Service report recommending that the term "psychopathic personality" be used to "specify such types of pathologic behavior as homosexuality or sexual perversion"). *See generally* Boutilier v. INS, 387 U.S. 118, 122 (1967) ("Congress used the phrase 'psychopathic personality' not in the clinical sense, but to effectuate its purpose to exclude from entry all homosexuals and other sex perverts."); Aleinikoff, *supra* note 86, at 47-54. *But cf.* Boutilier v. INS, 363 F.2d 488, 496-99 (2d Cir. 1966) (Moore, J., dissenting) (Congress intended to exclude only if homosexuality amounted to personality disorder in individual); Eskridge, *Gadamer/Statutory Interpretation*, 90 COLUM. L. REV. 609, 642-44 (1990) (discussing the difficulties of determining the legislative intent behind INA § 212(a)(4)).

171. *See Fleuti*, 374 U.S. at 463-68 (Clark, J., dissenting); *see also* T. ALEINIKOFF & D. MARTIN, *supra* note 15, at 335 ("Isn't *Fleuti* an embarassment to the United States Supreme Court? Isn't the dissent clearly correct that the majority rewrote the statute along the lines of proposals made to, but rejected by, Congress?"); Comment, *Exclusion and Deportation of Resident Aliens: The Re-Entry Doctrine and the Need for Reform*, 13 SAN DIEGO L. REV. 192, 197-200 (1975).

172. Boutilier v. INS, 387 U.S. 118, 123-24 (1967) (deportation ground not based on post-entry conduct, therefore "no necessity exists for guidance so that one may avoid the applicability of the law"; "Congress has plenary power to make rules for the admission of aliens."); *see also* Mahler v. Eby, 264 U.S. 32, 40-41 (1924); *cf.* Jordan v. De George, 341 U.S. 223, 229-32 (1951) (caselaw keeps deportation statute for crime of "moral turpitude" from being unconstitutionally vague).

173. At the time that the Court decided *Fleuti*, prevailing medical opinion had not yet changed in the way that later prompted the Ninth Circuit to interpret the statute not to permit exclusion of homosexuals without a medical certificate of excludability. *See* Hill v. INS, 714 F.2d 1470, 1481 (9th Cir. 1983), *aff'g* Lesbian/Gay Freedom Day Comm. v. INS, 541 F. Supp. 569, 579-80 (N.D. Cal. 1982); *see also infra* text accompanying notes 287-94.

174. The incomplete or underinclusive solution, which is a more general problem with phantom norm decisionmaking, is discussed more fully *infra* text accompanying notes 308-13.

175. 459 U.S. 21 (1982).

to Mexico. The INS considered her excludable rather than deportable as an "entering" alien on the ground that the *Fleuti* exception did not apply. The Service then excluded her for allegedly having aided six Mexican and Salvadoran nationals, who traveled with her in their attempt illegally to enter the United States.[176] After the Board of Immigration Appeals (BIA) dismissed Plasencia's appeal, she successfully filed for habeas corpus in the district court, which found that the INS had mistakenly placed her in the exclusion category for the purpose of deciding if her return was an "entry." A Ninth Circuit panel affirmed, relying heavily on *Chew*. The panel declared broadly: "[A] permanent resident alien does not lose the procedural protection to which he is otherwise entitled simply by making a brief journey abroad."[177] But the constitutional norm that would afford procedural due process to reentering permanent residents in exclusion hearings was just a phantom, and understandably the panel limited itself to interpreting the statute to afford Plasencia a deportation hearing to decide whether her return was an "entry."

The Supreme Court could have affirmed on the statutory ground, a course that would have been consistent with its approach in *Chew* and *Fleuti*. Instead, it construed the statute to label Plasencia as excludable rather than deportable. Reaching the constitutional issue, the Court held that even a returning permanent resident subject to exclusion "can invoke the Due Process Clause . . . ."[178] The Court distinguished *Mezei* as involving a much longer absence[179] and held that an alien's constitutional status does not depend strictly on the statutory exclusion or deportation categories. The Court openly recognized its holding's pedigree by relying heavily on the same constitutional considerations that had guided statutory interpretation in *Chew* and *Fleuti*. "[O]nce an alien gains admission to our country and begins to develop the ties that go with permanent residence, his constitutional status changes accordingly."[180] Justice O'Connor wrote that "[a]lthough the holding [in *Chew*] was one of regulatory interpretation, the rationale was one of constitutional law,"

---

176. INA § 212(a)(31), 8 U.S.C. § 1182(a)(31) (1988), excludes "[a]ny alien who at any time shall have, knowingly and for gain, encouraged, induced, assisted, abetted, or aided any other alien to enter or to try to enter the United States in violation of law."

177. Plasencia v. Sureck, 637 F.2d 1286, 1288 (9th Cir. 1980), *rev'd sub nom*. Landon v. Plasencia, 459 U.S. 21 (1982).

178. 459 U.S. at 32. The Court declined to say if the process afforded to Plasencia was sufficient and remanded on that issue.

179. *Id.* at 34.

180. *Id.* at 32 (citing Johnson v. Eisentrager, 339 U.S. 763, 770 (1950)). *See also* D. MARTIN, *supra* note 9, at 24 ("After *Plasencia*, due process protection no longer turns entirely on the distinction between exclusion cases and deportation cases—that is, on whether the alien is at the border or in the interior of the country. In some settings, at least, the courts may undertake a more sensitive inquiry into the alien's community ties in assessing procedural requirements under the Constitution."); *cf.* Rafeedie v. INS, 880 F.2d 506, 519-25 (D.C. Cir. 1989) (preliminary injunction against summary exclusion of returning permanent resident), discussed in greater detail *infra* text accompanying notes 322-28.

and she cited *Fleuti* as having acknowledged this aspect of *Chew*.[181] In *Chew* and *Fleuti* the constitutional norms were phantoms, and *Plasencia* transformed them into "real" constitutional immigration law.

III. CONTEMPORARY PLENARY POWER: CONSTITUTIONAL V. STATUTORY LAW

A. *Phantom Norm Review of Executive Action*

Plasencia is vital to this story because it shows that phantom norms can represent a transitional phase. But *Plasencia* addressed just one constitutional aspect of classical immigration law—procedural due process for returning permanent residents—leaving many others untouched. It said nothing about substantive constitutional claims by permanent residents, such as the claim that the Constitution prevents the government from deporting them for speech protected by the First Amendment. Nor did *Plasencia* say anything about claims asserted by those who were not permanent residents. In other recent cases, courts have recognized some of these claims through subconstitutional decisions guided by phantom constitutional norms, as in *Chew* and *Fleuti*. It remains to be seen whether these phantom norms will become part of the directly applied constitutional immigration law.

These recent phantom norm decisions have endorsed some degree of judicial review of immigration decisions committed by Congress to an agency of the executive branch. Of course, some of the earlier phantom norm decisions, certainly including both *Chew* and *Woodby*, also can be characterized as judicial review of agency decisions. But more recent cases vary from the early phantom norm decisions in one important sense. Their reasoning is distinctive—it relies heavily on first characterizing the decision as one of agency rather than legislative action, and then scrutinizing that agency action for subconstitutional defects.

The genesis of this decisionmaking pattern is complex; it reflects the emergence and combination of four views about the proper scope of judicial review. First, courts are more competent to review immigration decisions by agencies than by Congress. The first element is best explained through the Court's 1972 decision in *Kleindienst v. Mandel*, which upheld the denial of a

181. 459 U.S. at 33 ("Any doubts that *Chew* recognized constitutional rights in the resident alien returning from a brief trip abroad were dispelled by *Rosenberg v. Fleuti*, where we described *Chew* as holding 'that the returning resident alien is entitled as a matter of due process to a hearing on the charges underlying any attempt to exclude him.'"). *See also* Rafeedie v. INS, 880 F.2d 506, 520-21 (D.C. Cir. 1989) ("*Chew* was a constitutional decision"), *aff'g in part and rev'g in part* 688 F. Supp. 729, 743 (D.D.C. 1988) ("While the Supreme Court in *Chew* cast its conclusion in statutory terms, the decision itself and subsequent cases have made clear that the rationale was a constitutional one; even though the Court did not hold the regulation facially unconstitutional, it found that the statute could not constitutionally be applied to Chew.") (footnote and citations omitted).

nonimmigrant visa to Ernest Mandel, a Belgian journalist and self-described "revolutionary Marxist."[182] In 1968, Mandel received invitations from various universities throughout the United States. For two earlier visits, the government, without his knowledge, had exercised its statutory discretion to waive the exclusion ground for Communists, but this time it declined to do so.[183] Mandel and several of the United States citizen professors who had invited him sued, relying primarily on the argument that the First Amendment guaranteed them the right to meet and speak with Mandel.[184]

Mandel is important for our purposes because in that case the Court suggested a distinction between decisions made directly by Congress in an immigration statute and enforcement of those statutes by the executive branch. According to the Court, the plaintiffs raised a First Amendment challenge not to the plenary power doctrine itself, but just to the Attorney General's discretionary decision not to waive exclusion.[185] The government argued that the Attorney General is vested with "sole and unfettered discretion" to deny waiver by giving "any reason or no reason" at all,[186] but the Court found that it did not need to decide this question. Instead, it found that the Attorney General had given a "facially legitimate and bona fide reason" for his decision, and that this foreclosed further judicial review.[187] But by distinguishing the plenary power cases and conditioning its approval of discretionary decisions by requiring the government to offer a "facially legitimate and bona fide reason," the Court suggested some outer limits to executive discretion that might not apply to direct congressional decisions.

Mandel's suggested distinction between executive actions and legislative acts has found its way into lower court decisions that have scrutinized executive branch immigration decisions more closely than the plenary power doctrine would seem to allow.[188] The distinction reflects the view that courts intrude

182. 408 U.S. 753, 769-70 (1972).

183. INA § 212(d)(3)(A), 8 U.S.C. § 1182(d)(3) (1988), allows discretionary waiver of the ideological exclusion grounds applied to Mandel, which were INA § 212(a)(28)(D) & (G)(v), 8 U.S.C. § 1182(a)(28)(D) & (G)(v) (1988).

184. The original complaint also alleged violations of equal protection and procedural due process, see 408 U.S. at 760, but these issues were not before the Court.

185. *Id.* at 767. Justice Marshall, joined by Justice Brennan, dissented on the grounds that the constitutionality of § 212(a)(28) necessarily came into question, since no waiver would have been necessary without that subsection. *Id.* at 777-84.

186. *Id.* at 769.

187. *Id.* at 770 ("In the case of an alien excludable under § 212(a)(28), Congress has delegated conditional exercise of this power ['to make policies and rules for exclusion of aliens'] to the Executive. We hold that when the Executive exercises this power negatively on the basis of a facially legitimate and bona fide reason, the courts will neither look behind the exercise of that discretion, nor test it by balancing its justification against the First Amendment interests of those who seek personal communication with the applicant."). *Cf. id.* at 774 (Douglas, J., dissenting) (Attorney General exceeded his statutory powers because his discretion was limited to "matters commonly within the competence of the Department of Justice—national security, importation of drugs, and the like").

188. *See, e.g.*, Abourezk v. Reagan, 785 F.2d 1043, 1061 (D.C. Cir. 1986) ("The Executive has broad discretion over the admission and exclusion of aliens, but that discretion is not boundless. It extends only as far as the statutory authority conferred by Congress and may not transgress constitutional limitations.

on the legislative sphere more when they review decisions made by Congress directly than when they review immigration decisions made by the executive branch pursuant to a delegation of power by Congress.[189] This maxim is especially powerful when the executive operates under a broad grant of authority from Congress rather than under narrow or specific congressional direction.[190] With the addition of the three other elements, judicial scrutiny—at least in some cases—was to become less restrained than *Mandel*'s apparently deferential "facially legitimate and bona fide reason" formula would suggest.

The second view that contributed to this decisionmaking pattern was unremarkable and well established—that courts can review immigration decisions under a subconstitutional "arbitrary and capricious" or "abuse of discretion" test, to make sure the executive branch has stayed within congressionally

It is the duty of the courts, in cases properly before them, to say where those statutory and constitutional boundaries lie."), *aff'd by an equally divided Court*, 484 U.S. 1 (1987); Louis v. Nelson, 544 F. Supp. 973, 998 (S.D. Fla. 1982), *aff'd sub nom.* Jean v. Nelson, 711 F.2d 1455 (11th Cir. 1983), *rev'd*, 727 F.2d 957 (11th Cir. 1984) (en banc), *aff'd as modified*, 472 U.S. 846 (1985) (emphasizing importance of distinction between review of congressional and agency action); Haitian Refugee Center v. Civiletti, 503 F. Supp. 442, 452 (S.D. Fla. 1980) ("power of INS is more circumscribed [than Congress']. . . . Where it has been given discretion by statute or treaty, INS may not exercise that discretion arbitrarily or capriciously."), *aff'd as modified sub nom.* Haitian Refugee Center v. Smith, 676 F.2d 1023 (5th Cir. Unit B 1982). *But cf.* Schuck, *supra* note 6, at 15 ("Court has not hesitated to extend this 'special judicial deference to congressional policy choices in the immigration context' to administrative officials as well as to the Congress.") (footnote omitted).

Even after *Mandel*, courts still routinely refuse to review visa denials by consular officers overseas. *See, e.g.*, Li Hing of Hong Kong, Inc. v. Levin, 800 F.2d 970, 971 (9th Cir. 1986); Burrafato v. Department of State, 523 F.2d 554, 556 (2d Cir. 1975), *cert. denied*, 424 U.S. 910 (1976); Pena v. Kissinger, 409 F. Supp. 1182, 1185-87 (S.D.N.Y. 1976); *cf.* Wong v. Department of State, 789 F.2d 1380, 1384-85 (9th Cir. 1986) (judicial review of visa revocation, with alien in United States); Shimizu v. Department of State, CV 89-2741-WMB (C.D. Cal. May 31, 1990), *digested in* 67 INTERPRETER RELEASES 699, 702 (1990) (judicial review of visa revocation, with alien outside United States). For background and analysis regarding visa denials and their reviewability, see Recommendation of the Administrative Conference of the United States, 1 C.F.R. § 305.89-9 (1990); Martin, Mandel, Cheng Fan Kwok, *and Other Unappealing Cases: The Next Frontier of Immigration Reform*, 27 VA. J. INT'L L. 803, 809-14 (1987).

189. Executive decisions may be less appropriate for judicial review than direct legislative decisions because they are more "political," but the courts have gradually abandoned the political question doctrine as a basis for declining judicial review in immigration cases. *See, e.g.*, Orantes-Hernandez v. Meese, 685 F. Supp. 1488, 1503 (C.D. Cal. 1988); Harvard Law School Forum v. Shultz, 633 F. Supp. 525, 528 (D. Mass. 1986); *cf.* INS v. Chadha, 462 U.S. 919 (1983), which held unconstitutional the power of either house of Congress, by resolution, to overrule discretionary suspension of deportation by the Attorney General for reasons of "extreme hardship" under INA § 244(a)(1), (c)(2), 8 U.S.C. § 1254(a)(1), (c)(2) (1988). "The plenary authority of Congress over aliens under [the naturalization clause] is not open to question, but what is challenged here is whether Congress has chosen a constitutionally permissible means of implementing that power." 462 U.S. at 940-41. The decision seems confined to constitutional challenges based on separation of powers, instead of limiting plenary power generally, and thus seems akin to the cases that recognize more judicial competence to hear procedural due process claims than attacks on substantive immigration classifications. *Cf.* Fiallo v. Bell, 430 U.S. 787, 794 (1977) ("no occasion to consider in this case whether there may be actions of the Congress with respect to aliens that are so essentially political in character as to be nonjusticiable"). *See generally* Legomsky, *supra* note 5, at 255, 261-69.

190. *See, e.g.*, Narenji v. Civiletti, 481 F. Supp. 1132, 1140 (D.D.C. 1979) (less judicial review if case involves "a specific statutory enactment of Congress"), *rev'd on other grounds*, 617 F.2d 745 (D.C. Cir. 1979), *cert. denied*, 446 U.S. 957 (1980).

prescribed bounds.[191] Third, judicial review for an "arbitrary and capricious" decision or an "abuse of discretion" includes a look at any constitutional doubts about the agency decision. Put differently, an executive branch decision is an "abuse of discretion" or "arbitrary and capricious" if it relies on unconstitutional premises.[192]

The fourth and capstone element was that courts may broaden the grounds for an "arbitrary and capricious" or "abuse of discretion" finding to include conflict between agency action and a *phantom* constitutional norm. In other words, even if constitutional doubts about an agency decision are insufficient to justify finding it unconstitutional, courts still express those doubts at a subconstitutional level through an "arbitrary and capricious" or "abuse of discretion" finding. The reasoning might be that the agency abused its discretion by violating a phantom constitutional norm, or that the agency exceeded its delegated powers under a statute that would be interpreted narrowly using a phantom constitutional norm. In either case, the court effectively undermines what would seem to be the governing principles of constitutional immigration law.

B. Jean v. Nelson *and Other Phantom Norm Decisions*

Thanks to these four elements, in the past decade, phantom norm-based subconstitutional review of agency action—a variation on the phantom norm decisions of the 1950's and 1960's—has become a widespread method of deciding difficult immigration cases in many of the areas left open in *Plasencia*. These more recent immigration cases, like the early phantom norm decisions, represent a partial integration into immigration law of the values embodied in *Yick Wo*'s treatment of aliens outside of immigration law.[193]

Graham v. Richardson,[194] a 1971 Supreme Court decision, is a seminal case in *Yick Wo*'s modern legacy. *Graham* struck down a state-imposed residency requirement limiting lawfully admitted aliens' access to welfare benefits. The law pertaining to citizenship requirements is just one aspect of the question whether the Constitution protects aliens. Although *Yick Wo* involved race, not alienage, recent alienage cases suggest that *Yick Wo* remains alive and well. Beginning its equal protection analysis with that case, the near-unanimous

---

191. *See, e.g.*, Foti v. INS, 375 U.S. 217, 228-30 (1963) (abuse of discretion is standard for judicial review of denials of discretionary relief from deportation); Carlson v. Landon, 342 U.S. 524, 544 (1952) (applying "clear abuse of discretion" standard in upholding the Attorney General's refusal to release aliens on "bail" pending deportation proceedings, while noting "the executive judgment is limited by adequate standards"); *cf.* United States *ex rel.* Accardi v. Shaughnessy, 347 U.S. 260, 268 (1954) (Board of Immigration Appeals must exercise discretion conferred upon it by regulation).

192. *See, e.g.*, Wong Wing Hang v. INS, 360 F.2d 715, 719 (2d Cir. 1966) (agency decision is "abuse of discretion if . . . made without a rational explanation, inexplicably depart[s] from established policies, or rest[s] on an impermissible basis such as an invidious discrimination against a particular race or group").

193. These earlier cases are discussed *supra* text accompanying notes 94-108.

194. 403 U.S. 365 (1971).

Graham Court declared: "[C]lassifications based on alienage, like those based on nationality or race, are inherently suspect and subject to close judicial scrutiny. Aliens as a class are a prime example of a 'discrete and insular' minority . . . for whom such heightened judicial solicitude is appropriate."[195]

Plyler v. Doe,[196] a 1982 Supreme Court decision, has similarly broad implications. Relying heavily on *Yick Wo* and *Wong Wing*, the Court said that even undocumented aliens are entitled to constitutional protections and struck down a Texas statute that effectively barred children of undocumented aliens from attending public schools. Although the Court declined to say that undocumented aliens are a "suspect class,"[197] it found the state statute was not rational because it did not further "some substantial goal of the State."[198] More fundamentally, *Plyler* recognized a radically broader view of the constitutionally protected community than that implicit in the plenary power doctrine. *Plyler* recognized the membership of these undocumented children in American society as an accomplished fact, and further recognized that they could not be excluded by fiat from constitutional rights and privileges.[199]

The Court's reasoning in *Graham* and *Plyler* seemed broad enough to limit not only state laws, which were at issue in those cases and in most recent decisions invalidating citizenship requirements,[200] but the federal laws as well.[201] But in fact, the Court has recognized that the federal government's

195. *Id.* at 371-72 (8-0 decision with Harlan, J. concurring) (citing United States v. Carolene Prods. Co., 304 U.S. 144, 152-53 n.4 (1938)).

196. 457 U.S. 202 (1982).

197. *Id.* at 219 n.19.

198. *Id.* at 224.

199. *See* Schuck, *supra* note 6, at 54 (*Plyler* "may mark a fundamental break with classical immigration law's concept of national community and of the scope of congressional power to decide who is entitled to the benefits of membership."). Schuck supports this broad reading of *Plyler* in part by comparing it with De Canas v. Bica, 424 U.S. 351 (1976), which suggested that states may enact legislation that discourages immigration in ways that are generally consistent with federal policy. *Id.* at 57. On this broadening of the national community in general, see Aleinikoff, *Membership*, *supra* note 102, at 20-22; Bosniak, *supra* note 102, at 967-87; Rosberg, *The Protection of Aliens From Discriminatory Treatment by the National Government*, 1977 SUP. CT. REV. 275; Schuck, *supra* note 6, at 34-58; *cf.* Lopez, *Undocumented Mexican Migration: In Search of a Just Immigration Law and Policy*, 28 UCLA L. REV. 615, 700 (1981) ("If that set of attributes [heretofore assumed to be shared only by citizens] is now shared by people without citizenship, the interests of that group must be acknowledged and considered; any moral theory of community status that fails to take account of these interests will lose its purpose for being.").

200. *See, e.g.*, Bernal v. Fainter, 467 U.S. 216 (1984) (notaries public); Nyquist v. Mauclet, 432 U.S. 1 (1977) (recipients of financial assistance for higher education); Examining Bd. of Eng'rs v. Otero, 426 U.S. 572 (1976) (civil engineers); *In re* Griffiths, 413 U.S. 717, 718 n.3 (1973) (attorneys); Sugarman v. Dougall, 413 U.S. 634, 646 (1973) (civil service employees); Truax v. Raich, 239 U.S. 33 (1915) (employees in general).

201. In *Graham* the Court expressly refrained from deciding if Congress could enact a statute imposing on aliens a uniform nationwide residency requirement as a condition of federally funded welfare benefits. 403 U.S. at 382 n.14. But it discussed supremacy separately as an "additional," and thus, perhaps unnecessary reason for its holding. *Id.* at 376-77; *see also* Nyquist v. Mauclet, 432 U.S. 1, 12 (1977); *In re* Griffiths, 413 U.S. 717, 718 n.3 (1973); Sugarman v. Dougall, 413 U.S. 634, 646 (1973); *cf.* Toll v. Moreno, 458 U.S. 1, 10 (1982) (state university's policy of denying in-state status to holders of certain nonimmigrant visas is invalid under the supremacy clause; Court did not reach due process and equal protection issues); *Graham*, 403 U.S. at 383 (Harlan, J., concurring) (supremacy clause sufficient to invalidate state law); Hines v. Davidowitz, 312 U.S. 52, 68 (1941) (under supremacy clause, federal alien registration statute precludes

constitutional power over immigration and naturalization gives it greater freedom from judicial scrutiny when it classifies by alienage. *Hampton v. Mow Sun Wong*,[202] decided in 1976, involved a challenge to a Civil Service Commission regulation that made aliens ineligible for most federal employment, much like the state regulations that the Court had invalidated three years earlier in *Sugarman v. Dougall*.[203] The Court struck down the federal regulation as a deprivation of liberty without due process, because it was unwilling to find that the Commission had acted in furtherance of national interests connected with the immigration and naturalization power.[204] But the Court suggested that either Congress or the President would have the power to bar aliens from federal jobs; indeed, lower courts have consistently upheld the Executive Order issued by President Ford shortly after *Mow Sun Wong*.[205] On the same day that the Court decided *Mow Sun Wong*, it also decided *Mathews v. Diaz*,[206] which challenged a statute, similar to the state welfare rules struck down in *Graham*, that barred aliens from access to federal Medicare unless they were permanent residents who had lived in the United States for five years. The Court in *Diaz* upheld the statute, reasoning that "[i]n the exercise of its broad power over naturalization and immigration, Congress regularly makes rules that would be unacceptable if applied to citizens."[207]

The ultimate results in the *Mow Sun Wong* and *Diaz* litigation sustained the power of Congress and the executive branch, and even in the case law concerning state classifications, *Graham* and *Plyler* remain high-water marks. But the key lesson is that the reasoning of all of these alienage cases carried on the *Yick Wo* tradition.[208] The Court allowed constitutional scrutiny, squarely and seriously considering the rights of aliens, in contrast to its cavalier treatment of many constitutional claims in immigration cases. *Mow Sun Wong* said that distinctions between aliens and citizens merited constitutional scrutiny.[209] In *Diaz*, the Court began by citing *Wong Yang Sung*, *Wong Wing*, and *Russian Volunteer Fleet* for the proposition that the Constitution protects aliens from deprivations of life, liberty, or property without due process, even if their

enforcement of state alien registration statute). Rosberg, *supra* note 199, at 316-36, argues that the constitutional limits on classification by alienage should apply to federal as well as state government.

202. 426 U.S. 88 (1976).

203. 413 U.S. 634 (1973); *cf. In re* Griffiths, 413 U.S. 717 (1973) (state exclusion of aliens from practice of law is unconstitutional).

204. 426 U.S. at 105, 114-16.

205. *See, e.g.*, Mow Sun Wong v. Campbell, 626 F.2d 739, 744-45 (9th Cir. 1980), *cert. denied*, 450 U.S. 959 (1981); Jalil v. Campbell, 590 F.2d 1120, 1123 n.3 (D.C. Cir. 1978); Vergara v. Hampton, 581 F.2d 1281, 1284-85 (7th Cir. 1978), *cert. denied*, 441 U.S. 905 (1979).

206. 426 U.S. 67 (1976).

207. *Id.* at 79-80.

208. *See* Aleinikoff, *Federal Regulation*, *supra* note 102, at 865 ("The legacy of *Yick Wo* has also survived and flourished in the 20th century. Outside the immigration process, aliens receive most of the constitutional protections afforded citizens."). Yick Wo v. Hopkins, 118 U.S. 356 (1886), and its immediate progeny are discussed *supra* text accompanying notes 94-108.

209. 426 U.S. at 101-03.

presence in the country is unlawful, involuntary, or transitory.[210] *Diaz* also endorsed the view that the character of an alien's stake has constitutional significance.[211]

Even decisions that have upheld certain types of state alienage classifications began their analysis with *Yick Wo*.[212] And in the spirit of *Wong Wing*, the Court has also made it clear that the Fourth Amendment protects aliens in the United States, even if the actual holding favored the government.[213] Lower courts have held that aliens are entitled to *Miranda* warnings in any custodial interrogation that seeks information to be used against them in criminal proceedings,[214] and that excludable aliens are entitled under the due process clauses of the Fifth and Fourteenth Amendments to be free of gross physical abuse at the hands of state or federal officials.[215] And when the underlying issue is not alienage alone, but race as in *Yick Wo* itself, the Constitution protects persons whether they are citizens or aliens.[216] It is also significant that in several cases recognizing aliens' constitutional rights, the dissent's approach, rejected by implication, was to reason directly from the plenary power cases in immigration law.[217]

---

210. *Diaz*, 426 U.S. at 77. Wong Yang Sung v. McGrath, 339 U.S. 33 (1950), is discussed *supra* text accompanying notes 121-30; Wong Wing v. United States, 163 U.S. 228 (1896), is discussed *supra* text accompanying notes 97-99; and Russian Volunteer Fleet, 282 U.S. 481 (1931), is discussed *supra* text accompanying note 100.

211. 426 U.S. at 80 ("The decision to share that bounty ["that a conscientious sovereign makes available to its own citizens and *some* of its guests"] with our guests may take into account the character of the relationship between the alien and this country . . . .") (emphasis in original).

212. *See, e.g.*, Cabell v. Chavez-Salido, 454 U.S. 432, 436 (1982) (probation officers); Ambach v. Norwick, 441 U.S. 68, 72 (1979) (public school teachers). These cases permit denying aliens political rights and employment that are believed to be closely tied to the political functions of government; *see also* Foley v. Connelie, 435 U.S. 291 (1978) (state troopers).

213. *See* United States v. Martinez-Fuerte, 428 U.S. 543, 566-67 (1976) (Fourth Amendment permits Border Patrol's routine stop of vehicle at permanent checkpoint near border without individualized suspicion); United States v. Ortiz, 422 U.S. 891, 896-97 (1975) (Fourth Amendment bars warrantless search of vehicle without probable cause at permanent checkpoint near border); United States v. Brignoni-Ponce, 422 U.S. 873, 878-87 (1975) (Fourth Amendment bars roving Border Patrol stop of vehicle near border based only on appearance of occupants); Almeida-Sanchez v. United States, 413 U.S. 266, 273 (1973) (Fourth Amendment bars warrantless search of vehicle without probable cause by roving Border Patrol); *cf.* United States v. Verdugo-Urquidez, 110 S. Ct. 1056 (1990) (Fourth Amendment does not apply to search by United States authorities of Mexican residence of Mexican citizen and resident who had no voluntary attachment to United States).

214. United States v. Henry, 604 F.2d 908, 913-15 (5th Cir. 1979).

215. Lynch v. Cannatella, 810 F.2d 1363, 1372-74 (5th Cir. 1987); *cf.* Medina v. O'Neill, 589 F. Supp. 1028, 1040-42 (S.D. Tex. 1984) (harsh conditions of detention of excludable stowaway aliens violated due process).

216. *See* Takahashi v. Fish & Game Comm'n, 334 U.S. 410 (1948); Yick Wo v. Hopkins, 118 U.S. 356 (1886).

217. *See Mow Sun Wong*, 426 U.S. at 117-18, 121-22, 127 (Rehnquist, J., dissenting) (citing Kleindienst v. Mandel, 408 U.S. 753 (1972); Galvan v. Press, 347 U.S. 522 (1954); Harisiades v. Shaughnessy, 342 U.S. 580 (1952), and Fong Yue Ting v. United States, 149 U.S. 698 (1893)); *Almeida-Sanchez*, 413 U.S. at 291 (White, J., dissenting) (citing Kleindienst v. Mandel, 408 U.S. 753 (1972); Chae Chan Ping v. United States, 130 U.S. 581 (1889); Fong Yue Ting v. United States, 149 U.S. 698 (1893), and other plenary power cases).

Indeed, judges who have rendered phantom norm decisions have relied heavily on *Yick Wo* values. The most prominent example of recent phantom norm decisionmaking based on these values is *Jean v. Nelson*,[218] which concerned claims of racial and national origin discrimination in connection with detention of Haitian asylum applicants. Similar reasoning is evident in the lower courts. The discussion that follows attempts to identify and analyze some of the most important of those decisions. It is not intended to be a complete catalog of recent phantom norm decisions. Rather, it seeks to demonstrate the broad influence of this judicial practice by focusing on four important areas of immigration law in which phantom norms seem to explain a number of the most significant recent cases. They are: race and national origin discrimination claims, indefinite detention of Mariel Cubans, asylum procedures for excludable first-time entrants, and First Amendment challenges to exclusion grounds.

Much as *Chew* had done, some of these decisions have stopped short of declaring executive action unconstitutional and simply invalidated executive action solely on an articulated "abuse of discretion" or similar rationale. These decisions often openly express constitutional skepticism that lacks support in the immigration law precedents. This reasoning allows us to identify them as phantom norm decisions. We can identify other decisions as phantom norm decisions because those courts have also made alternative findings of unconstitutionality, again based on phantom norms with little support in immigration law. Appellate courts sometimes reverse these constitutional holdings, and sometimes not. In some settings, evidence suggests that phantom norms are evolving, as illustrated by *Plasencia*, toward directly applicable constitutional immigration law.

1. *Race and National Origin Discrimination: Haitians and Others*

Jean is the most prominent, but not the first, of several recent cases that raised the issue whether government policy toward aliens from certain countries unlawfully discriminates on the basis of race and national origin. A 1979 decision, *Narenji v. Civiletti*,[219] tested special reporting requirements for Iranian students after the takeover of the American Embassy in Teheran. The district court reached alternative findings. Subconstitutionally, it decided that the regulations establishing the requirements lacked statutory authority because the enabling statute was "drawn neutrally." The court found "no statutory basis for the discriminatory classification established by the regulation such that defendants could cloak their rule's discriminatory effect in the mantle of congressio-

218. 472 U.S. 846 (1985).

219. 481 F. Supp. 1132 (D.D.C.), *rev'd*, 617 F.2d 745 (D.C. Cir. 1979), *cert. denied*, 446 U.S. 957 (1980).

nal approval under its power over immigration and naturalization and thereby for practical purposes, exempt the regulation from judicial scrutiny."[220]

There is ample evidence that the subconstitutional holding in *Narenji* was based on a phantom constitutional norm barring national origin classifications. The court's discussion of the statute follows a citation to *Yick Wo* for the notion that mainstream equal protection analysis should apply to the students' claim of unconstitutional discrimination.[221] The court also relied on statutes from outside immigration law to establish that "Congress has indicated its disdain for discrimination based on national origin."[222] The court's thinking is also clear from the rest of its opinion, which relied on *Mow Sun Wong*[223] to hold in the alternative that the regulation was an unconstitutional violation of equal protection.[224] The District of Columbia Circuit subsequently reversed on both subconstitutional and constitutional grounds. In so rejecting the district court's constitutional reasoning, the appeals court confirmed that the lower court's reasoning was based on a background constitutional norm that was phantom, not real.

Other phantom norm decisions involving race and national origin discrimination claims arose in the context of migration of Haitians to the United States, most of whom sought political asylum. Until 1982, aliens at the border who appeared inadmissible were usually released on parole pending exclusion hearings. That year, the government, largely responding to the Haitian influx, announced that almost all aliens arriving without entry documents, as most Haitians did, would be detained instead.[225]

Haitian Refugee Center v. Civiletti,[226] a 1980 case from the Southern District of Florida involving claims of national origin discrimination and denial of due process, reached alternative subconstitutional and constitutional findings as did the court in *Narenji v. Civiletti*. First, the court found that the INS lacked statutory authority to discriminate on the basis of national origin or race where Congress had not employed those criteria.[227] That this subconstitutional hold-

220. 481 F. Supp. at 1141.

221. *Id.* at 1138 ("Plaintiffs are quite correct in their assertions that the classification in this instance is one founded upon national origin and that the constitutional guarantee of equal protection can be evoked in most circumstances to protect against the government's invocation of such a standard, even in its dealings with aliens.") (citing *Yick Wo*). The court continued: "[I]t has been long acknowledged that distinctions based on ancestry or national origin are 'odious to a free people whose insitutions are founded upon the doctrine of equality.'" *Id.* at 1139 (citing Hirabayashi v. United States, 320 U.S. 81, 100 (1943)).

222. 481 F. Supp. at 1141 (citing 42 U.S.C. § 2000a (1976) (public accommodations); 42 U.S.C. § 2000b (public facilities); *id.* § 2000e-2 (employment)).

223. 481 F. Supp. at 1144.

224. *Id.* at 1145.

225. *See* 47 Fed. Reg. 30,044 (1982) (as amended at 47 Fed. Reg. 46,493 (1982)) (amending 8 C.F.R. § 235.3(b)).

226. 503 F. Supp. 442 (S.D. Fla. 1980), *aff'd as modified sub nom.* Haitian Refugee Center v. Smith, 676 F.2d 1023 (5th Cir. Unit B 1982).

227. 503 F. Supp. at 453 ("INS has no authority to discriminate on the basis of national origin or race—except perhaps by promulgating regulations in a time of national emergency"); *cf. id.* at 452 ("Congress, if it so chooses, may discriminate against and among aliens on grounds which would violate the

ing rested on a constitutional foundation is evident from the opinion as a whole, especially in its overwhelming emphasis on the constitutional issues and its tendency to treat constitutional and subconstitutional analysis as virtually indistinguishable.[228] The nature of that constitutional foundation is clear from the court's alternative finding of unconstitutional national origin discrimination. To make this finding, the court borrowed norms from mainstream constitutional law—relying most prominently on *Yick Wo*[229] and *Graham*[230]—that remained phantoms in the immigration law context.[231] The Fifth Circuit did not reverse on the constitutional issue, but it expressed skepticism by noting: "[W]e do not address the equal protection contentions any more than to observe that we do not approve the sweeping conclusions of the district court."[232]

The plaintiffs in *Vigile v. Sava*,[233] a 1982 district court case, similarly attacked INS detention policy toward Haitian asylum applicants as an abuse of discretion and as race and national origin discrimination proscribed by the Constitution and international law. The court granted the plaintiffs' relief, but restricted its holding to a subconstitutional level. Recognizing that the plenary power doctrine barred direct attacks on racial or national origin, the court avoided a directly constitutional holding. Instead, it subsumed all substantive issues in the case under abuse of discretion analysis, declaring that "invidious racial or national origin based discrimination constitutes abuse of discretion when insinuated into a neutral grant of decision-making authority."[234]

A Second Circuit panel reversed on the ground that the district court's review for abuse of discretion had been overbroad.[235] But importantly, the panel agreed with the district court's use of subconstitutional abuse of discretion analysis to borrow constitutional norms from other areas of public law, even

---

Constitution if applied to American citizens.").

228. *Id.* at 456 (coming very close to equating violations of regulations and operating procedures with denials of procedural due process as a constitutional matter). The Fifth Circuit noted this on appeal. *Haitian Refugee Center*, 676 F.2d at 1036 n.29.

229. 503 F. Supp. at 471 n.50.

230. *Id.* at 454.

231. *Id.* at 532 (government policy "is offensive to every notion of constitutional due process and equal protection").

232. *Haitian Refugee Center*, 676 F.2d at 1041.

233. 535 F. Supp. 1002 (S.D.N.Y. 1982), *rev'd on other grounds sub nom.* Bertrand v. Sava, 684 F.2d 204 (2d Cir. 1982).

234. 535 F. Supp. at 1016; *id.* at 1007 ("review for abuse of parole discretion is the norm. While displaying total restraint on issues of exclusion, federal courts have examined the treatment of aliens awaiting that final order."; blanket denial of parole to almost all Haitians was abuse of discretion); *id.* at 1016 ("Even though Congress may employ race or national origin as criteria in determining which aliens to exclude from the country, a district director may not apply neutral regulations to discriminate on such grounds."). The court reconciled its subconstitutional holding with the plenary power doctrine as follows:

> Judicial review of the parole process, therefore, does not impinge upon the political judgment to exclude or accept nor interfere with the executive and legislative power to control our borders. It only insures that parole status, which Congress has determined does not necessarily interfere with such control, is conferred by district directors within the bounds anticipated by the delegation of discretion.

Id. at 1006.

235. Bertrand v. Sava, 684 F.2d 204, 213-18 (2d Cir. 1982).

when those norms had never been established in constitutional immigration law. The panel stated, for example, that "immigration policies based on criteria that are not acceptable elsewhere in our public life" would be inconsistent with the "facially legitimate and bona fide reason" required by *Kleindienst v. Mandel*.[236]

In *Jean v. Nelson*, a class of Haitian asylum applicants also claimed that the INS detention policy unconstitutionally discriminated on the basis of race and national origin. The district court first decided subconstitutionally that the government's new detention policy was invalid because it had not been adopted properly under the Administrative Procedure Act (APA).[237] The effect was to restore the old policy, under which the plaintiffs were entitled to be released.[238] The court also extensively discussed the constitutional equal protection issue. In constitutional reasoning that reflected a phantom equal protection norm, the court distinguished claimants seeking parole from those seeking admission, citing *Yick Wo*, *Wong Wing*, *Mathews v. Diaz*, as well as the lower court decision that the Supreme Court later affirmed in *Plyler v. Doe*.[239] Ultimately, however, the court found that plaintiffs had failed to show a constitutional violation, leaving the decision to stand on the subconstitutional ground alone.[240]

On review, an Eleventh Circuit panel concurred with the district court's constitutional dictum that the plenary power doctrine did not apply because plaintiffs sought parole, not admission.[241] The panel relied on *Yick Wo* and *Plyler v. Doe*, on *Plasencia*, as well as on a phantom norm decision, *Chew*.[242] Like the district court, the panel actually reached only a subconstitutional decision, but it went beyond simply affirming the district court's finding of

---

236. *Id.* at 212 n.12.

237. Louis v. Nelson, 544 F. Supp. 973, 993-97 (S.D. Fla. 1982), *aff'd in part sub nom.* Jean v. Nelson, 711 F. 1455 (11th Cir. 1983), *rev'd*, 727 F.2d 957 (11th Cir. 1984) (en banc), *aff'd as modified*, 472 U.S. 846 (1985).

238. 544 F. Supp. at 1006. The district court stayed its order for 30 days, during which the INS promulgated a new parole policy in compliance with the APA.

239. *See id.* at 998 ("[w]hen requesting release on parole, Plaintiffs are entitled to the same constitutional protections afforded all persons within the territorial jurisdiction of the United States"); *see also id.* at 990 (citing *Diaz*, 426 U.S. at 77, for proposition that plaintiffs "are persons within the territorial jurisdiction of the United States and they cannot be denied their liberty without due process of law").

240. 544 F. Supp. at 998-1002 (evidence shows no equal protection violation).

241. Jean v. Nelson, 711 F.2d 1455, 1484 (11th Cir. 1983), *rev'd*, 727 F.2d 957 (11th Cir. 1984) (en banc), *aff'd as modified*, 472 U.S. 846 (1985); *see also* Shaughnessy v. United States *ex rel.* Mezei, 345 U.S. 206, 220 (1953) (Jackson, J., dissenting) ("Realistically, this man is incarcerated by a combination of forces which keep him as effectually as a prison, the dominant and proximate cause of these forces being the United States immigration authority."); *id.* at 227 ("Confinement is no longer ancillary to exclusion; it can now be justified only as the alternative to normal exclusion. It is an end in itself."). *cf.* Bertrand v. Sava, 535 F. Supp. 1020, 1030 (S.D.N.Y. 1982) ("There comes a point at which 'temporary' detention takes on the aura of permanence."), *rev'd on other grounds*, 684 F.2d 204 (2d Cir. 1982).

242. 711 F.2d at 1483-85. The court also relied on Rodriguez-Fernandez v. Wilkinson, 654 F.2d 1382 (10th Cir. 1981), a phantom norm decision that is discussed in greater detail *infra* text accompanying notes 256-65.

noncompliance with the APA.[243] It reached a second subconstitutional holding based on a phantom equal protection norm—whether or not the plenary power cases barred plaintiffs from raising a constitutional equal protection claim, the statute prohibited discrimination.[244] The panel went on to decide that plaintiffs had proved unlawful discrimination.[245]

The Eleventh Circuit en banc also spoke subconstitutionally, but it was unwilling to use abuse of discretion analysis as a vehicle for expressing phantom equal protection norms not indigenous to immigration law.[246] Unsurprisingly, it found no defects, forcing it to reach the constitutional issue. Flatly disagreeing with the constitutional dicta in the district court and the panel opinions, the court en banc held that aliens have no greater right to seek parole than to seek admission.[247] Reasoning that *Mezei* compelled rejection of plaintiffs' constitutional claims,[248] it upheld the detention policy. Four judges dissented in an opinion that adopted the panel decision's view that the statute prohibited discrimination, whether the Constitution did or not.[249]

The Supreme Court, like all three lower courts, addressed the subconstitutional issue. Noting that the parole regulations were "facially neutral" and that the parties all believed them to require "parole decisions to be made without regard to race or national origin,"[250] the Court found that the statute and regulations barred any race or national origin discrimination against the Haitians. Unlike the lower courts, the Court declined to address the constitutional issue and criticized the Eleventh Circuit en banc opinion for

---

243. 711 F.2d at 1474-83.

244. *Id.* at 1485 ("Our standard of review under *Mezei* and *Knauff* is to assure aliens receive that process Congress determined was due; on the record before us we can but determine that Congress intended the statutes under which the INS acted to be applied in a non-discriminatory fashion."). For cases basing similar subconstitutional results on international law, see Haitian Refugee Center v. Civiletti, 503 F. Supp. 442, 453-54 (S.D. Fla. 1980), *aff'd as modified sub nom.* Haitian Refugee Center v. Smith, 676 F.2d 1023, 1038 n.35 (5th Cir. Unit B 1982); Sannon v. United States, 427 F. Supp. 1270, 1274-77 (S.D. Fla. 1977), *vacated and remanded without opinion*, 566 F.2d 104 (5th Cir. 1978). *But cf.* Bertrand v. Sava, 684 F.2d 204, 218-19 (2d Cir. 1982) (international law does not change status under domestic law); Pierre v. United States, 547 F.2d 1281 (5th Cir. 1977) (same), *vacated and remanded to consider mootness*, 434 U.S. 962 (1977).

245. 711 F.2d at 1501-02.

246. Jean v. Nelson, 727 F.2d 957, 975-79 (11th Cir. 1984) (en banc), *aff'd as modified*, 472 U.S. 846 (1985) ("both executive and congressional actions are to be regulated by the judiciary according to the same narrow standard of review"); *id.* (immigration officials were "acting within the scope of their delegated powers"); *see also* Bertrand v. Sava, 684 F.2d 204, 217 (2d Cir. 1982); Pierre v. United States, 547 F.2d 1281, 1289 (5th Cir. 1977), *vacated and remanded to consider mootness*, 434 U.S. 962 (1977). *Jean* is a variation on the standard abuse of discretion case because lower level INS officials may have failed to follow internal INS guidelines. The Eleventh Circuit en banc remanded this question to the district court. 727 F.2d at 978-79.

247. 727 F.2d at 963 ("decision to parole or detain an excludable alien is an integral part of the admissions process"); *see also id.* at 969 ("alien's legal status is not altered by detention or parole"); Bertrand v. Sava, 684 F.2d 204, 211-12 (2d Cir. 1982).

248. 727 F.2d at 969-72.

249. *Id.* at 987-89 (Kravitch, J., concurring in part and dissenting in part).

250. 472 U.S. 846, 852 (1984). The regulation at that time was 8 C.F.R. § 212.5 (1982). The current regulation is 8 C.F.R. § 212.5 (1990). *See* 47 Fed. Reg. 30,044 (1982), (as amended at 47 Fed. Reg. 46,493 (1982)); *see also* 8 C.F.R. §§ 212.12, 212.13, 235.3 (1990).

having done so.[251] It remanded the case to the district court to decide whether INS officials had made individualized parole determinations in exercising their statutory discretion, and whether they had acted without regard to race or national origin.[252]

In dissent, Justice Marshall correctly observed that nothing in the statute or regulations expressly prohibited consideration of race or national origin in parole decisions, notwithstanding any stipulation by the parties.[253] The parties to the litigation and all three lower court opinions had viewed any such possible restrictions as constitutional in nature, drawn from the *Yick Wo* tradition.[254] The Court could have expressed that reasoning directly in constitutional terms, but had it done so, the Court would have significantly limited the plenary power doctrine, especially as set out in *Knauff* and *Mezei*. Such a constitutional decision would have done for equal protection challenges by asylum applicants what *Plasencia* had done for procedural due process claims by reentering permanent residents. Of course, the Court could have taken that step, but as in *Chew*, it was unwilling to do so. As long as it would not decide *Jean* on constitutional grounds, the antidiscrimination norm that guided the Court's interpretation of the parole statute and regulations remained a phantom. This characterization fits not only the Supreme Court's reasoning in *Jean*, but also the reasoning of the lower courts—none of which had the power even to consider overruling or limiting *Knauff* or *Mezei*—in *Jean* and in the other discrimination cases discussed here.

---

251. 472 U.S. at 854-55.

252. *Id.* at 857.

253. *Id.* at 861 (Marshall, J., dissenting) ("[the regulations] do not, by their terms, prohibit the consideration of race or national origin"). Interestingly, Justice Marshall's dissent focused almost entirely on national origin discrimination, which he seemed to regard as sufficient to undermine the majority's position. *See, e.g., id.* at 864 ("[T]he Court's conclusion that, aside from constitutional constraints, the parole regulations prohibit national-origin distinctions draws no support from anything in the regulations themselves or in the statutory and administrative background to those regulations."). In a footnote, he observed without explanation that "the analysis would be different for race discrimination." *Id.* at 859 n.1. In fact, the statute and regulations address neither race nor national origin discrimination. I intend my analysis of *Jean* as a phantom norm decision to include both race and national origin, but the analysis would remain valid even based on only one of the two.

The statute makes parole highly discretionary. *See id.* at 863 (Marshall, J., dissenting) ("These regulations indicate that the INS believes that nationality-based distinctions are not necessarily inconsistent with congressional delegation of 'discretion' over immigration decisions to the Executive. That interpretation of the statutes is, of course, entitled to deference.") (citing *Chevron*); *see also id.* at 865; *cf.* Haitian Refugee Center v. Civiletti, 503 F. Supp. 442, 453 (S.D. Fla. 1980) (congressional conclusion that discrimination in immigration law is "improper" is evident in INA § 202(a), 8 U.S.C. § 1152(a): "No person shall receive any preference or priority or be discriminated against in the issuance of an immigrant visa because of his race, sex, nationality, place of birth, or place of residence . . . ."), *aff'd as modified sub nom.* Haitian Refugee Center v. Smith, 676 F.2d 1023 (5th Cir. Unit B 1982).

The rest of Marshall's dissent addressed the constitutional issue directly. 472 U.S. at 868-81 (aliens have constitutional right to parole decisions free of race or national origin discrimination) (citing *Yick Wo, Wong Wing, Russian Volunteer Fleet, Diaz*, and *Plyler*); *see also id.* at 877-79 (granting parole to aliens, at least while decision on their excludability is pending, differs from admitting them).

254. *See* Louis v. Nelson, 544 F. Supp. at 998; Jean v. Nelson, 711 F.2d at 1483-85; Jean v. Nelson, 727 F.2d at 967-75.

It is striking that the racial equality aspect of this norm was the same norm that had influenced the Court's interpretation of the tax exemption statute in *Bob Jones*.[255] Both *Bob Jones* and *Jean* arose because the agency had to decide whether, in enforcing the statute, to apply a racial equality norm that Congress had not expressly specified. In *Bob Jones*, the racial equality norm was real, but in *Jean* it was phantom; the Court has never held an immigration classification unconstitutional on the ground that it discriminates on the basis of race or national origin.

2. *Indefinite Detention of Mariel Cubans*

Another group of phantom norm cases involves Mariel Cubans, about 125,000 of whom came to the United States from Cuba between April and October 1980. While many were excludable, the Cuban government would not take them back, and the United States government then placed many in indefinite detention.[256] In *Rodriguez-Fernandez v. Wilkinson*,[257] one detainee sought release on parole, claiming that continued indefinite detention was cruel and unusual punishment proscribed by the Eighth Amendment and a violation of the due process clause of the Fifth Amendment. In 1980, the district court found that the petitioner could not claim rights under the Fifth or Eighth Amendments, and that denying parole offended no statute, but that such a denial nonetheless constituted an abuse of discretion because it violated international law principles forbidding arbitrary detention.[258]

In 1981, a Tenth Circuit panel affirmed on the subconstitutional ground that the INS lacked statutory authority for indefinite detention.[259] But the heart of the opinion's reasoning was dictum, based on a phantom norm of Fifth and Eighth Amendment protections for detained would-be entrants that would render indefinite detention of these excludable Cubans unconstitutional punishment.[260] The panel conceded that the plenary power cases, especially *Mezei*,

255. *Bob Jones* is discussed *supra* text accompanying notes 78-83.

256. For background, see 52 Fed. Reg. 48,799-802 (1987); Note, *Legal Fictions Mask Human Suffering: The Detention of the Mariel Cubans—Constitutional, Statutory, International Law, and Human Considerations*, 62 S. CAL. L. REV. 1733, 1735-43 (1989). Under the Cuban Review Plan adopted on May 25, 1987, Mariel Cubans still in custody receive an annual INS interview to determine whether they will be released on parole. The detainees must show that they are nonviolent and likely to remain so, not likely to commit future crimes, do not pose a danger to the community, and are not likely to violate parole conditions. *See* 8 C.F.R. § 212.12 (1990). After prison riots in Atlanta, Georgia, and Oakdale, Louisiana, in November 1987, the government agreed that Cubans in custody at that time also would receive a one-time written review by a special Department of Justice panel (from which INS personnel are excluded) if they were denied parole by the INS. *See* 8 C.F.R. § 212.13 (1990).

257. Fernandez v. Wilkinson, 505 F. Supp. 787 (D. Kan. 1980), *aff'd sub nom.* Rodriguez-Fernandez v. Wilkinson, 654 F.2d 1382 (10th Cir. 1981).

258. 505 F. Supp. at 790-800.

259. 654 F.2d at 1389-90 (construing INA § 233, 8 U.S.C. § 1223(b), *repealed by* § 206, Act of Oct. 30, 1986, Pub. L. No. 99-591, 100 Stat. 3341-56).

260. 654 F.2d at 1386 ("We dispose of the appeal by construing the applicable statutes to require Rodriguez-Fernandez' release at this time. Nevertheless, it seems important to discuss the serious constitutional

normally insulate exclusion decisions from constitutional challenges, but it concluded that indefinite detention was not just incident to exclusion, and instead that it closely resembled criminal punishment.[261] This analysis relied on *Yick Wo* and four of its most prominent progeny—*Wong Wing*, *Russian Volunteer Fleet*, *Mow Sun Wong*, and *Graham*.[262] The panel distinguished *Mezei* in four ways: it focused on denial of a hearing, it involved national security considerations, it involved confinement under less harsh conditions, and it raised a claim of not only parole but also admission.[263]

The court based its subconstitutional decision on constitutional reasoning. The opinion offered a questionable interpretation of the statute[264] and resulted in a surreal order that the INS release an excludable alien, but not necessarily within the United States. The court offered this reading with little discussion, but observed that the reading was consistent with the court's constitutional analysis.[265] In *Diaz v. Haig*,[266] the federal district court in Wyoming followed *Rodriguez-Fernandez* and relied on *Yick Wo* and *Wong Wing* in reaching alternative findings that indefinite detention of juvenile Cubans was abuse of discretion and a Fifth Amendment due process violation.[267] Other cases have reached similar subconstitutional holdings, though the extent to which they relied on constitutional reasoning varies.[268] That the subconstitutional analysis

---

questions involved if the statute were construed differently.").

261. *Id.* at 1385-86.

262. *Id.* at 1386-87. These cases are discussed *supra* text accompanying notes 95-100, 194-209.

263. *Id.* at 1388. The focus on denial of a hearing in *Mezei* was indeed a difference between the facts in *Mezei* and *Chew*, but one that would seem to strengthen it as a barrier against the Cubans' claims in *Rodriguez-Fernandez*.

264. For a critique of the court's interpretation of the statute, see Note, *Statutory and Constitutional Limitations on the Indefinite Detention of Excludable Aliens*, 62 B.U.L. REV. 553, 570-79 (1982).

265. 654 F.2d at 1390.

266. 594 F. Supp. 1 (D. Wyo. 1981).

267. *Id.* at 5-6 (abuse of discretion), 7-8 (due process).

268. *See* Moret v. Karn, 746 F.2d 989, 992 (3d Cir. 1984) (abuse of discretion to deny parole without dangerousness finding required by INS internal procedures); Fernandez-Roque v. Smith, 557 F. Supp. 690, 696 (N.D. Ga. 1982) (some sponsorship practices were abuse of discretion; other parole-related policies are not), *rev'd*, 734 F.2d 576, 583 (11th Cir. 1984) (applying same abuse of discretion test but reaching opposite result); Soroa-Gonzales v. Civiletti, 515 F. Supp. 1049, 1057-61 (N.D. Ga. 1981) (abuse of discretion for INS to revoke parole); *cf.* Gutierrez v. Ilchert, 702 F. Supp. 787, 789-94 (N.D. Cal. 1988) (denial of parole to IRCA legalization applicant lacks facially legitimate and bona fide reason); Gutierrez v. Ilchert, 682 F. Supp. 467, 471-72 (N.D. Cal. 1988) (denial of parole to IRCA legalization applicant is abuse of discretion and lacks facially legitimate and bona fide reason). *See generally* Note, *supra* note 256. *But cf.* Perez-Perez v. Hanberry, 781 F.2d 1477, 1481-82 (11th Cir. 1986) (broad discretion to grant or deny parole); Sidney v. Howerton, 777 F.2d 1490, 1491 (11th Cir. 1985) (vacating district court order to parole alien during pendency of habeas corpus petition); Note, *Indefinite Detention of Excluded Aliens: Statutory and Constitutional Justifications and Limitations*, 82 MICH. L. REV. 61, 65-76 (1983) (statute allows detention instead of parole).

At least one court relied on international law. A gravitational pull on statutes may be exercised by international law norms, which, like phantom constitutional norms, are inconsistent with prevailing constitutional law. *See, e.g.*, Fernandez v. Wilkinson, 505 F. Supp. 787, 795-800 (D. Kan. 1980), *aff'd on other grounds sub nom.* Rodriguez-Fernandez v. Wilkinson, 654 F.2d 1382 (10th Cir. 1981); Eskridge, *supra* note 74, at 1026-28. *But see* Amanullah v. Nelson, 811 F.2d 1, 16 n.10 (1st Cir. 1987); Singh v. Nelson, 623 F. Supp. 545, 552-58 (S.D.N.Y. 1985); Fernandez-Roque v. Smith, 567 F. Supp. 1115, 1122 n.2 (N.D. Ga. 1983), *rev'd on other grounds*, 734 F.2d 576 (11th Cir. 1984); Fernandez-Roque v. Smith, 622 F. Supp.

in these cases rested on a phantom norm is evident not only from *Mezei* itself, but also from other appellate decisions that have consistently rejected Fifth and Eighth Amendment protections for indefinitely detained Cubans. For example, in *Fernandez-Roque v. Smith*[269] an Eleventh Circuit panel squarely reversed a district court decision that started its analysis with *Yick Wo* and ended by adopting reasoning similar to the constitutional dictum in *Rodriguez-Fernandez*.[270] Relying heavily on the en banc decision in *Jean v. Nelson*,[271] the panel followed *Mezei* in holding that decisions regarding release from detention on parole are within the political branches' plenary power because they are part of the admissions process.[272]

3. *Asylum Procedures for Excludable First-Time Entrants*

Several recent lower court cases on procedures for deciding asylum applications by excludable first-time entrants are probably subconstitutional decisions guided by phantom constitutional norms. In *Chun v. Sava*,[273] decided in 1983, a Second Circuit panel interpreted the asylum statutes and regulations to require the INS to grant stowaways an exclusion hearing at which they could raise asylum claims. *Augustin v. Sava*,[274] a 1984 Second Circuit panel decision, interpreted the asylum statutes and regulations to require adequate translations. Both interpretations relied heavily on a constitutional norm of due process for excludable first-time entrants seeking asylum, and in particular on the view that the scheme for granting political asylum creates a constitutionally protectable liberty or property interest that brings asylum applicants within the procedural due process framework set out in *Goldberg v. Kelly*,[275] *Mathews v. Eldridge*,[276] and their progeny. *Chun*'s subconstitutional analysis relied heavi-

887, 901-04 (N.D. Ga. 1985), *aff'd in pertinent part and rev'd in part on other grounds sub nom.* Garcia-Mir v. Meese, 788 F.2d 1446, 1453-55 (11th Cir.), *cert. denied*, 479 U.S. 888 (1987); Ishtyaq v. Nelson, 627 F. Supp. 13, 17-26 (E.D.N.Y. 1983). On international law as it applies to exclusion, see Nafziger, *The General Admission of Aliens Under International Law*, 77 AM. J. INT'L L. 804 (1983).

269. 734 F.2d 576 (11th Cir. 1984), *rev'g* 567 F. Supp. 1115 (N.D. Ga. 1983).

270. *See* 567 F. Supp. at 1125-29 (detention was not merely incident to exclusion, and judicially ordered parole was provisional procedural remedy separate from substantive admission decisions).

271. 727 F.2d 957 (11th Cir. 1984) (en banc), *aff'd as modified*, 472 U.S. 846 (1985), discussed *supra* text accompanying notes 237-55.

272. *Fernandez-Roque*, 734 F.2d at 581-82; *see also Mezei*, 345 U.S. at 215 ("such temporary harborage, an act of legislative grace, bestows no additional rights"); Carlson v. Landon, 342 U.S. 524, 533 (1952) ("Detention is necessarily a part of this deportation procedure."); *Amanullah*, 811 F.2d at 9-11; Garcia-Mir v. Meese, 766 F.2d 1478, 1484 (11th Cir. 1985), *cert. denied*, 475 U.S. 1022 (1986); Palma v. Verdeyen, 676 F.2d 100, 103-05 (4th Cir. 1982); *Fernandez*, 505 F. Supp. at 790. *See generally* Schuck, *supra* note 6, at 71 ("As a practical matter, this ostensibly procedural decision amounts to granting a substantive right to remain in this country indefinitely to many aliens who are clearly excludable under our laws.") (emphasis in original); *id.* at 30 (classical immigration law treated exclusion and detention alike); *id.* at 28-30, 68-72 (detention power is key to classical immigration law).

273. 708 F.2d 869 (2d Cir. 1983).

274. 735 F.2d 32 (2d Cir. 1984).

275. 397 U.S. 254 (1970).

276. 424 U.S. 319 (1976).

ly on "the dictates of procedural due process" and the presence of a constitutionally "protectable interest."[277] Similarly, *Augustin* cited *Mathews v. Diaz* for the proposition that "an alien who is present in the United States, even illegally, is entitled to constitutional protections."[278] It then noted that its interpretation of the statute and regulations may very likely be the constitutional due process minimum, because of asylum applicants' "constitutionally protected liberty or property interests."[279]

The constitutional norm of due process for excludable first-time entrants seeking asylum was probably a phantom. *Plasencia* embraced the *Mathews v. Eldridge* procedural due process analysis, but only for returning permanent residents.[280] Lower courts have imposed some modest procedural due process requirements in asylum cases, but for deportable rather than excludable aliens.[281] The Eleventh Circuit's 1982 panel decision in *Haitian Refugee*

---

277. 708 F.2d at 876 ("our construction of the statute is aided to some extent, if not guided, by what we perceive to be the dictates of procedural due process"); *id.* at 877 & n.25 ("refugee who has a 'well-founded fear of persecution' in his homeland has a protectable interest recognized by both treaty and statute, and his interest in not being returned may well enjoy some due process protection not available to an alien claiming only admission").

278. 735 F.2d at 36.

279. *Id.* at 37; *see also id.* ("procedural protections may well be required not only by the pertinent statutes and regulations but also by the due process clause"); *id.* ("likely that some due process protection surrounds the determination of whether an alien has sufficiently shown that return to a particular country will jeopardize his life or freedom"); *id.* at 38 ("appellant was denied procedural rights protected by statute and INS regulations and very likely by due process as well"). Justice Marshall's dissent in *Jean* noted that "the Court of Appeals for the Second Circuit has suggested that unadmitted aliens can invoke the protections of the Constitution." 472 U.S. at 868 n.8 (Marshall, J., dissenting) (citing *Augustin* and *Chun*).

280. *See* 459 U.S. 21, 34 (1982). *Plasencia* is discussed in greater detail *supra* text accompanying notes 175-81.

281. *See* Orantes-Hernandez v. Meese, 685 F. Supp. 1488, 1506-08 (C.D. Cal. 1988) (under Constitution and *Eldridge* analysis, government must give notice of right to apply for asylum); *id.* at 1509-11 (government may not interfere with access to retained counsel in asylum cases); Orantes-Hernandez v. Smith, 541 F. Supp. 351, 374-78, 380-82 (C.D. Cal. 1982) (similar). The court recently found that the INS failed to comply with a permanent injunction ordering better access to legal services. Orantes-Hernandez v. Thornburgh, CV 82-1107 KN (C.D. Cal. Sept. 28, 1989), *digested in* 66 INTERPRETER RELEASES 1177-78 (1989); *see also* Nunez v. Boldin, 537 F. Supp. 578, 583-87 (S.D. Tex. 1982) (identifying the interests as "life and liberty," court issued temporary restraining order requiring government to notify detainees of right to apply for asylum), *appeal dismissed*, 692 F.2d 755 (5th Cir. 1982); *cf.* Haitian Refugee Center v. Nelson, 872 F.2d 1555, 1562-63 (11th Cir. 1989) ("entitlement interest exists"; Constitution requires certain procedures for Special Agricultural Worker legalization), *aff'g* Haitian Refugee Center v. Nelson, 694 F. Supp. 864, 878 (S.D. Fla. 1988) ("constitutionally protected right to seek SAW status"). *See generally* Note, *INS Transfer Policy: Interference With Detained Aliens' Due Process Right to Retain Counsel*, 100 HARV. L. REV. 2001 (1987).

The *Orantes-Hernandez* litigation also includes subconstitutional holdings that may reflect a constitutional due process norm, but for deportable aliens this norm may not be a phantom. *See* Orantes-Hernandez v. Meese, 685 F. Supp. 1488, 1506, 1509 (C.D. Cal. 1988); Orantes-Hernandez v. Smith, 541 F. Supp. 351, 374-76 (C.D. Cal. 1982) (similar). *But cf.* Azzouka v. Sava, 777 F.2d 68, 75-76 (2d Cir. 1985) (statute gives no right to asylum hearing if government finds alien is national security risk), *cert. denied*, 479 U.S. 830 (1986); *see also* Azzouka v. Sava, 820 F.2d 585 (2d Cir. 1987). In what may be a similar phantom norm decision, a district court held recently that the government's failure to interpret entire deportation and exclusion proceedings for non-English speakers violates their statutory rights to be present, examine evidence, confront and cross-examine witnesses, as well as their constitutional Fifth Amendment due process rights. El Rescate Legal Servs. v. Executive Office of Immigration Review, 727 F. Supp. 557, 560-63 (C.D. Cal. 1989); *cf.* Sannon v. United States, 427 F. Supp. 1270, 1273, 1275-77 (S.D. Fla. 1977) (relying on United Nations Convention and Protocol Relating to the Status of Refugees to require equal treatment of

Center v. Smith[282] recognized a constitutionally protected liberty or property interest in asylum applicants who are excludable first-time entrants.[283] But *Knauff* and *Mezei*[284] either squarely block a finding that a first-time entrant applying for asylum has a liberty interest in entry, or compel finding that she is entitled only to minimal due process under the *Eldridge* analysis. In fact, the Eleventh Circuit has severely limited *Haitian Refugee Center v. Smith*, if not effectively overruled it.[285]

4. *First Amendment Challenges to Exclusion Grounds*

Another important group of phantom norm cases involves First Amendment challenges to exclusion grounds.[286] In 1982, for example, a federal district

---

excludable and deportable aliens), *vacated and remanded without opinion*, 566 F.2d 104 (5th Cir. 1978). *But see* Pierre v. United States, 547 F.2d 1281, 1287-89 (5th Cir. 1977), *vacated and remanded to consider mootness*, 434 U.S. 962 (1977). For a discussion of *Sannon* and *Pierre*, see Note, *Due Process Rights for Excludable Aliens Under United States Immigration Law and the United Nations Protocol Relating to the Status of Refugees—Haitian Aliens, A Case in Point*, 10 N.Y.U. J. INT'L L. & POL. 203 (1977).

282. 676 F.2d 1023 (5th Cir. Unit B 1982).

283. *See id.* at 1038 ("constitutionally protected liberty or property interests may have their source in positive rules of law, enacted by the state or federal government and creating a substantive entitlement to a particular governmental benefit. In this case we conclude that Congress and the executive have created, at a minimum, a constitutionally protected right to petition our government for political asylum."), *aff'g and modifying* Haitian Refugee Center v. Civiletti, 503 F. Supp. 442, 455 (S.D. Fla. 1980) ("asylum applicant who fears to return to his homeland because of persecution has raised the specter of truly severe deprivations of life, liberty, and property"). For commentaries adopting this general approach to excludable aliens' due process rights, see Martin, *supra* note 103, at 187; Note, *The Right to Appointed Counsel in Asylum Proceedings*, 85 COLUM. L. REV. 1157 (1985); Note, *Penetrating the Entry Doctrine: Excludable Aliens' Constitutional Rights in Immigration Processes*, 98 YALE L.J. 639 (1989). For a much earlier case that similarly combined phantom constitutional reasoning with a subconstitutional holding, see United States *ex rel.* Paktorovics v. Murff, 260 F.2d 610, 613-15 (2d Cir. 1958) ("since a construction of [the statute] which requires a hearing on the subject of revocation of parole will remove serious doubt regarding the validity of the statute, we so construe the section").

284. *Plasencia* limited *Mezei*, but only with respect to returning permanent residents. *See supra* text accompanying notes 157-81.

285. Jean v. Nelson, 727 F.2d 957, 981-83 (11th Cir. 1984) (en banc) (no liberty or property interest in asylum; neither due process nor statute/regulations requires notice of right to seek asylum), *rev'g* 711 F.2d 1455, 1507-08 (11th Cir. 1983) (aliens in exclusion must be notified of right to seek asylum), *aff'd as modified*, 472 U.S. 846 (1985); *cf.* Ukrainian-American Bar Ass'n v. Baker, 893 F.2d 1374, 1380-82 (D.C. Cir. 1990) (government need not tell asylum applicants of offer of free legal advice); Perez-Perez v. Hanberry, 781 F.2d 1477, 1479-81 (11th Cir. 1986) (statute does not authorize appointed counsel in habeas corpus action attacking parole denial); Garcia-Mir v. Meese, 788 F.2d 1446, 1449-53 (11th Cir. 1986), (indefinitely detained Cubans have no constitutionally protected liberty interest in parole), *cert. denied*, 479 U.S. 889 (1986); Fernandez-Roque v. Smith, 734 F.2d 576, 582 (11th Cir. 1984) (same, citing 11th Circuit en banc opinion in *Jean*); Ramirez-Osorio v. INS, 745 F.2d 937, 944-47 (5th Cir. 1984) (even assuming that deportation asylum applicant has protectable liberty interest, due process does not require notice of asylum); Avila v. Rivkind, 724 F. Supp. 945, 947-48 (S.D. Fla. 1989) (alien seeking admission "is requesting a privilege, and has extremely limited constitutional protection"; Attorney General may exclude without further hearing an alien previously convicted for terrorist activities as security risk under INA § 235(c)); Note, *Protecting Aliens From Persecution Without Overloading the INS: Should Illegal Aliens Receive Notice of the Right to Apply for Asylum?*, 69 VA. L. REV. 901, 908-13 (1983) (even if immigration laws create liberty interest in asylum applicant, no due process right to notice of asylum).

286. American-Arab Anti-Discrimination Comm. v. Meese, 714 F. Supp. 1060 (C.D. Cal. 1989), an important recent case sustaining a First Amendment challenge to ideological deportation grounds, is discussed *infra* text accompanying notes 352-57.

court in *Lesbian/Gay Freedom Day Committee v. INS*[287] barred the government from enforcing the "psychopathic personality" statutory exclusion ground against homosexuals.[288] The opinion relied largely on constitutional analysis, specifically a norm of First Amendment limits on government enforcement of the immigration statutes. The court found that because medical authorities no longer considered homosexuality per se to be a mental disorder, applying this exclusion ground to homosexuals was unconstitutional under the First Amendment and *Mandel*'s "facially legitimate and bona fide" test.[289] But *Mandel* itself tells us that the district court in *Lesbian/Gay Freedom Day Committee* relied on a phantom norm. There, the Supreme Court dismissed a First Amendment claim by finding that the government had given a "facially legitimate and bona fide reason," without background inquiry of the sort in *Lesbian/Gay Freedom Day Committee*.[290] The only other federal court of appeals to consider a similar constitutional challenge upheld enforcement.[291] And because of the Supreme Court decision in *Boutilier v. INS*,[292] any constitutional challenge on vagueness grounds would be no less based on phantom norm reasoning.

Lesbian/Gay Freedom Day Committee merits description as a subconstitutional decision based on a phantom constitutional norm, because its constitutional reasoning also provided the basis for its alternative subconstitutional holding—that enforcement of the "psychopathic personality" exclusion ground against homosexuals was an abuse of discretion because it was inconsistent with the statute.[293] By transplanting the "facially legitimate and bona fide reason" test to subconstitutional ground, using it to ask if government action was "beyond the scope of delegated powers" or an "abuse of discretion," the court expressed its constitutional views in a much less audacious form and gave new, subconstitutional life to First Amendment challenges. Indeed, the Ninth Circuit affirmed on the statutory ground without reaching the constitutional issue.[294]

Several similar decisions involve First Amendment challenges to government decisions to exclude aliens on openly ideological grounds, as authorized

---

287. 541 F. Supp. 569 (N.D. Cal. 1982), *aff'd only on statutory ground sub nom.* Hill v. INS, 714 F.2d 1470, 1481 (9th Cir. 1983).

288. INA § 212(a)(4), 8 U.S.C. § 1182(a)(4) (1988), discussed *supra* text accompanying notes 158-73 in connection with Rosenberg v. Fleuti, 374 U.S. 449 (1963).

289. 541 F. Supp. at 585-88.

290. Kleindienst v. Mandel, 408 U.S. 753 (1972), is discussed in greater detail *supra* text accompanying notes 182-87.

291. Matter of Longstaff, 716 F.2d 1439, 1442-43, 1450 n.56 (5th Cir. 1983), *cert. denied*, 467 U.S. 1219 (1984).

292. 387 U.S. 118 (1967), discussed *supra* text accompanying note 172 in connection with Rosenberg v. Fleuti, 374 U.S. 449 (1963).

293. 541 F. Supp. at 580; *see also* Note, *The Immigration and Nationality Act and the Exclusion of Homosexuals:* Boutilier v. INS *Revisited*, 2 CARDOZO L. REV. 359 (1981). *But cf.* Matter of Longstaff, 716 F.2d 1439, 1445-51 (5th Cir. 1983) (no subconstitutional problem with enforcement of exclusion ground against homosexuals), *cert. denied*, 467 U.S. 1219 (1984).

294. Hill v. INS, 714 F.2d 1470, 1481 (9th Cir. 1983).

by statute.[295] *Allende v. Shultz*,[296] a 1985 district court case, challenged denial of a nonimmigrant visa to Hortensia Allende, widow of former Chilean President Salvador Allende. *Allende* resembles *Lesbian/Gay Freedom Day Committee* in that it also turned *Mandel*'s "facially legitimate and bona fide reason" test for constitutional challenges into a subconstitutional reason to deny the government's motion to dismiss or in the alternative for summary judgment. According to the court, "the explanation given must be 'facially legitimate and bona fide' not only in a general sense, but also within the context of the specific statutory provision on which the exclusion is based."[297] A later opinion in the same case reached the same conclusion after considering new evidence, and suggested strongly that its holding on both occasions was a matter of statutory interpretation.[298] *El-Werfalli v. Smith*,[299] a 1982 district court decision, is similar in reasoning though not result. The court applied the "facially legitimate and bona fide reason" test as part of an analysis that probably was subconstitutional, in that it measured whether the stated reason conformed to the exclusion ground as set forth in the statute. The court tentatively found that the government "arguably fail[ed] to establish a reasoned basis for action," but then later found for the government after *in camera* inspection of classified material.[300]

Abourezk v. Reagan,[301] the only recent federal appellate decision involving a First Amendment challenge to ideological exclusion, may also be an example of phantom norm decisionmaking. Without clearly separating its constitutional and subconstitutional analyses,[302] the district court first interpreted the statute

295. *See* INA § 212(a)(27), (28), (29), 8 U.S.C. § 1182(a)(27), (28), (29) (1988).

296. 605 F. Supp. 1220 (D. Mass. 1985), *aff'd*, 845 F.2d 1111 (1st Cir. 1988).

297. 605 F. Supp. at 1224, *aff'd*, 845 F.2d 1111 (1st Cir. 1988); *cf.* Gutierrez v. Ilchert, 702 F. Supp. 787, 789-95 (N.D. Cal. 1988) (denial of parole to IRCA legalization applicant lacks facially legitimate and bona fide reason); Gutierrez v. Ilchert, 682 F. Supp. 467, 471-72 (N.D. Cal. 1988) (denial of parole to IRCA legalization applicant is abuse of discretion and lacks facially legitimate and bona fide reason). *But cf.* Bertrand v. Sava, 684 F.2d 204, 210-13 (2d Cir. 1982) ("facially legitimate and bona fide reason," not abuse of discretion test, applies to attack on parole denial as impermissibly based on race and/or national origin).

298. *See* Allende v. Shultz, No. 83-3984-C, slip op. at 7 1987 LEXIS 2798, 15 (D. Mass. Mar. 31, 1987) ("This Court [earlier] ruled that under subsection 27, Mrs. Allende's mere membership in [certain communist organizations] was an insufficient reason for the denial of her visa application."); *see also id.* at 16 ("There is thus no lawful basis for defendants, proceeding under subsection 27."). In February 1990, Pub. L. No. 101-246, § 128, 104 Stat. 15, made permanent § 901 of the Foreign Relations Authorization Act, Pub. L. No. 100-204, which barred the government from deporting, excluding, or denying visas to nonimmigrants because of past or current political beliefs, statements or associations which, if engaged in by a citizen in this country, would be protected by the Constitution. Section 128 also makes it clear that entry by nonimmigrants protected under § 901 cannot be subject to conditions or restrictions.

299. 547 F. Supp. 152 (S.D.N.Y. 1982).

300. *Id.* at 154. The *Allende* opinion, 605 F. Supp. at 1224, *aff'd*, 845 F.2d 1111 (1st Cir. 1988), cited *El-Werfalli*; *see also* NGO Comm. on Disarmament v. Haig, No. 82 Civ. 3636, slip op. (S.D.N.Y. June 10, 1982), *aff'd mem.*, 697 F.2d 294 (2d Cir. 1982).

301. 592 F. Supp. 880 (D.D.C. 1984), *rev'd and remanded on other grounds*, 785 F.2d 1043 (D.C. Cir. 1986), *aff'd by an equally divided Court*, 484 U.S. 1 (1987).

302. *See* 592 F. Supp. at 884 n.11 (plaintiffs' arguments that statute either does not permit exclusion or if it does is unconstitutional "do not differ materially, and . . . therefore will be considered together").

to permit the visa refusals in question.[303] It then acknowledged the seriousness of the First Amendment challenge, but rejected it after *in camera* inspection of classified material.[304] A District of Columbia Circuit panel reversed on the ground that the record in the district court had been inadequate to support its interpretation of the statute.[305] It remanded the case, avoiding the constitutional issue altogether.[306] The district court then resolved the case by interpreting the statute as not authorizing the exclusions.[307] It may be fair to analyze the combined result of these two decisions as judicial review of visa refusals through narrow interpretation of the statutory exclusion grounds, based on a phantom norm of First Amendment limits on government enforcement of the immigration statutes.

IV. In Search of Constitutional Immigration Law

In spite of the Court's reluctance to provide guidance in *Jean v. Nelson*, or perhaps because of it, public debate about plenary power will intensify rather than abate. One problem with this debate is that it has focused almost exclusively on the doctrine's most immediate and obvious consequences, asking simply if it is fair or necessary to keep immigration law outside our system of public law. But it is just as important to examine the plenary power doctrine's less obvious, systemic effects on immigration law as a whole.

A. *The Problem of Awkward or Unpredictable Subconstitutional Solutions*

Phantom norm decisions have eliminated some troubling results produced by the plenary power doctrine; this is why courts have reached them. Courts that find acceptable subconstitutional solutions for individual cases will adopt them. Yet a subconstitutional decision can be a very awkward way to express a phantom constitutional norm, because statutes and other subconstitutional texts provide judges with limited interpretive possibilities. These might suffice to prevent injustice in a particular case but do little to find a coherent and fully satisfactory solution if the particular injustice is merely symptomatic of a more fundamental problem that is rooted in the alien's inability, attributable to the plenary power doctrine, to raise a constitutional claim directly.[308]

303. 592 F. Supp. at 884-86. The exclusion ground principally at issue was INA § 212(a)(27), 8 U.S.C. § 1182(a)(27) (1988); also important was the relationship between this ground and INA § 212(a)(28), 8 U.S.C. § 1182(a)(28) (1988).

304. 592 F. Supp. at 886-88.

305. 785 F.2d at 1053-60.

306. *Id.* at 1060 n.24; *cf. id.* at 1075-76 (Bork, J., dissenting) (*Mandel* lets government exclude based on content of political beliefs as constitutional matter).

307. Abourezk v. Reagan, Nos. 83-3739, 83-3741, 83-3895 1988 LEXIS 5203, 15-17 (D.D.C. June 7, 1988).

308. Legislative options are more promising. Congressman Mazzoli offered legislation in recent sessions that would have amended the INA to recognize a returning resident's ties to the United States, much as

Awkwardness includes both overbreadth and underinclusiveness. *Fong Haw Tan*[309] illustrates the former. Concern for permanent residents prompted the Court to adopt a generous rule of construction that has been applied to all deportation cases.[310] Both underinclusiveness and overbreadth are evident in *Fleuti*,[311] which created an exception to the reentry doctrine because the statute failed to reflect a returning permanent resident's "stake." *Fleuti*'s subconstitutional solution is underinclusive in that many permanent residents, like Maria Plasencia, have significant equities based on ties to this country, but do not benefit from *Fleuti*. Their problems required the direct constitutional solution in *Landon v. Plasencia*.[312] *Fleuti* is also overinclusive; at least one court has applied it to help nonimmigrants.[313] Underinclusiveness means some of the problem remains unsolved. Overbreadth represents waste; courts may compensate for the overbroad generosity of *Fong Haw Tan*, *Fleuti*, and similar decisions with a more begrudging attitude toward permanent residents on other issues, both constitutional and subconstitutional.

A related problem is unpredictability. The problem arises because sometimes, in *Fleuti* for example, a subconstitutional decision can incorporate a phantom constitutional norm only by going beyond reasonable readings of a subconstitutional text. Phantom norm decisionmaking may help some aliens when a case evokes sympathy, but we cannot predict those situations. The Supreme Court's 1984 decision in *INS v. Phinpathya*[314] illustrates the problem. The Attorney General has discretion to suspend deportation and confer permanent residency on an otherwise deportable alien of "good moral character" who "has been physically present in the United States for a continuous period of not less than seven years," if deportation would result in "extreme hardship" to the alien or his spouse, parent, or child.[315] Applying *Fleuti* by analogy, some lower courts disregarded "innocent, casual, and brief"[316] ab-

happens in constitutional immigration law after *Plasencia*. H.R. 4823, 99th Cong., 2d Sess. § 101 (1986). This bill passed the House, but the Senate did not act on it. 132 CONG. REC. H8570 (1986). Mazzoli introduced virtually the same measure the next term, see H.R. 2921, 100th Cong., 1st Sess. (1987). Another option is to eliminate the concept of "entry" altogether. *See* T. ALEINIKOFF & D. MARTIN, *supra* note 15, at 342-47; Martin, *supra* note 188, at 806-07. This would mean that a unitary type of immigration hearing would test an alien's claim to be in the United States, in effect combining the statutory deportation and exclusion categories for most but not necessarily all purposes.

309. 333 U.S. 6, 10 (1948). *See supra* notes 117-20 and accompanying text.

310. *See, e.g.*, Woodby v. INS, 385 U.S. 276, 286 n.19 (1966) ("This standard of proof applies to all deportation cases, regardless of the length of time the alien has resided in this country."); Squires v. INS, 689 F.2d 1276, 1280 (6th Cir. 1982); D. MARTIN, *supra* note 9, at 19 (Supreme Court applies "most lenient interpretation to ambiguous statutes or regulations, particularly in deportation cases, in light of the high stakes involved for the alien.").

311. Rosenberg v. Fleuti, 374 U.S. 449 (1963), discussed *supra* text accompanying notes 158-73.

312. 459 U.S. 21 (1982), discussed *supra* text accompanying notes 175-81.

313. *See* Joshi v. INS, 720 F.2d 799, 800-02 (4th Cir. 1983). The Board of Immigration Appeals has refused to apply *Fleuti* except in cases involving returning permanent residents. *See Matter of Torres*, Int. Dec. 3010 (BIA 1986); *Matter of Mundell*, 18 I. & N. Dec. 467 (BIA 1983).

314. 464 U.S. 183 (1984).

315. INA § 244(a)(1), 8 U.S.C. § 1254(a)(1) (1988).

316. *Fleuti*, 374 U.S. at 461.

sences for purposes of the seven-year continuous physical presence requirement.[317] *Phinpathya* struck down this practice as contrary to the statute's plain meaning, rejecting the aliens' principal argument, that suspension, like the *Fleuti* exception, was meant to be "equitable and ameliorative."[318] But the lower courts' broad view of continuous physical presence was no more contrary to plain meaning than the Court's own reading of "entry" in *Fleuti*, which the Court reached in spite of evidence that Congress tried to limit exceptions to the reentry doctrine. In fact, *Phinpathya* read "continuous physical presence" narrowly in spite of evidence that Congress wanted a broader reading.[319]

One possible explanation for *Phinpathya*—though Justice O'Connor deemed it "not essential" to the decision—lies in her contrasting characterizations of Fleuti as a "*lawful* resident alien" and Phinpathya as "an *unlawful* alien who could have been deported even had she remained in this country."[320] But these are just conclusory labels, and it is difficult to know how courts, in interpreting statutes, should distinguish between the Fleutis and the Phinpathyas of immigration law.[321] It is difficult to know when to adhere strictly to a statute's "plain meaning" in spite of a harsh result, and when to take greater interpretive liberties to reflect a phantom constitutional norm, for example, that certain hardships should not be visited on some aliens with ties to this country. We should be troubled when the range of possible "reasonable" readings of a statute broadens to the point that little will surprise us.

As judges become more willing and able to address constitutional issues directly, rather than through subconstitutional reasoning, they should find that they do not need to grope for these awkward and unpredictable subconstitutional solutions. Similarly, *Fleuti* and the strain that it placed on the statute are needed less now than in 1963, thanks to *Plasencia*'s expansion of judicial review to include openly constitutional due process challenges to exclusion procedures.

---

317. *See, e.g.*, Phinpathya v. INS, 673 F.2d 1013, 1017-18 (9th Cir. 1981), *rev'd*, 464 U.S. 183, 193 (1984); Kamheangpatiyooth v. INS, 597 F.2d 1253, 1257 (9th Cir. 1979); Wadman v. INS, 329 F.2d 812, 815-17 (9th Cir. 1964).

318. 464 U.S. at 192-96.

319. *Id.* at 196-205 (Brennan, J., concurring). According to the majority, Congress thought that the definition of "entry" was too broad while the "continuous physical presence" requirement was intended "specifically to limit the discretionary availability of the suspension remedy." *Id.* at 193. Indeed, Congress wasted little time after *Phinpathya* to enact legislation that abrogated its result. Immigration Reform and Control Act § 315(b), Pub. L. No. 99-603, 100 Stat. 3359 (1986), amended INA § 244, 8 U.S.C. § 1254(b)(2), to exclude "brief, casual, and innocent" absences that "did not meaningfully interrupt" continuous physical presence.

320. 464 U.S. at 194 (emphases in original). She added, "[s]uch an alien has no basis for expecting the Government to permit her to remain in the United States or to readmit her upon her return from foreign soil." *Id.* Her comment that this difference was "not essential" to the Court's decision, *id.*, suggests that the distinction had some effect.

321. *See also* Tovar v. INS, 612 F.2d 794, 797-98 (3d Cir. 1980); *cf.* INS v. Hector, 479 U.S. 85, 88-91 (1986) (per curiam), which reversed, as contrary to the statute's "plain meaning," a court of appeals decision that had considered hardship to two nieces in determining extreme hardship for suspension of deportation under INA § 244(a)(1), 8 U.S.C. § 1101(b)(2) (1988), because their relationships were functionally indistinct from a parent-child relationship.

The District of Columbia Circuit's 1988 decision in *Rafeedie v. INS*[322] illustrates this shift nicely. The INS tried to exclude Rafeedie, a permanent resident and allegedly a high-ranking member of a Palestinian terrorist group, without a hearing under the summary exclusion statute when he tried to enter the country after a trip overseas.[323] Rafeedie argued that that statute did not apply to permanent residents, and alternatively that the denial of a hearing resulting from its application violated due process. The district court rejected the statutory argument[324] but found that the due process concerns were serious enough to merit a preliminary injunction barring summary exclusion,[325] and the appeals court affirmed the grant of the injunction.[326]

In *Rafeedie*, a subconstitutional decision based on a phantom constitutional norm was unnecessary because *Plasencia* had recognized a returning permanent resident's due process rights as a constitutional matter. Both the district and the appeals courts addressed the constitutional issue directly without having to squeeze *Rafeedie* through the *Fleuti* loophole.[327] In fact, the appeals court held that *Fleuti*'s "meaningfully interruptive absence" test for statutory "reentry" would not decide if a returning permanent resident could claim procedural due process, even though some of the same factors would count.[328] What might have been a phantom norm decision before *Plasencia* instead became a candid application of constitutional doctrine.

With such changes in constitutional immigration law giving judges new freedom to address constitutional claims directly, it seems less objectionable when a court breaks the *Fong Haw Tan* rule, which is the least precise subconstitutional solution, by refusing to read deportation statutes in favor of aliens in *Phinpathya* and similar cases.[329] Nonetheless, a systemic problem with phantom norm decisionmaking is that even if these awkward or unpredictable solutions are needed less now than before, they remain with us after the transition from phantom to real is complete.

322. 880 F.2d 506 (D.C. Cir. 1989).

323. This, the INS argued, made him excludable under INA §§ 212(a)(27), (28)(F), 8 U.S.C. §§ 1182(a)(27), (28)(F) (1988). It sought to exclude him without a hearing under INA § 235(c), 8 U.S.C. § 1225(c) (1988).

324. 688 F. Supp. 729, 742-43 (D.D.C. 1988).

325. *Id.* at 743-51.

326. 880 F.2d at 519.

327. *See* 688 F. Supp. at 743 ("Plaintiff's real quarrel with Section 235(a) is a constitutional, not a statutory, one.").

328. 880 F.2d at 521-24.

329. *See, e.g.*, INS v. Rios-Pineda, 471 U.S. 444, 451 (1985); INS v. Stevic, 467 U.S. 407 (1984); INS v. Jong Ha Wang, 450 U.S. 139, 143-44 (1981) (per curiam) (overturning liberal construction of "extreme hardship" because "the Act commits their definition [of "extreme hardship"] in the first instance to the Attorney General and his delegates, and their construction and application of this standard should not be overturned by a reviewing court simply because it may prefer another interpretation of the statute.").

B. *The Problem of Misdirected Judicial Review*

Apart from awkwardness or unpredictability, subconstitutional phantom norm decisions, once established in response to problems of constitutional dimensions, set a precedent for excessive review of routine matters. By "excessive," I do not necessarily suggest "more" or "less" judicial review; explicit adoption of constitutional norms certainly would invite judicial scrutiny of agency decisions. Rather, I mean "excessive" in that the judicial habit of review not tied to real constitutional norms is open-ended and unbounded. As a result, courts may be least likely to intervene in agency decisionmaking when they can help most, and most likely to intervene when they can help least. The problem goes back as far as *Fong Haw Tan*, which by calling for interpretation of deportation statutes in the light most favorable to the alien, often asks courts to limit deference to agency interpretation of statutes. Much more recently, *Jean v. Nelson*[330] derived limits on INS discretion in parole decisions from phantom constitutional considerations, when the subconstitutional texts contained no express limitations.

Under a narrow reading of *Jean*, the INS abuses its parole discretion[331] if it considers factors that conflict with phantom constitutional norms, whether or not a court would squarely hold unconstitutional a statute that expressly made those factors pertinent.[332] Thus, it is an abuse of discretion to consider race in parole decisions, even if it is not directly unconstitutional. But it is difficult to avoid slipping into a slightly broader reading of *Jean*, especially (but not only) the reading that the INS abuses its discretion by considering factors not expressly authorized by statute or regulation. This reading is entirely understandable, since the prevailing view of *Jean* has not been to analyze it as a "phantom norm decision." Such a phantom norm analysis of *Jean* might help contain judicial review of agency action, but unless *Jean* is so limited, this slightly broader reading creates an open-ended and therefore troubling precedent for excessive judicial intrusion into agency decisionmaking. As Justice Marshall wrote in his dissent in *Jean*: "The Court's restrictive view of the Attorney General's discretionary authority with respect to parole decisions, adopted in the face of no authoritative statements limiting such discretion, will presumably affect the scope of his permissible discretion in areas other than parole deci-

330. 472 U.S. 846 (1985), discussed more fully *supra* text accompanying notes 237-55.

331. INA § 212(d)(5), 8 U.S.C. § 1182(d)(5) (1988).

332. As Deborah Anker has suggested, the holding in *Jean* could have an even narrower meaning, that courts limit an agency's discretion when its conduct shows it to be unworthy of judicial deference. *See* Anker, *supra* note 9, at 11. This might apply to the INS in general, and to its treatment of Haitian asylum applicants in particular. But, as Anker points out, the opinion did not rely on these factors. Rather than assume that the INS mistreated Haitians, the Court remanded to the district court to find if the INS had discriminated against Haitians at all. *Jean* might also mean that the INS may not consider factors not reasonably related to the discretionary relief sought, in this case parole, but national origin seems closely related to a decision to grant or deny parole in a pending asylum case, if only with regard to likelihood of success on the application.

sions. . . . This is indeed a costly way to avoid deciding constitutional issues."[333]

INS v. Rios-Pineda,[334] decided by the Court just six weeks before it decided *Jean*, reflects the more typical framework for judicial review of agency decisions. The Court said that it was not an abuse of discretion to deny certain motions to reopen suspension of deportation proceedings. The Court also emphasized that agencies must be free to base decisions on factors that relate generally to the law entrusted to it—in this case, "legitimate concerns about administration of the immigration laws."[335] These aspects of *Rios-Pineda* merely continued a tradition of judicial decisions that had established broad discretion for the INS.[336] Two prominent examples are *United States* ex rel. *Hintopoulos v. Shaughnessy*[337] and *Jay v. Boyd*,[338] both of which involved the discretionary denial of aliens' requests for suspension of deportation. *INS v. Abudu*, a 1988 Supreme Court decision, adopts a similar approach,[339] as have numerous lower court decisions.[340]

333. 472 U.S. at 867-68 (Marshall, J., dissenting); *cf.* Jean v. Nelson, 727 F.2d 957, 985-86 (11th Cir. 1984) (en banc) (Tjoflat, J., concurring in part and dissenting in part) (court may review agency action by Attorney General or by lower level officials to see if it is outside agency discretion, but it is impractical for court to ask if lower level officials' acts were consistent with internal agency directives), *aff'd as modified*, 472 U.S. 846 (1985). If Judge Tjoflat is correct, fear of constitutional questions may make INS operations unduly rigid.

334. 471 U.S. 444 (1985).

335. *Id.* at 451-52.

336. *See generally* Anker, *supra* note 9, at 11.

337. 353 U.S. 72, 78 (1957) (proper for INS to consider "present-day conditions and congressional attitudes" in exercising discretion).

338. 351 U.S. 345, 358 (1956) (The Court felt "constrained to construe the statute as permitting decisions based upon matters outside the administrative record, at least when such action would be reasonable.").

339. 485 U.S. 94, 104-10 (1988). The Court also noted in *Jean* that the INS enjoys "broad statutory discretion" to deny parole. 472 U.S. at 857; *cf.* INS v. Cardoza-Fonseca, 480 U.S. 421, 446-48 (1987) (courts decide pure questions of statutory construction).

340. *See, e.g.*, M.A. v. INS, 899 F.2d 304, 309-10 (4th Cir. 1990) (en banc). The Eleventh Circuit reached the same conclusion:

> As we stated in *Jean*, Congress has delegated remarkably broad discretion to executive officials under the Immigration and Nationality Act, and these grants of statutory authority are nowhere more sweeping than in the context of parole of excludable aliens. . . . After thoroughly exploring the broad discretion given the Attorney General in making parole decisions, we concluded [in *Jean*] that a federal court's scope of review in such instances is not the traditional abuse of discretion standard, but rather is limited to ascertaining whether he has advanced a 'facially legitimate and bona fide reason' for his decision.

Garcia-Mir v. Meese, 766 F.2d 1446, 1478 (11th Cir. 1986), *aff'd in part and rev'd in part*, 766 F.2d 1022, *cert. denied*, 475 U.S. 1022 (1987); *accord* Jean v. Nelson, 727 F.2d 957, 975-79 (11th Cir. 1984) (en banc), *aff'd as modified*, 472 U.S. 846 (1985); Narenji v. Civiletti, 617 F.2d 745, 747 (D.C. Cir. 1979) (INA "need not specifically authorize each and every action taken by the Attorney General, so long as his action is reasonably related to the duties imposed upon him."), *cert. denied*, 446 U.S. 957 (1980); Hotel & Restaurant Employees Union, Local 25 v. Smith, 594 F. Supp. 502, 506 (D.D.C. 1984) (Attorney General has discretionary power to take actions and to "develop standards, principles and rules" as long as those decisions are based upon "considerations rationally related to the statute he is administering.") (quoting Fook Hong Mak v. INS, 435 F.2d 728, 730 (2d Cir. 1970)), *aff'd in relevant part and rev'd in part on other grounds*, 804 F.2d 1256 (D.C. Cir. 1986), *aff'd by an equally divided court*, 846 F.2d 1499 (D.C. Cir. 1988) (en banc).

Courts that get into the habit of using expanded "abuse of discretion" and similar subconstitutional constructs to apply phantom constitutional norms indirectly are likely to succumb to the temptation to define "legitimate" so broadly that they in effect try to run the agency. While this may be understandable in light of the record of the INS,[341] judicial review still represents the commitment of a precious resource. Review of the wrong type is an uncertain improvement over no judicial review at all.[342] And the problem is compounded when judicial review is not only misdirected but also imposes cumbersome or unworkable procedures.[343]

Plenary power has prevented the growth of a coherent constitutional framework for immigration law, within which its subconstitutional levels—statutes, regulations, agency directives, and so forth—can develop and be administered fairly and predictably. There is a paradox here. On the one hand, the courts adopted the plenary power doctrine to insulate immigration decisions from constitutional judicial review. Judicial sensitivity to the need to maintain the flexibility to respond to unexpected contingencies, especially pertaining to foreign policy, may explain some of the plenary power doctrine's persistence—for example, the Supreme Court decided *Knauff*, *Mezei*, and *Harisiades* at the height of the nation's preoccupation with the perceived Communist threat. The irony is that the steady erosion of the plenary power doctrine through phantom norm decisionmaking may, precisely because no coherent body of constitutional norms exists to anchor and thus limit judicial review in immigration cases, lead to subconstitutional decisions that intrude into executive or legislative operations even more aggressively. There may be times when agency decisionmaking, to reach the best results, should be able to apply expertise, discretion,

341. *Cf.* Trillin, *Making Adjustments*, THE NEW YORKER (May 18, 1984) ("[W]hen people who deal regularly with the I.N.S. try to illustrate the depths of its inefficiency and obduracy, they often find themselves at a loss for American institutions to compare it with, and turn to foreign examples—the South Vietnamese Army, maybe, or the Bolivian Foreign Service.").

342. *See generally* Sofaer, *Judicial Control of Informal Discretionary Adjudication and Enforcement*, 72 COLUM. L. REV. 1293 (1972); Sofaer, *The Change-of-Status Adjudication: A Case Study of the Informal Agency Process*, 1 J. LEGAL STUD. 349 (1972).

343. *See* Aleinikoff, *Aliens, Due Process and "Community Ties": A Response to Martin*, 44 U. PITT. L. REV. 237, 259 (1983) (problem with plenary power doctrine is that it prevents continuing dialogue between courts and Congress on due process, so that when government conduct becomes "outrageous," "courts are forced to leap in with both feet, demanding costly and intrusive procedures that make control of the borders and deportation of aliens considerably more difficult."); Martin, *supra* note 103, at 171 ("Casting an eye over the procedural exuberance of the lower courts—procedural rulings theoretically applicable, without discernible limitation, to anyone in the world who presents himself at our borders—may well prompt a retreat to something like *Knauff-Mezei*."); Schuck, *supra* note 6, at 82-85 (discussing problems that could result from "active, intrusive judicial role in immigration cases," among them "ever more formalized, extensive hearing procedures whose effects, apart from increased cost and delays, are uncertain and quite possibly perverse"); Verkuil, *A Study of Immigration Procedures*, 31 UCLA L. REV. 1141, 1179-82 (1984) (parole-detention cases show that lack of administrative review of informal procedures may lead to judicial intervention, which in turn may lead to more procedural formality of discretionary decisions, more judicial review, or both).

and flexibility after considering the unusual and the unpredictable.[344] Since *Mandel*, the most negative effects of phantom norm decisions have been to impede the sound exercise of executive branch discretion. Tight supervision may correct short-run problems, but in the long run it also prevents immigration law from maturing and thus continues its traditional isolation—albeit isolation of a different character—from the mainstream of our public law.

C. *The Problem of Dialogue and the Future of Plenary Power*

Awkward or unpredictable solutions and misdirected judicial review are symptoms of a more basic condition. Immigration law after a century of plenary power has two conflicting sets of constitutional norms. In each of the four areas discussed in Part III.B, constitutional immigration law remains tied to the key plenary power cases. And the Court has continued to rebuff constitutional challenges to substantive, directly legislative choices. The most prominent recent case of *Fiallo v. Bell*,[345] decided in 1977, challenged the statute that recognized only an illegitimate child and mother, but not the father, as parent and child for immigration purposes.[346] The Court rejected an equal protection attack on two classifications in the statute—gender and legitimacy—that usually trigger something more than casual scrutiny.[347]

---

344. The type of case that may raise concerns about excessive judicial review is Hotel & Restaurant Employees Union, Local 25 v. Smith, 563 F. Supp. 157 (D.D.C. 1983), 594 F. Supp. 502, 507-09 (D.D.C. 1984), *aff'd in relevant part and rev'd in part on other grounds*, 804 F.2d 1256 (D.C. Cir. 1986), *aff'd by an equally divided court*, 846 F.2d 1499 (D.C. Cir. 1988) (en banc). The case involved "extended voluntary departure," which is the discretionary INS practice of letting some aliens, usually by nationality group, stay in the country for humanitarian reasons even though they may lack a meritorious case for political asylum or other formal status. The plaintiff union argued that the INS decision to deny extended voluntary departure to Salvadorans, while granting it to other nationalities, was "arbitrary and capricious and an abuse of discretion." Judge Richey asserted a power of limited judicial review, explaining that "abuse of discretion . . . can be shown by demonstrating a violation of constitutional, statutory, regulatory or other legal mandates." 563 F. Supp. at 162. Later, he entered summary judgment for the INS, finding no abuse of discretion because it had a "facially legitimate" reason for its decision, and the appeals court affirmed. 594 F. Supp. at 508. Had the Supreme Court already decided *Jean*, Judge Richey might well have inquired more deeply. *Cf.* American Baptist Churches v. Thornburgh, No. C 85-3255 FRP (N.D. Cal. 1989), *digested in* 66 INTERPRETER RELEASES 1061-63 (1989) (citing *American-Arab* to deny government's motions to dismiss and for summary judgment where plaintiffs claimed illegal discrimination against Salvadorans and Guatemalans in asylum applications and grants of extended voluntary departure); *see also* American Baptist Churches v. Meese, 712 F. Supp. 756, 767-74 (N.D. Cal. 1989). *See generally* Schuck, *supra* note 6, at 59-62; Note, *Extended Voluntary Departure: Limiting the Attorney General's Discretion in Immigration Matters*, 85 MICH. L. REV. 152 (1986).

345. 430 U.S. 787 (1977).

346. INA § 101(b)(1)(D), 8 U.S.C. § 1101(b)(1)(D) (1988), then provided that a "child" includes "an illegitimate child, by, through whom, or on whose behalf a status, privilege, or benefit is sought by virtue of the relationship of the child to its natural mother." *See Fiallo*, 430 U.S. at 797 (1977). Later, a 1986 amendment included fathers with "a bona fide parent-child relationship." Immigration Reform and Control Act § 315(a), Pub. L. No. 99-603, 100 Stat. 3359 (codified at INA § 101(b)(1)(D), 8 U.S.C. § 1101(b)(1)(D) (1988)).

347. *See, e.g.*, Trimble v. Gordon, 430 U.S. 762, 766-76 (1977) (legitimacy); Califano v. Webster, 430 U.S. 313, 317 (1977) (gender); Califano v. Goldfarb, 430 U.S. 199, 210-11 (1977) (gender); Craig v. Boren, 429 U.S. 190, 197 (1976) (gender); Jimenez v. Weinberger, 417 U.S. 628 (1974) (legitimacy); Gomez v. Perez, 409 U.S. 535, 538 (1973) (legitimacy).

Signs of change now appear on the horizon, in the form of expressly constitutional lower court decisions that have refused to accept the plenary power doctrine as controlling. While *Fiallo* clearly relied on the plenary power doctrine in rejecting the constitutional claim,[348] the Court left the door slightly ajar for judicial review by characterizing the statute as "'*largely* immune from judicial control.'"[349] Lower courts have taken this statement, sometimes together with *Mandel*'s "facially legitimate and bona fide reason" test, to support limited judicial review using a rational basis test. A few lower courts have held immigration classifications to lack a rational basis and therefore to violate equal protection. For example, the Second Circuit held in *Francis v. INS*[350] that, absent a rational basis, the BIA violated equal protection when it interpreted the statute to distinguish, for purposes of granting discretionary waivers, between permanent residents who had temporarily left the United States after becoming deportable and those who had not left.[351]

In 1989, a district court in *American-Arab Anti-Discrimination Committee v. Meese*[352] sustained a First Amendment challenge to several ideological deportation grounds as applied to two permanent residents and six nonimmigrant aliens, based on their alleged affiliation with a Marxist group, the Popular Front for the Liberation of Palestine.[353] The court reasoned that the provisions

348. 430 U.S. at 799. *See also* Landon v. Plasencia, 459 U.S. 21, 34-35 (1982) ("role of the judiciary is limited to determining whether the procedures meet the essential standard of fairness under the Due Process Clause and does not extend to imposing procedures that merely displace congressional choices of policy"). Interestingly, *Fiallo* cited *Mandel* more broadly, in support of the plenary power doctrine generally. *Fiallo*, 430 U.S. at 799.

349. 430 U.S. at 792 (quoting *Mezei*, 345 U.S. at 210) (emphasis added); *see also* 430 U.S. at 793 n.5 ("Our cases reflect acceptance of a limited judicial responsibility under the Constitution even with respect to the power of the Congress to regulate the admission and exclusion of aliens."); *id.* at 795-96 n.6 ("This is not to say, as we make clear in n.5, *supra*, that the Government's power in this area is never subject to judicial review."). The lower court in *Fiallo* similarly reasoned that distinguishing between unwed mothers and fathers was neither "wholly devoid of any conceivable rational purpose" nor "fundamentally aimed at achieving a goal unrelated to the regulation of immigration." Fiallo v. Levi, 406 F. Supp. 162, 166 (E.D.N.Y. 1975), *aff'd sub nom.* Fiallo v. Bell, 430 U.S. 787 (1977).

The plenary power cases in the early 1950's had hinted at some limited constitutional judicial review. *See* Harisiades v. Shaughnessy, 342 U.S. 580, 589-90 (1952) ("policy toward aliens" is "largely immune" from judicial review; issue is whether "congressional alarm" about "Communist conspiracy within the United States is either a fantasy or a pretense"). *Harisiades* rejected the view "that there were then or are now no possible grounds on which Congress might believe that Communists in our midst are inimical to our security." *Id.* at 590 (emphasis added); *see also* Carlson v. Landon, 342 U.S. 524, 544 (1952) ("no denial of . . . due process . . . where there is reasonable apprehension of hurt from aliens charged with a philosophy of violence against this Government").

350. 532 F.2d 268 (2d Cir. 1976)

351. *Francis* concerned interpretation of INA § 212(c), 8 U.S.C. § 1182(c) (1988). The Ninth Circuit followed *Francis* in Tapia-Acuna v. INS, 640 F.2d 223, 225 (9th Cir. 1981). *Cf.* Immigration Assistance Project v. INS, 709 F. Supp. 998, 1002-03 (W.D. Wash. 1989) (equal protection violation because no rational basis for distinguishing between groups of legalization applicants); *cf.* Elias v. Department of State, 721 F. Supp. 243, 247-50 (N.D. Cal. 1989) (no "facially legitimate and bona fide reason" to let citizen fathers, but not mothers, transmit citizenship to offspring born abroad).

352. 714 F. Supp. 1060, 1063 (C.D. Cal. 1989).

353. The deportation grounds at issue were INA §§ 241(a)(6)(D), (F)(iii), (G)(v), (H), 8 U.S.C. §§ 1251(a)(6)(D), (F)(iii), (G)(v), (H) (1988). 714 F. Supp. at 1062. The court did not reach plaintiffs' argument that sections 901(a) and (b) of the Foreign Relations Authorization Act, Fiscal Years 1988 and

were unconstitutional because the same activities, if engaged in by citizens, would be protected by the First Amendment.[354] The court relied heavily on *Chew* and Justice Murphy's concurrence in *Bridges*,[355] and also cited *Plyler v. Doe*[356] as authority in rejecting the government's argument that "a fundamentally 'different' Bill of Rights applies to aliens seeking to avoid expulsion from the United States."[357]

Harvard Law School Forum v. Shultz,[358] a 1986 district court decision, took the "facially legitimate and bona fide reason" language, which *Allende* and *El-Werfalli* took from *Mandel* and applied to heighten subconstitutional judicial scrutiny, and reintroduced it at the constitutional level to recognize a First Amendment challenge of the sort that *Mandel* had rejected. The case concerned the government's refusal to let a member of the Palestine Liberation Organization's mission to the United Nations travel to Harvard to participate in a political forum. The analysis began by relying heavily on *Allende*[359] but then moved quickly to the constitutional level. In reasoning that ran counter to *Mandel*, the court found that under the First Amendment the government's justification for refusing permission was "not facially legitimate because it is related to the suppression of protected political discussion."[360]

Another recent case, *Manwani v. INS*,[361] sustained substantive constitutional challenges to section five of the Immigration Marriage Fraud Amendments of 1986[362] on the grounds that it imposes an impermissible irrebuttable presumption and violates equal protection.[363] That section bars immigration

1989, Pub. L. No. 100-204, § 901, 101 Stat. 1331, 1399 (1987), were also unconstitutional.

354. In reaching this conclusion, the court read *Harisiades*, which rejected a First Amendment challenge, as having applied the same First Amendment standard to aliens and citizens. The district court reasoned that First Amendment law had changed in the meantime, since the precedent cited by the *Harisiades*, Dennis v. United States, 341 U.S. 494 (1951), had been superseded by later cases, most prominently Brandenburg v. Ohio, 395 U.S. 444 (1969). *See* discussion of *Harisiades supra* note 68.

355. 714 F. Supp. at 1074, 1076. Bridges v. Wixon, 326 U.S. 135 (1945), is discussed *supra* text accompanying notes 109-16.

356. 714 F. Supp. at 1075 n.11. Plyler v. Doe, 457 U.S. 202 (1982), is discussed *supra* text accompanying notes 196-99.

357. 714 F. Supp. at 1079.

358. 633 F. Supp. 525 (D. Mass. 1986).

359. *Id.* at 531.

360. *Id.* at 531. *cf.* United States v. Aguilar, 871 F.2d 1436, 1468-70 (9th Cir. 1989) (prosecuting sanctuary movement members for smuggling undocumented aliens does not abridge free exercise); American Friends Serv. Comm. v. Thornburgh, 718 F. Supp. 820, 822-23 (C.D. Cal. 1989) (IRCA employer sanctions do not abridge free exercise), *appeal docketed*, No. 89-56095 (9th Cir. 1989).

361. 736 F. Supp. 1367 (W.D.N.C. 1990).

362. Pub. L. No. 99-639, 100 Stat. 3537 (1986) (codified at INA § 204(h), 8 U.S.C. § 1154 (h) and INA § 245(e), 8 U.S.C. § 1255(e) (1988)).

363. 736 F. Supp. at 1386-92. The district court also found violations of procedural due process rights. *Id.* at 1383-86. In contrast, a panel of the District of Columbia Circuit relied on characterizing the statute as violating procedural due process because it delayed for two years an affected couple's opportunity to prove their marriage bona fide. *See* Escobar v. INS, 896 F.2d 564, 567-73 (D.C. Cir. 1990), *rev'g* 700 F. Supp. 609, 611-13 (D.D.C. 1988), *withdrawn and reh'g en banc granted* (Apr. 25, 1990). The Supreme Court rejected a similar argument in Fiallo v. Bell. *See* 430 U.S. 787, 791 (1977). The reasoning in *Escobar* illustrates an interesting variation on phantom norm decisionmaking. Rather than give a statute a particular interpretation, it casts the statute as "procedural," which makes the constitutional challenge more likely to

based on a marriage entered into while deportation proceedings are pending, unless the alien first leaves the country for two years.[364] Both *Yick Wo* progeny—*Wong Wing*, *Russian Volunteer Fleet*, *Mathews v. Diaz*, and *Plyler v. Doe*—and phantom norm decisions—*Chew* and *Wong Yang Sung*—figured centrally in the district court's reasoning.[365] This pattern—reliance upon both phantom norm decisions and the line of cases from *Yick Wo* to *Plyler*—is likely to be the reasoning in other decisions undermining or limiting the plenary power doctrine, just as it was, for example, in Justice Douglas' dissent in *Harisiades*[366] and Justice Marshall's dissent in *Jean*.[367]

These recent cases notwithstanding, the key Supreme Court constitutional precedents—*Knauff*, *Mezei*, and *Harisiades* being the most prominent among them—still severely limit judicial review. Almost always, substantive rules made directly by Congress have withstood constitutional challenge.[368] The Court may someday disavow the plenary power doctrine, but that step will take time. In the meantime, however, courts will continue to avoid this directly applicable constitutional doctrine through subconstitutional decisions that rely on phantom constitutional norms much more favorable to aliens. Cases that purport to speak only subconstitutionally have become the indirect expression of an alternative body of constitutional immigration law in which the plenary

---

succeed.

364. INA § 204(h), 8 U.S.C. § 1154(h) (1988).

365. *Manwani*, 736 F. Supp. at 1375-76, 1382-83, 1388-90.

366. 342 U.S. 580, 598-99 (Douglas, J., dissenting) (citing *Yick Wo*, *Wong Wing*, *Russian Volunteer Fleet*, and other cases).

367. 472 U.S. 846, 872-75 (Marshall, J., dissenting) (citing *Chew*, *Wong Wing*, *Russian Volunteer Fleet*, and *Yick Wo*). *Cf.* Note, *The Constitutional Rights of Excludable Aliens: History Provides a Refuge*, 61 WASH. L. REV. 1449 (1986) (proposing analytical framework based on *Wong Wing* and *Yamataya* to replace plenary power doctrine). *Jean* is discussed in greater detail *supra* text accompanying notes 237-55.

368. *See, e.g.*, Newton v. INS, 736 F.2d 336, 339-42 (6th Cir. 1984) (excluding exchange visitors from eligibility for suspension of deportation); Nademi v. INS, 679 F.2d 811, 814-15 (10th Cir. 1982) (treating Iranians differently after taking of Tehran hostages), *cert. denied*, 459 U.S. 872 (1982); Sadegh-Nobari v. INS, 676 F.2d 1348, 1351 (10th Cir. 1982) (same); Malek-Marzban v. INS, 653 F.2d 113, 116 (4th Cir. 1981) (same); United States v. Barajas-Guillen, 632 F.2d 749, 752-54 (9th Cir. 1980) (voluntary departure rules tending to exclude indigent aliens); Narenji v. Civiletti, 617 F.2d 745, 747-48 (D.C. Cir. 1979), *cert. denied*, 446 U.S. 957 (1980) (classifying aliens by nationality); Menezes v. INS, 601 F.2d 1028, 1034 (9th Cir. 1979) (treating fiances and spouses of United States citizens differently in inquiring into bona fides of marriage); Castillo-Felix v. INS, 601 F.2d 459, 467 (9th Cir. 1979) (applying different readings of INA section in different circuits); Alvarez v. District Director, 539 F.2d 1220, 1224 (9th Cir. 1976) (limiting special treatment of "commuter aliens" to aliens from contiguous countries); Guan Chow Tok v. INS, 538 F.2d 36, 38 (2d Cir. 1976) (deporting narcotics offenders); Noel v. Chapman, 508 F.2d 1023, 1028-29 (2d Cir. 1975) (different extended voluntary departure rules for spouse of citizens and of permanent residents); Dunn v. INS, 499 F.2d 856, 859 (9th Cir. 1974) (distinguishing Eastern and Western Hemisphere aliens for relief from deportation), *cert. denied*, 419 U.S. 1016 (1975).

Other recent cases have upheld the constitutionality of § 5 of the Immigration Marriage Fraud Amendments of 1986, Pub. L. No. 99-639, 100 Stat. 3537. *See* Almario v. Attorney Gen., 872 F.2d 147, 151 (6th Cir. 1989) ("*Kleindienst* and *Fiallo* require that we exercise a very narrow standard of review assuming the classification at issue in this case is supported by a facially legitimate and bona fide reason"; purpose is to deter marriage fraud); *see also* Azizi v. Thornburgh, 908 F.2d 1130 (2d Cir. 1990), *aff'g* 719 F. Supp. 86, 89-96 (D. Conn. 1989); Anetekhai v. INS, 876 F.2d 1218, 1221-24 (5th Cir. 1989), *aff'g* 685 F. Supp. 599 (E.D. La. 1988); Minatsis v. Brown, 713 F. Supp. 1056, 1060-61 (S.D. Ohio 1989); Smith v. INS, 684 F. Supp. 1113, 1116-20 (D. Mass. 1988).

power doctrine has lost much of its force. From this broader perspective, the practical demise of the plenary power doctrine seems closer in time.[369]

One basic problem that arises in this situation is that it is difficult accurately to identify and distinguish between constitutional and subconstitutional immigration law. Was *Woodby* a constitutional or subconstitutional decision? What about *Jean*? What are the limits of Congress' power to legislate different results in those situations? A related point is that the alternative body of phantom constitutional norms is relatively inaccessible. Litigants and judges know, in a loose way, about the gradual erosion of plenary power, but they generally do not distinguish between directly applied constitutional law and the alternative body of phantom norm decisions. Inaccessibility in turn limits the range of litigants' arguments and judges' decisions.

The inaccessibility problem forms part of a much larger and more fundamental difficulty. If the constitutional norms that courts use to interpret statutes to avoid constitutional doubt are phantoms, then those norms may not guide, but may misguide. In fact, the phantom norms actually turn on its head the practice of interpreting statutes to avoid constitutional doubt. If we heed the plenary power doctrine, there is no real doubt to be avoided, but courts, perhaps hearing the echo of *Yick Wo*, nonetheless assume the existence of a constitutional question and answer it by relying on a phantom norm. Concerns for legislative supremacy, which arise whenever courts bend statutes by relying on a constitutional norm,[370] are even greater when the norm is a phantom.[371] In this sense, mainstream public law decisions that avoid constitutional issues only by severely straining the statutory text[372] may be more defensible than *Chew* or *Fleuti*, because those mainstream decisions at least respond to real constitutional norms. After the gradual enhancement of the constitutional status of permanent residents, and especially after *Plasencia*, we can look back and believe that *Chew* was "correctly" decided. But for other norms that have yet to become "real," for instance in the four currently important areas of immigration law discussed in Part III.B, we lack the comfort of hindsight, and the problem of misguided statutory interpretation looms large.

---

369. *Cf.* Legomsky, *supra* note 5, at 303 ("We have entered a new phase in the life of the plenary power doctrine. This stage is characterized by a judicial willingness, so far episodic, to cut away at the notion of plenary Congressional power over immigration."); *id.* at 305 ("Constitutional review of immigration legislation will enter another, perhaps final, stage. This next stage will be marked by a return to general principles of constitutional law. It will be unnecessary for courts to distinguish immigration statutes from other federal statutes."); Schuck, *supra* note 6, at 34 ("Capitulation seems only a matter of time for some of these doctrines, while for others the outcome remains highly uncertain."); *id.* at 58 (increasing willingness of courts to review immigration decisions "essentially remains a lower court phenomenon").

370. *See* Eskridge, *supra* note 74, at 1065-73.

371. Thus, Judge Posner's critical observation that the canon leads to overenforcement of constitutional norms, *see* R. POSNER, *supra* note 75, at 285, has greater merit when directed toward immigration law's phantom norm decisions, even if, as Sunstein argues, Posner is incorrect as to constitutional law generally because those norms are real even though underenforced. *See* Sunstein, *supra* note 77, at 469.

372. *See, e.g.*, NLRB v. Catholic Bishop, 440 U.S. 490 (1979), discussed in Eskridge, *supra* note 74, at 1022, 1066.

One defense of phantom norm decisions is that they have been a useful testing ground for new constitutional ideas without the need to challenge prevailing doctrine. What began subconstitutionally—first with *Chew*, then *Fleuti*—may lead ultimately to an expressly constitutional change—such as in *Plasencia*—as phantom subconstitutional decisions undermine the existing constitutional structure to the point that it topples.[373] As I suggested in the Introduction, this process may be a healthy, perhaps preferred, and perhaps even inevitable form of constitutional change. Still, this pattern has taken an extreme form in immigration law. What seems aberrant is the persistence, beyond any reasonable definition of "transitional phase," of real, plenary power-based constitutional norms in immigration law, as well as the degree of conflict between them and the phantom norms from the *Yick Wo* tradition that guide statutory interpretation.

As a result, phantom constitutional norms and directly applicable constitutional immigration law provide little by way of dialogue en route to new constitutional doctrine. Instead, the tension between the plenary power doctrine and subconstitutional phantom norm decisions has caused considerable damage to the process of dialogue about the future of immigration law. Each body of constitutional immigration law has its adherents. Government briefs in immigration cases rely heavily on directly applicable constitutional immigration law, while advocates for aliens cite the subconstitutional phantom norm decisions. Each of these two groups of cases seems determined to speak in a louder voice, each hoping to drown out the other, but there is no conversation. This absence of dialogue is what convinces me that immigration law is an extreme case, where phantom norm decisionmaking has lingered beyond a period of honest, tentative groping and experimentation and into a phase in which it has become mere technique and perhaps even subterfuge.[374]

In spite of these many problems, any overall assessment of phantom norm decisionmaking should reflect a more basic ambivalence. It oversimplifies the history to identify only the problems and, with the acuity of hindsight, say that these judges were just renegades who disliked the prevailing constitutional

---

373. *Cf.* L. FULLER, LEGAL FICTIONS (1967) (legal fictions are ephemeral building blocks of legal communication, to be discarded when no longer useful).

374. *See generally* G. CALABRESI, A COMMON LAW FOR THE AGE OF STATUTES 178-81 (1982). In the short run, any movement away from phantom norm decisionmaking and toward more direct dialogue may deprive aliens of a strategy of argumentation that has helped them in many of the cases discussed in this Article. Certainly in the lower courts, judicial candor about the exact nature of the constitutional norms that inform interpretation of statutes may briefly strengthen the precedential force of the key plenary power cases such as *Mezei*, *Knauff*, and *Harisiades*. For a while, then, aliens may lose a few more immigration cases than before. But I believe that this movement will hasten the complete demise of the plenary power doctrine for two reasons. First, the case for abandoning the doctrine will become more compelling because courts no longer will be able to rely on the mechanism—phantom norm decisionmaking—through which they effectively have accomplished a partial integration of immigration law into mainstream public law. And second, when the extent of that integration becomes more widely known, express application of mainstream public law norms in immigration cases will seem less radical, and indeed sometimes consistent with precedent. In the long run, the end of plenary power should be worth the interim price.

immigration law and acted lawlessly in smuggling in constitutional rules more to their liking. Phantom norm decisions have been an understandable and perhaps even noble response by conscientious judges to the perceived anachronistic character of plenary power. We have these decisions to thank for the fact that the Constitution is not as far removed from immigration law as the plenary power doctrine would suggest. These decisions have been the only means by which immigration law has moved beyond the nineteenth-century world of classical immigration law and come to reflect, in some small part, the values, embodied in the *Yick Wo* tradition, that are more consistent with our modern sense of justice. The dissonance between immigration law and our mainstream public law has made some judges feel that they must find a way to harmonize the two. Some of those judges have been Justices of the Supreme Court of the United States, as *Chew*, *Jean*, and other cases amply demonstrate. They, like lower court judges, have felt constrained by precedent to render decisions that give them some considerable discomfort, but not enough discomfort to prompt the immediate creation of a new constitutional immigration law.

To move confidently into the second century of immigration law, we need to understand what went wrong in the first century. The development of a subconstitutional case law based on phantom constitutional norms is persuasive evidence that as long as judges have discretion to decide cases, no area of law, not even immigration law, can hope to insulate itself forever from the rest of our legal culture. This was the vain hope, and ultimately the failure, of the plenary power doctrine.

The future of immigration law requires considerable thought and discussion. To what extent should the analytical framework that applies in the mainstream of our public law apply to immigration cases? To what extent should the answers be the same or different because immigration is at issue? And ultimately, who are "we" as a community, and what do we demand of those who want to join? The most fundamental problem with the plenary power doctrine is that it has seriously impaired the process of dialogue on these questions, whatever the best answers may be. We will begin to restore dialogue only when courts, including the Supreme Court in cases like *Jean v. Nelson*, take the next step and begin to put their constitutional thinking in expressly constitutional terms.

347 U.S. 522
GALVAN v. PRESS.

No. 407.

Argued Jan. 11, 12, 1954.

Decided May 24, 1954.

Habeas corpus proceeding. The United States District Court for the Southern District of California, Southern Division, denied the petition, and the petitioner appealed. The Court of Appeals for the Ninth Circuit, 201 F.2d 302, affirmed. On certiorari granted, the United States Supreme Court, Mr. Justice Frankfurter, held that even if petitioner, an alien, had joined Communist Party without knowledge of its advocacy of violence he was deportable as a party "member", absent any claim that he had joined the party "accidentally, artificially, or unconsciously in appearance only."

Affirmed.

Mr. Justice Black and Mr. Justice Douglas dissented.

74 S.Ct.—47

1. Habeas Corpus ⚷114

Where issues of constitutionality and construction of 1950 Internal Security Act were raised for first time in habeas corpus proceeding, certiorari was granted for review of judgment affirming denial of writ.

2. Aliens ⚷53

An alien's membership in a proscribed organization may have been so nominal as to keep him out of the deportable class even though he did not join such organization (1) as a child, (2) by operation of law, or (3) to obtain necessities of life, since those instances, enumerated in 1951 corrective statute, were not intended to be exclusive. Immigration and Nationality Act, § 212(a) (28) (E), (I) (i), 8 U.S.C.A. § 1182(a) (28) (E), (I) (i).

3. Aliens ⚷53

For purpose of statutory provision making "membership" in Communist Party a basis for deportation of an alien, it is enough that alien join Party, aware that he was joining an organization known as Communist Party and operating as a distinct and active political organization, and that he did so of his own free will; cognizance of the Party's advocacy of violence not being a prerequisite to deportation. Immigration and Nationality Act, § 212(a) (28) (E), (I) (i), 8 U.S.C.A. § 1182(a) (28) (E), (I) (i); Internal Security Act of 1950, §§ 7, 13, 50 U.S.C.A. §§ 786, 792.

See publication Words and Phrases, for other judicial constructions and definitions of "Membership".

4. Aliens ⚷54(9)

On issue as to alien's deportability, evidence sustained finding that alien had in fact joined Communist Party. Immigration and Nationality Act, § 241(a) (6) (C), 8 U.S.C.A. § 1251(a) (6) (C).

5. Aliens ⚷53

Even if alien had joined Communist Party without knowledge of its advocacy of violence, he was deportable as a party "member", absent any claim that he had joined the party "accidentally, artificially, or unconsciously in appearance only." Immigration and Nationality Act, § 212(a) (28) (E), (I) (i), 8 U.S.C.A. § 1182(a) (28) (E), (I) (i); Internal Security Act of 1950, §§ 7, 13, 50 U.S.C.A. §§ 786, 792.

See publication Words and Phrases, for other judicial constructions and definitions of "Member".

6. Aliens ⇨39

The power of Congress over the admission of aliens and their right to remain is necessarily very broad, touching as it does basic aspects of national sovereignty, and more particularly our foreign relations and the national security; and formulation of policies pertaining to entry of aliens and their right to remain here is entrusted exclusively to Congress.

7. Constitutional Law ⇨252

An "alien" is a "person" entitled to the same protection for his life, liberty and property under the due process clause as is afforded to a citizen. U.S C.A.Const. Amend. 5.

See publication Words and Phrases, for other judicial constructions and definitions of "Person" and "Alien".

8. Constitutional Law ⇨251

Fair play is the essence of "due process." U.S.C.A.Const. Amend. 5.

See publication Words and Phrases, for other judicial constructions and definitions of "Due Process".

9. Constitutional Law ⇨251

Substantive due process is a limitation upon all powers of Congress, including the war power. U.S.C.A.Const. Amend. 5.

10. Constitutional Law ⇨197

Ex post facto clause of federal Constitution is applicable only to punitive legislation. U.S.C.A.Const. Amend. 5.

11. Constitutional Law ⇨197

Ex post facto clause of the federal Constitution has no application to deportation, even though deportation may deprive a man of all that makes life worth living and may at times be the equivalent of banishment or exile. U.S.C.A.Const. Amend. 5.

12. Constitutional Law ⇨318

Policies pertaining to entry of aliens and their right to remain here are peculiarly concerned with political conduct of government, and in enforcement of those policies, executive branch of government must respect procedural safeguards of due process. U.S.C.A.Const. Amend. 5.

13. Aliens ⇨40
Constitutional Law ⇨199, 255

Statute making present or former membership in Communist Party a basis for deportation of alien was not repugnant to either due process or ex post facto clauses of federal Constitution. Internal Security Act of 1950, § 2(1), 50 U.S.C.A. § 781(1); U.S.C.A.Const. Amend. 5.

14. Courts ⇨90(3)

Not being prepared to deem themselves wiser or more sensitive to human rights than their predecessors, especially those who had been most zealous in protecting civil liberties under Constitution, members of Supreme Court could not justifiably overturn a view of the Constitution which was deeply rooted and had been consistently adhered to. U.S.C.A. Const. Amend. 5.

523

Messrs. Harry Wolpin and A. L. Wirin, Los Angeles, for petitioner.

Mr. Oscar H. Davis, Washington, D. C., for respondent.

Mr. Justice FRANKFURTER delivered the opinion of the Court.

Petitioner, an alien of Mexican birth, first entered the United States in 1918 and has since resided here with only occasional brief visits to his native country. In the course of two questionings, in March 1948, by the Immigration and

Naturalization Service, he indicated that he had been a member of the Communist Party from 1944 to 1946. In March of 1949, the petitioner was served with a deportation warrant, and on the same day a preliminary deportation hearing was held to acquaint him with the charges against him—that after entry he had become a member of an organization which advocated the violent overthrow of the United States Government, and of an organization which distributed material so advocating. In December 1950, petitioner had a *de novo* hearing at which the transcripts of all earlier proceedings were, by agreement, made part of the record. Shortly after the hearing commenced, the examining officer lodged the additional charge against the petitioner that after entry he had been a member of the Communist Party, membership in which had been made a specific ground for deportation by the Internal Security Act of 1950, 64 Stat. 987, 1006, 1008.

At this final hearing the evidence against the petitioner was derived from two principal sources. The first was
524
his own testimony during the two interrogations by immigration authorities in 1948. During those interrogations, he had testified as to the time and place he had joined the Communist Party, talked freely about his membership in the Party, and indicated generally that the distinction between the Party and other groups was clear in his mind; he had explained that the reason he had not applied for naturalization was that he feared his former Party membership might be revealed, and had offered to make amends by rejoining the Party as an undercover agent for the Government. At the hearing in December of 1950, petitioner denied that in his prior hearing he had admitted joining the Party, insisting that at the time he thought the question related to labor union activities. In response to a question whether he had ever attended meetings of the Spanish Speaking Club, an alleged Communist Party unit, he replied: "The only meetings I attended were relating to the Fair Employment Practices Committee."

The second source of information was the testimony of a Mrs. Meza to the effect that she had been present when petitioner was elected an officer of the Spanish Speaking Club. Petitioner denied the truth of this and other statements of Mrs. Meza calculated to establish his active participation in the Communist Party and said: "She must have been under great strain to imagine all those things."

The Hearing Officer found that petitioner had been a member of the Communist Party from 1944 to 1946 and ordered him deported on that specific ground. He did not deem it necessary to make findings on the more general charges contained in the original warrant. The Hearing Officer's decision was adopted by the Assistant Commissioner and an appeal was dismissed by the Board of Immigration Appeals. A petition for a writ of habeas corpus was denied by the District Court, and the dismissal
525
was affirmed by the Court of Appeals for the Ninth Circuit. 201 F.2d 302.

[1] On certiorari, petitioner challenged the sufficiency of the evidence to sustain deportation under § 22 of the Internal Security Act of 1950 and attacked the validity of the Act as applied to him.[1] These are issues that raise the constitutionality and construction of the

1. In his petition, petitioner also contended that the procedure used against him was unfair because of the new charge lodged by the Examining Officer in the December 1950 hearing. Apart from the fact that this claim was not pressed in the argument or petitioner's brief, it is sufficient to note that there was no element of surprise in the additional charge, since it was simply in more specific terms the same ground for deportation that petitioner already knew he had to defend against, namely, membership in the Communist Party. Furthermore, petitioner declined the Hearing Officer's offer of a continuance to meet the new charge.

1950 Act for the first time and so we granted certiorari. 346 U.S. 812, 74 S. Ct. 53.

Petitioner's contention that there was not sufficient evidence to support the deportation order brings into question the scope of the word "member" as used by Congress in the enactment of 1950, whereby it required deportation of any alien who at the time of entering the United States, or at any time thereafter, was a "member" of the Communist Party.[2] We are urged to construe the Act as providing for the deportation only of those aliens who joined the Communist Party fully conscious of its advocacy

526

of violence, and who, by so joining, thereby committed themselves to this violent purpose.

But the Act itself appears to preclude an interpretation which would require proof that an alien had joined the Communist Party with full appreciation of its purposes and program. In the same section under which the petitioner's deportation is sought here as a former Communist Party member, there is another provision, subsection (2) (E),* which requires the exclusion or deportation of aliens who are "members of or affiliated with" an organization required to register under the Internal Security Act of 1950,[3] "unless such aliens establish that they did not know or have reason to believe at the time they became members of or affiliated with such an organization * * * that such organization was a Communist organization." 64 Stat. 1007. In describing the purpose of this clause, Senator McCarran, the Act's sponsor, said: "Aliens who were innocent dupes when they joined a Communist Front organization, as distinguished from a Communist political organization [such as the Communist Party], would likewise not ipso facto be excluded or deported." 96 Cong.Rec. 14180. In view of this specific escape provision for members of other organizations, it seems clear that Congress did not exempt "innocent" members of the Communist Party.

While the legislative history of the 1950 Act is not illuminating on the scope of "member," considerable light was shed by authoritative comment in the debates on the statute which Congress enacted in 1951 to correct what it regarded as the unduly expanded interpretation by the Attorney General of "member" under the 1950 Act. 65

527

Stat. 28.† The amendatory statute dealt with certain specific situations which had been brought to the attention of Congress and provided that where aliens had joined a proscribed organization (1) when they were children, (2) by operation of law, or (3) to obtain the necessities of life, they were not to be deemed to have been "members." In explaining the measure, its sponsor, Senator McCarran, stated repeatedly and emphatically that "member" was intended to have the same meaning in the 1950 Act as had been given it by the courts and administrative agencies since 1918, 97 Cong.Rec. 2368–2374. See S.

2. Section 22 of the Internal Security Act of 1950 provides that the Attorney General shall take into custody and deport any alien "who was at the time of entering the United States, or has been at any time thereafter, * * * a member of any one of the classes of aliens enumerated in section 1(2) of this Act * * *."
Subparagraph (C) of § 1(2) lists "Aliens who are members of or affiliated with (i) the Communist Party of the United States * * *." The substance of this provision was incorporated in the Immigration and Nationality Act of 1952, 66 Stat. 163, 205, 8 U.S.C. § 1251(a) (6) (C), 8 U.S.C.A. § 1251(a) (6) (C).

* Now 8 U.S.C.A. § 1182(a) (28) (E).

3. Under § 7 of the Internal Security Act of 1950, 50 U.S.C.A. § 786, "Communist-action" and "Communist-front" organizations are required to register as such with the Attorney General. Section 13, 50 U.S.C.A. § 792, provides that where such an organization fails to register the Attorney General may institute proceedings requiring such registration.

† Now 8 U.S.C.A. § 1182(a) (28) (I) (i).

Rep. No. 111, 82d Cong., 1st Sess. 2; H. R.Rep. No. 118, 82d Cong., 1st Sess. 2. To illustrate what "member" did not cover he inserted in the Record a memorandum containing the following language quoted from Colyer v. Skeffington, D.C., 265 F. 17, 72: "Congress could not have intended to authorize the wholesale deportation of aliens who, accidentally, artificially, or unconsciously in appearance only, are found to be members of or affiliated with an organization of whose platform and purposes they have no real knowledge." 97 Cong.Rec. 2373.

[2] This memorandum, as a weighty gloss on what Congress wrote, indicates that Congress did not provide that the three types of situations it enumerated in the 1951 corrective statute should be the only instances where membership is so nominal as to keep an alien out of the deportable class. For example, the circumstances under which the finding of membership was rejected in Colyer v. Skeffington, supra, would not have been covered by the specific language in the 1951 Act. In that case, the aliens passed "from one organization into another, supposing the change to be a mere change of name, and that by assenting to membership in the new organization they had not really changed their affiliations or political or economic activities." 265 F. at 72.

528

[3] On the other hand, the repeated statements that "member" was to have the same meaning under the 1950 Act as previously, preclude an interpretation limited to those who were fully cognizant of the Party's advocacy of violence. For the judicial and administrative decisions prior to 1950 do not exempt aliens who joined an organization unaware of its program and purposes. See Kjar v. Doak, 7 Cir., 61 F.2d 566; Greco v. Haff, 9 Cir., 63 F.2d 863; In the matter of O.—, 3. I. & N. Dec. 736.

It must be concluded, therefore, that support, or even demonstrated knowledge, of the Communist Party's advocacy of violence was not intended to be a prerequisite to deportation. It is enough that the alien joined the Party, aware that he was joining an organization known as the Communist Party which operates as a distinct and active political organization, and that he did so of his own free will. A fair reading of the legislation requires that this scope be given to what Congress enacted in 1950, however severe the consequences and whatever view one may have of the wisdom of the means which Congress employed to meet its desired end.

[4, 5] On this basis, the Hearing Officer's finding that petitioner here was a "member" of the Communist Party must be sustained. Petitioner does not claim that he joined the Party "accidentally, artificially, or unconsciously in appearance only," to use the words in Senator McCarran's memorandum. The two points on which he bases his defense against the deportation order are, first, that he did not join the Party at all, and that if he did join, he was unaware of the Party's true purposes and program. The evidence which must have been believed and relied upon for the Hearing Officer's finding that petitioner was a "member" is that petitioner was asked to join the Party by a man he assumed to be an organizer, that he attended a number of meetings and that he did not apply for citizenship because he feared his Party

529

membership would become known to the authorities. In addition, on the basis of Mrs. Meza's testimony, the Hearing Officer was entitled to conclude that petitioner had been active in the Spanish Speaking Club, and, indeed, one of its officers. Certainly there was sufficient evidence to support a finding of membership. And even if petitioner was unaware of the Party's advocacy of violence, as he attempted to prove, the record does not show a relationship to the Party so nominal as not to make him a "member" within the terms of the Act.

[6–14] This brings us to petitioner's constitutional attack on the statute. Harisiades v. Shaughnessy, 342 U.S. 580, 72 S.Ct. 512, 96 L.Ed. 586, sustained the constitutionality of the Alien Registration Act of 1940. 54 Stat. 670. That Act made membership in an organization which advocates the overthrow of the Government of the United States by force or violence a ground for deportation, notwithstanding that membership in such organization had terminated before enactment of the statute. Under the 1940 Act, it was necessary to prove in each case, where membership in the Communist Party was made the basis of deportation, that the Party did, in fact, advocate the violent overthrow of the government. The Internal Security Act of 1950 dispensed with the need for such proof. On the basis of extensive investigation Congress made many findings, including that in § 2(1) of the Act that the "Communist movement * * * is a worldwide revolutionary movement whose purpose it is, by treachery, deceit, infiltration into other groups (governmental and otherwise), espionage, sabotage, terrorism, and any other means deemed necessary, to establish a Communist totalitarian dictatorship", 50 U.S.C.A. § 781(1) and made present or former membership in the Communist Party, in and of itself, a ground for deportation. Certainly, we cannot say that this classification by Congress is so baseless as to be violative of due process and therefore beyond the power of Congress.

530

In this respect—the dispensation with proof of the character of the Communist Party—the present case goes beyond Harisiades. But insofar as petitioner's constitutional claim is based on his ignorance that the Party was committed to violence, the same issue was before the Court with respect to at least one of the aliens in Harisiades.

The power of Congress over the admission of aliens and their right to remain is necessarily very broad, touching as it does basic aspects of national sovereignty, more particularly our foreign relations and the national security. Nevertheless, considering what it means to deport an alien who legally became part of the American community, and the extent to which, since he is a "person," an alien has the same protection for his life, liberty and property under the Due Process Clause as is afforded to a citizen, deportation without permitting the alien to prove that he was unaware of the Communist Party's advocacy of violence strikes one with a sense of harsh incongruity. If due process bars Congress from enactments that shock the sense of fair play—which is the essence of due process—one is entitled to ask whether it is not beyond the power of Congress to deport an alien who was duped into joining the Communist Party, particularly when his conduct antedated the enactment of the legislation under which his deportation is sought. And this because deportation may, as this Court has said in Ng Fung Ho v. White, 259 U.S. 276, 284, 42 S.Ct. 492, 495, 66 L.Ed. 938, deprive a man "of all that makes life worth living"; and, as it has said in Fong Haw Tan v. Phelan, 333 U.S. 6, 10, 68 S.Ct. 374, 376, 92 L.Ed. 433; "deportation is a drastic measure and at times the equivalent of banishment or exile".

In light of the expansion of the concept of substantive due process as a limitation upon all powers of Congress, even the war power, see Hamilton v. Kentucky Distilleries & Warehouse Co., 251 U.S. 146, 155, 40 S.Ct. 106, 107, 64 L.Ed. 194, much could be said for the view, were we writing on a clean slate, that the Due Process

531

Clause qualifies the scope of political discretion heretofore recognized as belonging to Congress in regulating the entry and deportation of aliens. And since the intrinsic consequences of deportation are so close to punishment for crime, it might fairly be said also that the *ex post facto* Clause, even though applicable only to punitive legislation,[4] should be applied to deportation.

But the slate is not clean. As to the extent of the power of Congress under review, there is not merely "a page of history," New York Trust Co. v. Eisner, 256 U.S. 345, 349, 41 S.Ct. 506, 507, 65 L.Ed. 963, but a whole volume. Policies pertaining to the entry of aliens and their right to remain here are peculiarly concerned with the political conduct of government. In the enforcement of these policies, the Executive Branch of the Government must respect the procedural safeguards of due process. Kaoru Yamataya v. Fisher (The Japanese Immigrant Case), 189 U.S. 86, 101, 23 S.Ct. 611, 614, 47 L.Ed. 721; Wong Yang Sung v. McGrath, 339 U.S. 33, 49, 70 S.Ct. 445, 453, 94 L.Ed. 616. But that the formulation of these policies is entrusted exclusively to Congress has become about as firmly imbedded in the legislative and judicial tissues of our body politic as any aspect of our government. And whatever might have been said at an earlier date for applying the *ex post facto* Clause, it has been the unbroken rule of this Court that it has no application to deportation.

3 We are not prepared to deem ourselves wiser or more sensitive to human rights than our predecessors, especially those who have been most zealous in protecting civil

532

liberties under the Constitution, and must therefore under our constitutional system recognize congressional power in dealing with aliens, on the basis of which we are unable to find the Act of 1950 unconstitutional. See Bugajewitz v. Adams, 228 U.S. 585, 33 S.Ct. 607, 57 L.Ed. 978, and Ng Fung Ho v. White, 259 U.S. 276, 280, 42 S.Ct. 492, 493, 66 L.Ed. 938.

Judgment affirmed.

Mr. Justice REED concurs in the judgment of the Court and in the opinion as written, except as to the deductions drawn from Senator McCarran's citation of Colyer v. Skeffington, D.C., 265 F. 17, 72.

Mr. Justice BLACK, with whom Mr. Justice DOUGLAS concurs, dissenting.

Petitioner has lived in this country thirty-six years, having come here from Mexico in 1918 when only seven years of age. He has an American wife to whom he has been married for twenty years, four children all born here, and a stepson who served this country as a paratrooper. Since 1940 petitioner has been a laborer at the Van Camp Sea Food Company in San Diego, California. In 1944 petitioner became a member of the Communist Party. Deciding that he no longer wanted to belong to that party, he got out sometime around 1946 or 1947. As pointed out in the Court's opinion, during the period of his membership the Communist Party functioned "as a distinct and active political organization". See Communist Party of United States of America v. Peek, 20 Cal.2d 536, 127 P.2d 889. Party candidates appeared on California election ballots, and no federal law then frowned on Communist Party political activities. Now in 1954, however, petitioner is to be deported from this country solely because of his past lawful membership in that party. And this is to be done without proof or finding that petitioner knew that the party had any evil purposes or that he

4. First in Ogden v. Saunders, 12 Wheat. 213, 271, 6 L.Ed. 606, and again in Satterlee v. Matthewson, 2 Pet. 380, 686 (appendix), 7 L.Ed. 458, a characteristically persuasive attack was made by Mr. Justice Johnson on the view that the *ex post facto* Clause applies only to prosecutions for crime. The Court, however, has undeviatingly enforced the contrary position, first expressed in Calder v. Bull, 3 Dall. 386, 1 L.Ed. 648. It would be an unjustifiable reversal to overturn a view of the Constitution so deeply rooted and so consistently adhered to.

agreed 533 with any such purposes that it might have had. On the contrary, there is strong evidence that he was a good, law-abiding man, a steady worker and a devoted husband and father loyal to this country and its form of government.

For joining a lawful political group years ago—an act which he had no possible reason to believe would subject him to the slightest penalty—petitioner now loses his job, his friends, his home, and maybe even his children, who must choose between their father and their native country. Perhaps a legislative act penalizing political activities legal when engaged in is not a bill of attainder. But see United States v. Lovett, 328 U.S. 303, 315–316, 66 S.Ct. 1073, 1078, 1079, 90 L.Ed. 1252. Conceivably an Act prescribing exile for prior innocent conduct does not violate the constitutional prohibition of *ex post facto* laws. Cf. American Communications Ass'n v. Douds, 339 U.S. 382, 412–415, 70 S.Ct. 674, 690–693, 94 L.Ed. 925. It may be possible that this deportation order for engaging in political activities does not violate the First Amendment's clear ban against abridgment of political speech and assembly. Maybe it is not even a denial of due process and equal protection of the laws. But see dissenting opinions in Carlson v. Landon, 342 U.S. 524, 72 S.Ct. 525, 96 L.Ed. 547, and Harisiades v. Shaughnessy, 342 U.S. 580, 72 S.Ct. 512, 96 L.Ed. 586. I am unwilling to say, however, that despite these constitutional safeguards this man may be driven from our land because he joined a political party that California and the Nation then recognized as perfectly legal.

Mr. Justice DOUGLAS, with whom Mr. Justice BLACK concurs, dissenting.

As Mr. Justice BLACK states in his dissent, the only charge against this alien is an act that was lawful when done. I agree that there is, therefore, no constitutional basis for deportation, if aliens, as well as citizens, are to be the beneficiaries of due process of law.

534 The case might, of course, be different if the past affiliation with Communism now seized upon as the basis for deportation had continued down to this date. But so far as this record shows, the alien Galvan quit the Communist Party at least six years ago. There is not a word in the present record to show that he continued his affiliations with it *sub rosa* or espoused its causes or joined in any of its activities since he ceased to be a member of it.

I cannot agree that because a man was once a Communist, he always must carry the curse. Experience teaches otherwise. It is common knowledge that though some of the leading Socialists of Asia once were Communists, they repudiated the Marxist creed when they experienced its ugly operations, and today are the most effective opponents the Communists know. So far as the present record shows, Galvan may be such a man. Or he may be merely one who transgressed and then returned to a more orthodox political faith. The record is wholly silent about Galvan's present political activities. Only one thing is clear: Galvan is not being punished for what he presently is, nor for an unlawful act, nor for espionage or conspiracy or intrigue against this country. He is being punished for what he once was, for a political faith he briefly expressed over six years ago and then rejected.

This action is hostile to our constitutional standards, as I pointed out in Harisiades v. Shaughnessy, 342 U.S. 580, 598, 72 S.Ct. 512, 523, 96 L.Ed. 586. Aliens who live here in peace, who do not abuse our hospitality, who are law-abiding members of our communities, have the right to due process of law. They too are "persons" within the meaning of the Fifth Amendment. They can be molested by the government in times of peace only when their presence here is hostile to the safety or welfare of the Nation. If they are to be deported, it must be for what they are and do, not for what they once believed.

differences into an official doctrine of special judicial deference to Congress.

The precise degree of that special deference has varied with both the context and the era. The Court has described the congressional power to regulate immigration as "plenary."[3] It has said, "Over no conceivable subject is the legislative power of Congress more complete."[4] In the early years, this principle of plenary congressional authority over immigration was often expressed in absolute terms; the suggestion was that courts had literally *no* power to review the constitutionality of Congress's actions.[5] In later years the Court began to hedge, leaving open the possibility of *some* judicial role in assessing the constitutionality of federal immigration statutes.[6] In addition, by way of exception, the Court has generally guaranteed one particular constitutional right—procedural due process—in deportation cases.[7]

In the mid-1980's, I wrote two companion pieces—an article and a book chapter—on this plenary congressional power to regulate immigration.[8] The collective object of the two pieces was to study

3. *See, e.g.*, Kleindienst v. Mandel, 408 U.S. 753, 766, 768, 769 (1972); Boutilier v. INS, 387 U.S. 118, 123 (1967).

4. Oceanic Steam Navigation Co. v. Stranahan, 214 U.S. 320, 339 (1909), *quoted in* Fiallo v. Bell, 430 U.S. 787, 792 (1977), *and in* Kleindienst v. Mandel, 408 U.S. 753, 766 (1972).

5. *See, e.g.*, Lees v. United States, 150 U.S. 476, 480 (1893) (power to exclude aliens is "absolute" and is "not open to challenge in the courts"); Fong Yue Ting v. United States, 149 U.S. 698, 706 (1893) (Congress's decision is "conclusive upon the Judiciary"); Chae Chan Ping v. United States [Chinese Exclusion Case], 130 U.S. 581, 606 (1889) (same).

6. *See, e.g.*, Fiallo v. Bell, 430 U.S. 787, 792 (congressional power is "*largely* immune" from judicial review) (emphasis added), 793 n.5 (positing a judicial power, admittedly "limited", to review even congressional decisions excluding aliens) (1977); Harisiades v. Shaughnessy, 342 U.S. 580, 588-89 (1952) (immigration legislation "largely immune from judicial interference").

7. The leading case is Yamataya v. Fisher [the Japanese Immigrant Case], 189 U.S. 86, 100 (1903). *Accord*, Wong Yang Sung v. McGrath, 339 U.S. 33 (1950). I use the term "generally" because the leading early case had seemed to hold due process inapplicable even in deportation proceedings. *See, e.g.*, Fong Yue Ting v. United States, 149 U.S. 698 (1893).

More recently, the Court has extended this exception to those exclusion proceedings in which the aliens are lawful permanent residents seeking to return from brief visits abroad. *See, e.g.*, Landon v. Plasencia, 459 U.S. 21, 32-37 (1982). Under those limited circumstances, the individual interests at stake make exclusion the functional equivalent of deportation.

8. Stephen H. Legomsky, *Immigration Law and the Principle of Plenary Congressional Power*, 1984 SUP. CT. REV. 255 [hereinafter cited as Legomsky I); STEPHEN H. LEGOMSKY, IMMIGRATION AND THE JUDICIARY — LAW AND POLITICS IN BRITAIN AND AMERICA, ch. 3 (entitled *Immigration and the US Constitution*) (1987) [hereinafter cited as Legomsky II].

comprehensively the history, theory, soundness, politics, and likely future course of the plenary power doctrine. Since their publication, roughly ten years have passed and many crucial judicial decisions have been rendered. The present article examines those events, assessing which predictions have come true and which have not. In the process, I offer a set of new, improved predictions that will probably require similar humility and revision ten years hence.

Section I will summarize briefly the conclusions reached in the two prior publications. Section II will synthesize the subsequent judicial trends and reformulate the predictions in the light of those patterns.

I. The Mid-1980's Analysis[9]

In *Immigration Law and the Principle of Plenary Congressional Power*, my goal was to be comprehensive. I attempted to identify and to evaluate every legal theory that might be gleaned from a century's worth of Supreme Court jurisprudence in support of special judicial restraint in immigration cases.

Several such theories, some interrelated, have emerged over the years. The Court has suggested that the constitutionality of immigration legislation is inherently a political question[10] because immigration is inseparable from foreign affairs;[11] that an alien is merely a "guest" asserting a "privilege" rather than a "member" asserting a "right";[12] that allowing aliens to assert constitutional rights in addition to any remedies they enjoy under international law would give them an un-

9. In writing the two pieces discussed in this section, I drew liberally from several thoughtful articles that had already examined certain specific aspects of the plenary power doctrine. *See, e.g.*, Henry M. Hart, *The Power of Congress to Limit the Jurisdiction of Federal Courts: An Exercise in Dialectic*, 66 HARV. L. REV. 1362, 1387-96 (1953); Siegfried Hesse, *The Constitutional Status of the Lawfully Admitted Permanent Resident Alien: The Inherent Limits of the Power to Expel*, 69 YALE L.J. 262 (1959); Siegfried Hesse, *The Constitutional Status of the Lawfully Admitted Permanent Resident Alien: The Pre-1917 Cases*, 68 YALE L.J. 1578 (1959); James A.R. Nafziger, *The General Admission of Aliens Under International Law*, 77 AM. J. INT'L L. 804 (1983); Peter H. Schuck, *The Transformation of Immigration Law*, 84 COLUM. L. REV. 1, 14-30 (1984).

10. In the immigration context the Court has typically used language functionally equivalent to "political question." For examples, *see* Legomsky I, *supra* note 8, at 261 n.30; *see also* Reno v. Flores, 113 S. Ct. 1439, 1449 (1993) (describing alien regulation as "committed to the *political* branches of the Federal Government") (emphasis added). The semantics generate some difficult substantive questions that are thoughtfully examined in Louis Henkin, *Is There a "Political Question" Doctrine?*, 85 YALE L.J. 597 (1976).

11. Legomsky I, *supra* note 8, at 261-69.

12. *Id.* at 269-70.

fair advantage over United States citizens;[13] that aliens' lack of allegiance to the United States should correlate to lessened constitutional protection;[14] that the congressional power to regulate immigration is inherent in national sovereignty and therefore insulated from the usual constitutional constraints;[15] and that the Constitution[16] is inapplicable in exclusion proceedings because in the eyes of the law an excluded alien is outside the United States.[17]

In addition to those substantive rationales, the unrelenting stream of plenary power decisions lent overpowering force to *stare decisis*. In 1954, the Court lamented that "the slate is not clean" and that the principle of plenary congressional power over immigration had become too deeply ingrained to be overruled.[18] I argued that all of those theories—both substantive and procedural—were unconvincing reasons to preserve the plenary power doctrine. My stated aim, presumptuous I admit, was to "clean the slate."[19]

The same article explored some of the external influences on the plenary power decisions. It concluded that the social backgrounds and political attitudes of the Supreme Court Justices did not adequately explain the development of the plenary power doctrine. At the same time, the article suggested that the Justices' general conceptions of their appropriate roles, together with a combination of contemporary social and political forces, account in part for the evolution of plenary congressional power over immigration.[20]

The article then demonstrated that the lower courts, increasingly uneasy about sanctioning a virtually unreviewable congressional power, had begun to forge creative detours around the plenary power doctrine. Three such techniques were identified. One device was to draw esoteric distinctions, including distinctions between distinctions. A second was to carve out an exception for procedural due process in

13. *Id.* at 270-72.

14. *Id.* at 272-73.

15. *Id.* at 273-75.

16. In Sale v. Haitian Centers Council, 113 S. Ct. 2549 (1993), the Court similarly limited the reach of a treaty provision and a corresponding statutory provision that prohibited the forcible return of refugees to countries of persecution.

17. Legomsky I, *supra* note 8, at 275-77. A seventh theory, which I entitled the "poetic justice" theory, is more context-specific. It applies only to those aliens who are excluded or deported on the basis of their affiliation with the Communist Party. *Id.* at 277-78.

18. Galvan v. Press, 347 U.S. 522, 530-31 (1954), *followed in* Fiallo v. Bell, 430 U.S. 787, 792 n.4 (1977); Kleindienst v. Mandel, 408 U.S. 753, 766-67 (1972).

19. Legomsky I, *supra* note 8, at 306.

20. *Id.* § II.

deportation cases.[21] A third was to acknowledge the plenary power doctrine but to read it as if the word "plenary" were meaningless. I argued that all of these techniques, while well intentioned, were at least unsatisfying and in most cases logically indefensible.[22]

Finally, the article attempted to forecast the future of the plenary power doctrine. Analogizing to familiar legal traditions, the article predicted, first, that the various devices by which the lower courts have partially circumvented the doctrine would keep expanding. Second, the article said, once the exceptions are on the verge of swallowing the rule, the Supreme Court would candidly clean the slate. I thought the Court would

> describe the historical evolution of the rule, identify the exceptions, expose the analytical inadequacies of the exceptions, observe the inevitable direction of the tide, and conclude that the time has come to lay the general principle to rest.[23]

The second piece, written at the same time but published three years later as a book chapter, detailed the historical development of the plenary power doctrine.[24] It focused solely on the Supreme Court's use of precedent in creating, developing, and solidifying, but ultimately qualifying, this selective form of judicial restraint. My general thesis in that chapter was that the doctrine had unfolded inadvertently, through misplaced reliance on holdings far more modest than the existence of unreviewable congressional power. More specifically, I argued that the Court had unwittingly relied on federalism cases to reach conclusions about individual constitutional rights.[25]

In the past ten years, an army of commentators has sustained the shelling. Several have elaborated thoughtfully on specific aspects of the plenary power doctrine. Some have argued for First Amendment exceptions or qualifications.[26] Some have emphasized the international law constraints.[27] One writer has examined in detail the extent

21. A fine article by Professor Motomura later elaborated on the procedural due process exception. Hiroshi Motomura, *The Curious Evolution of Immigration Law: Procedural Surrogates for Substantive Constitutional Rights*, 92 COLUM. L. REV. 1625 (1992).

22. Legomsky I, *supra* note 8, § III.

23. *Id.* at 305.

24. Legomsky II, *supra* note 8.

25. *Id.* In the present article I spare the reader the tedious case-by-case details of that argument. For anyone who would like to experience that sort of tedium and monotony, however, I highly recommend the original source.

26. *See, e.g.*, John A. Scanlan, *Aliens in the Marketplace of Ideas: The Government, the Academy, and the McCarran-Walter Act*, 66 TEX. L. REV. 1481 (1988); Comment, *Ideological Exclusion, Plenary Power, and the PLO*, 77 CAL. L. REV. 831 (1989).

27. Louis Henkin, *The Constitution and United States Sovereignty: A Century of Chinese Exclusion and its Progeny*, 100 HARV. L. REV. 853, 863-85 (1987); Michael A. Scaper-

as long as the aliens are physically on United States soil.[41] The Ninth Circuit reached a similar result by declaring that either irrational or excessive detention can amount to punishment even without an intent to punish, and that such punishment cannot be imposed without due process.[42]

Welcome as such decisions are, they are hard to square with the long line of Supreme Court decisions holding due process inapplicable to excluded aliens.[43] The issues raised in these modern cases seem relevant only to the content of due process. Yet content becomes irrelevant if, as the Supreme Court has said, an excluded alien may not challenge the procedures authorized by Congress at all.

A second expansion of the procedural due process exception has occurred through broad characterizations of the word "procedural." In at least two lines of cases—one involving marriage fraud[44] and the other involving mandatory detention of certain criminal aliens[45]—some lower courts have managed to strike down congressional action that would probably have survived "substantive" constitutional attacks. In each of these lines of cases, courts that have characterized the constitutional challenges as procedural have done so by employing irrebuttable presumption reasoning. They have suggested, in effect, that the pertinent statutory distinctions reflected empirical assumptions that the aliens should have the opportunity to refute. Applying *Mathews v. Eldridge*,[46] these courts then invoked the same procedural due process standards that apply in other areas of administrative law.[47]

41. *See* Medina v. O'Neill, 838 F.2d 800, 803 (5th Cir. 1988) (but finding no gross physical abuse on the facts), *rev'g* Medina v. O'Neill, 589 F. Supp. 1028, 1032-33 (S.D. Tex. 1984) (holding due process constraints applicable to excluded aliens and ultimately finding gross physical abuse); Lynch v. Cannatella, 810 F.2d 1363, 1374 (5th Cir. 1987) (holding due process violated). For a thorough discussion of this line of cases, *see* Taylor, *supra* note 2, at 1143-51.

42. Alvarez-Mendez v. Stock, 941 F.2d 956, 962 (9th Cir. 1991).

43. In addition to *Mezei*, *see, e.g.*, United States ex rel. Knauff v. Shaughnessy, 338 U.S. 537 (1950); Nishimura Ekiu v. United States, 142 U.S. 651 (1892). Margaret Taylor argues that the "gross physical abuse" cases are not true "immigration" cases and thus not subject to the plenary power doctrine at all. Taylor, *supra* note 2. As she acknowledges, however, *id.* at 1153-56, federal statutes that regulate aliens but not immigration have similarly received judicial deference.

44. STEPHEN H. LEGOMSKY, IMMIGRATION LAW AND POLICY 94-95, 165-68 (1992); Motomura, *supra* note 21, at 1659-65.

45. Legomsky, *supra* note 43, at 95-96; Motomura, *supra* note 21, at 1669-73.

46. 424 U.S. 319 (1976).

47. I do not mean to imply that findings of due process violations have been the norm even in these cases. As the cited pages illustrate, the decisions have gone both ways.

3. *Carving out New Exceptions or Near-Exceptions*

Just as the courts have developed a procedural due process exception to the plenary power doctrine, so too there are signs—admittedly mixed[48]—of an emerging First Amendment exception. In *American-Arab Anti-Discrimination Committee v. Meese (ADC)*,[49] a district court held that the usual First Amendment standards apply in deportation cases. The district court based its conclusion on *Harisiades v. Shaughnessy*,[50] where the Supreme Court had rejected the First Amendment challenges of several deported aliens. In *Harisiades* the Court had summarily cited *Dennis v. United States*,[51] which at the time had defined the First Amendment limitations on the government's authority to curtail political speech. Because the Court in *Harisiades* had cited *Dennis* and had not mentioned the plenary power doctrine, the district court in *ADC* read *Harisiades* as an application of traditional First Amendment standards to deportation cases.

In *Rafeedie v. INS*,[52] another district court went a step further, applying customary First Amendment standards to aliens challenging their *exclusion* from the United States. The court held that aliens enjoy as much First Amendment protection as citizens do, and it saw no difficulty in applying that philosophy to exclusion cases.

Elsewhere I have questioned the assumption that the Supreme Court in *Harisiades* meant to apply the usual First Amendment standards in deportation cases.[53] My skepticism is, of course, even stronger with respect to exclusion. The *Dennis* case that the Court in *Harisiades* suggested it was applying had held that restrictions on political speech would be valid only when there was a "clear and present danger" of a sufficiently important evil.[54] Yet in *Harisiades* the Court neither stated nor applied the *Dennis* test. At no point did the Court in *Harisiades* consider whether the advocacy or associations that rendered the aliens deportable posed the required clear and present danger. Under those circumstances, while again I applaud the result, the district court's assumption about *Harisiades* might well have been wishful thinking.

48. *See* Price v. INS, 962 F.2d 836 (9th Cir. 1992) (according special judicial deference in naturalization cases).

49. 714 F. Supp. 1060 (C.D. Cal. 1989), *rev'd on other grounds*, 940 F.2d 445 (9th Cir. 1991).

50. 342 U.S. 580 (1952).

51. 341 U.S. 494 (1951), cited in *Harisiades*, 342 U.S. at 592 nn.18-19.

52. 795 F. Supp. 13 (D.D.C. 1992).

53. Legomsky, Immigration Law and Policy, *supra* note 44, at 87.

54. Dennis v. United States, 341 U.S. 494, 515 (1951).

As the cases discussed in this subsection demonstrate, the prediction that the courts would become increasingly inventive in their quests to avoid the principle of plenary congressional power has been borne out. Unfortunately, as the next subsection will show, my prophetic powers seem to end there.

B. The Supreme Court

As noted earlier, I also predicted that lower courts' expansion of the exceptions to the plenary power doctrine would eventually lead the Supreme Court to clean the slate in one fell swoop. I expected the dam to burst with a sudden, dramatic announcement that, henceforth, immigration cases would be treated just like any other cases. Whether involving exclusion or deportation,[55] the immigration-related subject matter would no longer affect either the formulation or the application of the pertinent constitutional standard.

Obviously, that has not happened. Since the making of those predictions, the Supreme Court has decided no cases in which the constitutionality of congressional[56] action in the field of immigration has been at issue. Perhaps the bursting of the dam is still on the horizon, waiting for the exceptions to grow fatter. But I now think the dismantling of the plenary power doctrine is unlikely to follow that script. For one thing, the composition of the Supreme Court has changed substantially; its members are more conservative, both politically and judicially. For another, a different scenario seems to be in progress already.

Under this revised scenario, the lower courts and the Supreme Court allow the plenary power doctrine to wear away by attrition. Little by little, exceptions and qualifications will reduce the doctrine to a shadow of its former self without an express overruling of contrary precedent. At the lower court level, this pattern is already evident in the cases that apply the rational basis test to substantive due process and equal protection challenges in the deportation setting; in the cases that apply the traditional *Mathews v. Eldridge* factors to procedural due process challenges; and in the cases that apply the then

55. My previous writings on the plenary power doctrine did not address those cases that involved aliens' rights outside the contexts of exclusion and deportation. Decisions like *Diaz* and *Nestor*, which involved federal restrictions on aliens' eligibility for welfare and social security, respectively, spoke in traditional rational basis terms but in the end accorded exceptional deference to Congress. *See supra* note 2. I would have expected the overruling of the plenary power doctrine to influence those cases as well.

56. The closest the Court has come to such a case is Reno v. Flores, 113 S. Ct. 1439 (1993), where the Court upheld the constitutionality of an INS regulation.

prevailing First Amendment standards to restrictions on political speech and associations.[57] Some sleight of hand might be necessary, but perhaps the Supreme Court will say something like this:

> As we have consistently observed, Congress has the plenary power to decide which classes of aliens will be excluded and which ones will be deported. The breadth of that power is considerable, and the courts may not disturb Congress's decision simply because they would have preferred a different resolution.
>
> But that is not to say the congressional power is unlimited. As when it legislates in any other area, Congress is constrained by the Constitution. And ever since *Marbury v. Madison*, 5 U.S. (1 Cranch) 137 (1803), we have recognized that, with some exceptions, the interpretation of those constitutional constraints is a judicial responsibility.
>
> We have never held otherwise. To be sure, some of our early opinions in this area contain broad language designed to emphasize the importance of appropriate judicial deference. But whenever we have found congressional decisions "conclusive upon the judiciary" or have otherwise suggested the absence of judicial power, the assumption was always that the particular congressional action was within the applicable constitutional constraints. The reason we could not intervene in those cases was simply that Congress had done nothing unconstitutional. That explanation especially fits our seminal decision in *Chae Chan Ping v. United States [Chinese Exclusion Case]*, 130 U.S. 581 (1889), as Professor Legomsky has convincingly demonstrated. See Stephen H. Legomsky, Immigration and the Judiciary — Law and Politics in Britain and America 192-94 (1987).[58]
>
> Our decision in *Fiallo v. Bell*, 430 U.S. 787 (1977), is illustrative. While upholding the particular statutory provision in question, we expressly acknowledged "a limited judicial responsibility under the Constitution even with respect to the power of Congress to regulate the admission and exclusion of aliens." *Id.* at 793 n.5. We proceeded to examine the substantive rationality of the challenged statutory distinction. At least one lower court has correctly read *Fiallo* as embodying the functional equivalent of the rational basis test — the same test that more generally governs substantive due process and equal protection challenges. *Azizi v. Thornburgh*, 908 F.2d 1130, 1133 n.2 (2d Cir. 1990).
>
> Our other immigration decisions follow the same pattern. In *Harisiades v. Shaughnessy*, 342 U.S. 580 (1952), we looked to *Dennis v. United States*, 341 U.S. 494 (1951), a non-immigration

57. See *supra* § II.A.

58. This last sentence is wishful thinking, but I have never been cited in a real Supreme Court opinion and will settle for being cited in a fictional one.

case, for the applicable First Amendment standard. In another deportation case, *Kaoru Yamataya v. Fisher*, 189 U.S. 86 (1903), we approved and applied traditional notions of procedural due process.

Thus, what some commentators have called the plenary power doctrine is in truth not a special "doctrine" at all. It is simply a convenient shorthand label for the common sense proposition that in immigration, as in most other areas of federal law, Congress has broad leeway to make sensitive policy decisions. Courts are ill equipped to second-guess those judgments. We will review federal immigration statutes for compliance with the Constitution, but, as in any other field, our role ends there.

If that form of denial strategy comes across as unlikely, there is a more plausible modified version. The Court could deny that special judicial deference is due in deportation cases (and possibly in other alien regulation cases when the aliens are formally within the United States) but adhere to either the existing plenary power doctrine or a scaled-down version of it in exclusion cases. If so inclined, the Court has many ways in which to distinguish deportation from exclusion.[59] Moreover, whatever difficulty the Court might face in trying to suggest that there has never been a plenary power doctrine for deportation cases, the barriers to credible denial would be even higher in exclusion cases, where both the rhetoric and the results of earlier opinions were often quite absolute.

Even with respect to exclusion, however, the modified approach would leave room for the Court to soften the plenary power doctrine. It could cite the *Fiallo* footnote and go from there. It could interpret *Kleindienst v. Mandel*, in which it applied a "facially legitimate and bona fide reason" test to the First Amendment rights of United States citizens to receive the ideas of excluded aliens,[60] as an all-purpose rational basis test for the constitutionality of speech-related exclusion grounds. Perhaps that kind of restricted plenary power doctrine—a new "PPD-lite"—is in the offing.

Thus, I still predict continued weakening of the special judicial deference to Congress in immigration (and other federal alien regulation) cases. But I now believe that the transformation will proceed

59. For a summary, *see* LEGOMSKY, IMMIGRATION LAW AND POLICY, *supra* note 44, at 44-47. The most compelling distinctions are those that rest on the presumed extraterritorial nature of exclusion proceedings or on assumptions about the differing magnitudes of the individual interests at stake, although even those distinctions are not free from doubt. *Id.*

60. 408 U.S. 753, 770 (1972).

only by steady erosion, rather than by a sequence of erosion followed by express overruling. Moreover, I no longer believe that in the foreseeable future the end result will be the complete abolition of this special deference. An emasculated version of the plenary power doctrine now seems more likely. Possible attributes of PPD-lite include (a) specific exceptions to the principle of special judicial deference; (b) justiciability of other constitutional challenges to immigration legislation but under narrower-than-usual standards of review; and (c) occasional references to the "plenary" nature of Congress's power or to the "special" deference owed in immigration cases, as makeweights to support independently reached decisions upholding congressional action.

III. Conclusion

Immigration commentators are well aware that our field has long been a constitutional oddity. For the most part, the Supreme Court has not applied to immigration cases the constitutional norms familiar in other areas of public law. That special judicial deference, known in immigration circles as the plenary power doctrine, has never been adequately explained on grounds of either policy or precedent. It is not enough to lament that "the slate is not clean."

In the mid-1980's, I argued that it was time to clean the slate and predicted with some confidence that the Court would do so. To be sure, there have been continued inroads. At this juncture, however, I must concede with regret that the slate is still dirty.

But the dirt has become no more encrusted, and the lower courts in fact have continued to chisel away at some of the more unsightly deposits. I believe that the Supreme Court will soon start to help out with the chiseling. Unlikely as the Court is to abolish the plenary power doctrine outright, it can, and probably will, give us PPD-lite.

Acknowledgments

Smith, James F. "A Nation that Welcomes Immigrants? An Historical Examinationof United States Immigration Policy," U.C. Davis Journal of International Law and Policy 1 (1995): 227–247. Reprinted with the permission of the University of California at Davis School of Law.

Act of Mar. 26, 1790, Statutes at Large 1: 103-104.

Chae Chan Ping v. *United States*, S.Ct. 623–32 (1889). Reprinted with the permission of the West Publishing Company.

Fong Yue Ting v. *United States*, 13 S.Ct. 1016–41 (1893). Reprinted with the permission of the West Publishing Company.

Hing, Bill Ong. "Immigration Policies: Messages of Exclusion to African Americans," *Howard Law Journal* 37 (1994): 237–82. Reprinted with the permission of the Howard University School of Law.

Aleinikoff, T. Alexander. "Federal Regulation of Aliens and the Constitution," *American Journal of International Law* 83 (1989): 862–871. Reprinted with the permission of the American Society of International Law.

Chin, Gabriel J. "Segregation's Last Stronghold: Race Discrimination and the Constitutional Law of Immigration," *UCLA Law Review* 46 (1998): 1–74. Copyright 1998, The Regents of the University of California. All rights reserved. Reprinted with permission.

Schuck, Peter H. "The Transformation of Immigration Law," *Columbia Law Review* 84 (1984): 1–90. Reprinted with permission.

Motomura, Hiroshi. "Immigration Law After a Century of Plenary Power: Phantom Constitutional Norms and Statutory Interpretation," *Yale Law Journal* 100 (1990): 545–613. Reprinted by permission of The Yale Law Journal Company and Fred B. Rothman and Company.

Galvan v. *Press*, 74 S.Ct. 737–45 (1954). Reprinted with the permission of the West Publishing Company.

Legomsky, Stephen H. "Ten More Years of Plenary Power: Immigration, Congress, and the Courts," *Hastings Constitutional Law Quarterly* 22 (1995): 925–937. Reprinted by permission. Copyright by University of California, Hastings College of Law.